Process Technology

Equipment and Systems

Process Technology

Equipment and Systems

Fourth Edition

Charles E. Thomas

Australia • Brazil • Canada • Mexico • Singapore • United Kingdom • United States

***Process Technology Equipment and Systems,* Fourth Edition**
Charles E. Thomas

VP, General Manager, Skills and Planning: Dawn Gerrain

Associate Product Manager: Nicole Sgueglia

Senior Product Development Manager: Larry Main

Senior Content Developer: Sharon Chambliss

Product Assistant: Maria Garguilo

Vice President, Marketing Services: Jennifer Ann Baker

Marketing Manager: Linda Kuper

Senior Production Director: Wendy Troeger

Production Manager: Sherondra Thedford

Art and Cover Direction, Production Management, and Composition: PreMediaGlobal

Cover photo credit: Steve Satushek/Getty Images

Library of Congress Control Number: 2014931118

ISBN-13: 978-1-285-44458-1

ISBN-10: 1-285-44458-2

Cengage
200 Pier 4 Boulevard
Boston, MA 02210
USA

Cengage is a leading provider of customized learning solutions with employees residing in nearly 40 different countries and sales in more than 125 countries around the world. Find your local representative at: **www.cengage.com.**

To learn more about Cengage platforms and services, register or access your online learning solution, or purchase materials for your course, visit **www.cengage.com.**

Notice to the Reader

Publisher does not warrant or guarantee any of the products described herein or perform any independent analysis in connection with any of the product information contained herein. Publisher does not assume, and expressly disclaims, any obligation to obtain and include information other than that provided to it by the manufacturer. The reader is expressly warned to consider and adopt all safety precautions that might be indicated by the activities described herein and to avoid all potential hazards. By following the instructions contained herein, the reader willingly assumes all risks in connection with such instructions. The publisher makes no representations or warranties of any kind, including but not limited to, the warranties of fitness for particular purpose or merchantability, nor are any such representations implied with respect to the material set forth herein, and the publisher takes no responsibility with respect to such material. The publisher shall not be liable for any special, consequential, or exemplary damages resulting, in whole or part, from the readers' use of, or reliance upon, this material.

Printed in the United States of America
Print Number: 13 Print Year: 2021

Dedication

To Kimberly

contents

preface

It should come as no surprise to anyone who knows the important role process technicians play in modern chemical manufacturing to discover the prominence that process technology programs have taken in U.S. and international community colleges and universities. Still, the controversial question remains, "what is process technology?" Process technology is a field of study offered at community colleges or universities that prepare students to take entry-level positions as technicians in the chemical processing industry. The curriculum can be categorized as the study and application of the scientific principles associated with the operation and maintenance of the CPI. Typically, there are 8 to 12 core classes that cover theory and hands-on operation, including safety, quality, equipment, instrumentation, physics, chemistry, math, process systems, unit operations, and troubleshooting. Hands-on training may include drawing P & IDs, operating bench-top units, starting-up, operating, and shutting down DCS simulations and small scale pilot plants. The first process technology program was offered at a Jr. college in Baytown Texas in 1938 by employees of Humble Oil, now Exxon Mobil.

This text is the product of many years of research in the field of process technology and operator training. It is unique and designed to enhance the learning sequence needed by adult learners. The materials in this text are the foundation topics for operator training.

Educators, of course, do their best to provide well-thought-out and illustrated textbooks, classroom lectures, computer-aided simulations and instruction, hands-on activities, bench-top and pilot units, CD/DVD materials, and the like. They take into consideration learning styles/strategies and find teaching approaches that work best—that is, how both an instructor and an individual approach the learning process and, given my experience in teaching process technology, there is a marked preference for teachers and students alike to emphasize self-study, of being responsible for this discipline on their own. This is the approach taken in the fourth edition of *Process Technology: Equipment and Systems.*

Recently, the industry has noticed a sharp increase in the number of women choosing process technology as an occupation. This text provides a balanced foundation for the rich diversity of students choosing to prepare for occupations in the chemical processing industry. This includes a much younger group of adult learners who will need to take the place of the baby boomer generation, as people from the latter retire in massive numbers.

As with the previous three editions, *Process Technology: Equipment and Systems* empowers the adult learner to accomplish the learning process. It covers the basic equipment and technology associated with the chemical processing industry.

What's New in This Edition?

The fourth edition includes new material on distillation, reactions, and process diagrams and equipment startups. This includes vapor liquid equilibrium diagrams, McCabe-Thiele method, detailed descriptions of plate and packed columns, calcium carbonate reactions, sodium phosphate reactions, ammonia reaction, reformers, and new detail on fluid catalytic cracking. The new edition also includes start-up procedures on the following systems: heat exchangers, cooling towers, boilers, furnaces, reactors, and distillation.

New features include:

- Process Diagrams: PFDs, PFiDs, P & IDds, and Control Loops
- Distillation: vapor-liquid phase diagrams & McCabe-Thiele method
- Distillation: reflux ratio, bubble point, dew point, mole fraction
- Distillation: plate and packed column detail, column loading
- Reactions: calcium carbonate, sodium phosphate, ammonia
- Reactions: fluid catalytic cracking, reformers
- Start-up Procedures: heat exchangers, cooling tower, boilers, furnace, reactor, distillation

Key concepts and learning features include:

- Valves, piping, tanks, and vessels
- Pumps, compressors, and simple systems
- Turbines and motors
- Heat exchangers, cooling towers, and simple systems
- Steam generation, fired heaters, and systems
- Process symbols, diagrams, and instrumentation
- Utility and separation systems
- Reactor, distillation columns, and systems
- Plastics systems

Each chapter in the text moves from simple to complex topics. Learning objectives are identified at the front of each chapter. Photographs and line drawings provide a rich visual documentation on which to *see* the concepts and equipment discussed in the narrative. A summary and a set of open-ended review questions end each chapter. This text also includes a short list of equipment symbols and diagrams discussed in a specific chapter. This allows new technicians to gradually build a good understanding of basic symbols and diagrams that are part of the visual lexicon of process technology around the world.

Acknowledgments

There are many individuals—too many to count—to thank for their contribution to process technology as an ongoing and developing discipline in higher education. I would like to extend my gratitude to the expert reviewers who read the early versions of this book and made recommendations to improve the final text, namely Gary Denson, Bryant Dyer, Paul Chance, Mike LaGrone, John Purdin, and Sharon Disspayne.

Charles E. Thomas, Ph.D.

Introduction to Process Equipment

Objectives

After studying this chapter, the student will be able to:

- Describe the basic hand tools used in industry.
- Explain the basic elements of rotary equipment.
- List the various types of stationary equipment.
- Identify the primary operation of a centrifugal pump.
- Explain the operation of a positive displacement pump.
- Describe dynamic and positive displacement compressors.
- Identify the purpose of a steam turbine.
- Describe the importance of equipment lubrication.
- Identify the various types of storage and piping equipment used in the chemical-processing industry.
- Describe the function of three types of industrial valves.
- Explain the purpose of a heat exchanger.
- Identify the primary function of a cooling tower.
- Identify the primary function of a boiler.
- Identify the purpose of a fired heater.
- Explain how a mixing reactor converts raw materials to useful products.
- Name the purpose of a distillation column.

Key Terms

Axial bearings—devices designed to prevent back-and-forth movement of a shaft; also called *thrust bearings*.

Basic hand tools—the typical tools process technicians use to perform their job activities.

Belt—used to connect two parallel shafts—the drive shaft and the driven shaft—each of which has a pulley mounted on the end; belts fit in the grooves of the pulleys.

Boiler—a type of fired furnace used to boil water and produce steam; also known as a *steam generator*.

Centrifugal force—the force exerted by a rotating object away from its center of rotation. Often referred to as a *center-seeking force*, centrifugal force is usually stated as the force perpendicular to the velocity of fluid moving in a circular path.

Chain drive—a device that provides rotational energy to driven equipment by means of a series of sprocket wheels that interlink with a chain; designed for low speeds and high-torque conversions.

Compressors—mechanical devices designed to accelerate or compress gases; classified as positive displacement or dynamic.

Cooling tower—a simple, rectangular device used by industry to remove heat from water.

Coupling—a device that attaches the drive shaft of a motor or steam turbine to a pump, compressor, or generator.

Distillation column—a cylindrical tower consisting of a series of trays or packing that provide a contact point for the vapor and liquid. The contact between the vapor and liquid in the column results in a separation of components in the mixture based on differences in boiling points.

Driven equipment—a device such as a compressor, pump, or generator that receives rotational energy from a driver.

Driver—a device designed to provide rotational energy to driven equipment.

Filter—a porous medium used to separate solid particles from a fluid by passage through it.

Fired heater—a high-temperature furnace used to heat large volumes of raw materials.

Gearbox—a power transmission mechanism consisting of interlocking toothed wheels (gears) inside a casing.

Heat exchanger—an energy-transfer device designed to transfer energy in the form of heat from a hotter fluid to a cooler fluid without physical contact between the two fluids.

Pumps—devices used to move liquids from one place to another; classified as positive displacement or dynamic.

Radial bearings—devices designed to prevent up-and-down and side-to-side movement of a shaft.

Reactor—a device used to combine raw materials, heat, pressure, and catalysts in the right proportions to form chemical bonds that create new products.

Rotary equipment—industrial equipment designed to rotate or move.

Rotor—the shaft and moving blades of rotary equipment or the moving conductor of an electric motor.

Seals—devices that prevent leakage between internal compartments in a rotating piece of equipment.

Stationary equipment—industrial equipment designed to occupy a stationary or fixed position.

Steam generator—see Boiler.

Steam turbine—an energy-conversion device that converts steam energy (kinetic energy) to useful mechanical energy; used as drivers to turn pumps, compressors, and electric generators.

Tanks and pipes—vessels and tubes that store and convey fluids.

Thrust bearing—see *Axial bearings.*

Torque—the turning force of rotating equipment.

Valve—a device used to stop, start, restrict (throttle), or direct the flow of fluids.

Viscosity—a measure of a fluid's resistance to flow.

Volute—the discharge chute of a centrifugal pump; a widening cavity that converts velocity to pressure.

Basic Hand Tools

The chemical processing industry is composed of refineries and petrochemical, paper and pulp, power generation, and food processing plants. Process technicians inspect and maintain equipment, place and remove equipment from service, complete checklists, control documentation, respond to emergencies, and troubleshoot system problems. To fulfill those responsibilities, the process technician must have a thorough understanding of tools, equipment, and systems.

Process technicians use hand tools to perform simple maintenance functions on operating units. The preventive maintenance role of process technicians is important because, in some cases, a little minor maintenance can prevent major equipment damage. **Basic hand tools** (Figure 1.1) used by process technicians include:

- Allen wrenches
- Wrenches: English and metric
- Pipe wrench
- Crescent wrench
- Valve wrench
- Channel locks
- Needle-nose pliers
- Pliers
- Vise grips
- File
- Tubing benders
- Broom or brush
- Electrical tape
- Measuring tape
- Water hose
- Hammer
- Ratchet and socket sets
- Wire brush
- Phillips screwdriver
- Flathead screwdriver
- Pencil or markers
- Wire cutter and strippers
- Utility knife

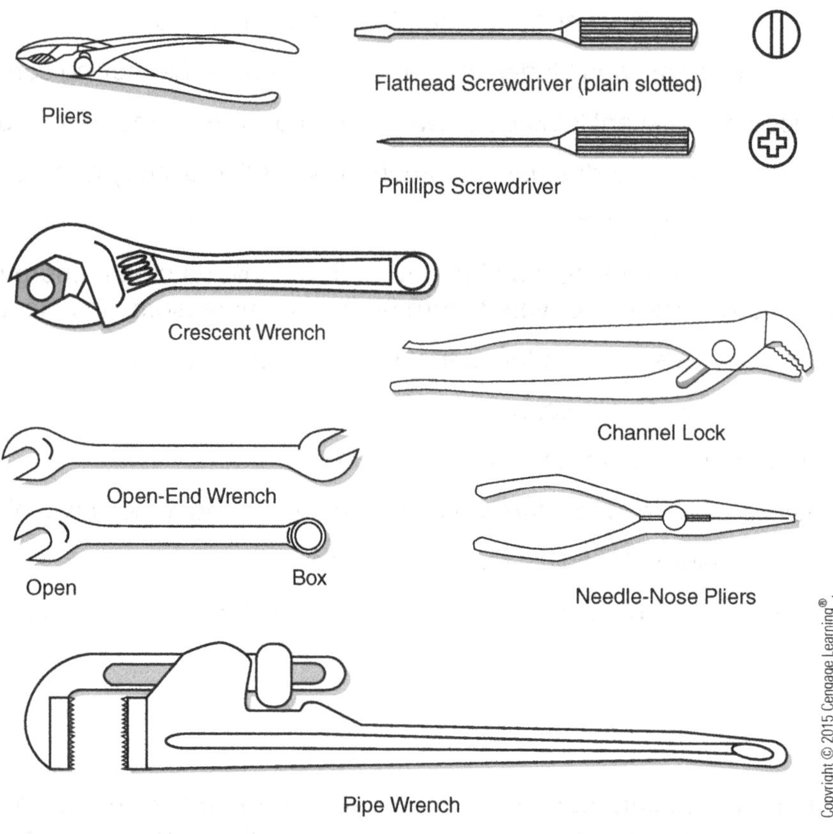

Figure 1.1 *Basic Hand Tools*

- Chisels
- Level
- Hacksaw
- Tubing cutters
- Dustpan
- Table vice
- Square
- Flash light

Rotary Equipment

The chemical processing industry has a wide variety of rotary, reciprocating, and **stationary equipment**. A circular movement characterizes **rotary equipment**. Examples can be found in motors, **pumps**, **compressors**, fans, blowers, turbines, conveyors, and feeders. Shaft rotation is typically measured in RPM, or revolutions per minute. Many motors run in excess of 3,000 RPM. A protective cover is typically placed over the moving shaft to minimize accidents.

CAUTION: *Severe injuries can result when loose clothing, shoelaces, jewelry, or long hair get tangled around rotating parts.*

Rotary equipment is composed of a driver, a connector, and the **driven equipment**.

Drivers and Driven Equipment

A **driver** is a device designed to provide rotational energy to another piece of equipment, the driven equipment. The most common drivers are electric motors and turbines. Examples of driven equipment include pumps, compressors, generators, fans, conveyors, and solids feeders. **Couplings**, **belts**, or chains connect drivers and driven equipment.

Couplings come in a variety of shapes and designs. The most common styles are fixed-speed couplings (rigid and flexible) and variable-speed couplings (hydraulic and magnetic). Figure 1.2 shows a rigid and a flexible coupling.

Belts are used to connect two parallel shafts: the drive shaft and the driven shaft. A pulley is mounted on the end of each shaft. Belts fit in the grooves of the pulley. The sizes of the pulleys allow the driver and driven equipment to operate at different speeds. When the drive pulley and the driven pulley are the same size, the speeds of the two shafts are virtually identical. Belts come in a variety of shapes and sizes and are made of durable material designed to withstand operating conditions. Belt drives (Figure 1.3)

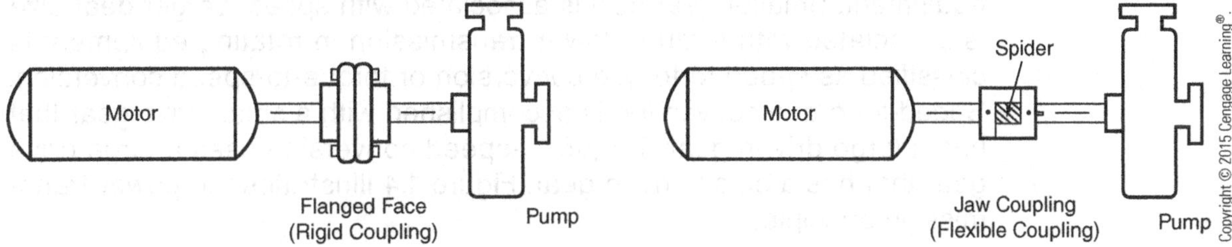

Figure 1.2 *Fixed Couplings: Rigid and Flexible*

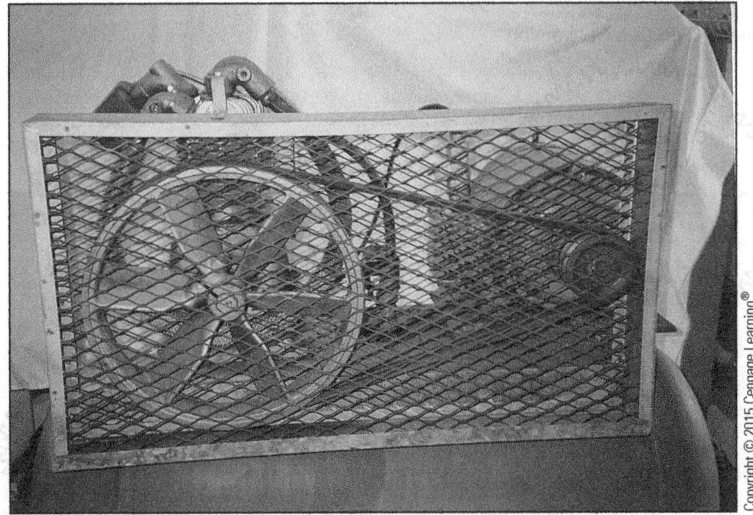

Figure 1.3 *Belt Drive*

require less space than fixed-speed or variable-speed couplings. A belt drive can make speed-to-torque and torque-to-speed conversions. (**Torque** is the turning force of rotating equipment.) Process technicians frequently inspect belts during rounds to ensure that safety guards are in place, that belt tension is correct, and that the belts are still mounted on the pulleys. Flopping, squealing, or smoking belts indicate wear, tension, or driven-equipment problems.

Chain drives are very similar to belt drives. Instead of using pulleys, however, a chain drive has a series of sprocket wheels that interlink with the chain. Chain drives are designed for low speeds and high torque conversions. In this type of system, slippage is minimal, chain replacement is rare, and temperature variations are not a factor as long as the chain is kept lubricated.

Gearboxes and Power Transmission

Gearboxes are often used between the driver and the driven equipment. A gearbox takes its name from the different-sized gears (toothed wheels) inside the casing. Inside the gearbox, the drive gear meshes with a larger or smaller gear, the driven gear. As the drive gear rotates, the interlocked gears in the box turn, transmitting power to the driven equipment. Smaller gear size is associated with speed. Larger gear size is associated with torque. Power transmission in rotating equipment is classified as speed-to-torque conversion or torque-to-speed conversion. Speed-to-torque conversion is accomplished with a small drive gear that has a large driven gear. Torque-to-speed conversion uses a large drive gear that has a small driven gear. Figure 1.4 illustrates the power transmission principle.

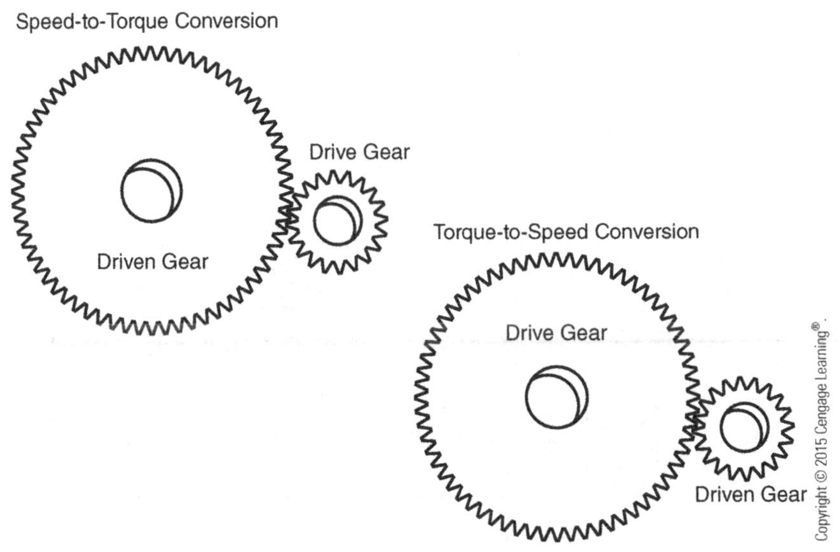

Figure 1.4
Power Transmission

Electric Motors

The process industry uses electric motors to operate pumps, generators, compressors, fans, blowers, and other equipment. Electric motors are either direct current (DC) or alternating current (AC). The operation of an electric motor is based on three principles: Electric current creates a magnetic field; opposite magnetic poles attract each other, and like magnetic poles repel each other; and current direction determines the magnetic poles. An electric motor consists of a stationary magnet (stator) and a moving conductor (rotor). A permanent magnetic field is formed by the lines of force between the poles of the magnet. When electricity passes through the conductor in a DC motor, the conductor becomes an electromagnet and generates another magnetic field. The twin fields increase in intensity and push against the conductor. The direction of rotation in a motor is determined by these strong magnetic fields.

The **rotor** in an AC motor (Figure 1.5) is a slotted iron core. Copper bars are fitted into the slots. Two thick copper rings hold the bars in place. Unlike the electric current in a DC motor, electric current in the AC motor is not run directly to the rotor. Alternating current flows into the stator, producing a rotating magnetic field. The stator artificially creates an electric current in the rotor, which generates the second magnetic field. When the two fields interact, the rotor turns.

Centrifugal Pumps

Centrifugal pumps (Figure 1.6) are devices that move fluids by **centrifugal force**. Centrifugal force is the force exerted by a rotating object away from its center of rotation; it is usually referred to as a "center-seeking force." The primary principle involves spinning the fluid in a circular motion that propels it outward and into a discharge chute known as a **volute**. The basic

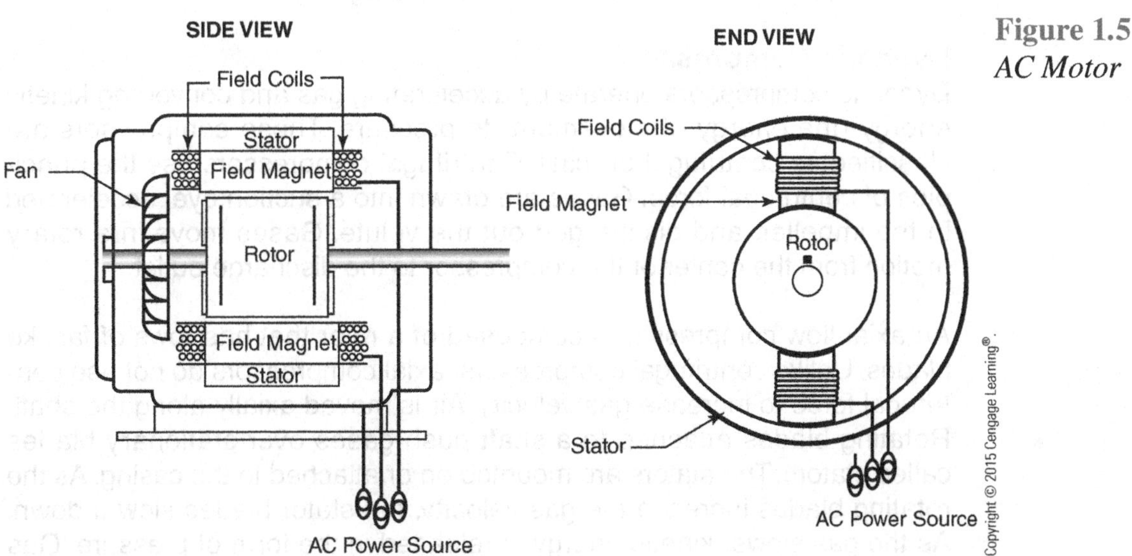

Figure 1.5
AC Motor

Figure 1.6
Centrifugal Pump

components of a centrifugal pump are the casing, motor or driver, coupling, volute, suction eye or inlet, impellers, wear rings, **seals**, bearings, discharge port, and suction and discharge gauges.

Positive Displacement Pumps

Positive displacement pumps displace a specific volume of fluid on each stroke or rotation. Positive displacement pumps can be classified as rotary or reciprocating. Rotary pumps displace fluids by means of rotating screws, gears, vanes, or lobes. Reciprocating pumps move fluids by drawing them into a chamber on the intake stroke and displacing them by means of a piston, diaphragm, or plunger on the discharge stroke. Reciprocating pumps are characterized by a back-and-forth motion. The basic components of a reciprocating pump are a connecting rod, a piston, plunger, or diaphragm, seals, check valves, motor, casing, and bearings.

Dynamic Compressors

Dynamic compressors operate by accelerating gas and converting kinetic energy (the energy of movement) to pressure. These compressors are classified as centrifugal or axial. Centrifugal compressors use the principles of centrifugal force. Gases are drawn into a suction eye, accelerated in the impeller, and discharged out the volute. Gases move in a rotary motion from the center of the compressor to the discharge outlet.

An axial flow compressor is composed of a rotor that has rows of fanlike blades. Unlike centrifugal compressors, axial compressors do not use centrifugal force to increase gas velocity. Air is moved axially along the shaft. Rotating blades attached to a shaft push gases over stationary blades called stators. The stators are mounted on or attached to the casing. As the rotating blades increase the gas velocity, the stator blades slow it down. As the gas slows, kinetic energy is released in the form of pressure. Gas

Figure 1.7
Blower

velocity increases as the gas moves from stage to stage until it reaches the discharge port. Jet engines and gas turbines contain axial flow compressors. Figure 1.7 Blower, illustrates how a centrifugal blower is used in a refrigeration system to evenly distribute ice on the exterior of tubes by bubbling air up from the bottom of the thermal ice bath to the top.

Positive Displacement Compressors

Positive displacement (PD) compressors operate by trapping a specific amount of gas and forcing it into a smaller volume. They are classified as rotary or reciprocating. PD compressors and PD pumps operate under similar conditions. The primary difference is that compressors are designed to transfer gases, whereas pumps move liquids. Rotary compressor design includes a rotary screw, sliding vane, lobe, and liquid ring. Reciprocating compressors include a piston and diaphragm. Figure 1.8 is an example of a typical PD compressor found in industry.

Steam Turbine

A **steam turbine** (Figure 1.9) is a device that converts kinetic energy to mechanical energy. Steam turbines have a specially designed rotor that

Figure 1.8 *PD Compressor*

Figure 1.9 *Steam Turbine*

rotates as steam strikes it. This rotation is used to operate a variety of shaft-driven equipment. Steam turbines are used primarily as drivers for pumps, compressors, and electric power generators.

Equipment Lubrication

One of the primary functions a process technician performs is periodic equipment checks. During these routine checks, equipment oil levels and operating conditions are closely inspected. High temperatures, unusual noises or smells, and erratic flows are all signs that a problem has developed. Proper lubrication must be maintained to ensure the good operation of process equipment. Lubrication protects the moving parts of equipment by placing a thin film of protection between surfaces that come into contact with each other. Under a microscope, the smooth surface of a gear may appear very rough. Without lubrication, a tremendous amount of friction would be developed. Lubrication helps remove heat generated by friction and provides a fluid barrier between the metal parts to reduce friction. Loss of lubrication will severely damage compressors, steam turbines, pumps, generators, engines, and so on. Most rotary equipment will require some type of lubrication.

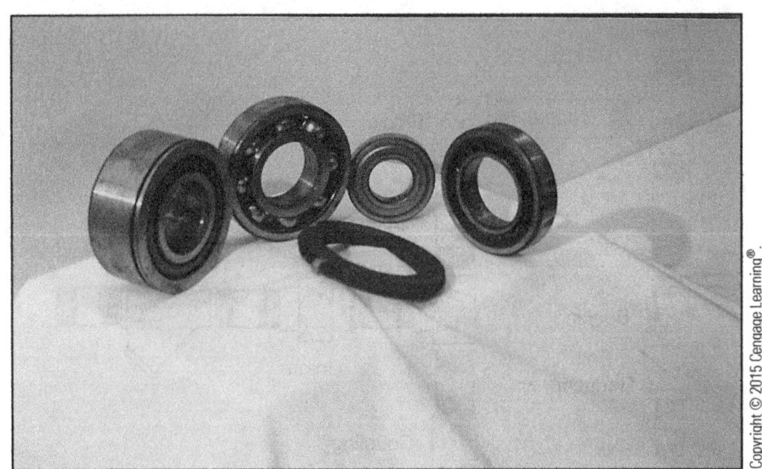

Figure 1.10
Seals and Bearings

Bearings

Radial and **axial bearings** can be found in most rotating equipment and require lubrication to operate properly. Radial bearings are designed to prevent up-and-down and side-to-side movement of the rotating shaft; axial bearings, often called **thrust bearings**, are designed to prevent back-and-forth movement of the shaft. Radial bearings can be found in a variety of designs such as ball bearings (Figure 1.10), friction or sleeve bearings, rolling element bearings, and shielded bearings.

Seals

Shaft seals (Figure 1.10) are designed to prevent leakage between internal compartments in a rotating piece of equipment. Shaft seals come in a variety of shapes and designs. Typical designs include labyrinth seals, carbon seals, packing seals, and mechanical seals. Labyrinth seals trap lubrication and fluids within a maze of ridges. Segmental carbon seals are mounted in a ring-shaped design around the rotating shaft. A spring holds the soft graphite seal in place and allows it to wear evenly. Mechanical seals come in a modular kit that is slid into place as one unit. Mechanical seals provide a stationary seat and a moving seal face. Mechanical seals can withstand high-pressure situations; carbon seals and labyrinth seals cannot. Shaft seals minimize air leakage into and out of the equipment; keep dirt, chemicals, and water out of the lubricant; and keep the clean lubricant in the chamber where the bearings and moving components are located.

Stationary Equipment

Piping and Storage Tanks

Industrial piping comes in a variety of shapes, designs, and metals to safely contain and transport chemicals. The engineering designer carefully

Figure 1.11
Pipe Fittings

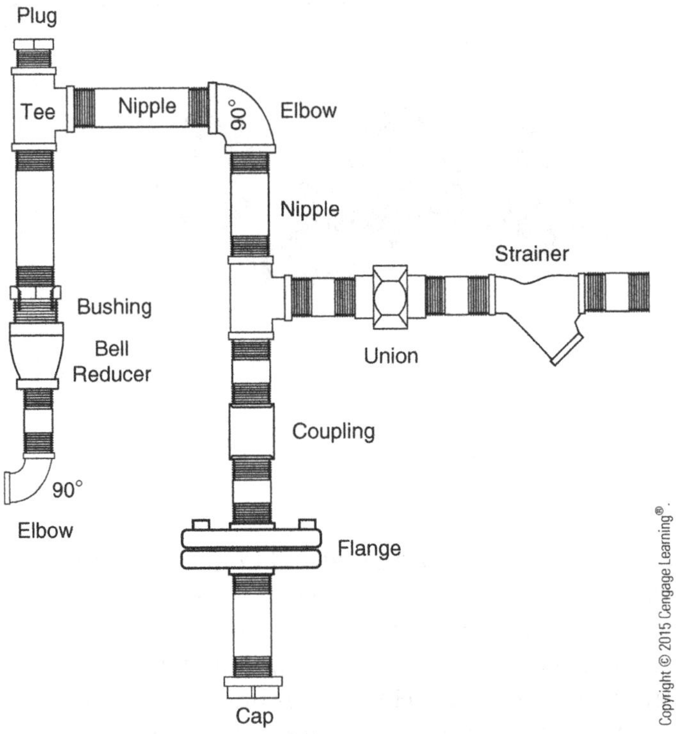

selects the types of materials that are compatible with the chemicals and operational conditions. Piping can be composed of stainless steel, carbon steel, iron, plastic, or specialty metals. Individual joints can be threaded on each end, flanged, welded, or glued. A wide array of fittings are used to connect the piping. The various types of fittings include couplings, unions, elbows, tees, nipples, plugs, caps, and bushings. Figure 1.11 illustrates the various types of fittings and piping. When the process fluid being transferred in the piping has a **viscosity value** (the measure of a substance's resistance to flow) that makes it difficult to transfer in cool weather, traced piping is often used (see Chapter 3).

The chemical processing industry uses tanks, drums, bins, and spheres to store chemicals. The materials used in these designs include carbon steel, stainless steel, iron, specialty metals, and plastic. A dome roof tank and a horizontal cylindrical vessel are shown in Figure 1.12.

Process technicians often inspect their equipment using the following methods: listen, touch, look, and smell. An experienced technician can identify a problem by listening for abnormal sounds and vibrations. Touching the equipment allows a technician to identify unusual heat patterns. Visually inspecting tank and sump levels allows a technician to look at and determine corrective action. An odor might indicate a leakage problem. Various tank designs are illustrated in Figure 1.13.

Figure 1.12 *Tank Storage*

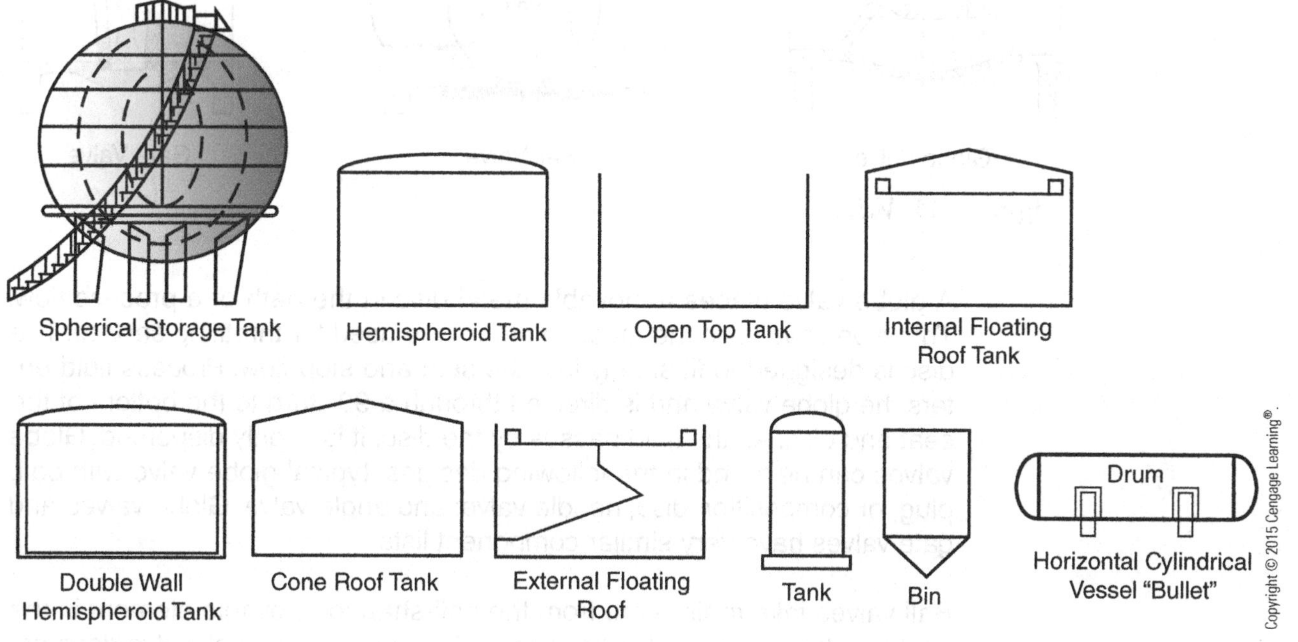

Figure 1.13 *Tank Designs*

Gate, Globe, and Ball Valves

Valves are used to stop, start, restrict (throttle), or direct the flow of fluids. Gate, globe, and ball valves are illustrated in Figure 1.14. A gate valve places a movable gate in the path of a process flow in a pipeline. The basic components of a gate valve include the body, bonnet, stem, handwheel, gate, packing, packing gland, and stuffing box. Gate valves come in two designs: rising and nonrising stems. Gate valves are designed for off/on control and not for restricting (throttling) the flow.

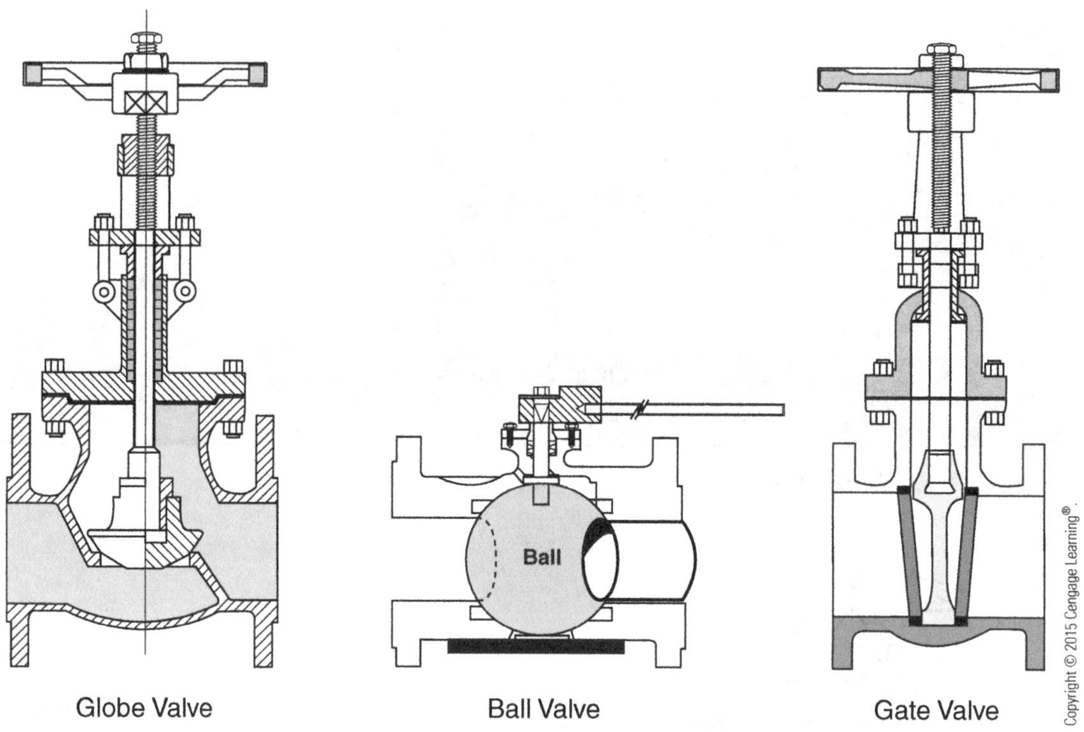

Figure 1.14 *Valves*

A globe valve places a movable metal disc in the path of a process flow. This type of valve is the most common one used for throttling service. The disc is designed to fit snugly into the seat and stop flow. Process fluid enters the globe valve and is directed through a 90° turn to the bottom of the seat and disc. As the fluid passes by the disc, it is evenly dispersed. Globe valves can be found in the following designs: typical globe valve with ball, plug, or composition disc; needle valve; and angle valve. Globe valves and gate valves have very similar component lists.

Ball valves take their name from the ball-shaped, movable element in the center of the valve. Unlike the gate and globe valves, a ball valve does not lift the flow control device out of the process stream; instead, the hollow ball rotates into the open or closed position. Ball valves provide very little restriction to flow and can be opened 100% with a quarter turn on the valve handle. In the closed position, the port is turned away from the process flow. In the open position, the port lines up perfectly with the inner diameter of the pipe. Larger valves require handwheels and gearboxes to be opened, but most only require a quarter turn on a handle.

Filters

Filters are used in the chemical processing industry to separate solid particles from a fluid by passage through a porous medium. Filters can be found in a variety of applications and services. The most common filters are cartridge filters. The replaceable filter medium is self-supporting and attached to a structural core. As material flows through the medium, suspended solids are separated from the fluid. As the pores in the medium fill up, fluid flow is restricted and delta pressure (ΔP; the difference between the pressure on one side and the pressure on the other side) across the body of the filter begins to increase. Redundant filter systems allow a technician to switch to a clean filter and safely remove the dirty cartridges.

One common use of large industrial filters is in the water-treatment area. Water-treatment requirements are linked to three factors: source water quality, how the water will be used, and environmental regulations. The chemical processing industry has adopted the practice of using surface water for industrial applications instead of well water. Water is initially brought into a large water basin where sediments are allowed to settle. Surface waters contain silt in suspended form and dissolved organic and inorganic impurities. Several large pumps take suction off the basin and transfer the water to filters designed to remove suspended solids. Industrial filters provide an important part of the water-treatment stage. Figure 1.15 illustrates a typical industrial filter.

Heat Exchangers

A **heat exchanger** allows a hot fluid to transfer energy in the form of heat to a cooler fluid without physical contact between the two fluids. A heat exchanger can provide heat or cooling to a process. Heat exchangers can be found in the following categories: shell and tube, plate and frame, spiral, and air cooled. A typical shell-and-tube heat exchanger is composed of a series of tubes surrounded by a shell. The tubes are typically connected

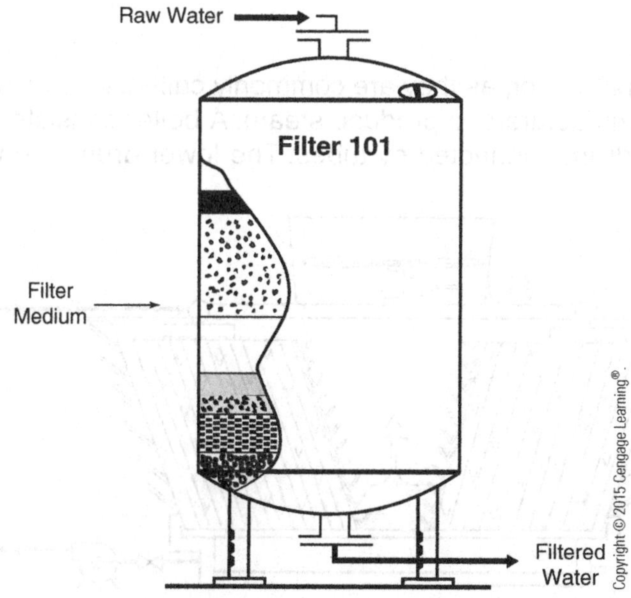

Figure 1.15 *Filter*

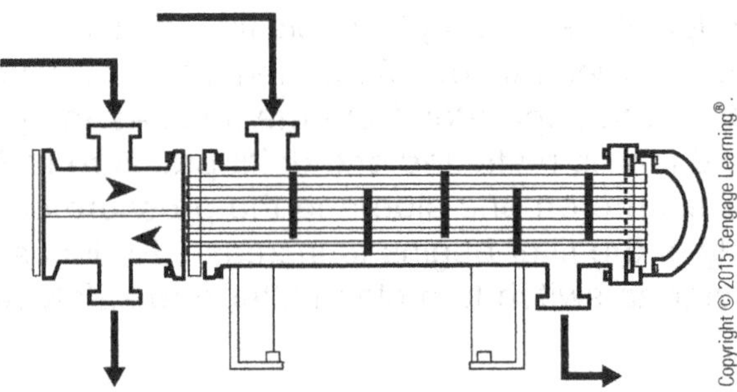

Figure 1.16
Shell and Tube Heat Exchanger

to a fixed tube sheet and are supported by a series of internal baffles. The shell-and-tube cylinder has a water box or head securely attached on the inlet and outlet side. A tube inlet and tube outlet admit fluid through the tubes. A shell inlet and outlet admit flow through the shell. Figure 1.16 illustrates the typical layout for a shell-and-tube heat exchanger.

Cooling Towers

Cooling towers are used by industry to remove heat from water. The internal design of the tower ensures good air and water contact. Hot water transfers heat to cooler air as it splashes on the boards inside the tower. The major portion of heat is stripped from the water by evaporation. The basic components of a cooling tower include a water basin, pump, and water makeup system at the base of the cooling tower. The internal frame is made of pressure-treated wood or plastic and is designed to support the internal components of the tower. Some of these components include the fill or splashboards and drift eliminators. Louvers on the side of the cooling water tower admit air into the device. A hot-water distribution system is typically located on the top of the cooling tower fill. A fan may be used to enhance airflow through the cooling tower. Figure 1.17 shows a typical cooling tower.

Boilers

Steam generators, or, as they are commonly called, **boilers**, are used by industrial manufacturers to produce steam. A boiler consists of an upper and a lower drum connected by tubes. The lower drum and water tubes

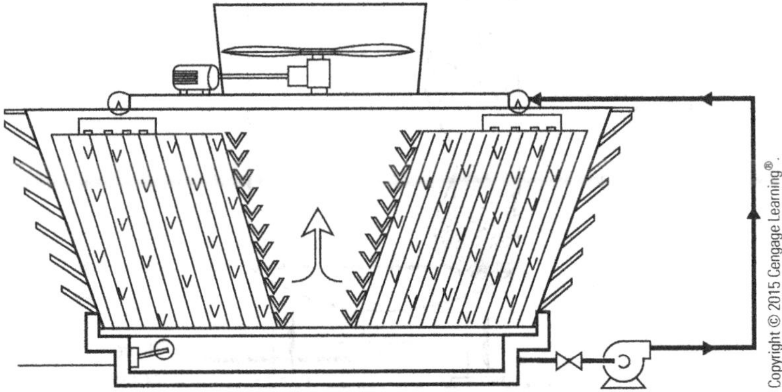

Figure 1.17
Cooling Tower

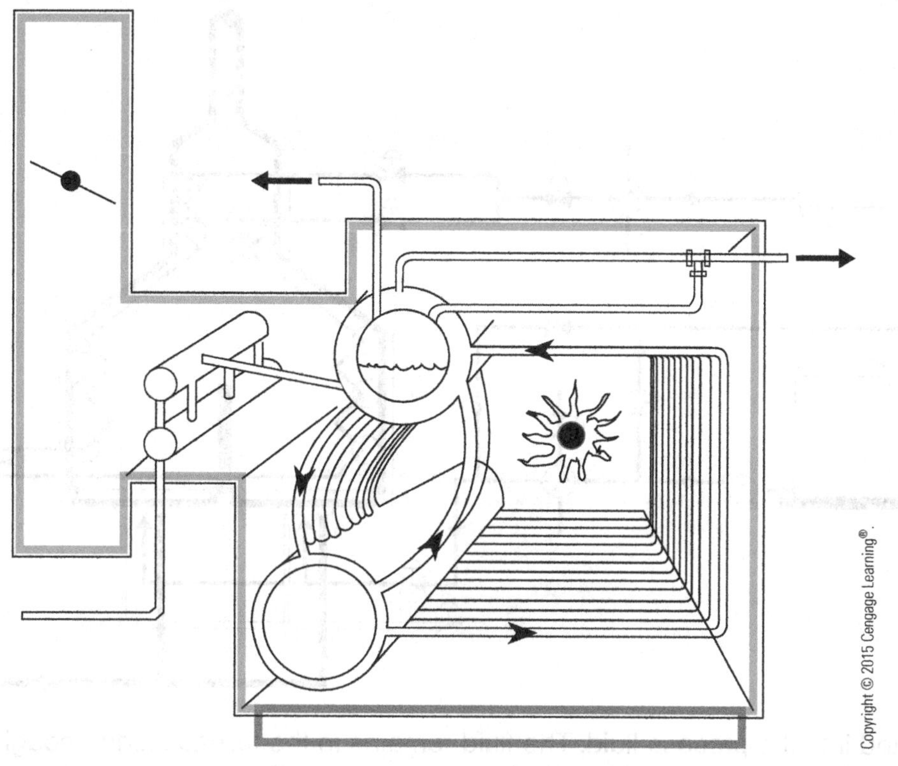

Figure 1.18
Water-Tube Boiler

are completely filled with water, while the upper drum is only partially full. This vapor cavity allows steam to collect and pass out of the header. Water is carried through tubes that flow into a heated chamber. As heat is applied to the water-generating tubes and drums, the water circulates around the boiler, down the down-comer tube, into the lower drum, and back up the riser tube and steam-generating tubes of the furnace. During normal operating conditions, steam rises into the upper drum and out the steam header. Steam generated in the boiler is replaced by makeup water. Figure 1.18 is an illustration of a water-tube boiler.

Furnaces

Fired heaters, or furnaces, are used to heat up raw materials so they can produce products such as gasoline, oil, kerosene, plastic, and rubber. Furnaces are used in many processes, including crude distillation, cracking, and olefins production. The basic components of a fired heater include a tough metal shell surrounding a firebox, a convection section, and a stack. The inside of the furnace is lined with a special refractory designed to reflect heat. A battery of tubes passes through the connection and radiant sections and into a common insulated header that passes out of the furnace. Fluid flow is balanced through the tubes to prevent equipment or product damage. Airflow and oxygen content are controlled through burner and damper adjustments. Figure 1.19 illustrates the basic layout of a furnace. The heat released by the burners is transferred through the tubes

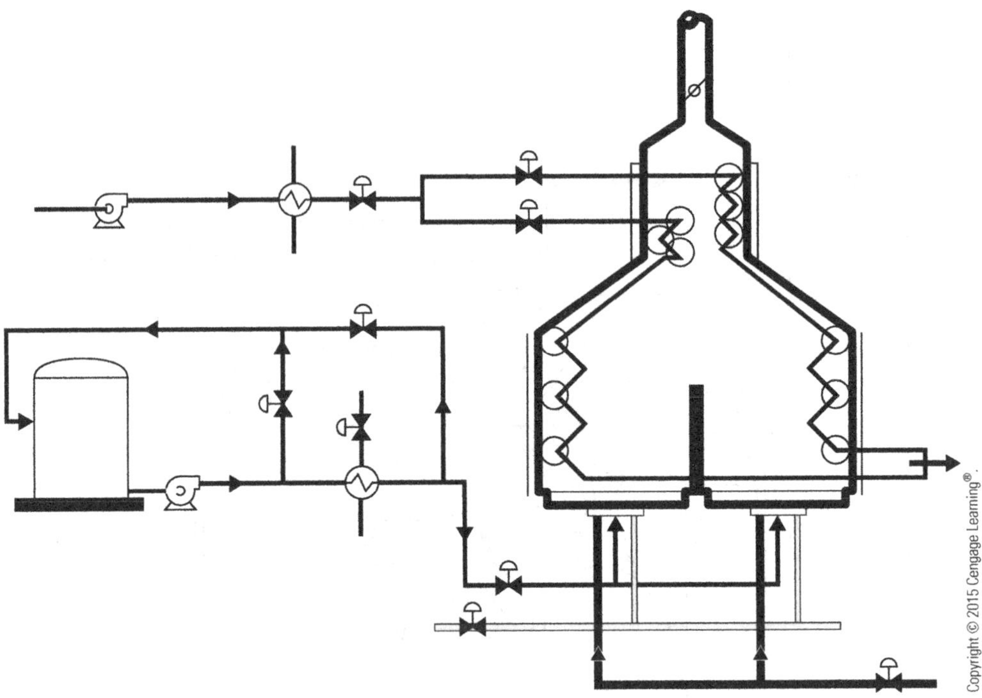

Figure 1.19
Fixed Heaters

and into the process fluid. The fluid remains in the furnace long enough to reach operating conditions before exiting for further processing. Furnace designs include cylindrical, cabin, and box.

Reactors

A **reactor** is a vessel used to convert raw materials into useful products through chemical reactions. Reactors are designed to operate under a variety of conditions. They combine raw materials with a catalyst, gases, pressure, or heat. A catalyst is a material designed to increase or decrease the rate of a chemical reaction without becoming part of the final product. The shape and design of a reactor are dictated by the application it will be used in. Reactors are used in a variety of processes and systems. The basic components of a reactor include a shell, a heating or cooling device, two or more product inlet ports, and one outlet port. A mixer may be used to blend the materials together. Figure 1.20 is an illustration of a simple mixing reactor.

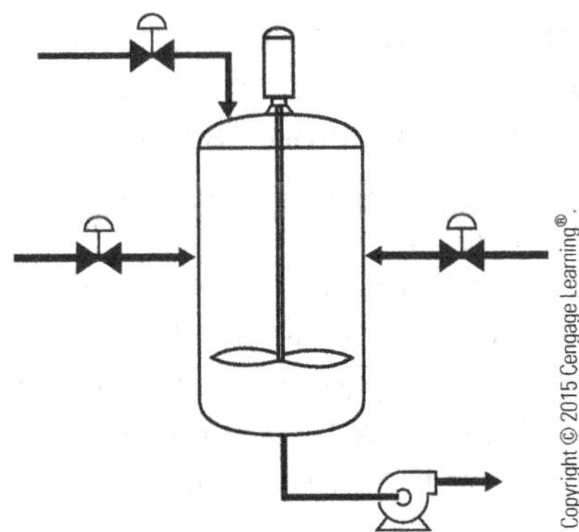

Figure 1.20
Mixing Reactor

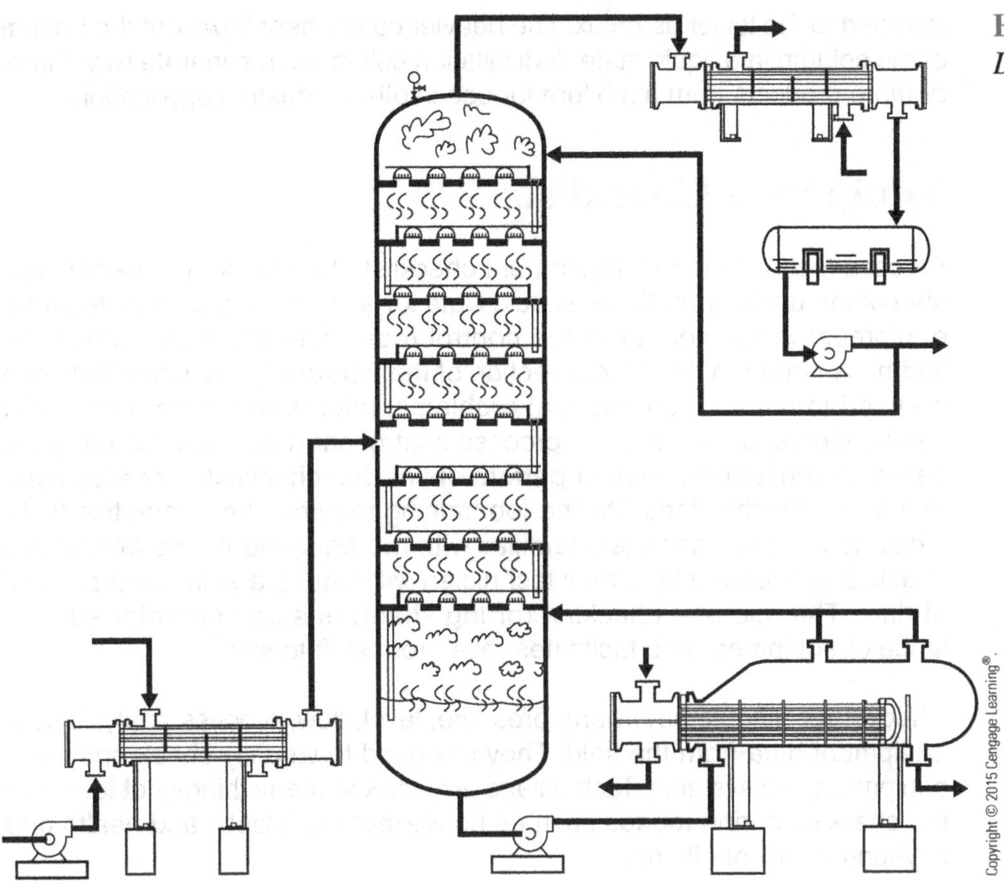

Figure 1.21
Distillation Column

Distillation Columns

A simple distillation system (Figure 1.21) is composed of a feed preheater, distillation tower, overhead condenser, accumulator, and reboiler. The preheated feed enters a feed tray or section in the tower. Part of the mixture vaporizes while the rest begins to drop into the lower sections of the tower. An exchanger known as a reboiler maintains energy balance on the column. **Distillation columns** or towers come in two designs: plate (tray) or packed.

As the vapor rises up the tower, it contacts the liquid descending the tower. This contact occurs on trays or the surface of packing and results in the mass transfer of material from the liquid phase to the vapor phase and from the vapor phase to the liquid phase. At each tray, the vapor leaving the tray has a higher concentration of the lower boiling material than the vapor entering the tray. Also, the liquid leaving the tray has a higher concentration of the higher boiling material than the liquid entering the tray.

The lightest component goes out the top of the column in a vapor state and is passed over the cooling coils of a shell and tube condenser. As the hot vapor comes into contact with the coils, it condenses and is collected in the overhead accumulator. Part of this product is sent to storage; the other is returned to the tower as reflux. The heavier component flows out the bottom of the column in a liquid state. A distillation column can separate two, three, or more products from a mixture for use in other industrial applications.

Equipment Checklists

Process technicians use equipment checklists to provide a current snapshot of the unit's operational status. Data for a checklist are collected from equipment in the field or in the control room. The checklist guides the technician through the critical section of the operating unit. Checklists can be used to pinpoint operational problems. Information from a checklist can be plotted on a statistical process control chart and used to compare past with present equipment performance. Unit checklists are also used to train new technicians. On-the-job training requires the apprentice technician to become intimately familiar with the equipment. The first time a checklist is filled out by a new technician will require a significant amount of time. The use of a checklist (or log sheet) ensures operator surveillance of equipment and facilitates relief from shift to shift.

Checklists collect temperature, pressure, level, flow, process analysis, and equipment data from the field. They are used to visually check operating equipment and systems. Technicians also look at many things not found on the checklist during rounds such as housekeeping, safety, and health and environmental conditions.

Summary

The chemical processing industry is composed of refineries and plants for petrochemical, paper and pulp, power generation, and food processing. Process technicians inspect and maintain equipment, place and remove equipment from service, complete checklists, control documentation, respond to emergencies, and troubleshoot system problems. Process technicians use a variety of hand tools to perform their jobs. In addition to having a knowledge of basic tools, process technicians must have a thorough understanding of equipment and systems.

Equipment can be classified as rotary or stationary. Steam turbines and electric motors are two of the most popular devices used by industry as drivers for rotary equipment.

Industrial drivers are connected to rotating equipment with a coupling. Rotating equipment uses seals and bearings to maintain operational integrity. Seals prevent leakage between internal compartments. Radial bearings are designed to prevent up-and-down and side-to-side movement of the rotating shaft; axial bearings are designed to prevent back-and-forth movement of the shaft.

Pumps transfer liquids from place to place. Positive displacement pumps can be classified as rotary or reciprocating. Reciprocating pumps are characterized by a back-and-forth motion, whereas rotary pumps move in a circular rotation. Dynamic pumps can be classified as centrifugal or axial. The centrifugal pumps use centrifugal force to move liquids; the axial pump pushes liquids along a straight line.

A compressor is designed to accelerate or compress gases. Compressors are closely related to pumps. They come in two basic designs, positive displacement and dynamic.

Tanks and pipes are stationary equipment designed to store and contain fluids. Filters are another type of stationary equipment; they remove suspended solids from fluids.

Valves are devices used to control the flow of fluids. Valves come in a variety of shapes, sizes, and designs that throttle, stop, or start flow. The more common designs include gate, globe, and ball.

Heat exchangers transfer energy in the form of heat between two fluids, which do not physically contact each other. Cooling towers remove heat from water.

Fired heaters, or furnaces, are used to heat large volumes of raw materials. Another type of furnace is a boiler. Boilers provide steam for industrial applications.

Reactors are used to combine raw materials, heat, pressure, and catalysts in the right proportions. Distillation towers separate chemical mixtures by the boiling points of the substances in the mixture.

Review Questions

1. Describe the primary purpose of valves.
2. Explain how a centrifugal pump operates.
3. Name the primary difference between a pump and a compressor.
4. What is the purpose of a steam turbine?
5. Explain the purpose of bearings and seals.
6. Describe the basic hand tools used by process technicians.
7. List four methods process technicians use to inspect equipment.
8. What are the three basic principles that allow an electric motor to operate?
9. What is the difference between rotary and stationary equipment?
10. How is power transmission in rotary equipment classified?
11. Explain the importance of lubrication for a pump, compressor, or turbine.
12. What is the difference between a rigid and a flexible coupling?
13. Explain the operation of a positive displacement pump.
14. What is the primary purpose of a mixing reactor?
15. What is the definition of a distillation column?
16. What is the purpose of a fired heater?
17. What is the primary purpose of a boiler?
18. What is the purpose of a cooling tower?
19. What are equipment checklists primarily used for?
20. Describe chain and belt drives.

Valves

Objectives

After studying this chapter, the student will be able to:

- Explain the purpose of valves in industrial manufacturing.
- Identify the basic components of a gate valve.
- Identify the basic components of a globe valve.
- Explain the operation and design of a ball valve.
- Identify the basic components of a check valve.
- Describe the design and operation of a butterfly valve.
- Examine the design and operation of a plug valve.
- Describe the design and operation of a diaphragm valve.
- Describe the design and operation of a relief valve and a safety valve.
- Describe the design and operation of automatic valves.

Key Terms

Accumulation—the pressure difference (ΔP) between initial lift pressure and full lift pressure on a relief valve.

Actuator—a device that controls the position of the flow-control element on a control valve by automatically adjusting the position of the valve stem.

Angle valve—a valve that operates by admitting fluid flow to the gate or plug and redirecting it 90° out the discharge port.

Antiseize compound—lubricant used on exposed valve stem threads.

Ball valves—named for the ball-shaped, movable element in the center of the valve.

Block valve—any valve that is intended to block flow; also called an *isolation valve*. The term generally refers to gate valves.

Bonnet—a bell-shaped dome mounted on the body of a valve.

Bridgewall markings—manufacturer information on the body of a valve.

Butterfly valves—characterized by their disc-shaped flow-control element, which pivots from its center.

Check valves—mechanical valves that prevent reverse flow in piping.

Control loop—a collection of instruments that work together to automatically control a process; usually consists of a sensing device, a transmitter, a controller, a transducer, and an automatic valve.

Control valves—automated valves used to regulate and throttle flow; typically provide the final control element of a control loop.

Diaphragm valve—a device that uses a flexible membrane to regulate flow.

Disc—a device made of metal or ceramic that fits snugly in the seat of a valve to control flow.

Flange—a device used to connect (bolt) piping to industrial equipment.

Flow-control element—the part of a valve that regulates flow; that is, the gate or the disc.

Fluid—of the three forms of matter—solids, liquids, and gases—liquids and gases are considered fluids.

Gate—the flow-control element of a gate valve.

Gate valve—a device that places a movable metal gate in the path of a process flow.

Globe valve—a device that places a disc in the path of a process flow.

Handwheel—attached to the valve stem and used to control the position of the flow-control element of a valve.

Multiport valve—has multiple inlets and/or outlets in specialized piping systems to divert flow direction, allowing fluid sources to be switched.

Needle valve—a type of globe valve that has a needle-shaped element that fits snugly into the seat.

Packing—a specially designed material used to stop fluids from entering or escaping; packed around the shaft (stem) of a valve, or shaft of a pump.

Packing gland—a mechanical device that contains and compresses packing.

Plug valve—a device that has a plug-shaped element; used for on/off service.

Safety/relief valve—device set to automatically relieve pressure in a closed system at a predetermined set point; relief valves are used for liquids; safety valves are used for gases.

Stem—a metal shaft attached to the handwheel and flow-control element of a valve.

Stuffing box—the section of a valve that contains packing.

Three-way valve—a valve with three ports (one inlet and two outlets) used to divert flow direction.

Throttling—reducing or regulating flow below the maximum output of a valve.

Trim—the flow-control element and seats in a valve.

Valve capacity—the total amount of fluid a valve will pass with a given pressure difference when it is fully open.

Warping—a term used to describe temperature changes and pipe expansion that cause a valve to seize, or "warp." Closing a valve too quickly can cause warping, and warping can cause a valve to stick.

Valve Applications and Theory of Operation

Process plants are a network of complex systems and processes. Just as arteries, veins, and the heart are vital to human life, pipes, valves, and pumps are indispensable in a process plant. The primary purpose of a valve is to direct and control the flow of **fluids** by starting, stopping, and **throttling** (restricting) flow to make processing possible. Valves are designed to withstand pressure, temperature, and flow and can be found in homes and industry across the world.

The common valves (Figure 2.1) found in the manufacturing environment are **gate valves, globe valves, ball valves, check valves, butterfly valves, plug valves, needle valves, three-way valves, diaphragm valves, relief** and **safety valves, angle valves**, and **multiport valves**. Valves normally are selected for a specific purpose. As you continue to read through this chapter, you will notice the variety of valve designs that exist. Operators need to be aware of how each valve works and the specific service for which it was designed.

Classification of Valves

Process operators classify valves by (1) **flow-control elements**, (2) function, and (3) operating conditions such as pressure, flow, or temperature. The most common way to classify valves is by the valve's flow-control

Figure 2.1
Valves

element design. This part of the valve controls or regulates the flow of fluid through the device. Some valves have movable metal gates, balls, plugs, diaphragms, **discs**, needles, or even butterfly-shaped elements. Most valves are named for the type or design of the flow-control element. Valves that are used for isolation are classified as **block valves**. While gate valves are the most common valve used for isolation, any valve can be used for this type of service. Another term associated with valve operation is valve capacity. **Valve capacity** is a term used to describe the total amount of fluid a valve will pass with a given pressure difference when it is fully open.

Bridgewall Markings

The body of the valve contains **bridgewall markings**, which provide manufacturer information specific to the valve. The bridgewall markings include the following letters and symbols:

W = Low-temperature water service
O = Low-temperature oil service
G = Low-temperature gas service
→ = Flow direction

Additional information on size and pressure rating is sometimes found on a plate attached to the **handwheel**.

Gate Valves

One of the more common valves found in industry is a gate valve. A gate valve places a movable metal gate in the path of a process flow in a pipeline. The gates are sized to fit the inside diameter of a pipe, so very little restriction occurs when it is in the open position. Valves vary in size from 0.125 inches to several feet. Gate valves typically are operated in the "wide

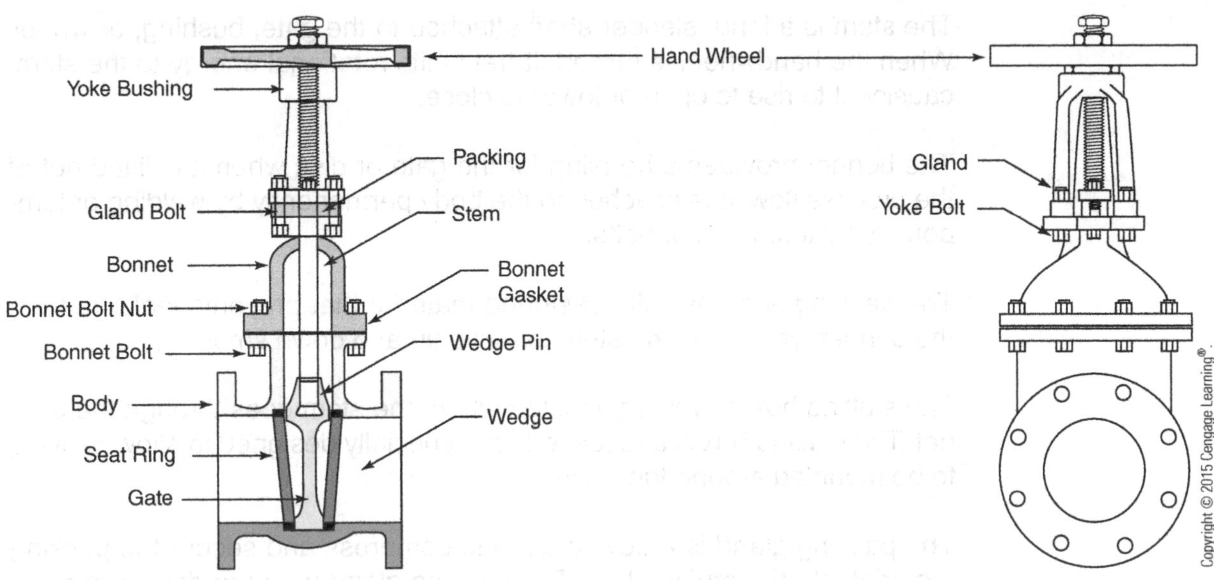

Figure 2.2 *Gate Valve Components*

open" or "completely shut" position. This type of valve is used where flow rates are not restricted. Gate valves should not be used to throttle flow for extended periods. Turbulent flow rates across the valve body will cause metal erosion, seat damage, and damage to the flow-control element, which can prevent the valve from blocking the flow completely.

The seats in a gate valve fall into two categories: replaceable and fixed. Smaller valves typically have fixed or cast seats because it is easier and more cost efficient to replace the valve than to replace the seats. In most cases, a gate valve has two vertical seating surfaces. The edges of the seats match up with the parallel discs or wedge.

The typical gate valve consists of a gate, body, seating area, **stem**, **bonnet**, **packing**, **stuffing box**, **packing gland**, and handwheel (Figure 2.2). The gate can be wedge shaped or may consist of parallel discs. It can be composed of a variety of materials. The gate is placed directly in the path of a process flow when it is shut and is lifted completely out of the way when open.

The body is the largest part of the valve. The body can be connected to the process piping in three ways: **flanges**, threaded connections, or welding. The rest of the valve is attached to the body.

The seating area consists of two fixed surfaces or rings inside the body of the valve that the gate closes against to stop flow. The seating area falls into two categories: replaceable or fixed. Seats must provide a clean mating surface for the gate to seal properly. The seat can be fabricated or cast as part of the valve, press-fit, threaded, or welded into place. Note that high-temperature and high-pressure situations may require a combination of threading and welding.

The stem is a long, slender shaft attached to the gate, bushing, or wheel. When the handwheel is turned, it transmits rotational energy to the stem, causing it to rise to open or lower to close.

The bonnet provides a housing for the gate or disc when it is lifted out of the process flow. It is attached to the body permanently by welding or temporarily by threading or bolts.

The packing is a specially designed material that prevents leakage from the bonnet, yet allows the stem to move up and down smoothly.

The stuffing box is typically located where the stem goes through the bonnet. The stuffing box is a recessed area specially designed to allow packing to be mounted around the stem.

The packing gland is a device used to compress and secure the packing material into the stuffing box. The packing gland nuts are designed to be evenly tightened by a technician to stop leaks.

The handwheel is attached to the valve stem. The handwheel transfers rotational energy to the stem. This rotational energy controls the movement of the flow-control element. Turning the handwheel clockwise closes the valve. When the handwheel is turned counterclockwise, it is opened.

Common Gates

The gate on a gate valve comes in a variety of shapes and sizes. The most common designs include the solid wedge, solid split gate, and parallel discs (Figure 2.3).

The solid wedge gate forms a positive seal when the solid metal gate slips into the seat. A set of guides keeps the disc aligned. Because the solid wedge gate is designed to fit snugly into the seat, it should never be overtightened. Excessive force will damage the seats, so the technician should shut the valve until it seats and then back off slightly.

Figure 2.3
Gates

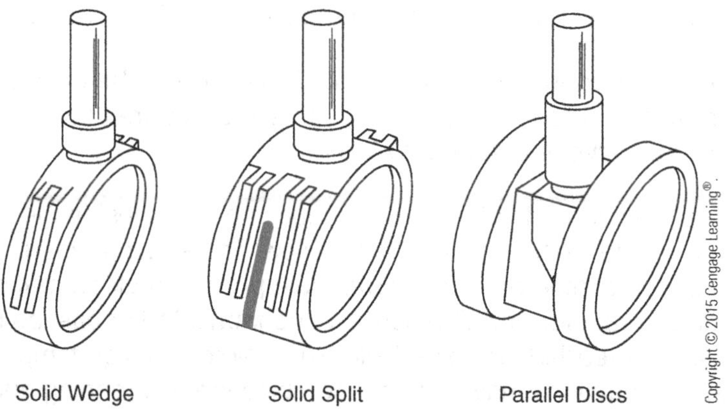

The solid split gate forms a positive seal when the split metal gate slips into the seat. The solid split gate is designed like the solid wedge but has a split element feature that makes it operate like a double disc. The split, nonjamming feature lets this type of gate handle higher temperatures than the solid wedge. Fluid pressure is used to seat this type of gate.

The parallel, or double, discs gate is composed of two separate discs mounted to a shaft. Some parallel discs have a spring located between the two discs, whereas others are attached to a solid metal core. The double discs gate valve is designed for high-temperature service because of its nonjamming design. As fluid enters this type of valve, it pushes against one of the discs, compressing the spring or flow-control element and forcing the opposite disc snugly into its seat. System pressure helps position the gate in its seat. The higher the pressure, the more tightly the disc fits into the seat.

Gate Valve Materials

Gate valves (Figure 2.4) are designed to be used in a variety of process conditions. The specific condition dictates what type of material the valve will be made of. For example, stainless steel gate valves are used in corrosive, high- and low-temperature services. Specialty alloy gate valves are used in high-pressure, high-pressure service. Bronze gate valves are used in low-temperature, low-pressure service. Brass gate valves are used in low-temperature, low-pressure service. Cast iron gate valves are used in water, lubrication, and some low-pressure steam services.

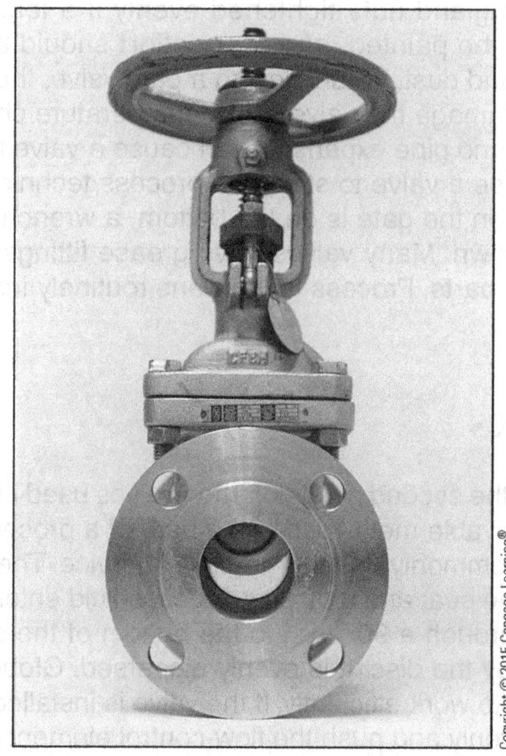

Figure 2.4
Stainless Steel Gate Valve

Gate valves are considered excellent devices for process services that are used intermittently. It is not uncommon to find gate valves that have not been operated for many years.

Rising Stem and Nonrising Stem

The stem on a gate valve comes in two designs: rising stem and nonrising stem. Located at the top of the gate valve is the handwheel. The handwheel is attached to a bushing, which is attached to the threaded stem. As the handwheel is turned counterclockwise, the stem in the center of the handwheel rises. As the stem rises, the gate is lifted out of the valve body, allowing fluid to flow. A process technician can look at this type of valve and tell if it is open or closed by checking the position of the stem. Another type of rising stem valve is threaded at the bottom of the stem. In this type of valve, the handwheel is firmly attached to the stem and rises with it as the valve is opened.

A nonrising stem gate has a collar that keeps the stem from moving up or down. The handwheel is attached firmly to the stem of a nonrising gate. Turning the handwheel screws the stem into or out of the gate. You cannot look at this type of valve and tell if it is open or closed.

Maintenance

Maintaining equipment is an important part of an operator's job. Valve stem threads exposed to weather need to be lubricated with **antiseize compound** to keep them operating properly. The packing should be inspected, and the gland nuts tightened evenly if a leak is found. Valve stems should not be painted, and every effort should be made to keep them free of dirt and dust. When closing a gate valve, the operator should take care not to damage the valve seats. Temperature changes, closing a valve too quickly, and pipe expansion can cause a valve to warp. Because **warping** can cause a valve to stick, the process technician should close valves slowly. When the gate is on the bottom, a wrench should never be used to snug it down. Many valves have grease fittings, which admit lubricant to moving parts. Process technicians routinely inspect, clean, and lubricate valves.

Globe Valves

Globe valves are the second most common valves used in industry. A globe valve places a movable metal disc in the path of a process flow. This type of valve is most commonly used for throttling service. The disc is designed to fit snugly into the seat and stop flow. Process fluid enters the globe valve and is directed through a 90° turn to the bottom of the seat and disc. As the fluid passes by the disc, it is evenly dispersed. Globe valves must be installed properly to work efficiently. If the valve is installed backward, it will tend to wear unevenly and push the flow-control element down.

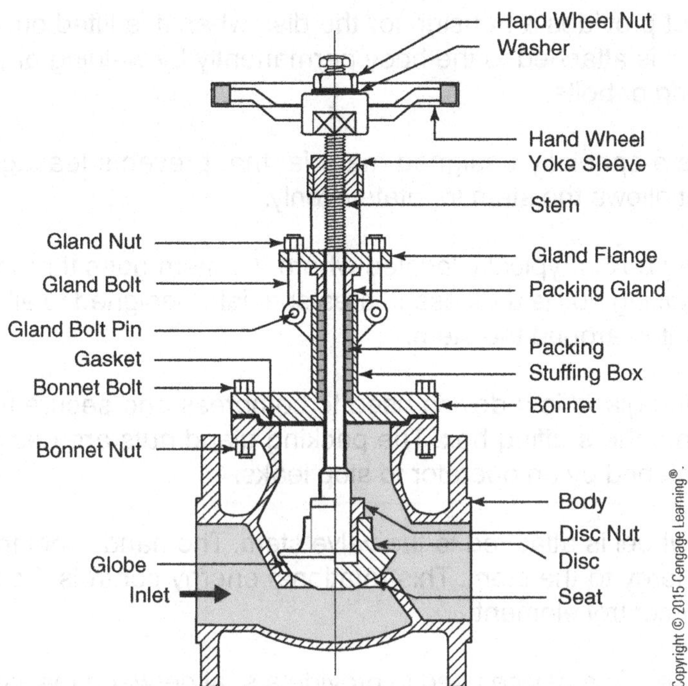

Figure 2.5
Globe Valve Components

The typical globe valve consists of a disc, body, seating area, stem, bonnet, packing, stuffing box, packing gland, handwheel, and back seat (Figure 2.5). The disc is attached to the stem in three ways: slip joint, threading, or one-piece manufacturing. The disc can be classified as plug, ball, composition, or needle shaped. It is composed of a variety of materials. The disc, or flow-control element, rests in the seat, directly in the path of a process flow when it is shut. Unlike the gate valve, the globe valve is designed to be used for throttling. Flow can be regulated by the percentage of opening of the flow-control element.

The body is the largest part of the valve. The body can connect to the process piping in three ways: flanges, threaded connections, or welding. The rest of the valve is attached to the body.

The seating area comes in four designs: cone-shaped, beveled flat surface, O-ring or washer, and tapered or needle-shaped cone. The seating area is where the flow-control element closes against the body of the valve to stop flow. The seating area can be replaceable or fixed. Seats must provide a clean mating surface to the flow-control element to seat properly. The seat can be fabricated or cast as part of the valve, press-fit, threaded, or welded into place. High-temperature and high-pressure situations may require a threading and welding combination.

The stem is a long, slender shaft attached to the disc, bushing, or wheel. Turning the handwheel transmits rotational energy to the stem.

The bonnet provides a housing for the disc when it is lifted out of the process flow. It is attached to the body permanently by welding or temporarily by threading or bolts.

Packing is a specially designed material that prevents leakage from the bonnet yet allows the stem to rotate evenly.

The stuffing box is typically located where the stem goes through the bonnet. The stuffing box is a recessed area specially designed to allow packing to be mounted around the stem.

The packing gland is a device used to compress and secure the packing material into the stuffing box. The packing gland nuts are designed to be evenly tightened by an operator to stop leaks.

The handwheel is attached to the valve stem. The handwheel transfers rotational energy to the stem. This rotational energy controls the movement of the flow-control element.

The back seat is a device used to provide a seal between the stem and the bonnet and to protect the packing from excess pressure inside the valve. It is used in steam service. The back seat is manufactured as part of the stem. When the stem is in the full-open position, the disc-shaped back seat firmly connects with the mating surface of the bonnet seat.

Four Common Disc Designs

The disc on a globe valve comes in a variety of shapes and sizes. The four most common designs are plug, ball, composition, and needle (Figure 2.6). The plug-type disc is used for throttling. It is equipped with renewable seat rings on a cone-shaped mating surface. It is the best device for throttling and in situations with wide temperature and pressure variations. The ball element seats against a beveled, ball-shaped, or flat surface. It is designed for on/off applications that have minimal throttling. The composition disc can be adapted for use with a variety of temperatures and flow rates. The composition disc is renewable. Mating surfaces use a rubber O-ring or washer. The needle disc is used for microthrottling service.

Globe Valve Materials

Globe valves are designed to be used in a variety of process conditions. The specific condition dictates the type of material the valve will be made of. For example, the stainless steel globe valve is used in corrosive, high- and low-temperature services. The specialty alloy globe valve is used in high-temperature and high-pressure services. Some common alloys used are nickel and iron or steel and titanium. The bronze globe valve is used in low-pressure and low-temperature systems. The brass globe valve is used

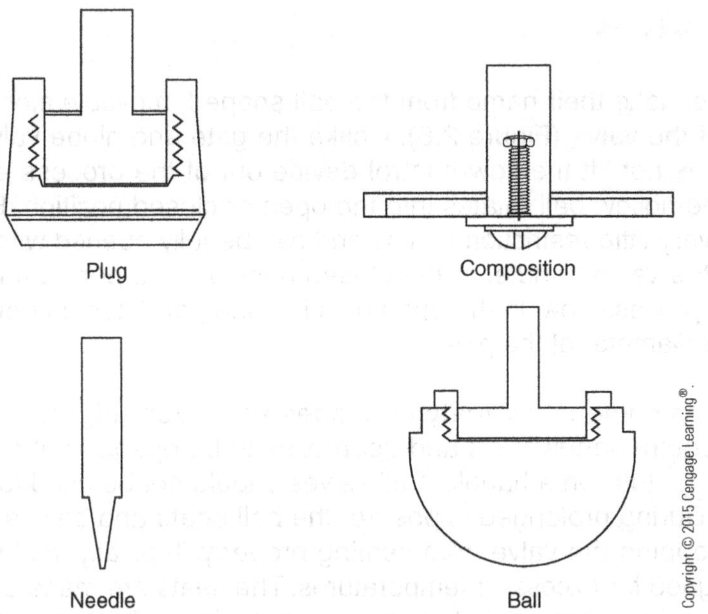

Figure 2.6
Globe Valve Disc Designs

in low pressure and low temperature. The cast iron globe valve is used in water lubrication and some low-pressure steam systems.

When compared to gate valves, globe valves (Figure 2.7) have a much greater pressure drop across the valve. Globe valves are designed to be installed in high-use areas. If a globe valve is installed in a low-use area, it tends to plug up even though it has a self-cleaning type design. Because of the 90° turns in a globe valve, it cannot be unplugged with a straight rod device.

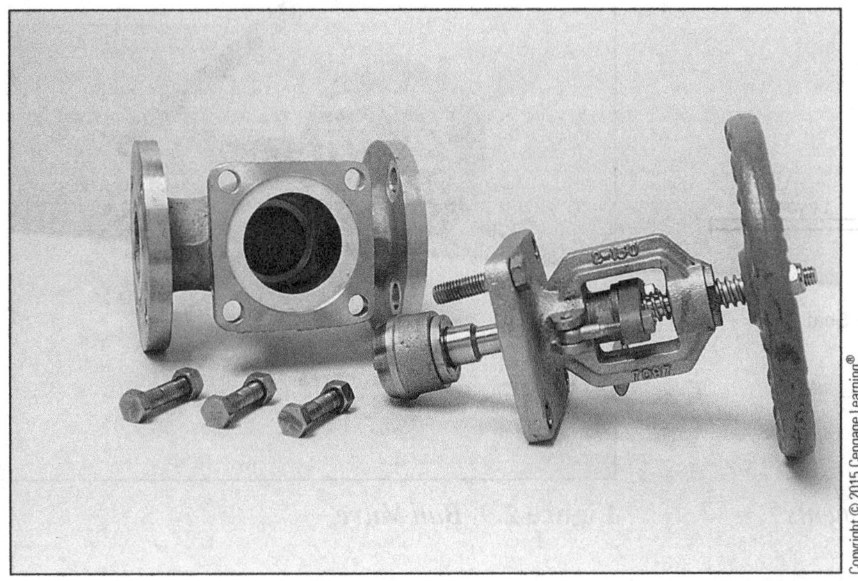

Figure 2.7
Globe Valve

Ball Valves

Ball valves take their name from the ball-shaped, movable element in the center of the valve (Figure 2.8). Unlike the gate and globe valves, a ball valve does not lift the flow-control device out of the process stream. Instead, the hollow ball rotates into the open or closed position. Ball valves provide very little restriction to flow and can be fully opened with a quarter turn on the valve handle. In the closed position, the port is turned away from the process flow. In the open position, the port lines up perfectly with the inner diameter of the pipe.

Ball valves come in a variety of shapes and sizes (Figure 2.9). Larger valves require handwheels and gearboxes to be opened but only require one-quarter turn on a handle. Ball valves should not be used for throttling service. During prolonged exposure, the ball seats and ball can be damaged, stopping the valve from sealing properly. Typically, ball valves are not designed for excessive temperatures. The seats are made of a plastic-coated material that tends to break down under high temperatures. Process technicians should be familiar with the temperature specifications of the valves.

Ball valves do not generally seal as well as globe valves in high-pressure service. Some ball valves (multiport valves) are designed with multiple ports, so an operator can switch fluid sources without stopping flow.

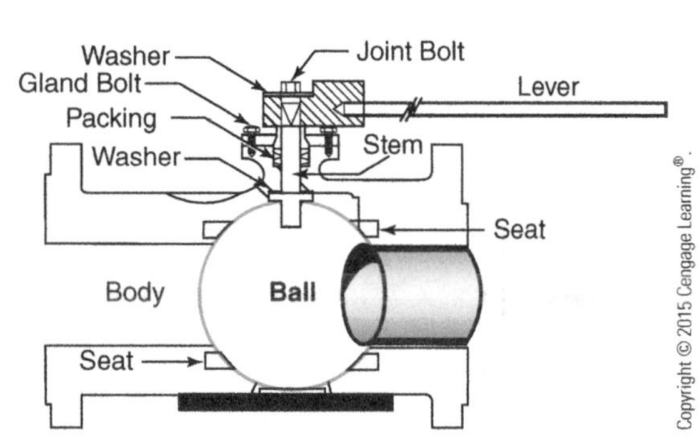

Figure 2.8 *Ball Valve Components*

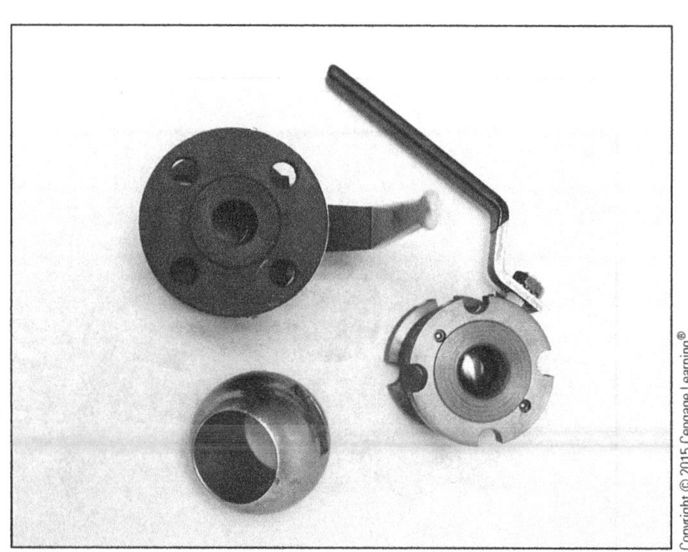

Figure 2.9 *Ball Valve*

Check Valves

A check valve is designed to prevent reverse flow and to avoid possible contamination or damage to equipment. The check valve limits backflow but is not considered a tight shutoff. Check valves come in a variety of designs and applications. Check valves have flow direction stamping on the valve body.

A typical check valve design is the swing check, which has a hinged disc that slams shut when flow reverses (Figure 2.10). Flow lifts the disc and keeps it lifted until flow stops or reverses. The body of the check valve has a cap for easy access to the flow-control element. Another design is the lift check, which has a disc that rests on the seat when flow is idle and lifts when flow is active (Figure 2.11). Special guides keep the disc in place. Like the swing check, it is designed to close when flow reverses. Lift checks are ideal for systems in which flow rates fluctuate. The lift check is more durable than the swing check. In the horizontal or vertical lift check design, a piston or ball is lifted up and out of the seat by process flow. A third design is the ball check design, which has a ball-shaped disc that rests on a beveled, round seat (Figure 2.12). The ball is down when flow is idle and up when flow is active. Special guides keep the ball disc in place. Like the swing check, it is designed to close when flow reverses. Ball checks are ideal for systems in which flow rates fluctuate or the fluid contains some solids. The ball check is as durable as a lift check and more durable than a swing check. A fourth design is the stop check design, which has characteristics of a lift check and a globe valve (Figure 2.13). In the closed position, the stop check disc is firmly seated. In the open position, the stem rises out of the body of the flow-control element and acts as a guide for the disc. In the open position, the stop check functions like a lift check with one exception. The degree of lift can be controlled.

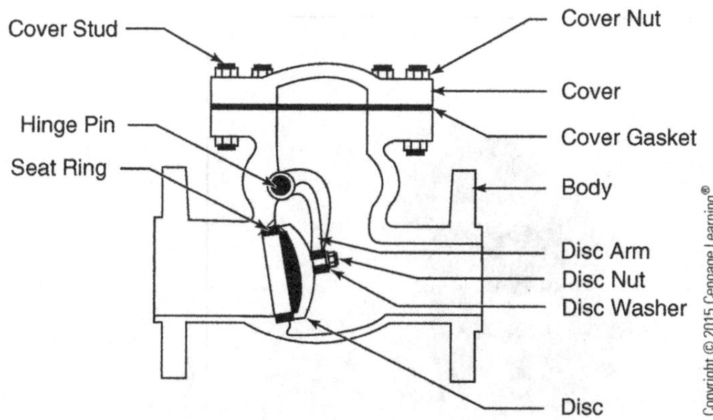

Figure 2.10 *Swing Check Valve*

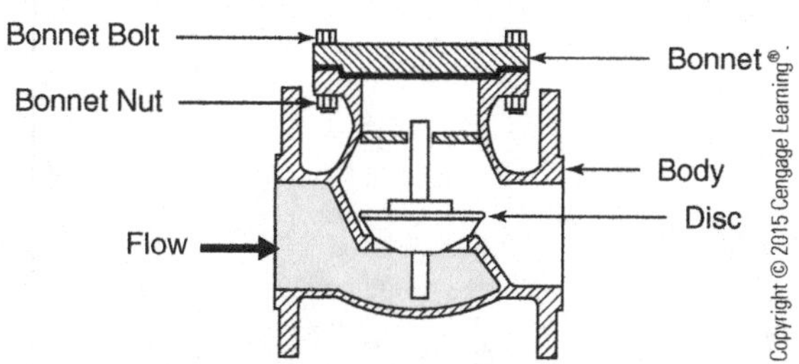

Figure 2.11 *Lift Check Valve*

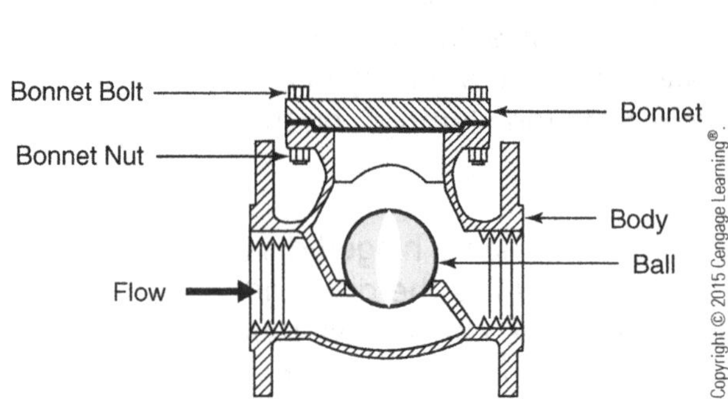

Figure 2.12 *Ball Check Valve*

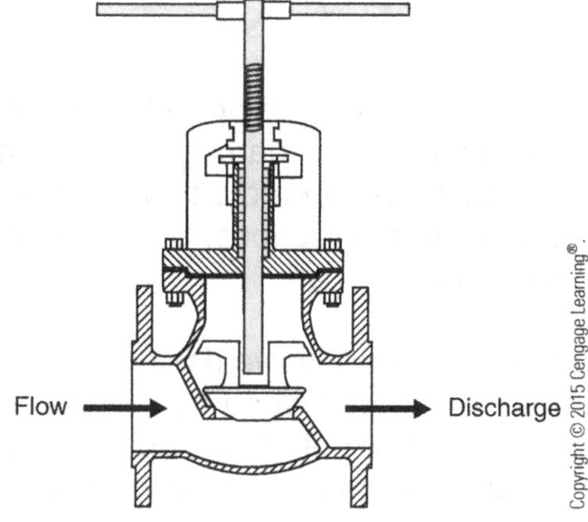

Figure 2.13 *Stop Check Valve*

Check valves operate by gravity and must be installed correctly. Ball, lift, swing, and stop checks come in horizontal and vertical designs.

Check valves are designed to be used in a variety of process conditions. The specific condition dictates the type of material the valve will be made of. For example, the stainless steel check valve is used on corrosive, high- and low-temperature services. The specialty alloy check valve is used in high-temperature and high-pressure services. Some common alloys used are nickel and iron or steel and titanium. The bronze check valve is used in low pressure and low temperature (Figure 2.14). The brass check valve is used in low pressure and low temperature. The cast iron check valve is used in water lubrication and some low-pressure steam services.

Figure 2.14
Bronze Swing Check

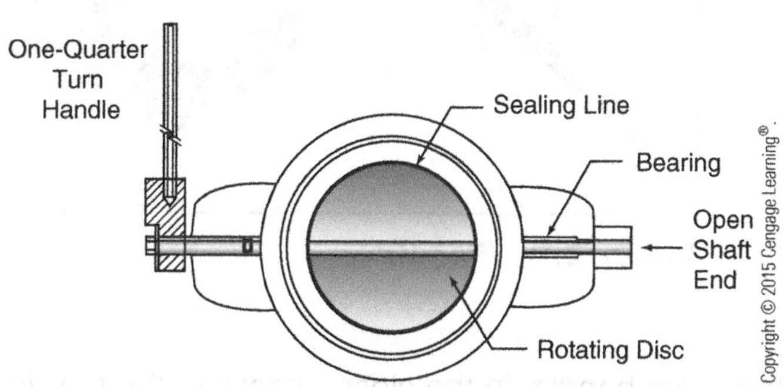

Figure 2.15 *Butterfly Valve Components*

Figure 2.16 *Butterfly Valve*

Butterfly Valves

Butterfly valves are commonly used for throttling and on/off service. The body of this type of valve is relatively small when compared with other valves and, therefore, occupies much less space in a pipeline. The flow-control element resembles a disc. A metal shaft extends through the center of the disc and allows it to rotate one-quarter turn (Figures 2.15 and 2.16). A one-quarter turn is all it takes to fully open or close the valve.

Technicians should be aware that butterfly valves are 100% open shortly after the valve handle passes one-eighth of a turn. During a throttling operation, a butterfly valve handle should be carefully secured. As flow enters the body of the valve, it contacts the disc and will cause it to open if the handle is not latched into position. Butterfly valves are used for throttling; however, it should be noted that they have nonuniform flow characteristics. Fifty percent open may provide near-maximum flow.

Butterfly valves are designed to be operated at low temperature and low pressure. They are commonly found in cooling water and heat exchanger systems throughout the process industry. The seats in a butterfly valve can be made of natural gum rubber or suitable plastics.

Plug Valves

Quick-opening, one-quarter turn plug valves are very popular in the process industry. The plug valve takes its name from the plug-shaped flow-control element it uses to regulate flow (Figures 2.17 and 2.18). Plug valves provide very little restriction to flow and can be opened 100% with

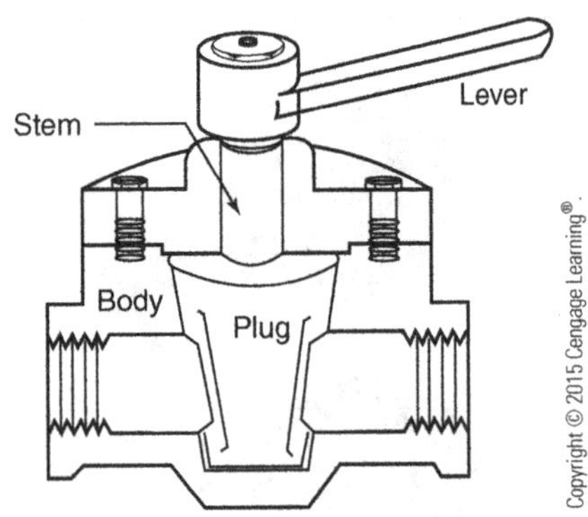

Figure 2.17 *Plug Valve Components*

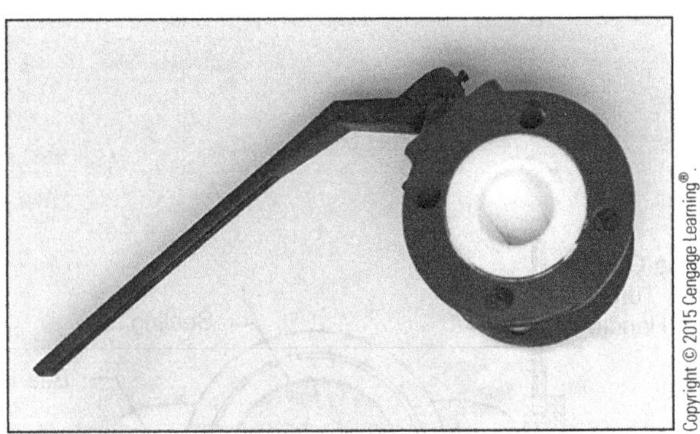

Figure 2.18 *Plug Valve*

a one-quarter turn on the valve handle. In the closed position, the port is turned away from the process flow. In the open position, the port lines up with the inner diameter of the pipe.

Plug Valve Design

Plug valves come in a variety of shapes and sizes. The plug valve is designed for fuel gas piping systems, low-pressure situations, slurry and lubrication service, on/off service, low-temperature service, and multiport operation. In on/off service, prolonged exposure can damage the seats. In low-temperature service, the seats are made of a plastic-coated material that tends to break down during higher temperatures.

Maintenance

With proper care and maintenance, a plug valve can last indefinitely. Lubrication plays a big part in the life of a plug valve. Plug valves depend on a lubricant inside the body of the valve. This lubricant helps the valve provide a leak-free seal.

A plug valve should never be used for throttling. Excessive wear will result if a plug valve is throttled, and the valve's ability to obtain a positive seal will be jeopardized. Most plug valves should be limited to temperatures under 480°F (248.9°C). Higher temperatures displace the grease and cause the plug and valve body to seize ("warp").

Diaphragm Valves

In a chemical plant, a variety of slurries, corrosive or sticky substances, are transferred from place to place. Standard valves would have a difficult time with this type of product, but diaphragm valves are specifically designed for the job. Diaphragm valves use a flexible membrane and seat to regulate flow. The handwheel operates just like the handwheel on a gate or globe valve. The stem is attached to a device called the *compressor*. The compressor

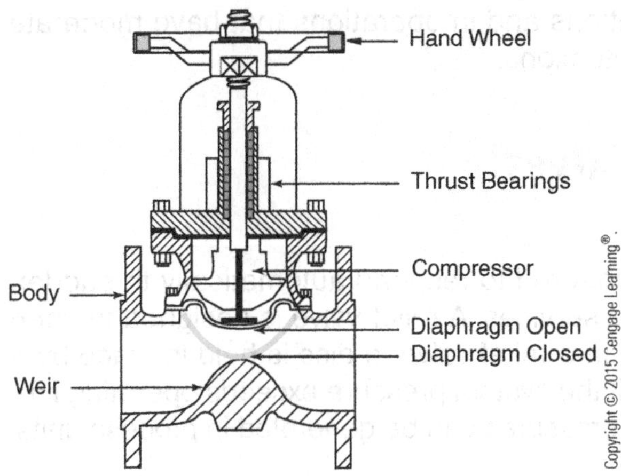

Figure 2.19 *Diaphragm Valve Components*

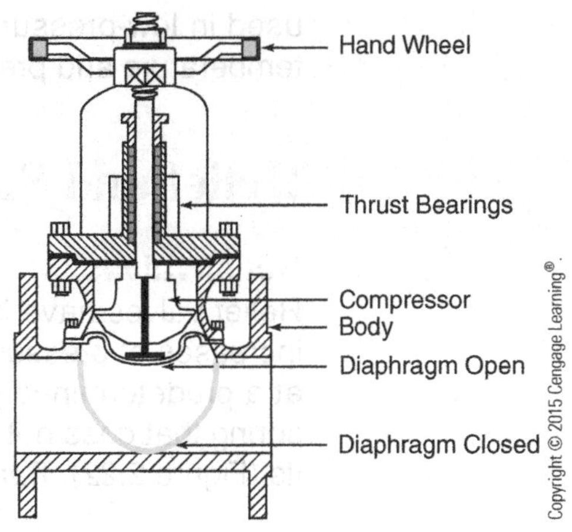

Figure 2.20 *Straight-Through-Diaphragm Valve*

pushes on a flexible diaphragm. The internal parts of the valve never come into contact with the process material. The diaphragm conforms to the setting on the handwheel. Diaphragm valves typically are used in low-pressure applications. The diaphragm valve seats are made of chemical-resistant plastic, rubber, or neoprene. This type of valve does not use packing.

Diaphragm valves come in two designs: weir and nonweir. The weir diaphragm valve has a weir (a dam) located in the body of the valve (Figure 2.19). Fluid must go over the top of the weir and under the diaphragm in order to exit. There is a large pressure drop across the body of the valve. It uses thick, durable diaphragm materials. The nonweir, or straight-through-diaphragm valve, has a flexible membrane that extends across the pipe (Figure 2.20). Pressure drop across the valve depends on the position of the diaphragm.

Diaphragm valves (Figure 2.21) can handle corrosive fluids and fluids that are ultrapure such as demineralized boiler feed water. They must be

Figure 2.21 *Weir Diaphragm Valve*

used in low-pressure applications and in operations that have moderate temperature and pressure fluctuations.

Relief and Safety Valves

Relief Valves

Relief valves have been engineered to respond automatically to sudden increases of pressure in liquid services. A relief valve is designed to open at a predetermined pressure. In a relief valve, a disc is held in place by a spring that does not open until the system pressure exceeds operating limits (Figure 2.22). Tremendous pressures can be generated in process units.

Relief valves are designed for pressurized liquid service. They do not respond well in gas service. Relief valves are designed to open slowly. This is a poor feature for gas service. Another reason relief valves are not used in gas service is because of the damage high-velocity gas would do to the seats and disc. Steam cut and wire drawn are two terms applied to damaged **trim**. Relief valves vary in design and style. Relief valves have a specific amount of travel, or lift, which varies from fractions of an inch to several inches. When a relief valve is lifted until it completely compresses the spring, it is said to be in "the fully open position." The difference between the initial lift-off pressure and the pressure of the fully open position is called **accumulation**.

A relief valve consists of the following components: a cap that protects the adjusting screw, the lock nut, and the internal components of the relief valve; an adjusting screw and lock nut; a casing or body; a spring and spring washer that hold a constant tension against the disc; inlet and outlet ports; a disc or plug; a spindle or stem that is threaded and screwed into the flow-control element to guide vertical movement of the disc; and a seat.

Figure 2.22 *Relief Valve Components*

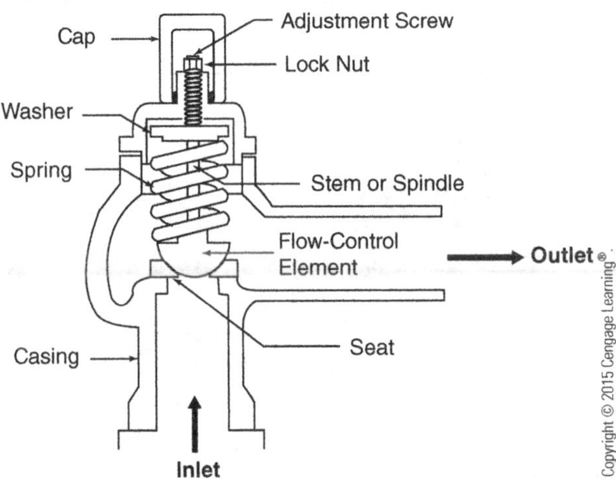

When a relief valve opens, it is relatively easy to detect flow through it. Cold-service relief valves ice up around both the inlet and discharge ports of the valve. Warm-service valves are hot on both sides of the valve. Most relief valves discharge to the cold or warm flare drum and ultimately to the flare. Some relief valves, such as those in steam service, discharge to the atmosphere. Some relief valves have a handle on the side to check the function of the valve. This is a maintenance function, however, and the technician should not use this lever.

The two advantages of relief valves are that they will reseat as soon as the pressure drops below the relief pressure and the spring tension is adjustable.

Safety Valves

Safety valves are considered to be a process system's last line of defense. They are designed to respond quickly to excess vapor, or gas, pressures. When a system overpressures, safety valves respond to allow excess pressure to be vented to the flare header or to the atmosphere. This venting prevents damage to equipment and personnel. This type of valve is very similar in design to a relief valve. The three major differences between a relief and a safety valve are liquid versus gas service, the pressure response time, and a larger exhaust port. Relief valves are designed to lift slowly, whereas safety valves tend to pop off. Because the exhaust port is much larger in a safety valve, it can release more flow at much lower velocities. This feature keeps the trim from being damaged.

Safety valves are classified by how they open and the number of adjusting rings used. Some safety valves reseat after lifting; others do not. Valves that do not reseat must be taken to the shop and mechanically reset.

A safety valve consists of the following components (Figures 2.23 and 2.24): a cap that protects the adjusting screw, the lock nut, and the internal

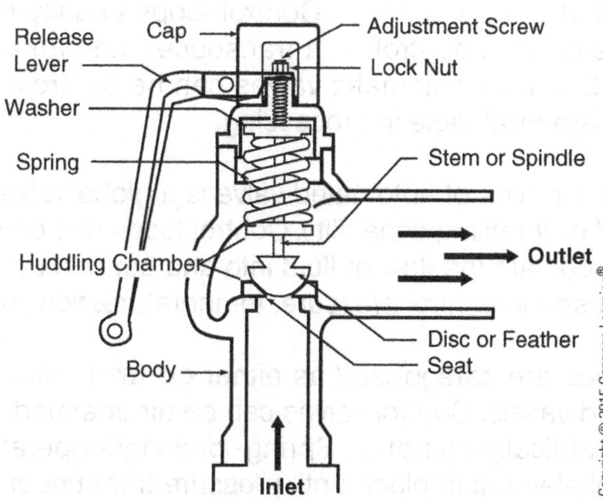

Figure 2.23 *Safety Valve Components*

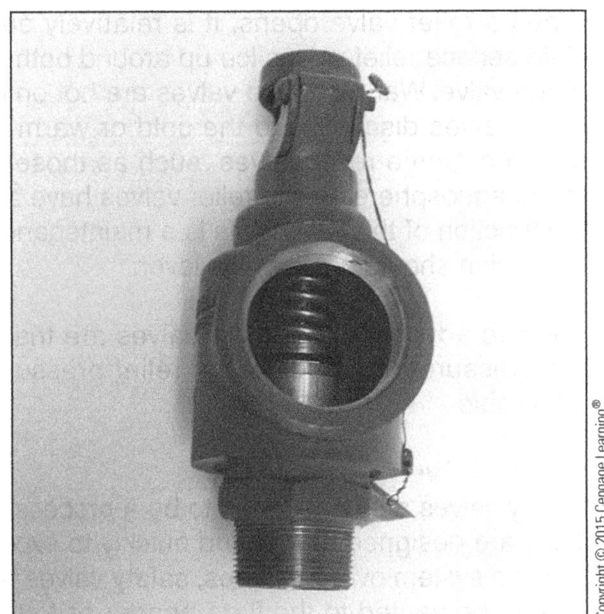

Figure 2.24 *Safety Valve*

components of the safety valve; a spring adjusting screw and lock nut; a casing or body; a spring and two-spring washers that hold a constant tension against the disc; an inlet and a large outlet; a disc or feather and disc guide; a spindle or stem that is threaded and screwed into the flow-control element to guide vertical movement of the disc; a seat and huddling chamber; an adjusting ring and ring pin; lifting levers; and pivot pins.

Automatic Valves

The chemical processing industry uses a complex network of automated instrument systems to control its processes. The smallest unit in this network is called a **control loop**. Control loops usually have a sensing device, a transmitter, a controller, a transducer, and an automatic valve (Figure 2.25). Because automatic valves can be controlled from remote locations, they are invaluable in processing.

The most common type of automated valve is a globe valve because of its versatile, on/off or throttling capability. Control loops use on/off or throttling-type valves to regulate the flow of fluid into and out of a system. Automatic valves can be used to control pressure, temperature, flow, or level.

Automatic valves are categorized as either **control valves** or spring- or weight-operated valves. Control valves can be air operated, electrically operated, or hydraulically operated. Spring- or weight-operated valves hold the flow-control element in place until pressure from under the disc grows

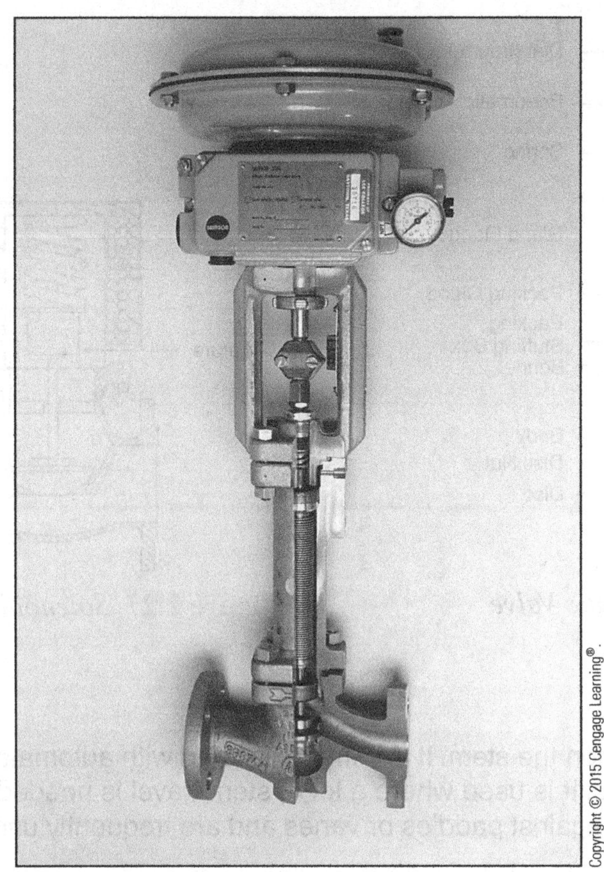

Figure 2.25
Pneumatic Automatic Valve

strong enough to lift the element from the seat; a check valve would fall into this category.

Any of the valves in this chapter can be automated by the installation of a device known as an **actuator**. The actuator controls the position of the flow-control element by moving and controlling the position of the valve stem. Actuators come in three basic designs: pneumatically (air) operated, electrically operated, and hydraulically operated.

Pneumatic Actuators

Pneumatically operated actuators are the most common type. They come in three designs: diaphragm, piston, and vane. Each design converts air pressure to mechanical energy. The diaphragm actuator is a dome-shaped device that has a flexible diaphragm running through the center (Figure 2.26). It typically is mounted on the top of the valve. The center of the diaphragm in the dome is attached to the stem. The valve position (on or off) is held in place by a powerful spring. When air enters the dome on one side of the flexible diaphragm, it opens, closes, or throttles the valve, depending on the design. The piston actuator uses an airtight cylinder and piston to

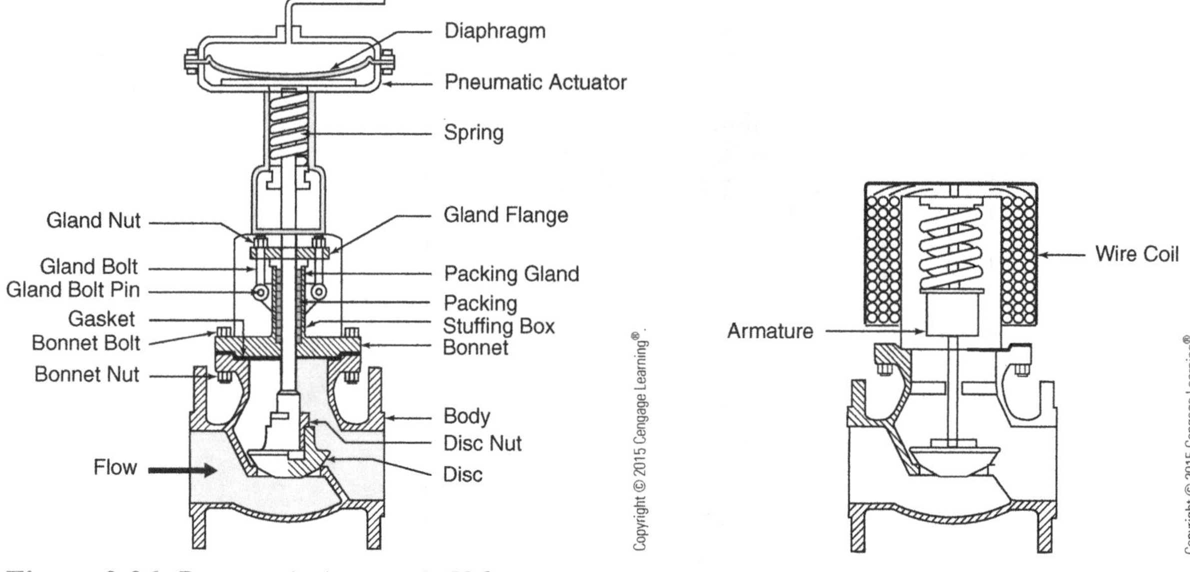

Figure 2.26 *Pneumatic Automatic Valve Components*

Figure 2.27 *Solenoid Valve*

move or position the stem. It is commonly used with automated gate valves or slide valves. It is used where a lot of stem travel is needed. Vane actuators direct air against paddles or vanes and are frequently used on quarter-turn valves.

Common terms for pneumatic actuators include:

- Air to open; spring to close. The valve fails in the closed position if the air system goes down. The air line typically is located on the bottom of the dome.
- Air to close; spring to open. The valve fails in the open position if the air system goes down. The air line is typically located on the top of the dome.
- Double-acting, no spring. Air lines are located on both sides of the dome.

Electric Actuators

Electrically operated actuators convert electricity to mechanical energy. Examples are the solenoid valve and the motor-driven actuator. Solenoid valves are designed for on/off service. The internal structure of a solenoid resembles that of a globe valve (Figure 2.27). The disc rests in the seat, stopping flow. The stem is attached to a metal core or armature that is held in place by a spring. A wire coil surrounds the upper spring and stem. When the wire coil is energized, a magnetic field is set up, causing the armature to lift and compress the spring. The armature is held in place until the current stops. A motor-driven actuator is attached to the stem of a valve by a set of gears. Gear movement controls the position of the stem.

Mechanical devices limit the movement of the stem by stopping the motor when open and when closed.

Hydraulic Actuators

Hydraulically operated actuators are very strong. They convert liquid pressure to mechanical energy. The hydraulic actuator uses a liquid-tight cylinder and piston to move or position the stem. It is commonly found in use with automated gate valves or slide valves. It is used where a lot of stem travel is needed.

Valve Symbols

Each type of valve can be represented by a symbol. Valve symbols are shown in Figure 2.28.

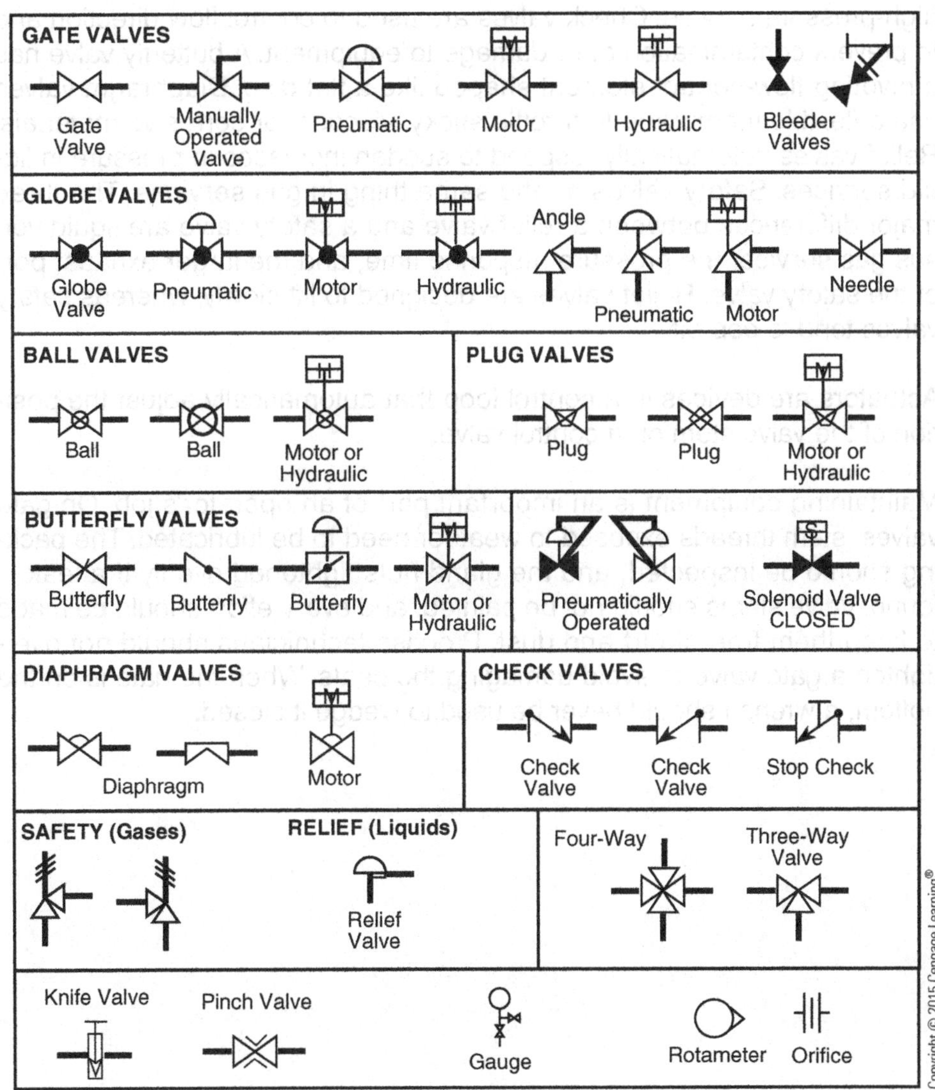

Figure 2.28
Valve Symbols

Summary

A valve is a device used to control (stop, start, or restrict or direct) the flow of fluids. Process technicians classify valves by flow-control element (the part of the valve that regulates or controls), function, and operating conditions (such as pressure, flow, or temperature).

A gate valve places a movable metal gate in the path of a process flow in a pipeline. Gate valves are considered excellent devices for process services that are used infrequently. A globe valve places a movable metal disc in the path of a process flow. This type of valve is the most common one used for throttling service. Ball valves and plug valves have a hollow ball- or plug-shaped movable element in the center of the valve that rotates into the open or closed position. These valves provide very little restriction to flow; they should not be used for throttling service or in high-temperature or high-pressure service. Check valves are used to control flow direction and to prevent contamination of or damage to equipment. A butterfly valve has a pivoting flow-control element shaped like a flat disc. Diaphragm valves use a flexible membrane to throttle sticky, viscous, or corrosive materials. Relief valves automatically respond to sudden increases in pressure in liquid services. Safety valves do the same thing in gas services. The three major differences between a relief valve and a safety valve are liquid versus gas service, the pressure response time, and the larger exhaust port of the safety valve. Relief valves are designed to lift slowly, whereas safety valves tend to pop off.

Actuators are devices in a control loop that automatically adjust the position of the valve stem on a control valve.

Maintaining equipment is an important part of an operator's job. On gate valves, stem threads exposed to weather need to be lubricated. The packing should be inspected, and the gland nuts tightened evenly if a leak is found. Valve stems should not be painted, and every effort should be made to keep them free of dirt and dust. Process technicians should not overtighten a gate valve to avoid damaging the seats. When the gate is on the bottom, a wrench should never be used to wedge it closed.

Review Questions

1. List the three basic types of actuators and explain how each works.
2. List the information found on the bridgewall markings.
3. List the basic components of a gate valve.
4. Compare a gate valve and its function to a globe valve.
5. Identify the three most common gate (flow element) designs.
6. List the two gate valve designs and explain how each works.
7. Globe valves are primarily used for what type of service?
8. Describe the flow of fluid through a globe valve.
9. Describe quick opening or quarter-turn valves.
10. Describe the operation and design of a diaphragm valve.
11. Compare a relief valve to a safety valve and describe how each works.
12. List the four types of check valves and explain how the device works?
13. A butterfly valve can be used for what type of service?
14. Explain a process technician's role in maintaining and operating valves.
15. List the four common disc designs used with globe valves.
16. Compare and contrast the operation of a weir and non-weir diaphragm valve.
17. Describe the term, *accumulation*.
18. List the basic components of a control loop.
19. Describe the three types of pneumatic actuators.
20. Describe the design and operation of a solenoid valve.

Tanks, Piping, and Vessels

OBJECTIVES

After studying this chapter, the student will be able to:

- Describe the different types of process piping.
- List the different types of aboveground storage tanks found in a tank farm.
- Describe the various vessels found in a process plant.
- Describe the effects of corrosion and cathodic protection.
- Explain the factors that influence the selection of materials used to construct a vessel.
- Define the term *alloy*.
- Describe the various inspection procedures used in a process plant.
- Identify the information found on a vessel sketch.
- Describe a vessel specification sheet.

Key Terms

Alloy—a material composed of two or more metals or a metal and a nonmetal.

Blind—a device used in piping to gain complete shutoff.

Bonding—is described as physically connecting two objects with a copper wire.

Bullet—cylindrical shaped tank with rounded ends that are classified as high pressure.

Butt-welded piping—pipe on which the parts to be joined are the same diameter and are simply welded together.

Cone-roof tank—an enclosed tank with a conical-shaped roof with vertical walls mounted on a circular concrete pad or directly on the ground.

Corrosion—electrochemical reactions between metal surfaces and fluids that result in the gradual wearing away of the metal.

Cryogenic tank—has been designed to store liquids below −100°F (−73.33°C).

Datum plate—a reference point on the bottom of a tank used to measure liquid level.

Dike—a containment wall or ditch that extends around a tank to prevent product loss.

Flanges—used to connect piping to equipment or where piping may have to be disconnected; consist of two mating plates fastened with bolts to compress a gasket between them.

Flat face flanges—generally used to mate against cast equipment, where bending from tightening bolts might break the flange; gasket should cover the entire face of the flange.

Floating-roof tank—has an open top and a pan-like structure that floats on top of the liquid and moves up and down inside the tank with each change in liquid level.

Gauge hatch—a door in the roof of an atmospheric tank that enables the contents to be measured and that provides some emergency pressure relief.

Grounding—is described as a procedure designed to connect an object to the earth with a copper wire and a grounding rod.

Gutted piping—see *Jacketed and gutted piping*.

Hemispheroid tank—has a rounded or dome-shaped top and vertical walls mounted on the ground or a concrete pad.

Jacketed tank—an insulated system designed to hold in heat or cold.

Jacketed and gutted piping—two concentric (one inside the other) pipes used when the conveyed fluid must be kept hot. In jacketed piping, the fluid is conveyed through the inner pipe and a heating medium is conveyed through the jacket. Gutted piping is the reverse.

Manway—a hatch or port used to provide open access into a tank.

Pig—a cylindrical device used to clean out pipes. Most pigs utilize a *pig launcher* to propel it through the line and into a pig trap.

Pipe size—the nominal (named) size of a pipe; usually close to the outside and inside diameters of the pipe but identical to neither. The outside diameter of a certain size pipe is constant. The inside diameter will change with the pipe wall thickness (schedule).

Pipe thickness—thickness of pipe wall, designated by a schedule number. Schedules 10 (thin walled), 40, 80, and 160 (heavy walled) are common. The schedule indicates a specific wall thickness for one pipe size only; a 3" schedule 40 pipe will have a different thickness than a 4" schedule 40. The pipe wall thickness increases as the schedule number increases.

Radiographic inspection—use of X-rays to locate defects in metals in much the same manner as an X-ray is taken of a broken bone.

Raised face flange—uses a gasket that fits inside the bolts.

Ring joint flange—uses only a metal ring for gasketing.

Slop tank—or off-spec tank is used to store product that does not meet customer expectations.

Socket-welded piping—type of piping in which the pipe is inserted into a larger fitting before being welded to another part.

Sphere—a circular-shaped tank with legs designed to contain high-pressure liquids or gases.

Spheroid—a circular tank with a flat bottom resting on a concrete pad or ground.

Stress-corrosion cracking—a mechanical-chemical type of deterioration associated with steel.

Tank farm—a collection of tanks used to store and transport raw materials and products.

Traced piping—used when the conveyed fluid must be kept hot; usually has a copper tubing containing steam or hot oil.

Vessel design sheets—identifies the factors entering into the selection, use, and need for periodic inspection of materials used to make vessels.

Tank Farm

A **tank farm** is best described as a collection of tanks designed to safely store and transport raw materials and products. These materials can be brought in from pipelines, barges, ships, or trucks. Aboveground storage tanks (ASTs) come in a variety of designs that can be classified as low, medium, or high pressure. Tank farms can safely store liquids or gases. Manufacturer code stamps on each tank will provide detailed information about the design specifications; pressures, temperatures, etc., that the tank should be operated at. Some tank farms include underground salt domes, caverns, and other belowground storage systems. Process technicians are required to safely operate and maintain each of the complex storage and transfer systems in a tank farm. Figure 3.1 shows a typical tank farm.

Figure 3.1
Tank Farm

Every tank farm will have a list of the chemicals stored on site and a corresponding material safety data sheet (MSDS). Safely handling and storing chemicals requires structured training and financial resources to maintain the integrity of the tanks, pipes, valves, pumps, and instrumentation. New technicians train from three to six months before being assigned to operate a complex system. During the training process, the trainee works with a senior technician to learn basic line-ups, standard operating procedures, safety rules and regulations, and sampling techniques. Process tanks and storage systems are equipped with the latest in modern process control. Process variables include flow rate, pressure, temperature, composition, and level.

Pigging

Technicians use specialized equipment to clean residues out of pipelines. The basic components used in this procedure include a pig launcher, pig, and pig trap. The pig launcher utilizes fluid pressure to launch a projectile called a **pig** through the pipe. A *pig trap*, designed to catch the dirty pig, is placed at the end of the pipe. Figure 3.2 illustrates the different type of pigs utilized in this procedure.

Tank Designs and Categories

Common names for tanks include cone roof, floating roof (internal or external), spheres, spheroid, **bullets**, hemispheroid, bins, silo, open top,

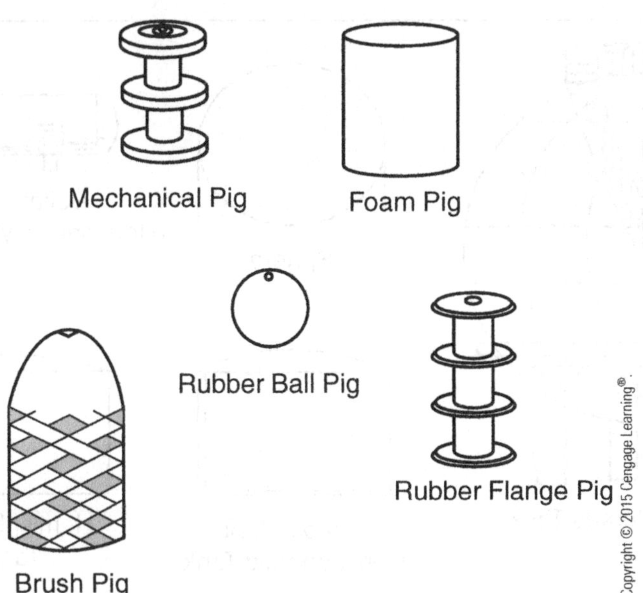

Figure 3.2
Pigs

or double wall. Technicians also refer to tanks as the feed tank, vaulted tank, elevated tank, recovery tank, surge tank, blend tank, cryogenic tank, **jacketed tank**, or blanketed tank. Process technicians use strapping tables to calculate the volume in a tank. A 55-gallon barrel typically holds about 42 gallons. If a tank is rated as a 5,000-barrel tank, it can safely store 210,000 gallons.

Tanks of various types are used for the storage of raw materials and finished products. Since these tanks (called *tankage*) represent a large concentration of value, the protection and safe operation of storage tanks are important. It is necessary that the operators utilizing storage tanks be completely familiar with the tankage and related equipment. **Cryogenic tanks** are designed to store liquids below −100°F (−73.33°C). Tanks used to store off-specification product are referred to as "**slop tanks**," or off-spec tanks.

There are various types of tanks, each of which has its own advantages and disadvantages. The type of tank to be used is generally determined by the product to be stored and pressure, measured as pounds per square inch gauge (psig). Tanks can be divided into four general categories: atmospheric tanks, low-pressure tanks (0 to 2.5 psig), medium-pressure tanks (2.5 to 15 psig), and high-pressure tanks (above 15 psig.) Figure 3.3 shows the different types of tanks found in the chemical processing industry.

Atmospheric Tanks

Atmospheric tanks can have a cone roof or a floating roof. Process technicians refer to a tank as being *atmospheric* when it is properly vented, or

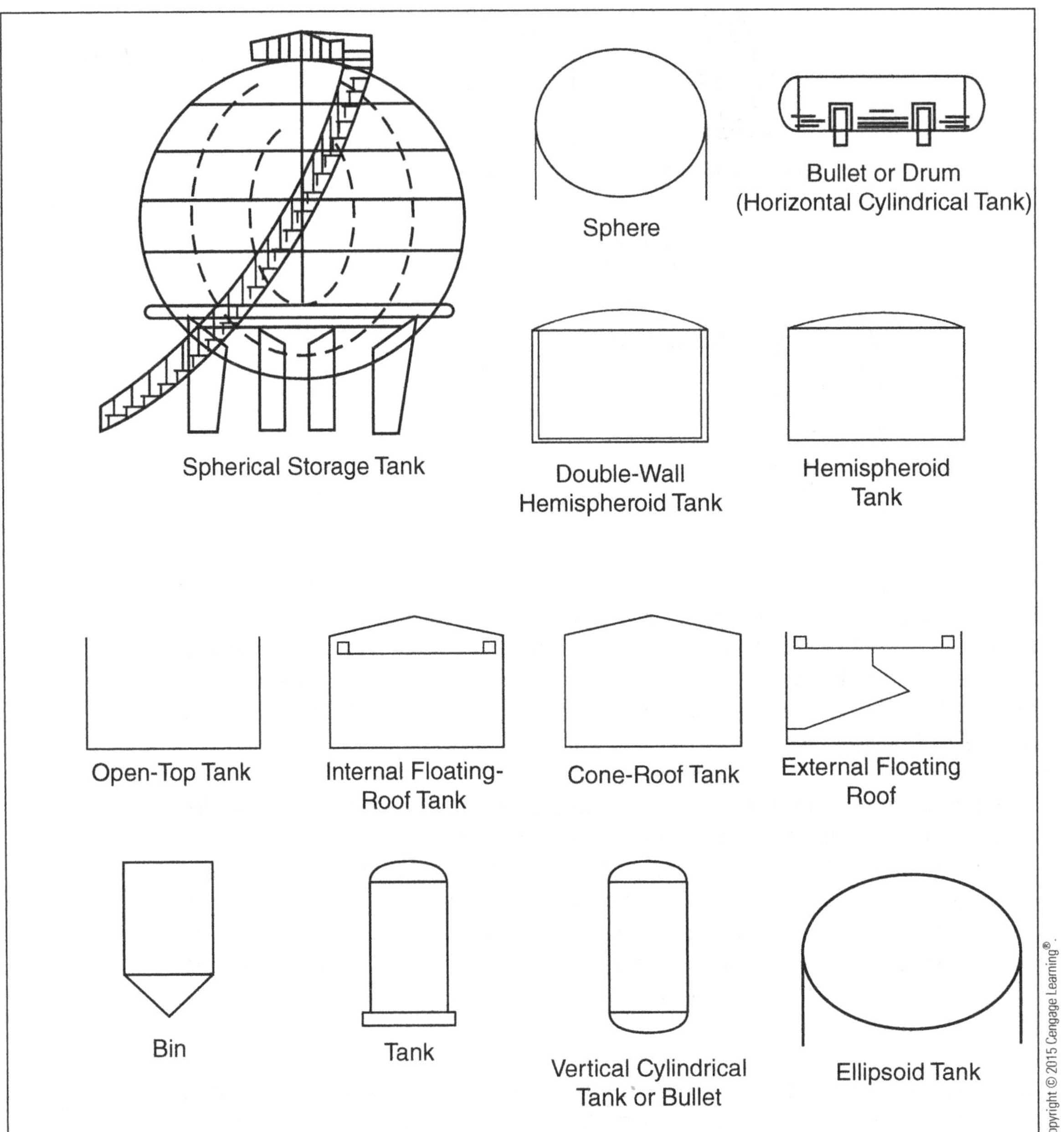

Figure 3.3 *Tank Designs*

designed to be run at 14.7 psia (pounds per square inch absolute) or zero gauge pressure (i.e., 0 psig).

Floating-Roof Tanks

A **floating-roof tank** has an open top and a pan-like structure that floats on top of the liquid and moves up and down inside the tank with each change in liquid level (see Figure 3.4). A close clearance is maintained between

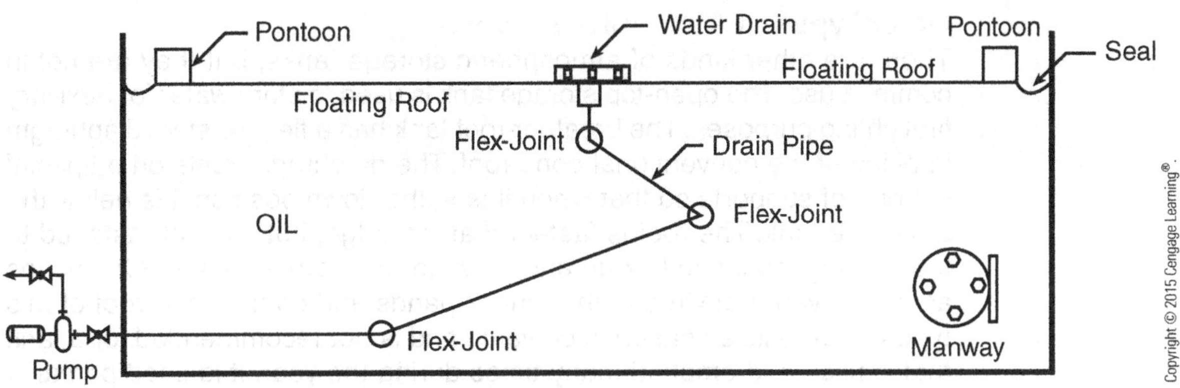

Figure 3.4 *Floating-Roof Tank Components*

the roof and the shell of the tank. The opening is sealed by means of a flexible curtain-like fabric attached to the roof and to steel bearing surfaces called *shoes*. The shoes slide on the shell and are kept in contact with the shell by means of a suitable mechanism.

There are three basic types of floating roofs: pan type, pontoon type, and double deck. John H. Wiggins invented and built the first practical floating roof in 1921. A pan-type, it featured a deck of a single thickness with a vertical cylindrical rim at the periphery or outer edge. The deck is coned slightly toward the center and is provided with radial rafters and trusses to give it stiffness. Today, the pan roof is used only in areas of low rainfall because the roof will tip and sink when loaded unevenly with water or snow.

The pontoon-type of floating roof has an annular (ring-shaped) pontoon around the outer edge and a deck of single thickness at the center. The annular pontoon at the outer edge provides air space insulation for a large area of the liquid surface, which is helpful in retarding boiling of the product. The pontoon roof is quite stable because rainfall will run to the center and cannot make an uneven load near the edge of the roof. Likewise, a leak in the center section of the roof will not sink the roof or make it unstable. The pontoon is divided into compartments so that a leak in one compartment is confined to a small area and the remaining compartments will be buoyant and support the roof. The area of the pontoon will vary between 25% and 55% of the roof area. The pontoon roof is the most common type of floating roof in use today.

The double-deck floating roof has a double deck over the entire liquid surface. The space between the decks is divided into compartments so that a leak will not sink the entire roof. Of course, the double-deck roof is more buoyant than the other types of floating roof, and the air space between the decks provides an insulation barrier over the whole roof. This type of roof is the most expensive of the three types.

Other Types of Atmospheric Tanks

There are other kinds of atmospheric storage tanks, but they are not in common use. The open-top storage tank is used to store water for auxiliary firefighting purposes. The breather-roof tank has a flexible steel diaphragm in place of the conventional cone roof. The diaphragm rests on a special set of roof supports so that when it is in the down position it is below the top of the tank. The roof is fastened at the edge, but it is not fastened to the framing, so it can flex up and down for a distance of about 20 inches as the air-vapor mixture in the tank expands and contracts. A roof of this type has very little conservation value and is not recommended for a tank that is filled and emptied many times during the year. It is used primarily for standing storage. The vapor-dome roof looks like a cone-roof tank with a hemisphere located at the center of the roof. Inside the hemisphere is a membrane of the same shape attached by its outer edge to the equator of the vapor dome. This membrane is free to hang downward in the form of a hemisphere. Hence, the movement of the membrane is equal to twice the volume of the hemisphere. Umbrella-roof tanks are very similar to cone-roof tanks except that the roof is rounded to a convex shape. Beams may support the roof, although internal supports are also used. Umbrella-roof tanks typically have small diameters.

Pressure Tanks

Pressure storage tanks are used to store volatile liquids, which have a Reid vapor pressure greater than 18 pounds per square inch (psi). There are three types of pressure storage vessels: drums, spheres, and spheroids. Drums are cylindrical vessels with ellipsoidal or hemispherical ends built to withstand a given internal pressure. Usually a drum is supported in the vertical position on a concrete foundation or in the horizontal position on two or more concrete piers.

Spheres, as we use the term, are pressure vessels shaped like a sphere and supported above grade on large tubular columns. A sphere 65 feet in diameter will have a volume of 25,000 barrels. A sphere has a more economical shape than a drum for the storage of liquid under relatively high pressure. **Spheroid** tanks are similar but have a somewhat flattened top and bottom.

Cone-Roof Tanks: Low to Medium Pressure

A **cone-roof tank** has a fixed, slightly conical roof, one or more inside support columns, and a flat bottom. Cone-roof tanks are used to store low-vapor pressure stocks. Cone-roof tanks are designed to operate within a range of about 1 inch of water pressure to 1 inch of water vacuum. The welded joint where the roof joins the shell is purposely made weaker than other joints so that it will burst and relieve pressure without spilling the tank contents. This design helps confine the fluid should a fire or explosion occur inside a tank.

Hemispheroid or Dome Tank: Low to Medium Pressure

Hemispheroidal tanks can be classified as medium-pressure tanks; 2.5 to 15 psig. This type of tank is typically used for the storage of higher volatility products. For this reason, hemispheroid tanks are a popular choice in the chemical processing industry. Figure 3.5 is an example of a hemispheroidal tank. A **hemispheroidal tank** or dome shaped tank has a rounded or dome shaped top and vertical walls mounted on the ground or a concrete pad or foundation.

Breathing

As a fixed-roof tank is filled, the air or vapor in the tank is expelled through a vent. As fluid is withdrawn, air enters the tank through the vent to replace the volume of liquid being withdrawn. To a lesser extent, this "breathing" action also takes place when the vapor in the tank expands or contracts from heating and cooling. Sunlight and warm days are sufficient to cause some expansion of vapor, and cooling at night or during a rainstorm will cause contraction of the vapors.

Flame Arrestors

In tanks that store flammable materials, the vapor expelled by filling or heating is sometimes mixed with air (oxygen) in the proper proportions to be ignited. The vents are equipped with a flame arrestor to prevent the possibility of fire reaching the contents of the tank. Flame arrestors are not designed to prevent flame passage indefinitely, and it is important to extinguish any fire at a flame arrestor immediately. Since the small passages in a flame arrestor element may plug from corrosion or foreign objects, the elements are cleaned on a regular schedule.

Manways and Manholes

The chemical processing industry uses **manways** as access hatches or ports into and out of tanks and vessels. These are used for visual inspection and access for cleaning. Manways are typically hinged for easy access. Gaskets are used to provide a positive seal and a series of bolts and nuts

Figure 3.5
Hemispheroid Tank

are used to secure the door to the vessel. Opening, blinding and confined space entry permits are required for entry into a vessel. The term manhole is used to describe a circular access port into below grade systems like sewers or tanks. Manhole covers are typically not hinged. Frequently the terms, manway and manhole are interchanged.

Conservation Vents

Fixed-roof tanks that store volatile fluids are often equipped with a conservation vent (Figure 3.6). A typical conservation vent is equipped with two valves having weighted discs to regulate pressure during operation. The exhaust valve will not open until a slight positive pressure is reached in the tank, and the intake valve will not open until the tank is under a slight vacuum. Controlling the pressure in the tank reduces loss of vapors.

Gauge Hatches

Gauge hatches (Figure 3.7) are provided in the roofs of atmospheric tanks to enable the contents to be measured. A secondary function of a gauge hatch is to provide some emergency pressure relief. Except when they are in use, gauge hatches should be kept closed to prevent loss of vapors, fire hazards, and entry of rainwater. Hatches should not be weighted or otherwise restricted from opening because restricting their ability to open eliminates their function as a pressure relief device. The **Datum plate** directly below the open hatch is used as a reference point when a technician is measuring liquid level with a gauge line.

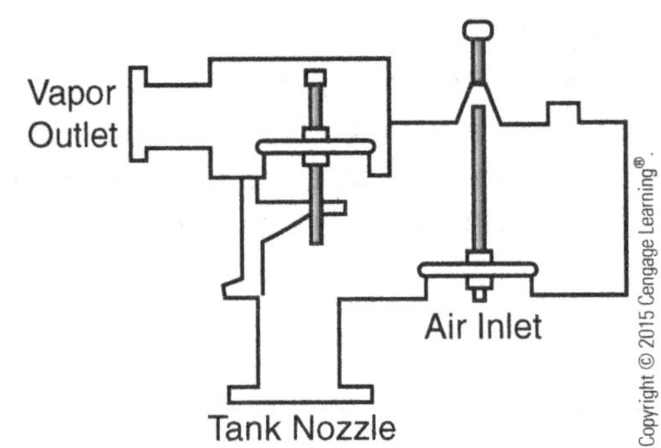

Figure 3.6
Conservation Vent

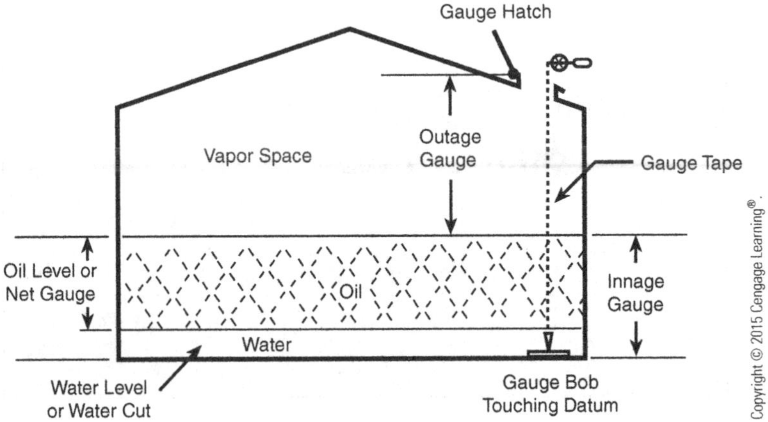

Figure 3.7
Tank Terminology

Water Draws

Water draw valves are provided at the lowest point in the tank bottom. They are used to remove water that has settled to the bottom of the tank and may be used to completely drain the tank.

Gas-Blanketed Tanks

Depending on the vapor pressure and temperature of the stock in an atmospheric tank, the vapor space may be filled with varying mixtures of vapor and air. The vapor space in tanks storing materials having a low vapor pressure at the storage temperature is usually too lean to explode. The vapor space in tanks storing very volatile materials is usually too rich to explode. In some tanks, however, the vapor space would be nearly always in the explosive range if air were allowed to enter.

Gas-blanketed tanks are used to store these hazardous feedstocks. They are also used for other stocks when contact with air or moisture would be harmful to the product. In general, gas-blanketed tanks are similar to other types of fixed-roof tanks except that they are equipped with a supply line for the gas blanket and a regulator to control the pressure.

Traditional and Modern Diking Techniques

A **dike** is best described as a containment wall or ditch that extends around a tank to prevent product loss. A variety of safety designs have been proposed. Examples of these can be found in Figure 3.8. Dikes are composed of earth, concrete, or metal. Fire walls and trenches are also used in diking designs.

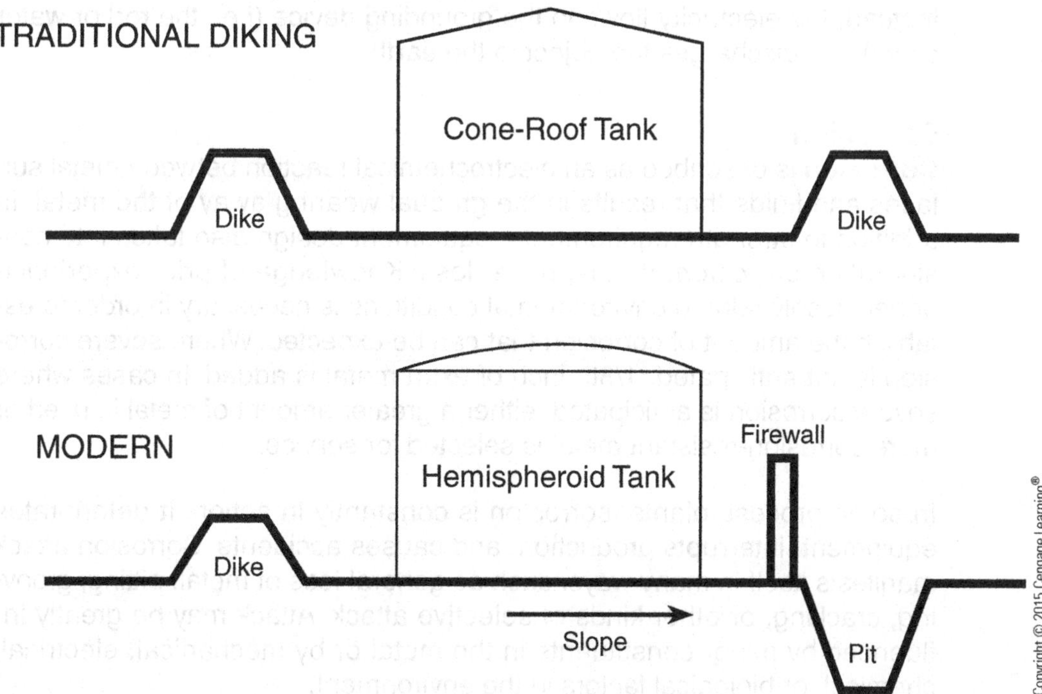

Figure 3.8 *Traditional and Modern Diking Techniques*

Piping

Piping in a chemical plant is used to convey all kinds of fluid materials. It constitutes approximately 30% of the initial investment for a new process plant. The materials used in piping construction are chosen to withstand the temperature, pressure, and other properties of the fluids being conveyed. Some other factors to be considered are codes and specifications, stress factors, layout or routing, and expansion flexibility. Commonly used materials include steel of different **alloys**, cast iron, aluminum, copper, and plastic compositions. Since metal piping, particularly steel, is the most common, we need to know something about its characteristics.

Bonding and Grounding Tanks, Pumps, and Piping

A static electric spark can be an ignition source and cause fire or explosion. This takes place when an electric spark discharges across a certain distance between a charged body and an uncharged body. Flammable liquid containers can build up static charges as the material is pumped in. Fluid movement of any type can produce a similar effect. The chemical industry has found two methods to prevent fire hazard from occurring—bonding and grounding. **Bonding** is achieved by physically connecting two objects together with a copper wire. **Grounding** is a procedure designed to connect an object to the earth with a copper wire and a grounding rod or grounding device. Underground water pipes can also function as a grounding device. Grounding provides an alternate path for the electricity to flow. When two objects are connected a spark cannot jump between them. Instead, the electricity flows to the grounding device (i.e., the rod or water pipes) and discharges the object to the earth.

Corrosion

Corrosion is described as an electrochemical reaction between metal surfaces and fluids that results in the gradual wearing away of the metal. In addition to strength requirements, equipment design also takes into consideration *corrosion*, that is, metal loss. Knowledge of prior experience under closely related environmental conditions is necessary in order to establish the amount of corrosion that can be expected. Where severe corrosion is not anticipated, 1/8th inch of extra metal is added. In cases where severe corrosion is anticipated, either a greater amount of metal is used or more corrosion-resistant metal is selected for service.

In some process plants, corrosion is constantly in action. It deteriorates equipment, interrupts production, and causes accidents. Corrosion attack manifests itself in many ways, such as general loss of metal, pitting, grooving, cracking, or other kinds of selective attack. Attack may be greatly influenced by minor constituents in the metal or by mechanical, electrical, chemical, or biological factors in the environment.

Improper operation also can affect corrosion rates. Typical examples are increasing temperature above design; using an incorrect amount of neutralizers and inhibitors; failing to drain water from equipment during shutdowns; using improper mixing in treating processes; and failing to drain water draws.

Cathodic Protection

Electrochemical corrosion control in the chemical processing industry is best taken care of using cathodic protection. Electrochemical corrosion causes tanks and pipes to prematurely corrode and fail. Cathodic protection utilizes a direct current device from an external source to counter the discharge of a metal submerged in a conducting medium. Typical conducting mediums include soil and water. The base of any aboveground storage tank is vulnerable to electrochemical attack and corrosion.

Cathodic protection utilizes two different methods:

- Use of sacrificial anodes
- Use of impressed current anodes

In both cases, the anodes are placed under or around the tank. Each anode is attached securely to the tank. Sacrificial anodes corrode in the place of the tank and will eventually need to be replaced, while the impressed current anodes do not corrode. When current flow is open the system can protect the tank indefinitely.

Steel and Other Types of Pipe

Most piping used in process units is carbon steel, primarily because it is fairly economical and has a wide temperature range. Carbon steel is used from −20°F (−28.88°C) to around 800°F (426.66°C). Specially heat-treated carbon steel is used in a temperature range of −21°F (−29.44°C) to −50°F (−45.55°C).

Farther down the temperature range, type 304 stainless steel or 3" nickel is normally used. The temperature range of 3" nickel is −51°F (−46.11°C) to −150°F (−101.11°C). Type 304 stainless steel is used in services from −151°F (−101.66°C) to −320°F (−195.55°C). The stainless steels at very low temperatures do not become brittle as regular carbon steel does. Stainless steel is also used in some high-temperature applications such as tube supports in furnaces. Stainless steel when coupled to carbon steel can cause problems because of its expansion rate, which is approximately 150% that of carbon steel.

Some low alloys (carbon-, moly-, and chrome alloys) are used in high-temperature service such as furnace tubes. These alloys are made for operations over 800°F (426.66°C). An *alloy* is a material consisting of two or more metals or a metal and a nonmetal. A low alloy is one that has a relatively small amount of the secondary material.

Steel pipe is manufactured in various diameters and wall thicknesses. **Pipe sizes** are identified by the pipe's nominal size (nominal means in name only), which is usually different from their actual inside and outside diameters (Figure 3.9). For example, a 3½" pipe has an outside diameter

Pipe Size Inches	Outside Diameter Inches	Identification: Steel: Iron Pipe Size	Identification: Steel: Sched. Number	Identification: Stainless Steel Sched. Number	Wall Thickness (1) Inches	Inside Diameter (d) Inches	Area of Metal Square Inches	Transverse Internal Area (a) Square Inches	Transverse Internal Area (A) Square Feet	Weight Pipe Pounds per foot	Weight Water Pounds per foot	External Surface Sq. Ft. per foot of pipe
				5S	.083	4.334	1.152	14.75	.10245	3.92	6.39	1.178
			40	10S	.120	4.260	1.651	14.25	.09898	5.61	6.18	1.178
		STD	80	40S	.237	4.026	3.174	12.73	.08840	10.79	5.50	1.178
4	4.500	NS	120	80S	.337	3.826	4.407	11.50	.07986	14.98	4.98	1.178
			160		.438	3.624	5.595	10.31	.07160	19.00	4.47	1.178
					.531	3.438	6.621	9.28	.06450	22.51	4.02	1.178
		XXS			.674	3.152	8.101	7.80	.05420	27.54	3.38	1.178
				5S	.109	8.407	2.916	55.51	.3855	9.93	24.06	2.258
				10S	.148	8.329	3.941	54.48	.3784	13.40	23.61	2.258
			20		.250	8.125	6.57	51.85	.3601	22.36	22.47	2.258
		STD	30		.277	8.071	7.26	51.16	.3553	24.70	22.17	2.258
			40	40S	.322	7.981	8.40	50.03	.3474	28.55	21.70	2.258
		XS	60		.406	7.813	10.48	47.94	.3329	35.64	20.77	2.258
8	8.625		80	80S	.500	7.625	12.76	45.66	.3171	43.39	19.78	2.258
			100		.594	7.437	14.96	43.46	.3018	50.95	18.83	2.258
			120		.719	7.187	17.84	40.59	.2819	60.71	17.59	2.258
		XXS	140		.812	7.001	19.93	38.50	.2673	67.76	16.68	2.258
					.875	6.875	21.30	37.12	.2578	72.42	16.10	2.258
			160		.906	6.813	21.97	36.46	.2532	74.69	15.80	2.258
				5S	.156	12.438	6.17	121.50	.8438	20.98	52.65	3.338
				10S	.180	12.390	7.11	120.57	.8373	24.17	52.25	3.338
			20		.250	12.250	9.82	117.86	.8185	33.38	51.07	3.338
			30		.330	12.090	12.87	114.80	.7972	43.77	49.74	3.338
		STD		40S	.375	12.000	14.58	113.10	.7854	49.56	49.00	3.338
			40		.406	11.938	15.77	111.93	.7773	53.52	48.50	3.338
12	12.75	XS		80S	.500	11.750	19.24	108.43	.7528	65.42	46.92	3.338
			60		.562	11.626	21.52	106.16	.7372	73.15	46.00	3.338
			80		.688	11.374	26.03	101.64	.7058	88.63	44.04	3.338
			100		.844	11.062	31.53	96.14	.6677	107.32	41.66	3.338
		XXS	120		1.000	10.750	36.91	90.756	.6303	125.49	39.33	3.338
			140		1.125	10.500	41.08	86.59	.6013	139.67	37.52	3.338
			160		1.312	10.126	47.14	80.53	.5592	160.27	34.89	3.338
				5S	.188	21.624	12.88	367.25	2.5503	43.80	159.14	5.760
				10S	.218	21.564	14.92	365.21	2.5362	50.71	158.26	5.760
			10		.250	21.500	17.08	363.05	2.5212	58.07	157.32	5.760
		STD	20		.375	21.250	25.48	354.66	2.4629	86.61	153.68	5.760
		XS	30		.500	21.000	33.77	346.36	2.4053	114.81	150.09	5.760
22	22.00		60		.875	20.250	58.07	322.06	2.2365	197.41	139.56	5.760
			80		1.125	19.750	73.78	306.35	2.1275	250.81	132.76	5.760
			100		1.375	19.250	89.09	291.04	2.0211	302.88	126.12	5.760
			120		1.625	18.750	104.02	276.12	1.9175	353.61	119.65	5.760
			140		1.875	18.250	118.85	261.59	1.8166	403.00	113.36	5.760
			160		2.125	17.750	132.68	247.45	1.7184	451.06	107.23	5.760

Figure 3.9 *Pipe Data*

of 4" and an inside diameter somewhere between 3" and 4". Most pipe used in plants is ½", ¾", 1", 1½", 2", 3", 4", 6", 8", 10", 12", and 14" and higher. When a pipe has an outer diameter of 14" or over, the outer diameter is the same as the nominal pipe size. Sizes of ⅕", 1¼", 2½", 3½", and 5" are not usually used.

The **pipe thickness** of the pipe wall is designated by a schedule number. Schedules 10 (thin walled), 40, 80, and 160 (heavy walled) are common. The schedule indicates a specific wall thickness for one pipe size only; a 3" schedule 40 pipe will have a different thickness than a 4" schedule 40. The pipe wall thickness increases as the schedule number increases.

Cast Iron Piping

Cast iron pipe and fittings are used to convey nonflammable fluids in some areas. The sizes and thicknesses of cast iron pipe are similar to those of steel pipe.

Because iron becomes brittle when it is exposed to fire and then sprayed with water, there is a good possibility that the cast iron would break under those conditions. For this reason, cast iron piping is not used in hazardous areas, nor is it used to handle flammable materials in any area.

Joining Pipes

Almost all piping in critical services is joined together by welded joints, which provide much more strength and less chance for leaks than do threaded joints. For piping that is not critical, as far as pressure or contents is concerned, piping with threaded joints is generally much cheaper and easier to install than is piping with welded joints.

Screwed Piping

Small pipes are commonly joined by the use of tapered pipe threads. The threads are cut into both male and female parts of the joint and are tapered to provide a tight fit. Usually a thread compound, or Teflon tape, is applied to the threads to aid in sealing the joint and for lubrication in connecting the joint. Since metal is removed in cutting threads, the weakest part of screwed piping is usually the joints. Screwed piping (Figure 3.10) is used in sizes up to 2" for handling nonhazardous materials.

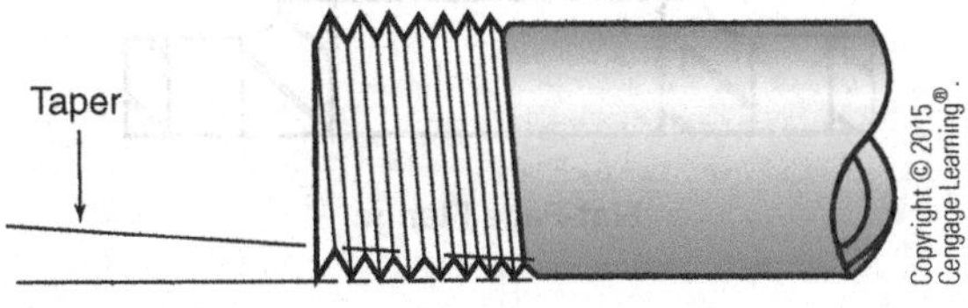

Figure 3.10
Screw Pipe

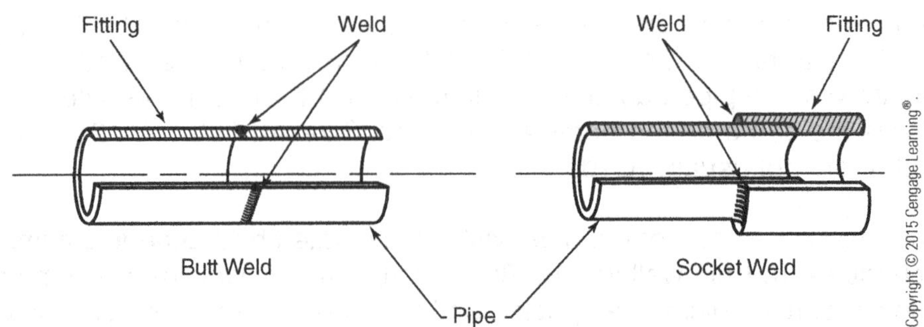

Figure 3.11
Welded Pipe

Welded Piping

Two types of welding may join piping (Figure 3.11). In **butt-welded piping**, the parts to be joined are the same diameter and are simply welded together. In **socket-welded piping**, the pipe is inserted into a larger fitting before being welded.

Socket-welded fittings are usually used in 2" size and smaller because there is less possibility that stray weld metal will obstruct the flow area. Butt-welding is used in all sizes, but particularly in 2" size and larger.

Flanges

Flanges are used to connect piping to equipment or where the piping may have to be disconnected. They consist of two mating plates fastened with bolts to compress a gasket between them. The three common types are shown in Figure 3.12. **Flat face flanges** are generally used to mate against cast equipment, where bending from tightening bolts might break the flange. A gasket should cover the entire face of the flange. **Raised face flanges** use a gasket that fits inside the bolts, and **ring joint flanges** use only a metal ring for gasketing.

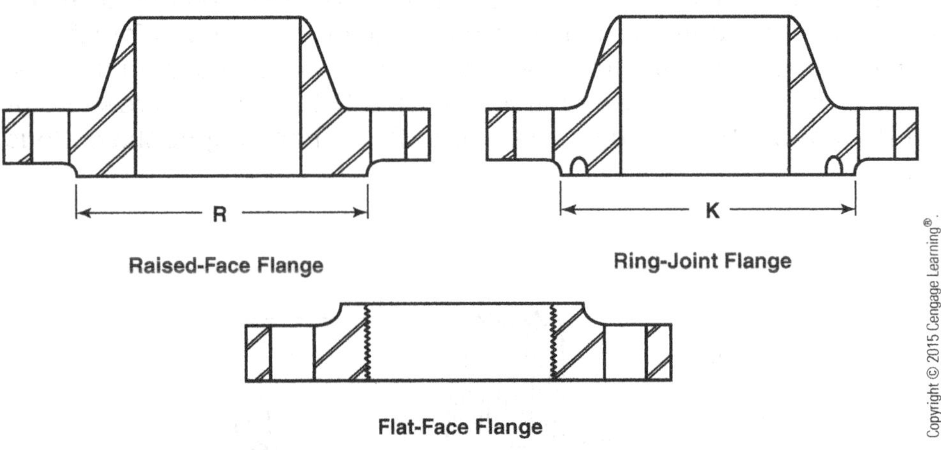

Figure 3.12
Flange Types

Flanges are made in various thicknesses and for various bolt sizes according to the pressure and temperature of the service. Ratings of 150 lb., 300 lb., and 600 lb. are common in chemical plants.

Fittings

Elbows, tees, flanges, valves, and other piping components are made to mate with screwed, flanged, and welded piping discussed in the previous sections. These items are also made in various weights and are constructed for certain pressure and temperature ratings.

Jacketed, Gutted, and Traced Piping

A special type of piping used where it is necessary to keep the conveyed fluid hot is jacketed, gutted, or **traced piping**. Both **jacketed and gutted piping** have two concentric (one inside the other) pipes. In jacketed piping, the fluid is conveyed through the inner pipe and a heating medium is conveyed through the jacket (the outer pipe). **Gutted piping** is the reverse; the fluid is conveyed through the outer pipe and the heating medium is conveyed through the inner pipe.

The general practice involved with steam tracing includes wrapping copper tubing around the process pipe and covering it with heat transfer cement or insulation. Hot oil tracing systems are used when the process fluid is hotter than the plant's steam system. Low-pressure steam or hot oil is passed through the copper tubing during operation. This procedure is often used to winterize a unit when temperatures are expected to drop below freezing. A negative aspect of steam tracing occurs when the copper tubing sweats under the insulation and, as a result, traps moisture next to the piping.

Paddle and Figure-Eight Blinds

Blinds are one of the main means used to gain a complete shutoff in piping. Two types of blinds are generally used in plants. The first, called a *paddle blind*, is nothing more than a piece of metal thick enough to be subjected to and withstand a specific pressure. The paddle blind is inserted between two flanges, with a gasket on each side, and tightened. The second type of blind, the *figure-eight blind*, is designed to be installed inside the piping. On one end of the blind is an opening to be used as a spacer between two flanges when the blind portion is not in use. The figure-eight blind has an advantage in that it is always at the location of use, whereas the paddle blind, when not in use, can be misplaced.

Double Block and Bleed

A double-block-and-bleed system is frequently used to stop flow; this is not considered a complete shutoff, however. The double block and bleed consists of two valves in line with a smaller valve opening to the atmosphere between the two line valves.

Vessels

It is important for operating people to understand the limitations of equipment in order to ensure uninterrupted operations. This section outlines some of the factors entering into the selection, use, and need for periodic inspection of materials used to make vessels and other plant equipment.

Vessel Documentation and Design

Each vessel will include a code stamp that will indicate high-pressure and high-temperature ratings, manufacturer, date, type of metal, storage capacity, and special precautions. Most vessel documentation includes strapping tables that will allow a technician access to data that can be used to identify capacity. For aboveground storage, the ASME (American Society for Mechanical Engineers) Code, Section VIII governs vessels that have pressures greater than 15 psig. Common storage designs include spheres, spheroids, horizontal cylindrical tanks (drums), bins, and tanks with fixed and floating roofs. Tanks, drums, and vessels are typically classified as low pressure, high pressure, liquid service, gas service, insulated, steam traced, or water cooled. Wall thickness and shape often determine the service a vessel can be used in. Some tanks are designed with internal or external floating roofs, double walls, dome or cone roofs, or open top. Earthen or concrete dikes often surround a tank and are designed for containment in the event of a spill.

Spherical and spheroidal storage tanks are designed to store gases or pressures above 5 psig. Spheroidal tanks are flatter than spherical tanks. Horizontal cylindrical tanks or drums can be used for pressures between 15 and 1,000 psig. Floating-roof storage tanks are used for materials near atmospheric pressure. In the basic design, a void forms between the floating roof and the product, forming a constant seal. The primary purposes of a floating roof are to reduce vapor losses and to contain stored fluids. In areas of heavy snowfall, an internal floating roof is used with an external roof because the weight of the snow would affect the seal.

Vessel Thickness

If the original design thickness of all pressure-retaining parts in a plant could be maintained and the process conditions held constant, the unit would not require periodic inspection and operation could be continuous. Since this is rarely possible, it is important to know what factors affect the initial required thickness of operating equipment and one factor, corrosion, that can reduce the thickness of equipment.

Essentially, the thickness of pressure-retaining equipment depends on the diameter of the pipe, vessel, exchanger, or other equipment; pressure; temperature; strength of material used; and anticipated corrosion rates (a 1/8" corrosion allowance is normally provided). Of these, the operator has control of pressure, temperature, and process changes that might affect the amount of corrosion.

Pressure

The equipment is designed for normal operating pressures plus an incremental increase in pressure to allow for operating upsets. The relief valve on the equipment or in the system is set to relieve when the design pressure is exceeded and is provided for equipment protection and safety of personnel. Blocking in equipment unprotected by relief valves, unofficially increasing the relief valve set pressure, and blocking in cooling water in exchangers while in service are only several possibilities of creating hazardous conditions in the unit. The operator should be aware that such unsafe practices can exceed design conditions and may cause failure.

Temperature

In general, the strength of metals decreases as temperature increases. Reduction in strength usually starts at 650°F (343.33°C) and becomes critical in the range of 950°F (510°C) to 1,000°F (537.55°C) in the case of carbon steel. For example, the strength of low carbon steel is reduced 22% when temperature is raised from 950°F to 975°F (510°C to 523.88°C). Similarly, because of this decrease in strength as temperatures increase, it is important that pressure in equipment be reduced and exposed metal surfaces cooled with water during fire conditions. Besides affecting strength, temperature also has a profound effect on corrosion rates. For example, in streams containing sulfuric acid, a 20°F (6.66°C) increase in temperature can double corrosion rates.

Materials: Carbon Steel, Alloys, and Nonferrous Alloys

The metals described in this section are those most commonly used in chemical plants. There are, of course, a great many other metals that are used, but they are not covered owing to space limitations.

Low-Carbon Steel

Fortunately, low-carbon steel, which is familiar to everyone, is a very satisfactory material for most plant applications. It is relatively inexpensive yet provides the strength, workability, and welding properties required. Most of the equipment used in a plant is made of this versatile material. The steel used for equipment is low in carbon (0.3% or less), sulfur, and phosphorus and contains sufficient manganese to offset the effect of sulfur. It may also contain small quantities of silicon or aluminum. The low-carbon content promotes ductility and weldability.

Although low-carbon steel is suitable for the majority of services, a number of other materials have been developed to cope with the severe conditions encountered as new processes were developed. However, none of these materials are suitable for all services.

Low-Alloy Steels

As the operating temperature of equipment increases above 650°F (343.33°C), the strength of low-carbon steels decreases. This decrease in strength becomes very pronounced between 950°F and 1,000°F (510°C and 537.77°C). For example, at 950°F and 1,000°F (510°C and 537.77°C), the strength of low-carbon steel is about one-third and one-sixth, respectively, of its room temperature strength. As a result, the strength of this type of steel becomes so low that it is not a satisfactory material. The strength and resistance to oxidation (rushing) required for these conditions are secured by adding small amounts of alloying elements. Molybdenum in quantities as small as 0.5% greatly increases the strength above 900°F (482.22°C). Chromium is added in amounts up to 9% to combat the tendency to oxidize at high temperatures and to resist corrosion from materials that contain sulfur. These steels still retain much of the ductility, toughness, and weldability of low-carbon steel but require more extensive and careful heat treatment when welded. The chrome alloys are used in pressure vessels, piping, furnace tubes, and exchangers operating at high temperatures and pressures.

Some of the processes used in refining and chemical plants employ hydrogen at high temperature or high pressure or both. Low-carbon steel normally becomes brittle in this service above 500°F (260°C). Embrittlement is prevented by using steels that contain small amounts of chromium or molybdenum or both.

When it is necessary to operate equipment at very low temperatures, low-carbon steel becomes an unsatisfactory material. Operating above −20°F (−28.88°C), it is a tough and ductile material. Below it, the steel begins to lose its ability to resist sudden shock. Most of the plain carbon steels are ductile to −20°F (−28.88°C), and suitable heat treatment could be used to −50°F (−45.55°C). Adding small amounts of alloying elements lowers the temperature at which steel becomes brittle. Nickel is the most common metal added to steel for this purpose. The addition of 3% to 5% nickel will produce steels that remain tough to −150°F (−101.11°C).

High-Alloy Steels

The properties of steel can be varied widely by small additions of other elements to produce steels that are satisfactory for most services. In some cases, however, it is not possible to produce steel that is satisfactory for a particular service by adding small amounts of other elements, and larger quantities of alloying elements are necessary to produce the desired characteristics. Steels that contain 10% or more of alloying metals are generally called high-alloy steels. The members of this group most often used in plants are chromium steel and austenitic (that is, stainless) steel.

Chromium Steels

Chemical components containing appreciable amounts of sulfur compounds become quite corrosive to steel at temperatures ranging from about 550°F to 850°F (287.77°C to 454.44°C). Chromium steels withstand this type of attack very well, but in some cases the low chromium alloys previously described are not resistant enough to be economically attractive. In these cases, alloys containing from 12% to 17% chromium are used.

The 17% chrome steels were used rather extensively initially for severe sulfur corrosion, but they had a tendency to become brittle after extended heating cycles in the 700°F to 1,000°F (371.11°C to 537.77°C) range. Their primary use is now largely confined to pump and compressor parts. The 12% chromium materials are widely used as protective linings in steel equipment, thermowells, and valve trim subject to this type of sulfur corrosion.

Austenitic (Stainless) Steels

When both nickel and chromium are added to steel in amounts totaling somewhat over 20%, the microscopic structure undergoes a pronounced change. The small grains of which steel is composed solidify in a form known as *austenite*, which behaves in many respects quite differently from the steels previously described.

The most common composition of stainless steel is commonly referred to as *18-8*. This name comes from the fact that this stainless steel contains about 18% chromium and 8% nickel. Other members of the family contain higher amounts of either or both of these elements, and some contain small amounts of other elements. Molybdenum is an additional element commonly used.

An outstanding characteristic of this group of steels is resistance to rusting when exposed to the atmosphere as well as resistance to corrosion by a wide variety of chemicals. They also retain much of their strength and have excellent resistance to oxidation at extremely high temperatures. In process units, they are widely used for brick hangers and tube supports in furnaces. Here, the materials flowing through them cool the tubes, but their supports are not and must maintain adequate strength and oxidation resistance at firebox temperatures.

In other applications, these steels are used when it is necessary to process materials at very low temperatures. They remain tough and ductile at temperatures far below those at which low-carbon steel becomes brittle.

Although the stainless steels are extremely useful implants for many severely corrosive and high-temperature services, they have some limitations that make them impractical for certain applications. Two conditions that cause these steels to deteriorate are **stress-corrosion cracking** and a high coefficient of expansion. Stress-corrosion cracking is a mechanical-chemical type of deterioration. Many materials, even low-carbon steel, are subject to it in particular critical environments. In plants, the most familiar occurrence is the cracking of stainless steels in chloride environments. The cracking is usually across the metal grains, and there is little metal loss from corrosion. The other undesirable characteristic of stainless steel is its high coefficient of expansion. When stainless steel is heated, it expands at a rate approximately 150% of that of steel. This expansion becomes a problem whenever stainless steel is used in close contact with other metals. At high temperatures, great internal strains can be produced because the two materials expand at different rates.

Nonferrous Alloys

A metal or alloy that contains little or no iron is called a nonferrous material. There are a great many elements other than iron that are metals in their pure form, and the combinations of these as alloys are almost limitless. Some of these alloys are rather widely used.

Nickel Alloys

In a few locations around a chemical plant where extreme resistance to chemicals is required and the stainless steels are unsatisfactory, a group of alloys containing large amounts of nickel are used. These alloys usually contain additions of iron, copper, aluminum, chromium, cobalt, and molybdenum. Some typical examples of these alloys are Monel, Hastelloy, and Inconel. These alloys are used in a variety of services that involve acids and caustics. For example, Monel is used in hot sodium hydroxide or hydrochloric acid service.

Copper Alloys

Brass is the term used to describe a family of alloys of copper and zinc. The copper content ranges from 90% to about 60%, with the balance being zinc. Some brasses have small amounts of other elements such as lead, tin, antimony, arsenic, and phosphorus.

Brasses are widely used because of their resistance to corrosion from water containing various impurities that are corrosive to steel. They are weaker than steel and lose much of their strength when heated. They are not normally used at temperatures above 450°F (232.22°C). Brass is most commonly used in condenser or cooler tubing when water is the cooling medium. Some brasses, notably those containing lead and antimony, have good antifriction properties and are widely used as bearing materials in pumps and compressors.

The bronzes are a second family of copper alloys. These alloys contain 90% or more copper. Aluminum, tin, or silicon makes up the balance. Aluminum and silicon bronzes are more resistant to salt water than brass and are widely used as condenser tubing when salt water is the cooling medium.

There are a number of copper-nickel alloys. One of these, called *cupronickel*, contains 70% copper and 30% nickel. Cupronickel is used in condenser tubing when the cooling water has extreme concentrations of salt.

Aluminum Alloys

The outstanding characteristics of aluminum are its good resistance to corrosion from sulfur compounds and its resistance to continuous oxidation when exposed to the atmosphere. Because of its resistance to corrosion by sulfur, aluminum is used for internal parts of equipment processing high-sulfur stocks. It is often used in sheet form to protect and weatherproof insulation on pipe and towers because of its resistance to atmospheric corrosion.

There are many alloys of aluminum, which contain small amounts of other metals that greatly increase its room temperature strength. This strength in most aluminum alloys decreases rapidly with increasing temperature.

Aluminum coatings over iron-base alloys have been used rather extensively in recent years to protect equipment from high-temperature sulfur and hydrogen sulfide corrosion as well as high-temperature oxidation.

Lead Alloys

Lead is a heavy, extremely ductile, relatively weak material that melts at a rather low temperature. It is used as a lining material in sulfuric acid-treating equipment.

Inspection

Prolonged and safe operation depends upon good inspection practices for assurance that equipment is being maintained in a safe condition and that off-stream time is reduced to a minimum by anticipation of necessary repairs. In general, the scope of work includes all pressure vessels, heat exchangers, storage tanks, process piping, pumps, relief valves, furnace tubes, fittings, breechings, stacks, and tube supports. Any equipment subjected to pressure or temperature extremes must be inspected periodically.

Power boilers and auxiliaries are subject to state regulations and inspection. Representatives of an insurance company may also inspect the boilers. Plant inspectors make joint inspections with state and insurance company inspectors and keep records for reference. Inspecting pumps and compressors is an important function of operators as well as maintenance and engineering personnel.

The inspector studies the condition peculiar to each piece of equipment. The nature of the material contained, the pressure, temperature, flow conditions, and other factors may cause or contribute to deterioration of equipment. Familiarity with operating conditions and knowledge of the materials of construction are essential. A study of conditions and materials leads to planning and actual performance of inspection, at which time the true condition of the equipment is determined. The scope of inspection work also includes keeping records, reporting results, recommending repairs and methods of repair, assisting in planning turnarounds, and determining safe working limits for equipment. When inspection reveals the need for replacement parts, it is important that the new parts be designed in accordance with recognized codes and specifications.

Inspection Frequency and Extent

The frequency and extent of inspection depend on factors such as pressure, temperature, corrosive action of the materials handled, and materials of construction, corrosive allowance, and past experience with the equipment involved. Equipment in high-pressure, high-temperature service subject to corrosion is, of course, inspected frequently. On the other hand, some equipment may require complete inspection only once in five years. The frequency and extent of inspection are established independently for each item and are subject to change with changes in operating conditions. In practically all cases, only part of the lines and equipment constituting a unit are inspected. Inspections are scheduled so that complete inspection of the unit will extend over several inspections.

Inspection Methods and Equipment

Visual inspection is the method most generally used and requires no explanation. Experienced inspectors use hammer testing to estimate the metal thickness. The equipment or line is tapped with a hammer and the feel and the sound are noted. The hammering sets up a vibration, and the sound depends on the thickness of the point struck. The feel of the hammer and the extent of denting also give an indication of thickness. Hammering can be used to determine doubtful areas. Other types of inspection, such as drilling or calipering, can be used to obtain an accurate reading in the thin area found by hammering. Transfer or direct reading calipers are used for measuring thickness when the areas being inspected are accessible. A variety of remote-reading instruments are available for measuring internal diameters of furnace and exchanger tubes.

Measuring through drilled holes, called *trepanning*, is the most accurate method of determining wall thickness when transfer calipers cannot be used. The thin area is first determined by visual inspection or a hammer test, and a hole is drilled completely through the wall. The thickness is measured through the hole. Holes are closed by threading the opening with a tapered thread and screwing in a tapered plug. Welding may also close holes.

Trepanning is used to inspect the welding on new storage tanks or similar equipment. It is also used at times to investigate the nature and extent of defects in plates or welds discovered by previous visual inspection. This method of inspection is no longer used extensively because it has quite generally been replaced by nondestructive radiographic (X-ray) techniques similar to those used to identify broken bones. Radiography is also used to determine pipe and tube wall thickness.

Weld probing is done by a special machine that removes boat-shaped samples from plates. These samples are generally taken to check welding or the condition of the material sampled.

Several electronic-sonic devices are available for measuring metal thickness; they are used mainly for determining the shell thickness of pressure vessels, storage tanks, piping, and thick-walled equipment. They have the ability to measure only the thickness of the material contacting the crystal probe or coupled in some manner to the probe. Multiple layers of metal, coke, or other deposits on the opposite side from the probe are excluded from the thickness readings obtained. These instruments are quite reliable if the opposite surface is not too severely pitted or if the material is not less than 1/8" thick.

An electronic-radium-source device is also used for measuring metal thickness. This type of instrument gives a rapid examination because no surface preparation is required, as in the case of most electronic-sonic devices, and readings are obtained directly in about 30 seconds. Metal temperatures up to 1,000°F (537.77°C) will not damage the instrument or seriously affect the accuracy of the measurements. The range is from a maximum of ¾" to zero, with its accuracy increasing near the zero end of the range. This characteristic makes it ideal for examining piping. However, it has its limitations, as it will include in the measurement any coke, liquid, extra layers of metal, or foreign deposits in the pipe or vessel. Also, in severely pitted areas, an average thickness reading will be obtained.

Crack or imperfection detectors use a dye penetrant to locate surface cracks in the metals. The technique consists of applying a dye penetrant to the suspected area, washing the surface, and then applying a developer solution. If a crack is present, a bright red line will appear in the white developing coating, locating its position.

Magnetic particle inspection is used to detect surface or near-surface flaws in equipment that can be magnetized. A magnetic field is induced and an iron powder is dusted on the piece to be inspected. The iron powder adheres to the piece at any discontinuity in the magnetic field, thus outlining such defects as cracks, porosity, and inclusions (embedded foreign material).

Hardness testers of various types are used in the shops and field to determine the hardness of metals. These hardness readings indicate the approximate strength and ductility of material.

Vessel Design Sheets

Vessel design sheets are sketches that include information necessary for the selection, use, and need for periodic inspection of materials used to make vessels. Figures 3.13 through 3.21 illustrate vessel design sheets of typical vessels found in process units.

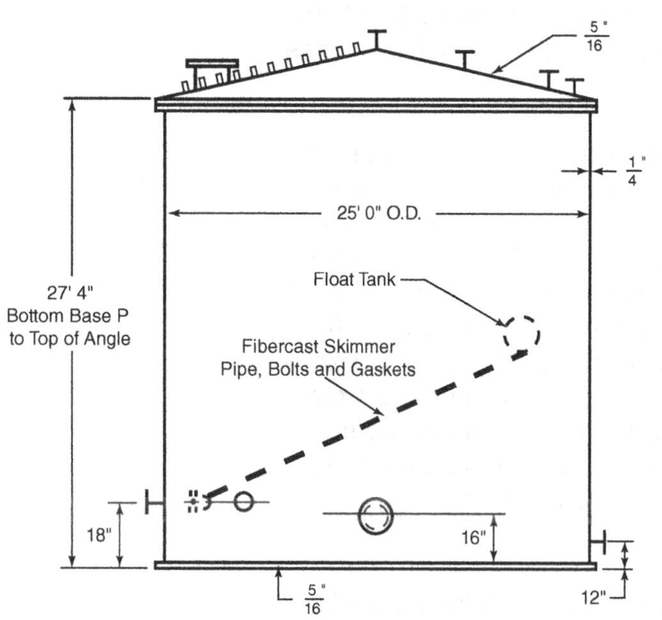

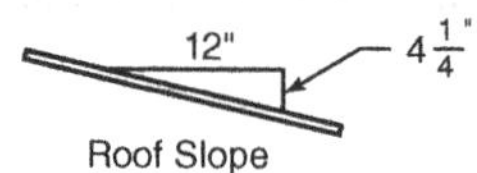

MATERIAL NOTES	DESIGN NOTES
Shell, Roof and Bottom Nozzle Necks Nozzle Flanges Manholes Gasket Bolting Lining Liquid Densities	Operating Temperature and Pressure – Full Liquid @ 120°F Design Temperature and Pressure – Full of Liquid @ 175°F Corrosion Allow Construction Inspection Test Painting
ADDITIONAL NOTES	
Tank to be lined infield with reinforced epoxy resistant to acid solution of AlCl, water, and aromatic HC (concentration 10% to 25% weight). Design temperature is limited by lining.	

Figure 3.13 *Storage Tank with Internal Skimmer*

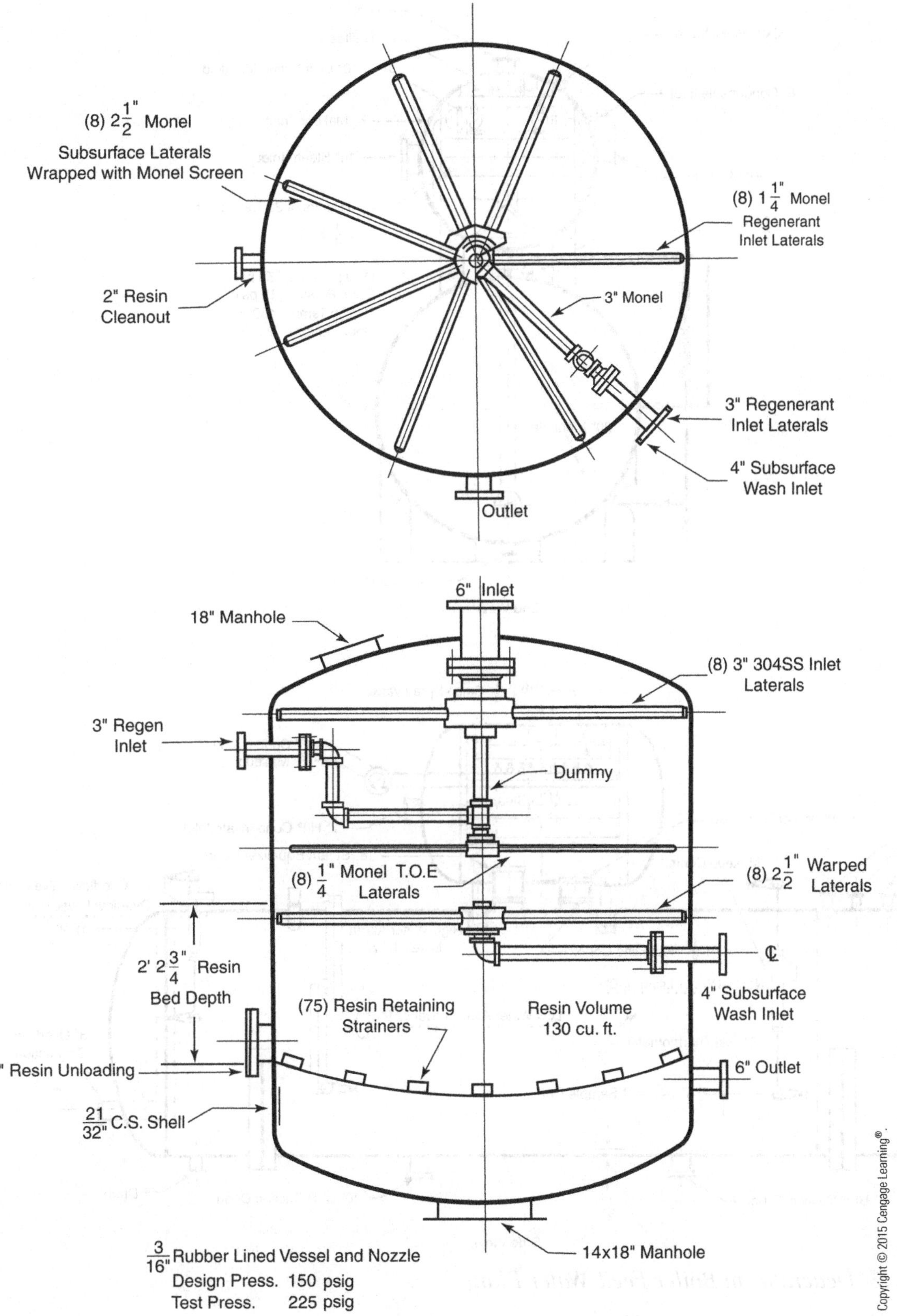

Figure 3.14 *Storage Tank*

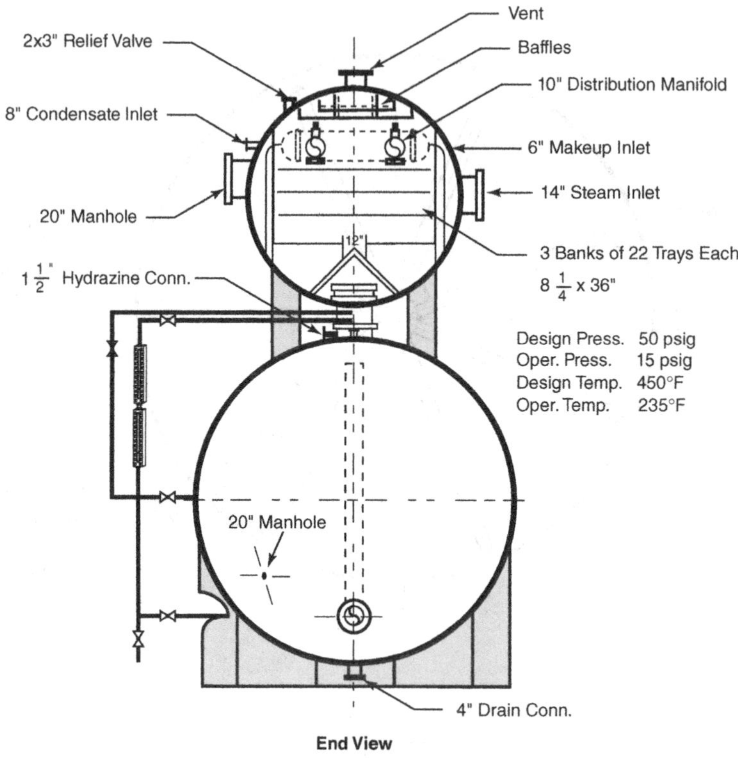

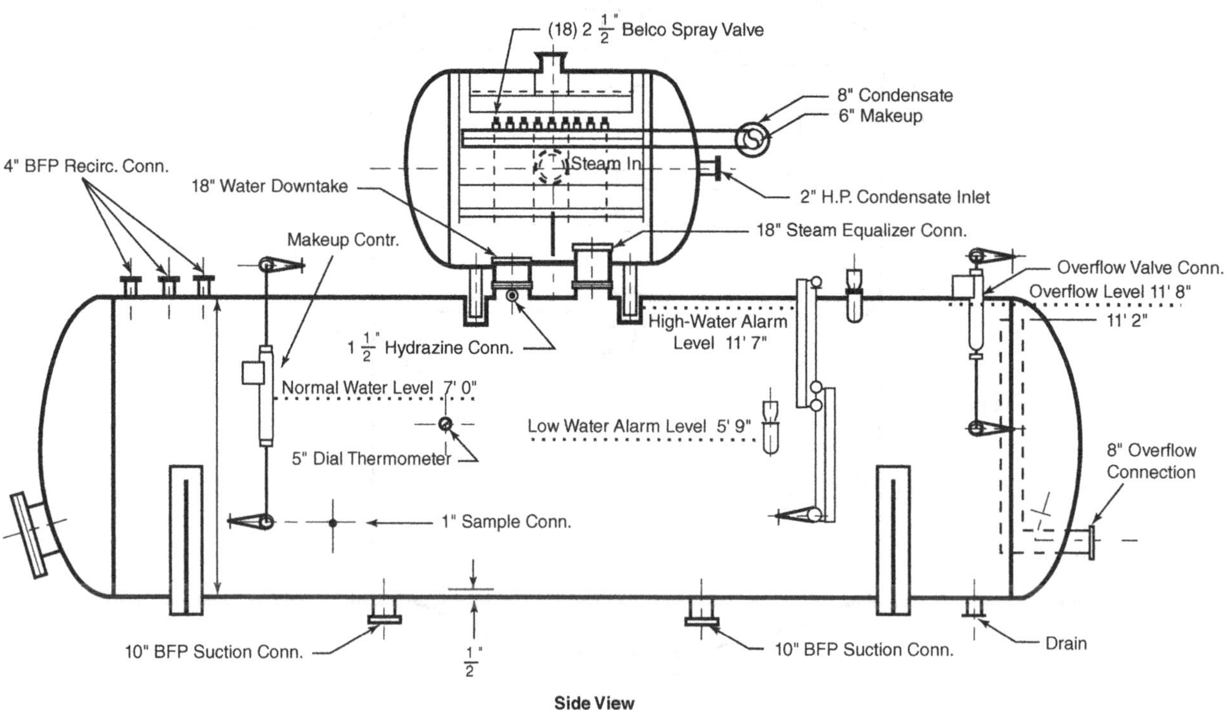

Figure 3.15 *Deaerator in Boiler Feed-Water Plant*

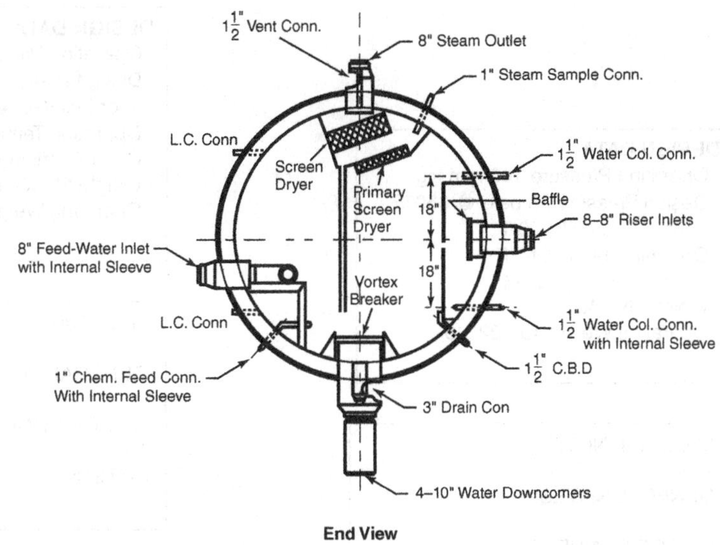

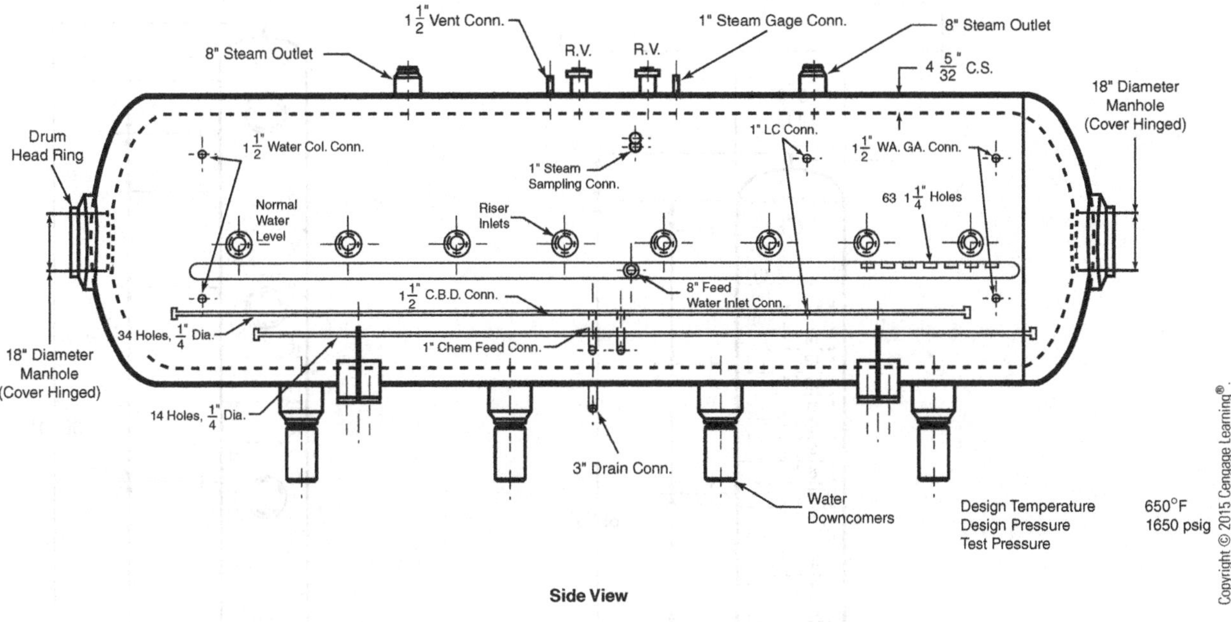

Figure 3.16 *High-Pressure Steam Drum*

DESIGN DATA
Operating Pressure 160 mm Hg
Design Pressure 25 psig @ 650°F
or Full Vac. @ 300°F
Operating Temp. 214°F
Code Stamping ASTM
Weight 69,140 lb.
Operating Weight 152,020 lb.

MATERIAL NOTES:

GENERAL NOTES:

LIST OF CONNECTIONS:

NOTES:

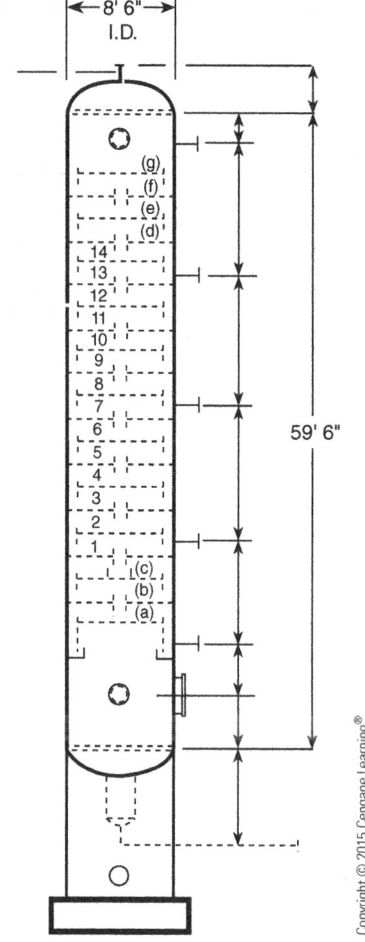

Figure 3.17 *Distillation Column*

DESIGN DATA
Operating Pressure 50 psig
Design Pressure 100 psig @ 150°F
or Full Vac. @ 300°F
Operating Temp. 120°F
Code Stamping ASTM
Weight 29,300 lb.
Operating Weight 114,300 lb.

MATERIAL NOTES:

GENERAL NOTES:

LIST OF CONNECTIONS:

NOTES:

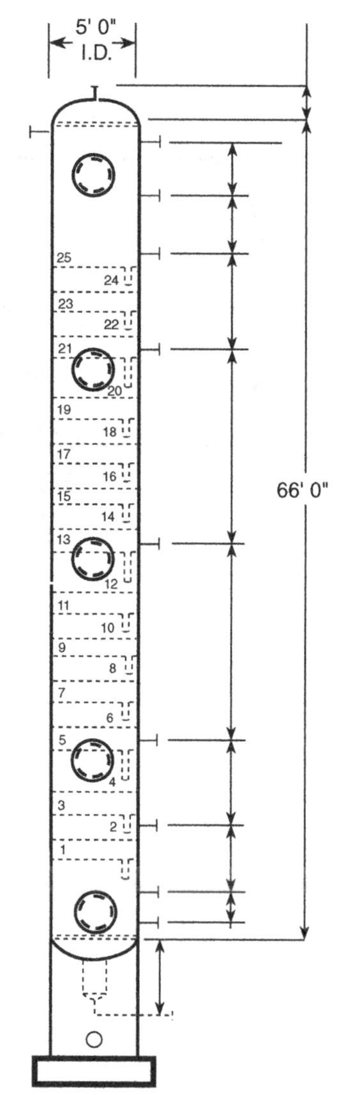

Figure 3.18 *Extraction Column*

Figure 3.19
Separator

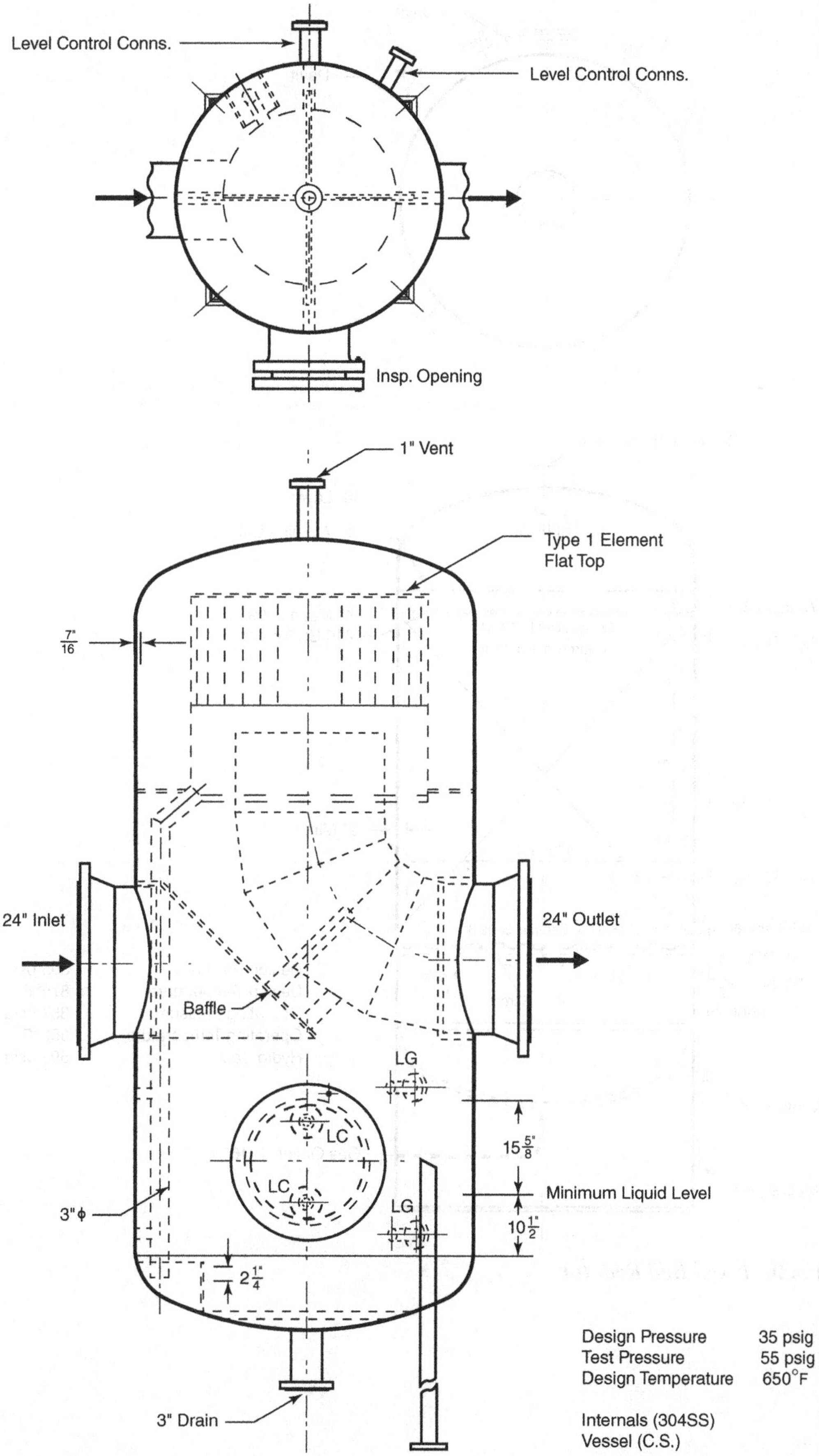

Level Control Conns.
Level Control Conns.
Insp. Opening
1" Vent
Type 1 Element
Flat Top
$\frac{7"}{16}$
24" Inlet
24" Outlet
Baffle
LG
LC
LC
$15\frac{5"}{8}$
Minimum Liquid Level
LG
$10\frac{1"}{2}$
3"ϕ
$2\frac{1"}{4}$
3" Drain

Design Pressure 35 psig
Test Pressure 55 psig
Design Temperature 650°F

Internals (304SS)
Vessel (C.S.)

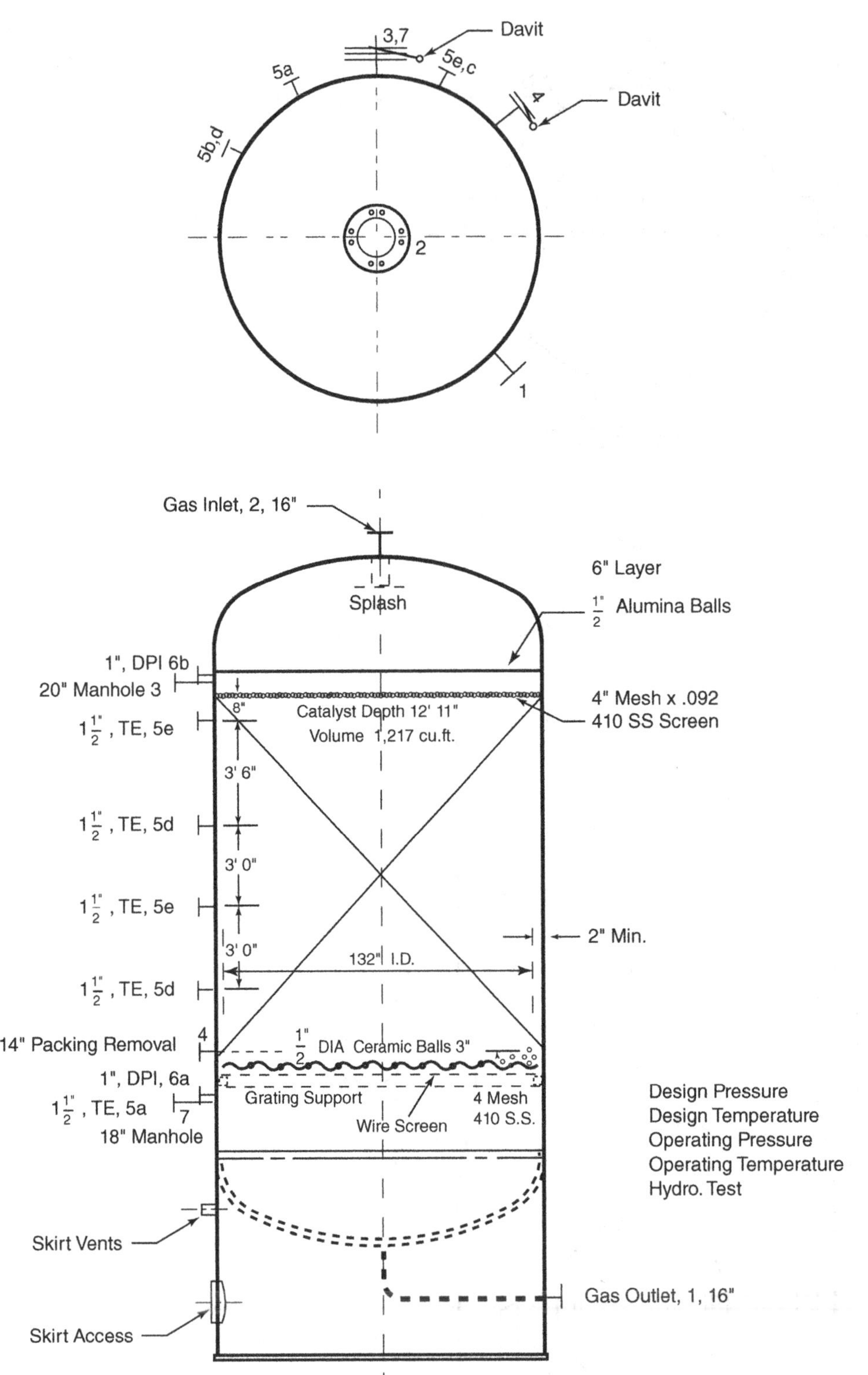

Figure 3.20 *Fixed Bed Reactor*

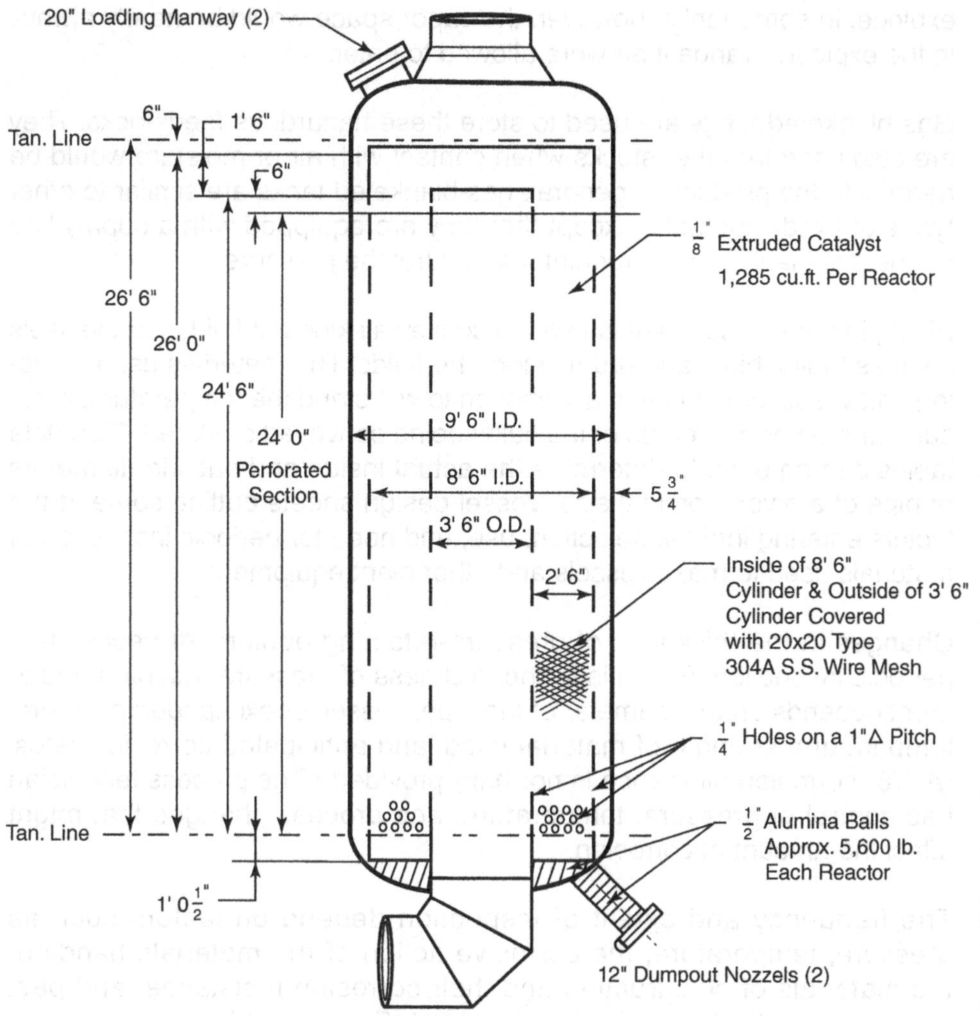

Figure 3.21
Radial Flow Reactor

Summary

Common names for tanks include cone roof, floating roof (internal or external), spheres, spheroid, bullets, hemispheroid, bins, silo, open top, or double wall. Technicians also refer to tanks as the feed tank, vaulted tank, elevated tank, recovery tank, surge tank, blend tank, cryogenic tank, jacketed tank, or blanketed tank. Tanks can be divided into four general categories: atmospheric tanks, low-pressure tanks (0 to 2.5 psig), medium-pressure tanks (2.5 to 15 psig), and high-pressure tanks (above 15 psig).

Depending on the vapor pressure and temperature of the stock in an atmospheric tank, the vapor space may be filled with varying mixtures of vapor and air. The vapor space in tanks storing materials having a low vapor pressure at the storage temperature is usually too lean to explode. The vapor space in tanks storing very volatile materials is usually too rich to explode. In some tanks, however, the vapor space would be nearly always in the explosive range if air were allowed to enter.

Gas-blanketed tanks are used to store these hazardous feedstocks. They are also used for other stocks when contact with air or moisture would be harmful to the product. In general, gas-blanketed tanks are similar to other types of fixed-roof tanks except that they are equipped with a supply line for the gas blanket and a regulator to control the pressure.

Piping in a chemical plant is used to convey all kinds of fluids, and vessels such as tanks, bins, and drums store the fluids. The materials used in piping and vessel construction are chosen to withstand the temperature, pressure, and other properties of the fluids being conveyed or stored. Pipe data tables can be used to determine the actual inside and outside diameters of pipe of a given nominal size. Vessel design sheets outline some of the factors entering into the selection, use, and need for periodic inspection of materials used to make vessels and other plant equipment.

Changes in the thickness of pressure-retaining equipment necessitate periodic inspection. Essentially, the thickness of pressure-retaining equipment depends on the diameter of the pipe, vessel, or exchanger; pressure; temperature; strength of material used; and anticipated corrosion rates. (A 1/8" corrosion allowance is normally provided.) The process technician has control of pressure, temperature, and process changes that might affect the amount of corrosion.

The frequency and extent of inspection depend on factors such as pressure, temperature, the corrosive action of the materials handled, the materials of construction and their corrosion resistance, and past experience with the equipment involved. Equipment in high-pressure, high-temperature service subject to corrosion is inspected frequently. Some equipment may require complete inspection only once in five years.

Visual inspection is the method most generally used. Experienced inspectors use hammer testing to estimate the metal thickness. Transfer or direct reading calipers are used for measuring thickness when areas being inspected are accessible. A variety of remote-reading instruments are available for measuring internal diameters of furnace and exchange tubes. Measuring through drilled holes, trepanning, is the most accurate method of determining wall thickness when transfer calipers cannot be used. Weld probing is done by a special machine that removes boat-shaped samples from plates. Several electronic-sonic devices are available for measuring metal thickness; they are used mainly for determining the shell thickness of pressure vessels, storage tanks, piping, and thick-walled equipment. An electronic-radium-source device is also used for measuring metal thickness. Crack or imperfection detectors use a dye penetrant to locate surface cracks in metals. Magnetic particle inspection is used to detect surface or near-surface flaws in equipment that can be magnetized. Inducing a magnetic field and dusting an iron powder on the piece to be inspected causes the iron powder to adhere to the piece at any discontinuity in the magnetic field. **Radiographic inspection** is used to locate defects in metals and pipe and tube wall thickness in much the same manner as an X-ray is taken of a broken bone. Hardness testers are used to determine the hardness of metals. These hardness readings indicate the approximate strength and ductility of material.

Review Questions

1. What type of shutoff does a paddle blind provide?
2. What two types of blinds do we use?
3. How does increased temperature affect the performance of metals?
4. Why are brass tubes used instead of carbon steel in many heat exchangers that use water as the cooling medium?
5. What metal would you use for a vessel containing noncorrosive material at 100 psig and 300°F (148.88°C)?
6. What is a nonferrous alloy?
7. Would you use brass, stainless steel, or Hastelloy in a hot, extremely corrosive process?
8. What information do process technicians find on vessel design sheets?
9. What is an alloy steel?
10. What is corrosion, and how is it manifested?
11. What factors determine how thick a vessel's walls should be?
12. What factors determine which material is best for vessel construction?
13. What is corrosion allowance?
14. What is the normal operating temperature range for vessels constructed of carbon steel?
15. What metals are used for extremely low temperatures (below −150°F or −101.11°C)?
16. Describe the various inspection procedures used in a process plant.
17. List the various types of vessels found in a process unit.
18. List the important facts associated with pipe size and diameter.
19. List the four distinct categories for aboveground storage tanks.
20. List the different types of tanks utilized in the chemical processing industry.
21. Explain how process technicians clean residue out of a pipeline.
22. Describe cathodic protection.
23. List the basic components of a floating-roof tank.
24. Compare modern and traditional diking methods.
25. Compare the terms "breathing" and "gas-blanketed tank."

Pumps

Objectives

After studying this chapter, the student will be able to:

- Review the history and design of pumps.
- Describe the scientific principles associated with centrifugal pump operation and identify key components.
- Describe the operation and maintenance of positive displacement pumps.
- List the various types of rotary pumps.
- Describe the basic components and operation of screw pumps.
- Explain how rotary gear pumps operate.
- Describe the basic components and operation of sliding and flexible vane pumps.
- Describe the basic components of a lobe pump.
- List the various types of reciprocating pumps.
- Explain the operation of diaphragm pumps.
- Describe the operation and design of piston pumps.
- Describe the scientific principles associated with the operation of plunger pumps.
- Start up and shut down a positive displacement pump.
- Start up and shut down a centrifugal pump.
- Troubleshoot typical problems associated with the operation of centrifugal and reciprocating pumps.

Key Terms

Acceleration head—the fluctuations of suction pressure created by the intake stroke of a reciprocating pump.

Axial pump—a dynamic pump that accelerates fluid in a straight line.

Cavitation—the formation and collapse of gas pockets around the impellers during pump operation; results from insufficient suction head (or height) at the inlet to the pump.

Centrifugal pump—a dynamic pump that accelerates fluid in a circular motion.

Diaphragm pump—a reciprocating pump that uses a flexible diaphragm to positively displace fluids.

Discharge head—the resistance or pressure on the outlet side of a pump.

Dynamic—class of equipment such as pumps and compressors that convert kinetic energy to pressure; can be axial or centrifugal.

Head—is described as Pressure (at suction) × 2.31 ÷ Specific gravity; 1 psi is equal to 2.31 feet of head.

Impeller—a device attached to the shaft of a centrifugal pump that imparts velocity and pressure to a liquid.

Lobe pump—a rotary pump that uses kidney-bean-shaped lobes to displace and transfer fluid.

Mechanical seal—provides a leak-tight seal on a pump; consists of one stationary sealing element, usually made of carbon, and one that rotates with the shaft.

Net positive suction head (NPSH)—the head (pressure) in feet of liquid necessary to push the required amount of liquid into the impeller of a dynamic pump without causing cavitation.

Net positive suction head available (NPSHa)—a term used to indicate the required pump suction pressure so the pump can operate properly. It is defined as atmospheric pressure (converted to head) 1 static head 1 surface pressure head 2 vapor pressure 2 frictional losses.

Net positive suction head required (NPSHr)—the minimum NPSH necessary to avoid cavitation. The NPSHa must be greater than or equal to the NPSHr, expressed as NPSHa ≥ NPSHr. It is the reduction in total head as the liquid enters the pump.

Piston pump—a reciprocating pump that uses a piston and cylinder to move fluids.

Positive displacement—class of equipment such as pumps and compressors that move specific amounts of fluid from one place to another; can be rotary or reciprocating.

Pressure relief valve—used to relieve excessive pressure on the discharge of a positive displacement pump.

Priming—becoming filled with fluid.

Pulsation dampener—a device installed close to a pump, in the suction or discharge line, to reduce pressure variations.

Reciprocating pump—a positive displacement pump that uses a plunger, piston, or diaphragm moving in a back-and-forth motion to physically displace a specific amount of fluid in a chamber.

Rotary pump—a positive displacement pump that uses rotating elements to move fluids.

Screw pump—a rotary pump that displaces fluid with a screw.

Slip—the percentage of fluid that leaks or slips past the internal clearances of a pump over a given time.

Specific gravity—is described as the ratio between the density of a given liquid to the known density of water, or the density of gases to the density of air. The specific gravity of water or air is one.

Vane pump—a rotary pump that uses flexible or rigid vanes to displace fluids.

Vapor lock—condition in which a pump loses liquid prime and the impellers rotate in vapor.

Pump Applications and Classification

Refineries and chemical plants use pumps to move liquids. Pumps are used in a variety of applications and processes, including refrigeration, automobiles, home heating systems, and water wells. The liquids moved by a pump vary from liquid sodium and liquid potassium for cooling nuclear reactors to domestic drinking water systems. In some situations, pumps are not needed to transfer liquid. In Figure 4.1, gravity transfers the liquid into tank 4.02B. In Figure 4.2, a pump is needed to move the liquid.

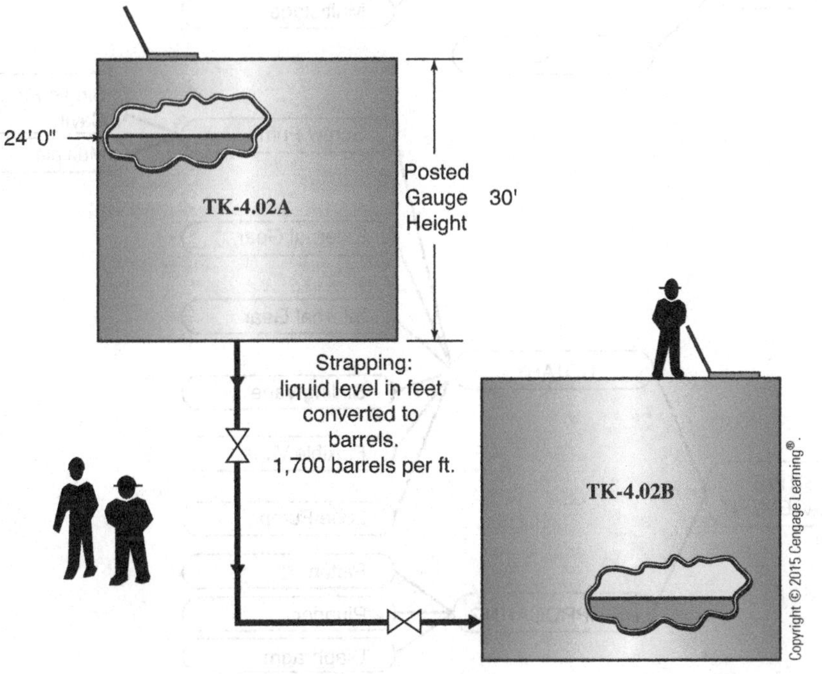

Figure 4.1
Gravity Flow

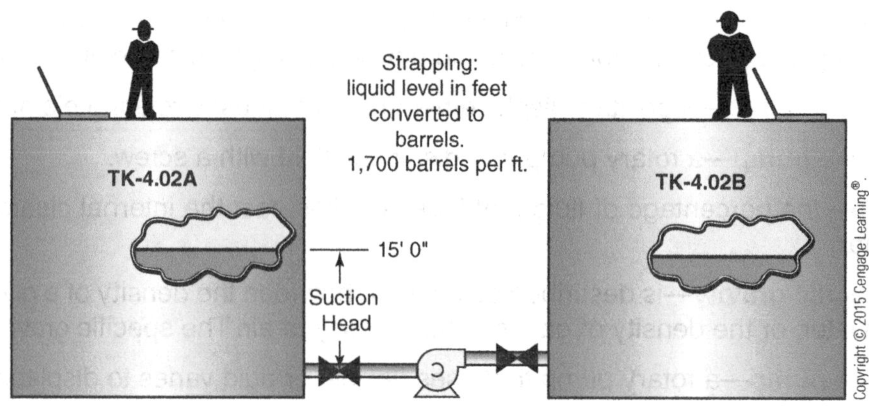

Figure 4.2
Pump Transfer

In general, pumps can be classified as **dynamic** or **positive displacement** (Figure 4.3). Both classes are designed to transfer liquids, but the way the transfer is accomplished is different.

Dynamic pumps accelerate liquids axially (in a straight line) or centrifugally (in circles). They are operated at high speeds to generate large flow rates at low discharge pressures. Pressure moves the liquid through the piping and equipment system.

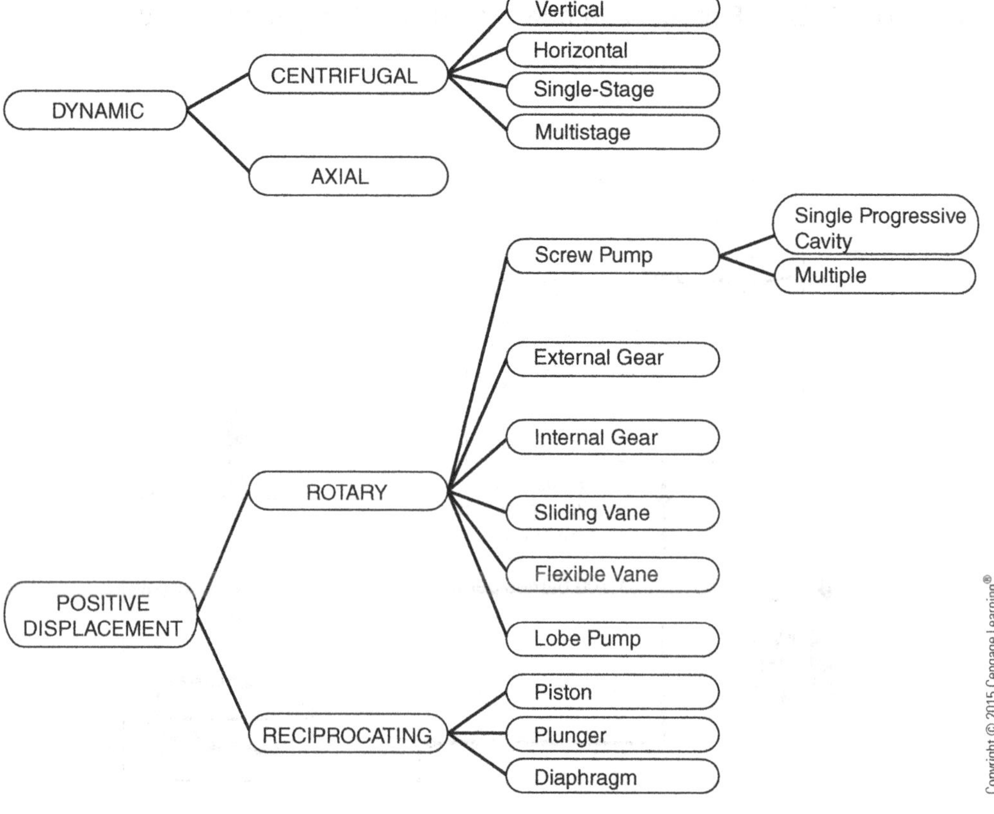

Figure 4.3
Pump Family Tree

Positive displacement (PD) pumps transfer liquids by using a rotary or reciprocating motion that displaces liquid on each rotation or stroke. They are used in processes that require specific amounts of fluid to be delivered. The operation of positive displacement pumps is significantly different from that of dynamic pumps. Positive displacement pumps transfer specific amounts of fluid no matter what the discharge pressure is, whereas the amount of fluid transferred by dynamic pumps is greatly affected by discharge pressure. **Rotary pumps** deliver a specific amount of fluid with each rotation of screws, gears, vanes, or similar elements. **Reciprocating pumps** move fluids by drawing them into a chamber on the intake stroke and pushing them out of the chamber with a piston, diaphragm, or plunger on the discharge stroke.

The first reciprocating pump, invented by a Greek, Ctesibius, in about 200 BCE, was used to pump water. Around this same time, Archimedes, a Greek mathematician, invented the first screw (rotary) pump. The first true **centrifugal pumps** were not invented until the 1600s by the French inventor Denis Papin. Papin's straight-vane centrifugal pump design was improved in 1851, when a British inventor, John Appold, designed a curved-**vane pump**.

Internal Slip

Slip is defined as the percentage of fluid that leaks or slips past the internal clearances of a pump over a given time. Slip also can be defined as the difference between how much liquid a pump can move and how much it actually does move.

Because of differences in design, positive-displacement pumps and centrifugal pumps respond differently to slip. A centrifugal pump can have as much as 100% slip if the discharge valve is closed. This principle can be illustrated by following the path of liquid as it enters a centrifugal pump with an open discharge valve. As fluid moves from the suction inlet into the **impeller**, flow is accelerated into a discharge chute known as a *volute*. The volute is a specially designed chamber that widens within the pump. Shutting the discharge valve stops flow to the process unit, but circulation within the pump continues. As the impeller turns, fluid is accelerated into the large volute discharge. Circular motion (clockwise) is sustained in the volute and discharge pipe up to the discharge valve (Figure 4.4). Fluid friction within this area begins to heat up the liquid. Operators should keep this principle in mind if the liquid being pumped is close to its boiling point. When the liquid vaporizes, it expands and gets hot; this process can create tremendous pressure that will damage the pump.

Positive-displacement (PD) pumps are designed to have minimal slip characteristics. There are physical laws that state that two separate bodies

Figure 4.4
Internal Slip

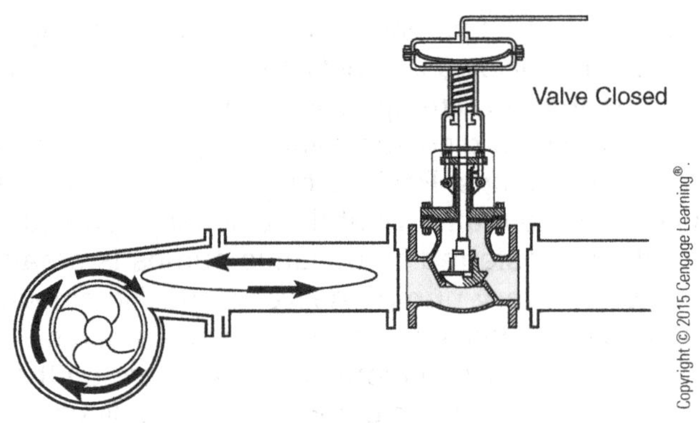

cannot occupy the same space at the same time. Positive-displacement pumps are designed to displace exact fluid volumes with solid objects, such as pistons or gears. When the discharge valve is closed on a PD pump, the following conditions exist:

- Very little (if any) slip is occurring within the body of the pump.
- Fluid pressure increases with every stroke.
- Fluid pressure is transferred equally to all isolated parts.
- The pump or discharge pipe can be damaged if a relief valve is not provided.

Dynamic Pumps

Dynamic pumps are classified as centrifugal or axial. Centrifugal pumps operate on the principle of centrifugal force. A spinning impeller inside a shell casing propels liquid outward. Fluid velocity is accelerated inside the shell of the pumps and liquid quickly moves toward the discharge port. Figure 4.5 illustrates the different types of impeller arrangements in a centrifugal pump.

Figure 4.5
Impeller Arrangements

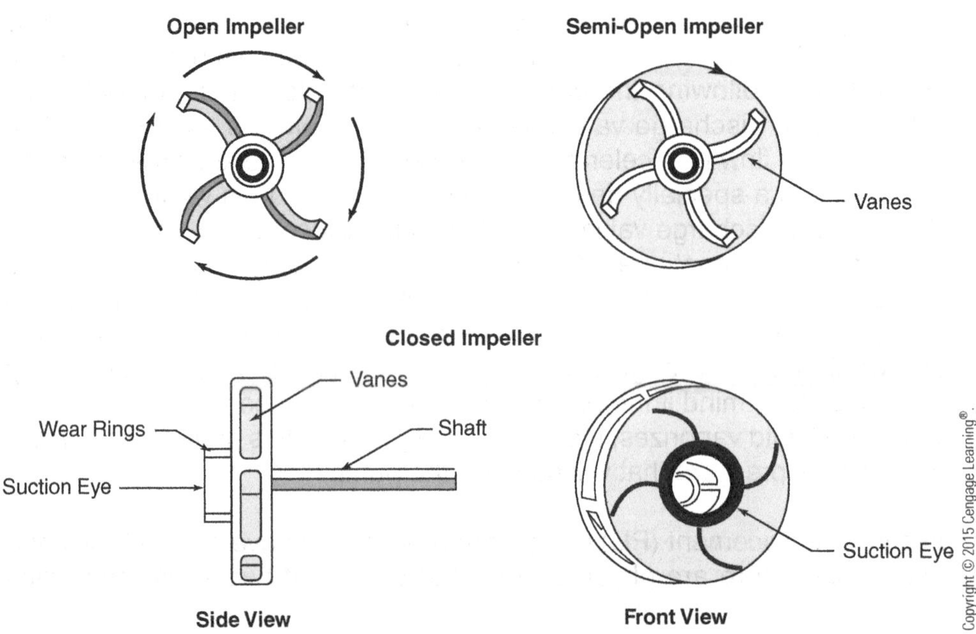

The **axial pump** utilizes a similar spinning motion to propel liquid, but the liquid moves in a straight line. This motion is directionally different from the centrifugal pump's outward movement.

Centrifugal Pumps

Centrifugal pumps are used widely in chemical processing plants and refineries. The primary principle used by centrifugal pumps is centrifugal force. As liquid enters the suction eye of a centrifugal pump, it encounters the spinning impeller (Figure 4.6). The liquid is propelled in a circular rotation that forces it outward and into the volute. Centrifugal force and volute design convert velocity energy to pressure. As the liquid leaves the volute, it slows down, building pressure. Diffuser plates also can be added to the impeller and volute area to slow down or change the velocity of the liquid.

Basic Components

The basic components of a centrifugal pump are shown in Figure 4.6. The outer casing is designed to enhance fluid flow. It contains the fluid, forms a volute, and holds external and internal parts together. Typically, it comes in two parts separated by a gasket and bolted together. An axially split casing is split parallel to the pump shaft. A radially split casing is split perpendicular to the pump shaft. A drain on the bottom of the casing allows liquid to be drained from the pump. The volute is a gradually widening cavity inside the casing of a pump. The inlet is sometimes referred to as the *suction eye*. Fluid comes into the center of the impeller and is spun toward the outside of the volute. The inlet line is designed to run when primed (full of liquid). The outlet line receives the discharge from the pump volute. Because the discharge line is wider than the inlet volute, fluid velocity slows,

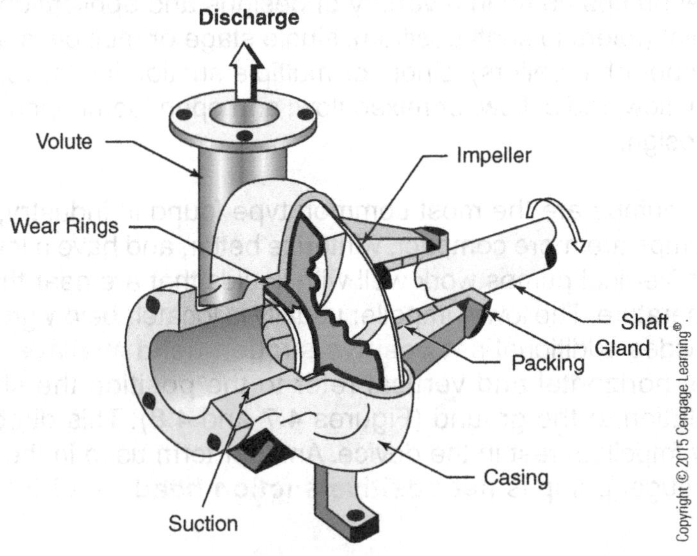

Figure 4.6 *Centrifugal Pump Components*

creating pressure. The impeller is a circular device attached to the shaft. The impeller resembles a wheel of curved blades that rotates around the shaft, spinning liquid from the eye (center of the pump) to the outer casing of the pump. This liquid enters the tapered neck of the volute before exiting the pump into the discharge line. The driver is an electric motor (fixed speed or variable speed) or a steam turbine. A manufacturer's nameplate lists the important information specific to the motor: phase, horsepower, manufacturer, and type. Radial bearings along the shaft minimize side-to-side movement. Thrust bearings eliminate axial movement along the shaft. Between the bearings and seals, a loose-fitting flinger ring mounted on the shaft conveys lubricant onto the bearings. The ring rotates on the shaft but at a lower speed. **Mechanical seals** installed where the rotating shaft enters the casing minimize or stop leakage from the internal components of the pump. Mechanical seals are composed of a stationary face and rotating face that fit closely. Seal faces may require lubrication or cooling. They require very low maintenance, have a reasonably long life, and have very little product leakage.

The stuffing box and packing gland hold packing firmly against the shaft and casing. A slight bit of internal leakage keeps the packing lubricated and cool. The packing material, located between the casing and the shaft, consists of rings composed of flexible material coated in graphite or Teflon®. The packing needs to be replaced frequently. Wear rings minimize leakage between the internal discharge and suction of the pump. Because the rotating impeller does not come into contact with the stationary casing, a void exists between the suction and discharge areas. Properly positioned wear rings minimize this leakage. If the wear rings are mounted on the pump's impeller, they turn with the impeller and are referred to as *impeller wear rings*. If the wear rings are mounted to the casing, they are stationary and are referred to as *casing wear rings*.

Pump Design

Centrifugal pumps come in a variety of designs and applications: vertical or horizontal (refers to shaft position); single stage or multiple stage (refers to the number of impellers); single or multiple suction inlets; volute or diffuser; axial flow, radial flow, or mixed flow; and open, semi-open, or closed impeller design.

Horizontal pumps are the most common type found in industry; however, vertical pumps are more compact, winterize better, and have a lower installation cost. Vertical pumps work well with liquids that are near their bubble point temperature. The lower impeller usually is located below ground level, which provides additional **net positive suction head available (NPSHa)**. The terms horizontal and vertical refer to the position the shaft occupies in relation to the ground (Figures 4.7 and 4.8). This direction indicates how impellers rest in the device. Another term used in the operation of a centrifugal pump is **net positive suction head required (NPSHr)**.

Figure 4.7
Horizontal Centrifugal Pump

Net positive suction head required is the minimum NPSH necessary to avoid cavitation. The NPSHa must be greater than or equal to the NPSHr, expressed as NPSHa $ NPSHr. It is the reduction in total head as the liquid enters the pump.

Most centrifugal pumps have a single suction inlet. In applications where the pumping volume is very high or source parameters vary significantly from destination parameters, pumps that have more than one suction inlet are used. In applications where the difference is minimal, single suction inlet pumps are used.

Figure 4.8
Vertical Centrifugal Pump

Impeller Design

The simplest type of centrifugal pump, a single-stage pump, has only one impeller. Multistage pumps have more than one. The impeller design comes in three basic types (see Figure 4.5). On the open impeller, vanes are connected only to the shaft. The open impeller is self-cleaning but does not have the structural support of a semi-open or closed impeller and is less efficient at producing pressure. On the semi-open impeller, the vanes are horizontally attached to a plate for structural support. On the closed impeller, the vanes are sandwiched between two plates, or shrouds. This is the strongest and most efficient design, but it is designed for use with clear liquids only. It is the most common type of impeller in industry.

Advantages and Disadvantages

The chemical process industry typically uses centrifugal pumps that operate between 1,200 and 8,000 RPM. In modern manufacturing, centrifugal pumps are used more than positive-displacement pumps because they are cheaper and require less maintenance and space. Another attractive feature of centrifugal pumps is that they will operate with a constant head pressure over a wide capacity range. In addition, it is easier to change the element (impeller versus piston) on a centrifugal pump than on a positive-displacement pump, and it is easier to change the driver. A final advantage is the adaptability of the selected driver—variable horsepower and fixed or variable speed.

Centrifugal pumps do have some disadvantages. They are not self-**priming** and respond poorly to viscous materials or variations in suction pressures. Figure 4.9 illustrates how a centrifugal pump transfers liquid from one vessel to another.

Head (Pressure)

In a centrifugal pump, fluids must be pushed (not sucked or pulled) into the impellers. Suction **head** is a term used to describe the pressure required to force liquid into a pump. Most processes are designed so that the suction pressure is sufficient to run the pump without **cavitation** (the formation and collapse of gas pockets around the impellers). A centrifugal pump must be primed (full of liquid, or liquid full) before the pump can be started. During operation, a centrifugal pump will artificially create a low-pressure area in

Figure 4.9
Fluid Transfer

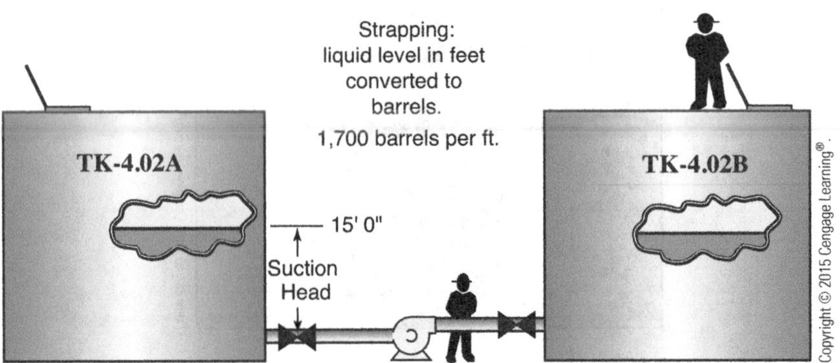

the suction eye. If the suction pressure is not carefully controlled, the low pressure could cause the liquid to boil. Boiling creates the condition called *cavitation*. Net positive suction head (NPSH) usually is calculated in feet of liquid. For example, if a pump took suction off the bottom of a 20-foot open tank with a liquid level of 10 feet, the static head would be 10 feet. This same principle can be applied to the **discharge head** on a pump. If the tank is closed, the vapor pressure of the liquid must be taken into consideration. Vapor pressure is closely related to the boiling point of a liquid. Heat affects vapor pressure by increasing molecular activity. NPSH is the minimum rating at which a centrifugal pump operates. Another problem associated with centrifugal pump operation is vapor lock. **Vapor lock** is a condition in which a pump loses liquid prime and the impellers rotate in vapor. Bled valves are located on the pump to remove air from the system. The pump must be shut down and air bled out before it will operate properly.

Factors that affect suction head pressure are temperature, viscosity (a fluid's resistance to flow), level of liquid in the suction system, restriction in the suction line, and flow rate through the line. Solutions to insufficient NPSH are smaller horsepower, lower speed, lower NPSH requirements, larger-diameter suction line, greater feed tank level or pressure, and cooler feed.

Determining Differential Head

A centrifugal pump requires a certain amount of back-pressure (differential head) in order to operate efficiently. The following exercise should illustrate the relationship between NPSH and the required differential head. It will also illustrate how piping changes and equipment modifications will affect the operation of a pump. A centrifugal pump should be viewed as having a set of conditions on the suction side and a wide array of variables on the discharge side. In the following depropanizer example, liquid propane is pumped from the reflux drum to the depropanizer column (Figure 4.10). On the suction side of the pump, a liquid level of 16 feet plus an operating pressure of 180 psia exists on the suction eye of the pump. The discharge of the pump is lined up through a series of gate valves, a check valve, a filter, an orifice, and a control valve. The discharge line contains a liquid leg of propane, 68 feet high. The depropanizer column operates at a pressure of 200 psia at a temperature of 90°F (32°C). The flow rates are 300 gallons per minute (GPM) normal and 340 GPM maximum. The required differential pressure of 318.7 psi takes into consideration all of the variables associated with the operation of the pump. Under these conditions, the centrifugal pump will operate efficiently. Pressure changes on the suction or discharge side of the pump will affect unit operation.

Pressure is defined as force or weight per unit area: Force (in pounds) ÷ Area (in square inches) = Pressure (in pounds per square inch). Atmospheric pressure is produced by the weight of the atmosphere as it presses down on an object resting on the surface of the earth. Pressure is directly proportional to height: the higher the atmosphere, gas, or liquid, the greater the pressure. At sea level, atmospheric pressure equals 14.7 psi. Pressure

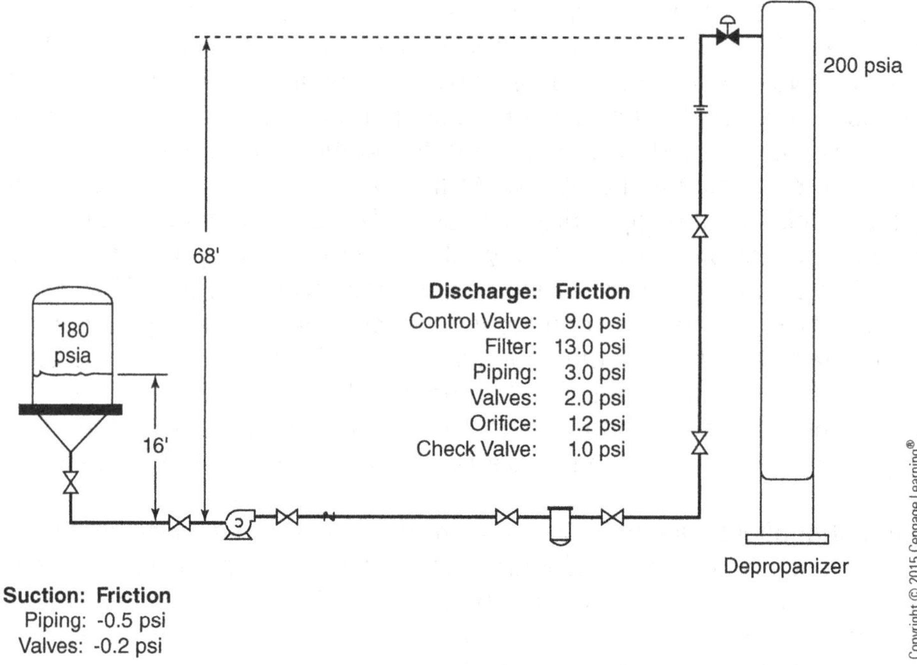

Figure 4.10 *Centrifugal Pump System*

calculations are easily performed using common standards. The primary standard for liquid calculations is water. For example, the weight of water can be used to determine the **specific gravity** of a liquid. Specific gravity is the ratio of the density of a solid or a liquid to the density of water or the ratio of the density of a gas to the density of air. Density is weight per unit volume. To calculate the pressure produced by 1 ft.^{3} of water, we need to use the equation Pressure = Force ÷ Area. One cubic foot of water weighs 62.4 pounds. The surface area of 1 ft.^{3} is 12 in. × 12 in. = 144 in. For water, 62.4 ÷ 144 = 0.433 psi. For each additional foot of water, an additional 0.433 psi can be added. Expressed another way, psi × 2.31 = Head (ft.), or Head (ft.) ÷ 2.31 = psi. A common equation for determining pressure is:

$$\text{Height} \times 0.433 \times \text{Specific gravity} = \text{Pressure}$$

To calculate the required differential head (h), we need to know the specific gravity of propane: specific gravity of propane = 0.485. The unknown factor is ΔP (differential pressure). To solve for ΔP, we must pay close attention to the variables that contribute to pressure changes in a process system.

$$\text{Head } (h) = \frac{(\Delta P)(2.31)}{\text{specific gravity (s.g.)}}$$

The absolute suction pressure factors that are involved are:

Reflux drum	180.00 psia
Elevation (16 ft.)	3.36 psi (16 ft. × 0.433 × 0.485)
Friction: piping	−0.50 psi
valves	−0.20 psi
	182.66 psia − 14.7 = 167.96 psig

The absolute discharge pressure factors are:

Depropanizer tower	200.00 psia
Elevation (68 ft.)	14.29 psi (68 ft. × 0.433 × 0.485)
Friction: piping	3.00 psi
Valves	2.00 psi
Control Valve	9.00 psi
Check Valve	1.00 psi
Orifice	1.20 psi
Filter	13.00 psi
	243.49 psia − 14.7 = 228.79 psig

Now we can solve for (ΔP = 228.79 psig − 167.96 psig = 60.83 psig) and plug that value into our equation for head.

$$h = \frac{(\Delta P)(2.31)}{\text{s.g.}}$$

$$h = \frac{(60.83)(2.31)}{0.485}$$

Multiply in a standard safety factor of 1.1:

289.73 × 1.1 = 318.7 ft.

The required differential head is 318.7 feet.

Total Head

Look again at Figure 4.2. The fluid in TK-4.02A appears to be at the same level as the fluid in TK-4.02B. The pump inlet and discharge are full, and the pump has been bled down and is liquid full. With both the suction and discharge valves open (indicated by the bow tie-shaped symbols), fluid levels equalize in the tanks. A pump is needed to transfer additional liquid into TK-4.02B. As the level increases in TK-4.02B, the discharge head (pressure in the discharge line) increases proportionally to the liquid level in the tank. Total head is equal to the discharge head minus the suction head. Operators should be aware that a centrifugal pump stops moving liquid if the discharge pressure gets too high. As discharge pressure increases, the liquid velocity slows down and even stops, and centrifugal force decreases, perhaps reaching zero.

Pump Curves

Centrifugal pumps are designed to work in specific services at specific rates. The best operating condition for a pump usually is indicated on the pump's efficiency curve. The efficiency curve includes several values: flow rate (in GPM), total head in feet (discharge head minus suction head; sometimes called differential head), pump efficiency, required pump horsepower, and pump NPSH.

Figure 4.11
Pump Curve

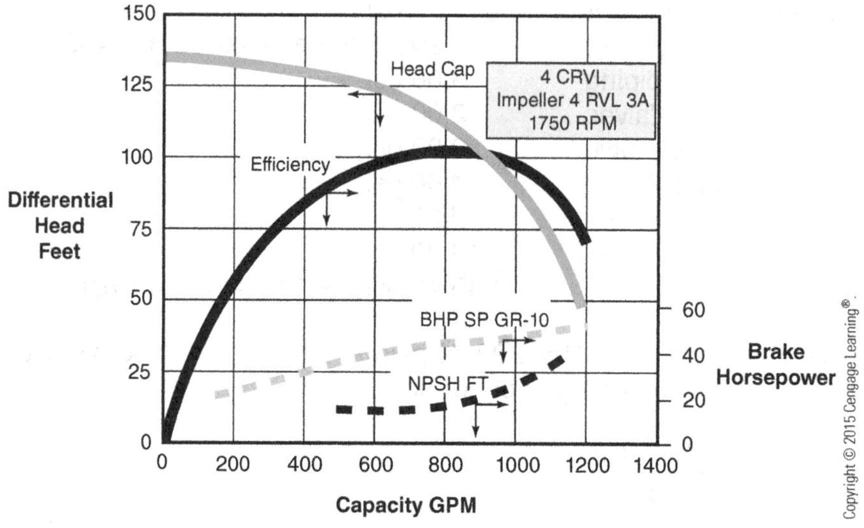

Because pumps are not always operating under their optimal conditions, they are designed to work across a range of rates and liquid properties. The operating parameters (driver horsepower, impeller size, liquid properties, pump efficiency, pump capacity and head, NPSH requirements, and so on) are all shown on a graph known as the pump curve (Figure 4.11). When there is a problem with a pump's performance or a change in the service, the pump curve is one of the first documents consulted. Manufacturers typically include pump curves with their products so that engineers and operators can refer to them before changing or redesigning a pump system. If the pump is operated at higher rates, efficiency decreases and the pump could be damaged.

Cavitation

Cavitation occurs when suction pressure drops below NPSH. At this point, the liquid vaporizes and forms gas pockets inside the pump casing. As these pockets form and collapse, the pump can be severely damaged. Operators easily can identify a pump that is cavitating. The sound of a cavitating pump closely resembles the noise you would hear if steel ball bearings were dumped into a pump's suction line. The operator also can expect to see rapid swings on the discharge pressure. Cavitation sends slugs of liquid and vapor through the pump. Each slug has an impact on the internal components of the pump. Serious damage will occur if this problem is not resolved quickly.

Affinity Laws

Relationships, called *affinity laws*, exist among a centrifugal pump's speed, capacity, head, and power. The capacity of a centrifugal pump is linked directly to its speed, or RPM. Total head is proportional to the square of the speed. Consumed power is proportional to the cube of the speed.

Pump Selection Chart

Modern industrial manufacturers use a pump selection chart when they are engineering a new process. The chart is set up with a horizontal axis that identifies capacity (in GPM) and two vertical axes that identify head (in feet of liquid) on the left and pressure (in pounds per square inch) on the right.

Construction Materials

Centrifugal pumps typically are manufactured with cast iron internal components and cast steel external cases. They can be made of most alloys as needed to cope with the nature of the liquid being pumped.

Factors That Affect Suction and Discharge

Operators need to be aware of the number of process variables that exist inside a system, including upstream and downstream vessels, valves, filters, piping, discharge head, and suction head. Each of these factors affects how efficiently a pump will operate. Figure 4.12 illustrates a tank-to-ship transfer.

Single- and Multistage Pumps

The terms single-stage and multistage are used to refer to the number of impellers a centrifugal pump uses. As flow enters a multistage pump, it enters the suction eye of the first-stage impeller. As the spinning impeller accelerates the fluid, it graduates into the suction of the next stage, where it is accelerated even more. At each stage, the pressure and fluid flow increase until the fluid reaches the discharge chamber.

When a designer builds a multistage pump, the concept of thrust must be considered. The suction side of an impeller exerts a small force on the shaft. The discharge of the impeller exerts a significant amount of thrust on the rotor. Collectively, the sum of the forces along the shaft is tremendous

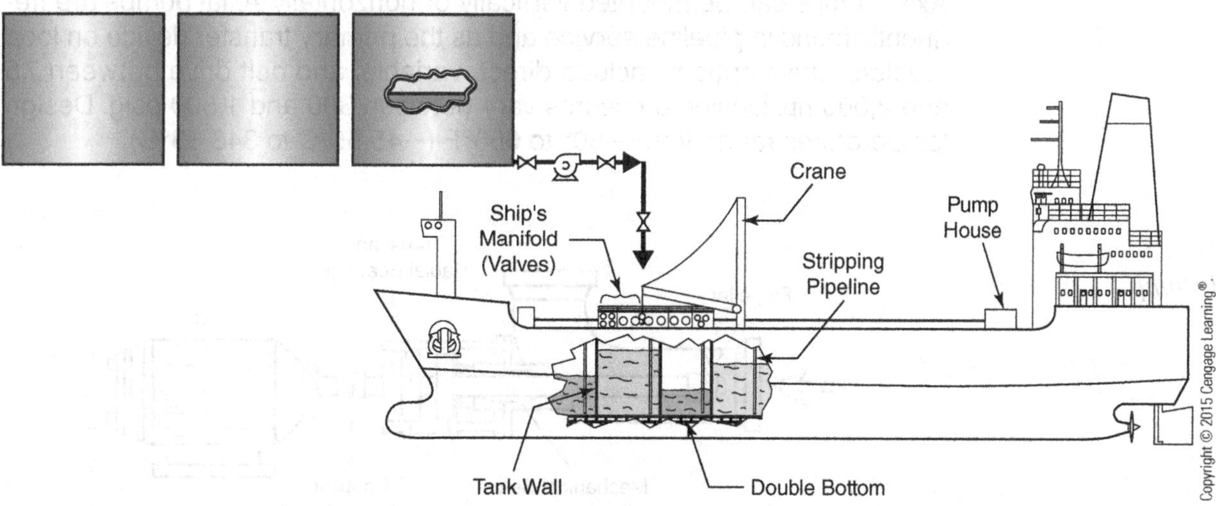

Figure 4.12 *Tank-to-Ship Transfer*

in a multistage pump. There are two ways to minimize thrust. One way is to use a balance device, a hydraulic system that connects the pump suction to the discharge end of the pump. The system is composed of a discharge side balancing drum, a balance line, and a low-pressure reservoir behind the drum. Another way is to align the impellers in opposite directions. Opposite alignment maintains even thrust along the shaft.

Multistage pumps must never be operated with the discharge valve closed. Usually, a minimum-flow system is designed into the piping so that there is always sufficient flow through the stages to protect them from damage.

Axial Pumps

Another way to accelerate and transfer fluid is to push it axially, or in a straight line. Axial pumps (Figure 4.13) are designed to provide this special feature. A common example of this principle is a boat motor. The motor turns a set of blades, forcing water to accelerate along a straight line. An axial pump operates using this same principle. Most axial pumps are located in an elbow on a piping run. The driveshaft extends through the elbow and into the process flow. A propeller is located on the process end of the driveshaft. The propeller is sized to fit the inside diameter of the pipe. The blading is engineered to pull fluid axially down the shaft. A mechanical seal prevents leakage where the shaft penetrates the pipe elbow. A specially designed thrust bearing prevents axial movement of the shaft. Heavy-duty radial bearings support the pump shaft and prevent radial movement. Some axial pumps have thrust-bearing oil coolers. An optional safety seal oil system or thrust-bearing lube system is available for some models. The motor is mounted just outside the elbow on a pad that allows for exact driveshaft lineup. A coupling securely connects the motor to the pump.

Axial pumps can be mounted vertically or horizontally. Axial pumps are frequently found in pipeline service and as the primary transfer device on loop reactors. Drive options include direct, variable, and belt drive between 7.5 and 2,000 hp. Design pressures vary between 300 and 1,500 psig. Design temperatures range from −50° to 650°F (−45.55°C to 343.33°C).

Figure 4.13
Axial Pump

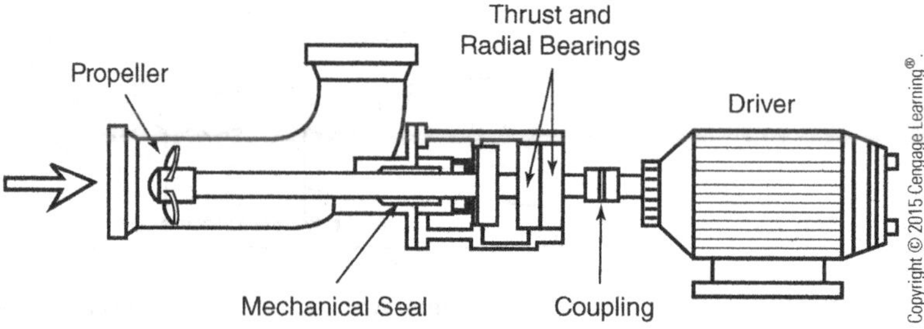

Jet Pumps

Jet pumps (Figure 4.14) take their name from the way they transfer liquids. A specially designed jet is engineered to utilize the venturi effect. Jet pumps are frequently used to lift water from wells over 200 feet deep. When a water well is set, a hole is drilled down to the water table. A polyvinylchloride (PVC) screen is connected to a PVC water pipe that is pushed down the open hole. The pipe is allowed to extend a few feet above ground level. Water passes through the screen and into the pipe. Pressure in the water pocket causes the water level to pass through the screen and up the pipe. The water level will typically stop before it reaches the top of the pipe. The process of lifting the water out of the pipe is the function of the jet pump. The different parts of the jet assembly and drop pipe include a foot valve, nozzle, venturi, suction, and discharge. During installation, a foot valve (check valve) is securely attached to the bottom of the jet. A leather seal is located between the check and the jet. The jet is secured to the drop pipe and lowered into the water pipe casing. A specially designed adapter is used to connect the center drop pipe to the outer casing. The adapter provides a base upon which the vertical centrifugal pump can be mounted. During operation, water is forced back down the void between the center

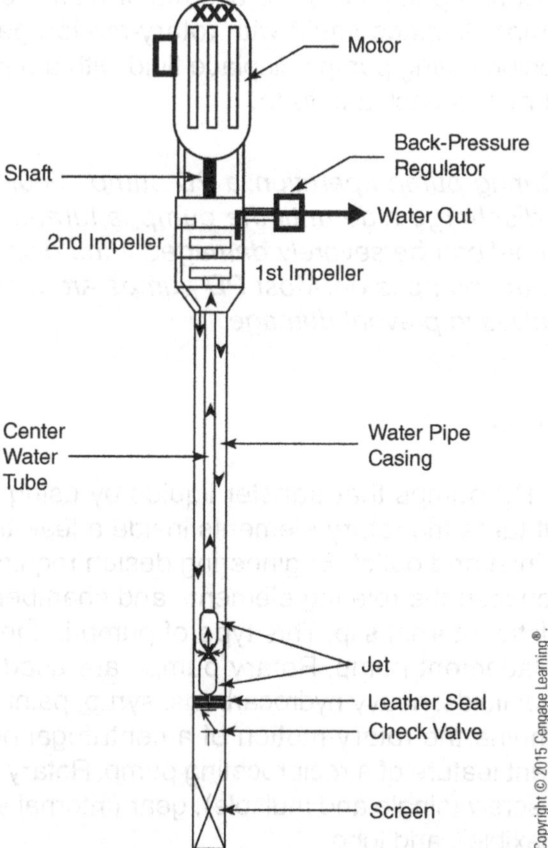

Figure 4.14
Jet Pump

drop pipe and the outer casing. As water is pumped back into the well casing, the foot valve slams shut. A small opening in the jet provides access back to the suction of the pump. A venturi effect occurs at the jet as pressure builds up in the casing. As pressure increases in the casing, velocity increases across the jet. A low-pressure zone is established inside the drop pipe as water quickly flows up toward the pump. A back-pressure regulator holds pressure inside the pump until it reaches operating conditions. When pressures reach operating conditions, water flow is divided as some water recirculates down the casing and the excess flows to a storage tank.

Another common jet pump arrangement includes a jet assembly and a centrifugal pump. The jet assembly forms a suction chamber that creates a vacuum when a stream of high-velocity water flows through a jet. The jet assembly is composed of two major parts: a nozzle and a venturi tube. The nozzle directs high-velocity water into the venturi tube or diffuser. As the high-velocity water exits the nozzle and enters the diffuser, it slows down, creating pressure. The primary purpose of the diffuser is to convert water velocity to pressure.

Positive Displacement Pumps

Rotary and reciprocating pumps are the two major classifications of PD pumps. Rotary pumps displace liquid with rotary-motion gears, screws, vanes, or lobes. Reciprocating pumps displace fluid with a diaphragm, piston, or plunger that moves back and forth.

> **CAUTION:** *During pump operation, a PD pump should never be blocked on the discharge side until the pump is turned off. Equipment and personnel can be severely damaged if the discharge side is blocked while the pump is on. Most PD pumps are provided with pressure relief valves to prevent damage.*

Rotary Pumps

Rotary pumps are PD pumps that transfer liquids by using a rotary motion. The driveshaft turns the rotary elements inside a leak-tight chamber that has a defined inlet and outlet. Engineering design requires close running clearances between the rotating elements and chamber wall. Rotary pumps have very little internal slip. This type of pump is the most widely used positive displacement pump. Rotary pumps are used to move the more viscous type of fluids: heavy hydrocarbons, syrup, paint, and slurries. Rotary pumps combine the rotary motion of a centrifugal pump and the positive displacement feature of a reciprocating pump. Rotary pumps come in four main types: screw (single and multiple), gear (internal and external), vane (sliding and flexible), and lobe.

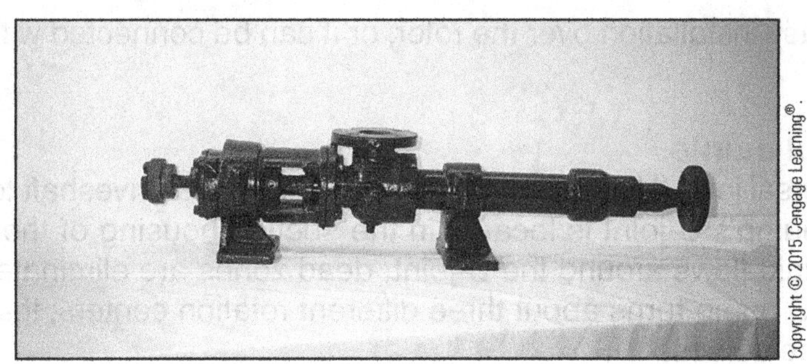

Figure 4.15
Progressive Cavity Pump

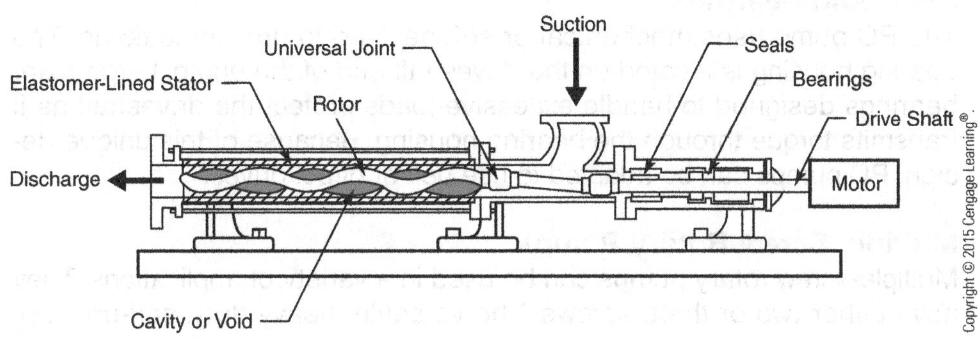

Figure 4.16
Progressive Cavity Pump Components

Single-Screw Rotary Pumps (Progressive Cavity Pumps)

A progressive cavity, or PC, pump (Figures 4.15 and 4.16) consists of only one moving part, the rotor. The rotor turns inside an elastomer-lined stator. When the self-priming rotor turns, cavities, or voids, are formed between the rotor and the stator. These voids progress axially from the suction casing to the discharge outlet. During operation, the cavities fill with fluid. Progressive cavity pumps provide high suction, extremely low shear, and smooth pulsation-free operation. These features are important where turbulence affects fluid composition. The PC pump is ideally suited for metering operations. Typically, this type of pump is used for heavy or viscous fluid service. The solid content of the process fluid does not affect the effectiveness of a PC pump. Progressive cavity pumps can be found in a variety of applications. Operating conditions include pressures up to 1,000 psi, flow rates to 950 GPM, and viscosities of over 1,000,000 cP (centipoises).

Rotor and Stator

The rotational speed of the rotor and the total volume of the cavities determine PC pump flow rate. The rubbing velocity determines rotor and stator life. If the space between the rotor and stator is small, a high degree of rubbing velocity is created. Rubbing velocity is a major factor in the life of the stator. The stator can be replaced after it is worn out. The rotor is typically made of high-chrome tool steel that resists particle abrasion. The stator is manufactured by extruding molten elastomer under high pressure into a heavy metal stator tube with an inner rotor core. When the molten elastomer cools, the core is removed. The stator tube can be threaded on each end for easy installation over the rotor, or it can be connected with four tension rods.

Universal Joints

The universal joint (U-joint) transmits torque from the driveshaft to the rotor. The PC pump's U-joint is located in the suction housing of the pump. As process fluid flows around the U-joint, dead zones are eliminated. The rotor in a PC pump turns about three different rotation centers; that is, it has about three curves.

Seals and Bearings

The PC pump uses mechanical or soft packing to prevent leakage. The bearing housing is located on the driveshaft end of the pump. Large roller bearings designed to handle excessive loads protect the driveshaft as it transmits torque through the bearing housing. Because of this unique design, PC pumps can be adapted to use belt or direct drives.

Multiple-Screw Rotary Pumps

Multiple-screw rotary pumps can be used in a variety of applications. They have either two or three screws. The versatile, heavy-duty, self-priming, two-**screw pump** has been in service since 1934. The special design of the pump elements enables the two-screw pump to provide high flow rates and excellent suction and to pump virtually any fluid. A two-screw pump has two rotors: a power rotor and an idler rotor. A set of external timing gears and bearings allows the screws (the rotors) to turn in unison without making contact with each other. This feature allows the pump to transfer any fluid regardless of abrasiveness, lubricity, or viscosity. In addition, because the screws do not touch, the pump can run empty without damaging the system. This feature makes it an ideal choice in tank-stripping operations or in operations in which fluid suction pressure will vary. The twin-screw pump has a driving and a driven shaft, two screws, an external set of timing gears and heavy-duty bearings, and divided-entry flow (Figure 4.17).

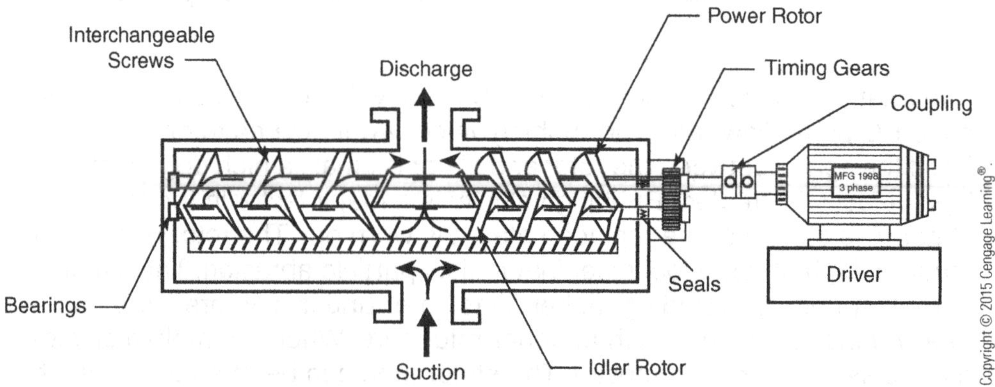

Figure 4.17
Two-Screw Pump

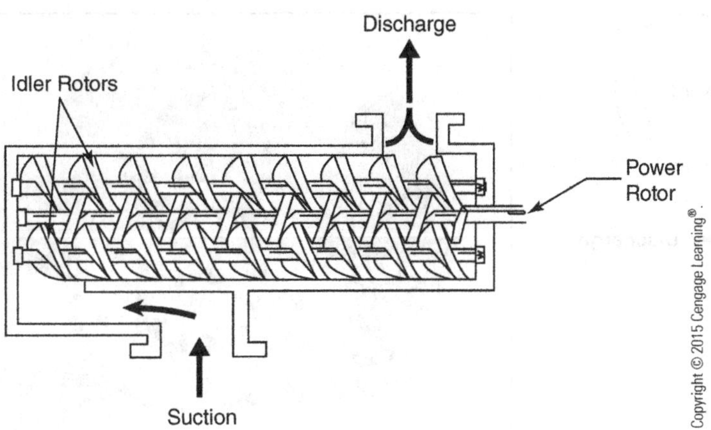

Figure 4.18
Three-Screw Pump

As fluid enters the pump, it is divided into two equal streams and directed to the two ends of the shaft. The pumping action of the screws moves the two streams of process fluid in a straight line between the closely spaced rotors until they combine at the discharge port. In a two- or three-screw design, the direction of rotation of the power rotor will determine whether the inlet and outlet ports are located on the top or bottom of the pump. Because the two streams have equal, simultaneous flow paths, the rotating rotors are balanced. This unique design reduces bearing wear. Most two-screw pumps require mechanical seals instead of soft packing on the suction side of the pump because of the large sealing area. Two-screw pumps typically operate at a speed of 3,500 RPM, flow rate of 8,800 GPM, inlet pressure up to 100 psi, discharge pressure up to 725 psi, and temperature up to 500°F (260°C).

The three-screw pump consists of a power or driver rotor and two idler rotors (Figure 4.18). The power screw meshes with the idler screws during operation. The three-screw pump is unlike the two-screw pump in that the three screws touch. Each screw rotates easily on a set of heavy bearings. During operation, the self-priming screws rotate, creating voids that transfer fluid in a continuous, pulsation-free flow. A typical three-screw pump does not need thrust bearings because of the balanced system. The engineering specifications allow this type of pump to be operated at speeds of up to 6,000 RPM, with pressures up to 750 psig, and flow rates around 600 GPM.

Gear Pumps

Gear pumps are similar to screw pumps in that they can be used in viscous service. Gear pumps typically can be found in two common types: external and internal.

External Gear Pumps

External gear pumps have two interesting gears that rotate parallel to each other, allowing fluid to be picked up by the gears and transferred out of the pump (Figure 4.19). One of the gears is an idler gear; the other is attached

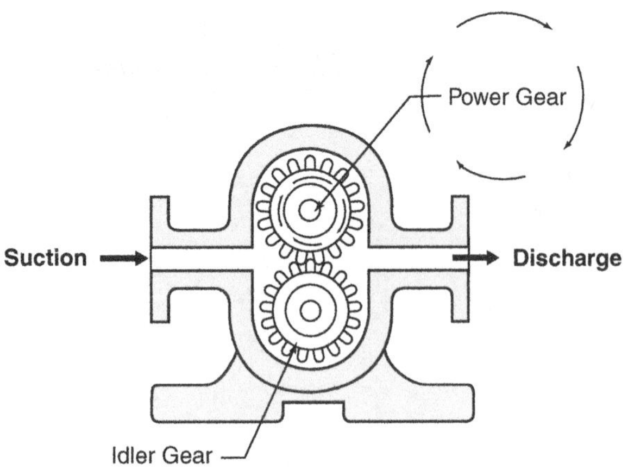

Figure 4.19 *External Gear Pump*

to a driver and is referred to as the power gear. External gear pumps consist of two mating gears that rotate inside a casing. The rotation of the driver gear turns the idler, or follower gear, trapping fluid and displacing it. Typically, in this type of operation, the driver gear is mounted on top. Most operators refer to external gear pumps as constant displacement pumps. The discharge of an external gear pump remains constant unless the shaft speed is changed.

The suction and discharge ports of an external gear pump are located on the opposite ends of the casing. When the pump is first started, air is forced out and into the discharge line. This process creates a low-level vacuum on the suction side. This vacuum causes fluid to enter the pump and be trapped between the gears. As the gears rotate, the fluid is swept around the housing and out of the discharge port.

Internal Gear Pumps

A Danish-American inventor, Jens Nielsen, invented the internal gear (IG) pump in 1915. Process technicians refer to the internal gear pump as the "gear-within-a-gear pump." Internal gear pumps operate with only two moving parts: a power gear driving an internal idler gear (Figures 4.20 and 4.21). When the power gear rotates, liquid enters the pump through the suction line. Since the pump is self-priming, the voids between the teeth of the power gear and the off-center idler gear fill with liquid. During rotation, liquid is separated by a crescent-shaped spacer. Liquid is pressed into the spaces above and below the crescent. As the gears rotate around the circular pump casing, the liquid is discharged out of the pump. The main components of an internal gear pump are a power gear or rotor; an idler gear; an idler pin; a driveshaft; a circular casing; the crescent, axial, and radial bearings; seals; and a relief valve. The idler gear rotates freely on a cylindrical idler pin. A main bearing and a second bearing

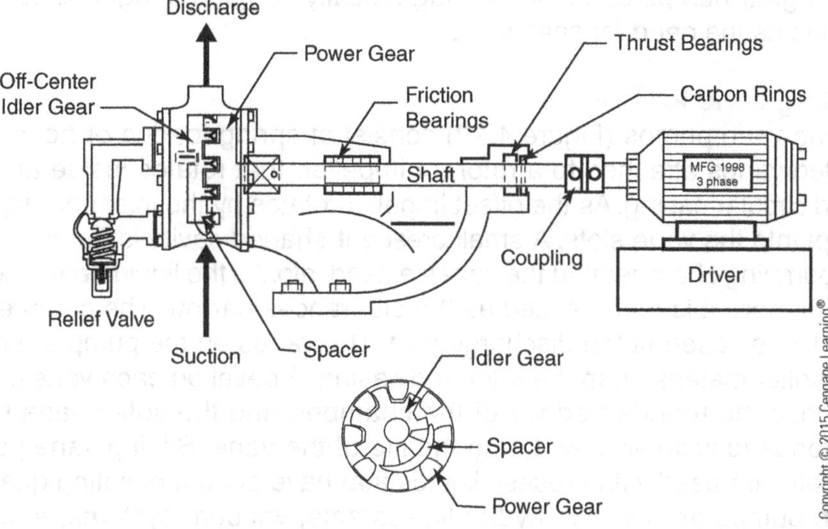

Figure 4.20
Internal Gear Pump

Figure 4.21
Internal Gear Pump

on the free end of the shaft support the driveshaft. Soft packing or mechanical seals can be used on the pump. Internal gear pumps require very little maintenance. Design parameters on internal gear pumps include flow rates from 1 to 750 GPM, speeds up to 1,750 RPM, temperatures up to 500°F (260°C), suction lift during operation up to 24 inches Hg vacuum, and differential pressures up to 250 psi. These pumps can be constructed of stainless steel for corrosive environments or of carbon steel or cast iron where applicable. The chemical processing industry uses internal gear pumps for chemicals such as acetone, acids, alcohol, alkalis, ammonium hydroxide, butadiene, polymers, resins, solvents, waxes, and xylenes. Internal gear pumps can also be magnetically coupled; magnetic coupling eliminates the need for shaft seals.

Sliding Vane Pumps

Sliding vane pumps (Figure 4.22) consist of spring-loaded or nonspring-loaded vanes attached to a rotor, or impeller, that rotates inside an oversized circular casing. As the offset impeller rotates by the inlet port, liquid is swept into the vane slots. A small crescent-shaped cavity is formed inside the pumping chamber that the vanes extend into. As the liquid nears the discharge port, it is compressed as the clearances narrow. The compressed liquid is released at the discharge port. The vanes on the pump are made of a softer material than the rotor and casing. A bevel on each vane closely matches the rounded edges of the chamber, and the softer vane material tends to wear evenly during the life of the vane. Sliding vane pumps typically are used with process liquids that have good lubricating qualities. Vane pumps are used in hydraulic systems, vacuum systems, and low-pressure oil systems. The typical vane pump has a capacity of 380 GPM and operates with a pressure differential of 50 psig.

Figure 4.22
Sliding Vane Pump

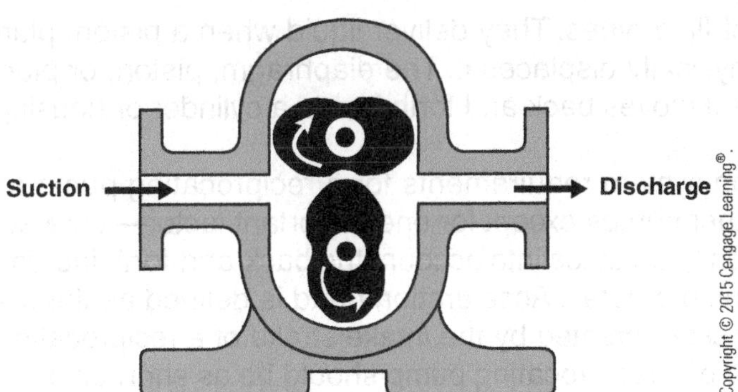

Figure 4.23
Lobe Pump

Flexible Vane Pumps

In a flexible pump system, the rotor is composed of a soft elastomer impeller, keyed to fit over the driveshaft that penetrates the pumping chamber. The pumping chamber is designed to provide good contact between the impeller and the inner chamber. Speeds are typically low since the rubbing velocity between the flexible vanes and chamber wall is significant. The impeller is centered in the pumping chamber. Flexible vane pumps are frequently used in vacuum service.

Lobe Pumps

Lobe pumps (Figure 4.23) have two rotating lobe-shaped screws that mesh during operation. As the lobes turn, voids are created that compress liquids around the outside of the pumping chamber. In a lobe pump, a set of external timing gears and bearings allows the lobes to turn in unison without making contact with each other. This feature allows the pump to transfer a wide variety of fluids. Because the lobes do not touch, the pump can run empty without damaging the system. A lobe pump has a driving and a driven shaft, two lobes, an external set of timing gears, and bearings. As fluid enters the pump, it is divided into two equal streams. The pumping action of the lobes moves the process fluid in two streams around the lobes in the close tolerances between the casing and the lobes. The streams combine at the discharge port. The direction of rotation of the driver will determine the locations of the inlet and outlet ports on the pump. Lobe pumps are designed to provide high flow rates at low pressures; they have excellent suction and pump a variety of fluids.

Reciprocating Pumps

Reciprocating pumps, especially in small volume sizes, are positive displacement pumps very commonly used in the petrochemical industry. Reciprocating pumps are engineered to transfer small volumes of liquid at relatively high pressures. Most reciprocating pumps are self-priming and are operated at relatively low speeds because of the back-and-forth motion and the effects of inertia on internal components. Reciprocating pumps can deliver consistently high volumetric efficiencies even when applied to a variety of fluid types. They deliver liquid when a piston, plunger, or diaphragm physically displaces it. The diaphragm, piston, or plunger pushes the fluid as it moves back and forth inside a cylinder or housing.

The suction system requirements for a reciprocating pump are similar to those of other pumps except for one important factor—**acceleration head**. Acceleration head takes into account the back-and-forth inertia that a reciprocating pump creates. Acceleration head is defined as the fluctuations of suction pressure created by the intake stroke of a reciprocating pump. The suction line of a reciprocating pump should be as short as possible. Excessive valves, bends, and fittings should be eliminated. The line should be large enough to deliver fluid velocities of 3 feet per second. Booster pumps typically can be added to the suction line to increase suction pressure.

Diaphragm Pumps

A **diaphragm pump** (Figure 4.24) uses a flexible sheet (diaphragm) to displace fluid. This type of pump has a crankshaft or eccentric wheel attached to a connecting rod. The connecting rod is anchored firmly to the center of the diaphragm. The outer edge of the diaphragm is bolted or secured to the exterior casing. As the eccentric wheel (referred to simply as the *eccentric*) starts its rotation, the diaphragm connecting rod goes up and down. This reciprocating motion creates a pumping action that displaces fluid. The pumping chamber below the diaphragm is connected to suction and discharge lines. Spring-loaded valves open or close, depending on the pressure in the chamber.

Diaphragm pumps have several advantages compared with most other types of pumps. They completely seal off the area between the diaphragm

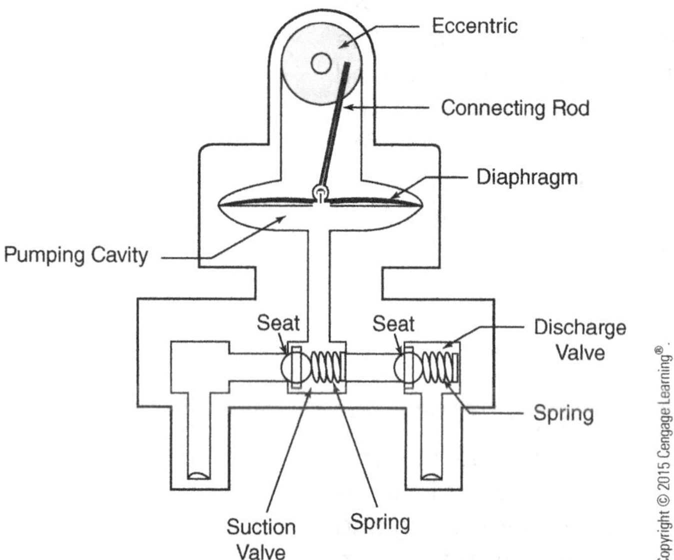

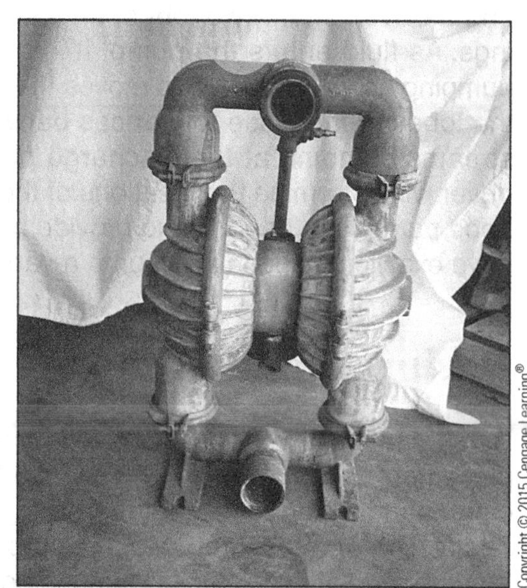

Figure 4.24 *Diaphragm Pump*

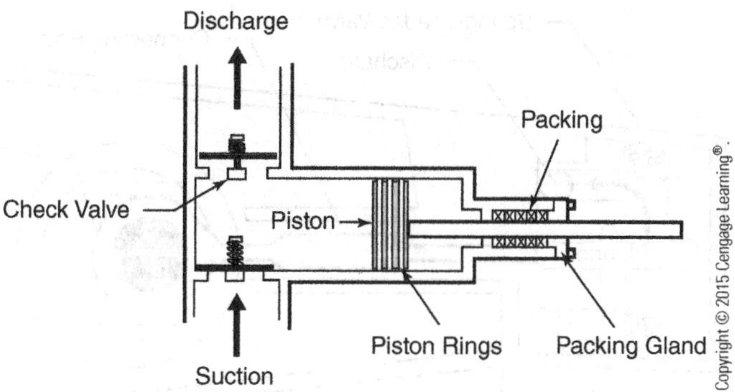

Figure 4.25
Simple Piston Design

and the pumping cavity; they can be used to pump a variety of chemicals; and they can be used with low or negative suction head.

Piston and Plunger Pumps

Reciprocating **piston pumps** (Figure 4.25) use a piston and a back-and-forth motion to displace fluid. During normal operation, the piston pump has a suction stroke and a discharge stroke. The suction stroke occurs when the piston pulls out of the cylinder. This motion creates a low- pressure vacuum in the cylinder, causing the discharge valve to close and the suction line to open, filling the cylinder. On the return stroke or discharge stroke, the suction valve slams shut, and the fluid is forced out the discharge valve. This type of pump continues to operate no matter how high the discharge head is.

The typical piston pump has a piston, piston ring, connecting rod, suction and discharge valves, casing and cylinder, and relief valve. Piston pumps are sealed internally and externally. Internally, the piston rings form a seal on the piston that can be seen only by tearing down the pump. The external packing seal is located where the piston rod enters the casing.

Reciprocating plunger pumps (Figure 4.26) operate with a back-and-forth motion and a device called a plunger to displace controlled amounts of liquid. The primary difference between a plunger pump and a piston pump is in the shape of the piston or plunger element and the way they seal. A piston pump has rings mounted on the piston that form a seal. The plunger on a plunger pump does not have moving rings. The plunger moves in and out of an O-ring or packing medium to form its own stationary seal. A major advantage of this type of sealing system is that the pump seals easily and can be replaced without major breakdown of the equipment. The basic components of a plunger pump are a plunger, crankshaft, connecting rod, pumping chamber, suction inlet valve, and discharge outlet valve.

Double-Acting Design

The back-and-forth motion of a piston or plunger pump limits the smooth flow control characteristic of other pumps that do not have long suction intake and

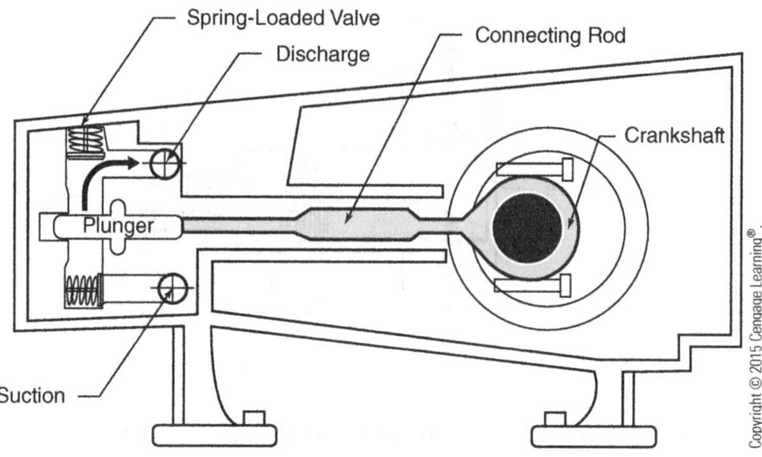

Figure 4.26
Plunger Pump

discharge strokes. Piston and plunger pumps tend to have pulsing or spurting flow rates. Using a double-acting design can compensate for this feature. A double-acting piston (Figure 4.27) or plunger pump discharges and pulls in fluid on each side of the chamber, on a single stroke. Because the double-acting pump discharges on each stroke, a more uniform flow rate can be maintained. Some engineering designs include **pulsation dampeners** to offset the problem of pulsing or spurting flow rates. The dampener should absorb cyclical flow variations. The pulsation dampener is a small pressure vessel with a diaphragm and a cushion of air to offset surges.

Duplex Design

Single-cylinder piston and plunger pumps have severe capacity limitations. Adding another cylinder compensates for this design flaw and doubles the flow capacity.

Vacuum Pumps

Operating systems that require pressures below atmospheric pressure use vacuum pumps. The typical vacuum pump falls under a variety of reciprocating, rotary, and dynamic categories. Vacuum pumps are frequently used in distillation operations, refrigeration, and air conditioning. Vacuum pumps are connected to the system to be evacuated with hoses, piping,

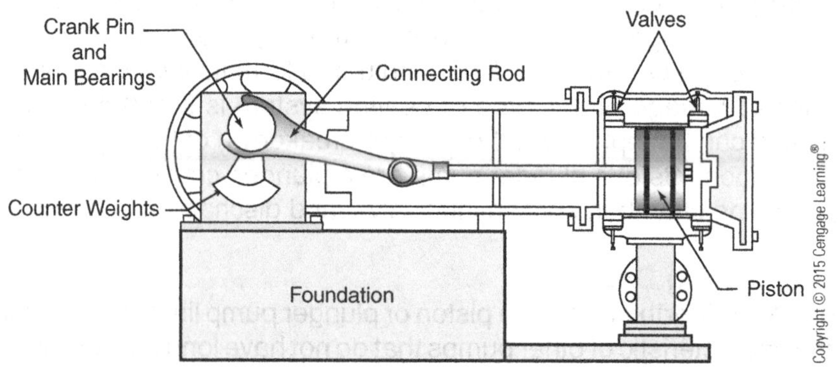

Figure 4.27
Double-Acting Piston Pump

or manifold systems. The connection is typically made in a location where liquid is not present. A knockout system is used to trap liquids before they can enter the pump. Some vacuum pumps have an oil seal inside the casing of the pump. During the evacuation process, the process flow is drawn into the pump and discharged. Contaminants are trapped in the oil, which may require frequent changing. The primary cause for pump failure is contaminated oil. When the pump is first started, it makes a gurgling sound as gas or vapors are drawn into it. A distinct rapping sound can be heard once the pumpdown is complete. The suction side of a vacuum pump should be closed before the pump is turned off because vacuum oil will be drawn into the internal pumping chamber. This problem could cause the pump to fail on the next startup. Vacuum pumps may also be classified as compressors, since the primary medium being transferred is gas or vapors. A variety of compressors—such as sliding vane, piston, liquid ring, and lobe compressors—can be adapted to serve as vacuum-type devices.

Startup, Shutdown, and Troubleshooting

The basic components of both dynamic and positive displacement pump systems are a pump, piping, valves, instruments, and process equipment (Figure 4.28). The heart of the system is the pump. It is used to accelerate, or add energy to, the process fluid. In most operating systems, spare (redundant) pump arrangements are used to permit pump maintenance without shutting down the process. Each plant has a standard operating procedure that should be followed closely during startup and shutdown. Tables 4.1 and 4.2 illustrate typical startup and shutdown procedures. Table 4.3 illustrates troubleshooting charts for identifying and solving pump problems.

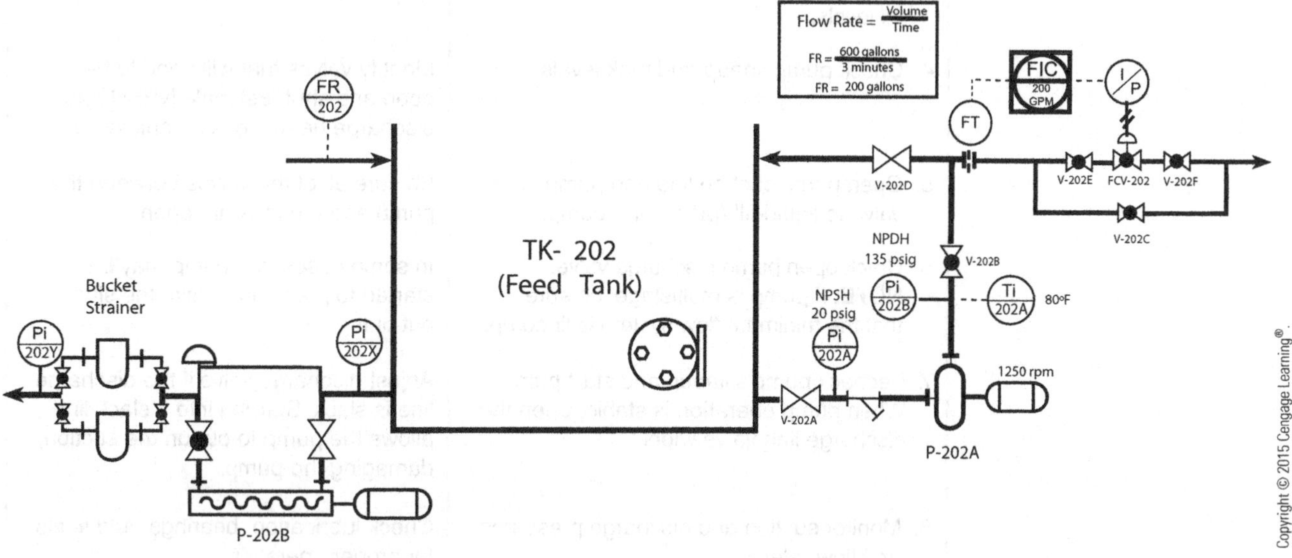

Figure 4.28 *Simple Pump System*

Table 4.1 *Start Up and Shut Down a Positive Displacement Pump*

Procedure	Comments
1. Review the standard operating procedure for the equipment.	Each unit will have its own operating procedure for equipment startup.
2. Check the pump driver oil level, cooling requirements, and other specifications.	Look for possible equipment damage.
3. Check pump lineup and tank levels.	Identify valves that need to be open and shut.
4. Open pump suction valves and fill pump.	Be sure all of the valves between the pump and the tank are open.
5. Open pump discharge valves.	PD pumps should never be started with the discharge line closed.
6. Recheck pump lineup and start pump.	
7. Monitor suction and discharge pressures and flow rates.	Be sure pump has adequate suction tank level.

Table 4.2 *Start Up and Shut Down a Centrifugal Pump*

Procedure	Comments
1. Review the standard operating procedure for this piece of equipment.	Each unit will have its own operating procedure for equipment startup.
2. Check lineups in auxiliary sealing, cooling, and flush systems.	
3. Check the pump and driver and the oil levels.	Look for possible equipment damage.
4. Check pump lineup and tank levels.	Identify valves that will need to be open and shut. Estimate NPSH and discharge head. Look at tank levels.
5. Open pump suction line and pump vent valve to liquid-fill (prime) the pump.	Be sure all of the valves between the pump and the tank are open.
6. Crack open pump discharge valve. **NOTE:** If pump is multistage, be sure that the minimum flow system is lined up.	In some cases, the pump may be started to press up or take the slack out of the line.
7. Recheck pump's lineup and start pump. When pump operation is stable, open the discharge line valve wide.	Adjust discharge valve if the discharge line is slack. Starting into a slack line allows the pump to outrun the suction, damaging the pump.
8. Monitor suction and discharge pressures and flow rates.	Check lubrication, bearings, and seals for proper operation.

Table 4.3 *Troubleshoot Typical Pump Problems*

Problem	Possible Cause	Solution
Pump is cavitating	• High suction temperature • Viscosity • Low suction pressure • Restriction in line • Too much horsepower • Pump speed too high • Too high NPSH • Small suction	• Lower temperature • Agitate suction tank • Increase level or pressure • Shut down and clear • Decrease horsepower • Lower RPM • Lower NPSH requirements • Increase diameter suction line
Pump vapor locked	• Pump not vented before startup • Variable-speed pumps	• Shut down and bleed off • Use a turbine or variable-speed motor drive
Specific gravity of product changes	• Different product composition	• Leave alone; will not affect pump capacity
Excessive vibration	• Starved suction • Bearings worn • Caused by the formation of vapor pockets • Pump misaligned • Rotor out of balance; usually intermittent • Shaft bent • Loose foundation bolts • Driver vibrating • Instrument malfunctions	• Pinch down discharge valve on pump • Shut down and replace • Vent pump to reestablish full capacity • Shut down and have realigned • Remove rotating element, check impeller; if passages clogged, remove foreign material; if impeller is damaged, a new one will need to be installed • Shut down and repair • Secure pump to foundation • Disconnect coupling and check driver • Locate and correct
Fails to deliver liquid	• Pump not primed • Wrong rotation • Suction line not filled with liquid • Air/vapor in suction line • NPSH insufficient • Low level in suction tank • Total head required greater than available	• Prime it • Reverse • Fill suction line • Bleed off • Increase NPSH • Increase level • Increase pump size

Pump Symbols

Each type of pump can be represented by a symbol. Pumps symbols are illustrated in Figure 4.29.

Summary

Pumps can be classified as dynamic or positive displacement. The two most common types of pumps are centrifugal and positive displacement. Centrifugal pumps are dynamic pumps that use centrifugal force and the design of the volute to add energy or velocity to the liquid. Positive-displacement (PD) pumps displace a specific volume of fluid (gas or liquid)

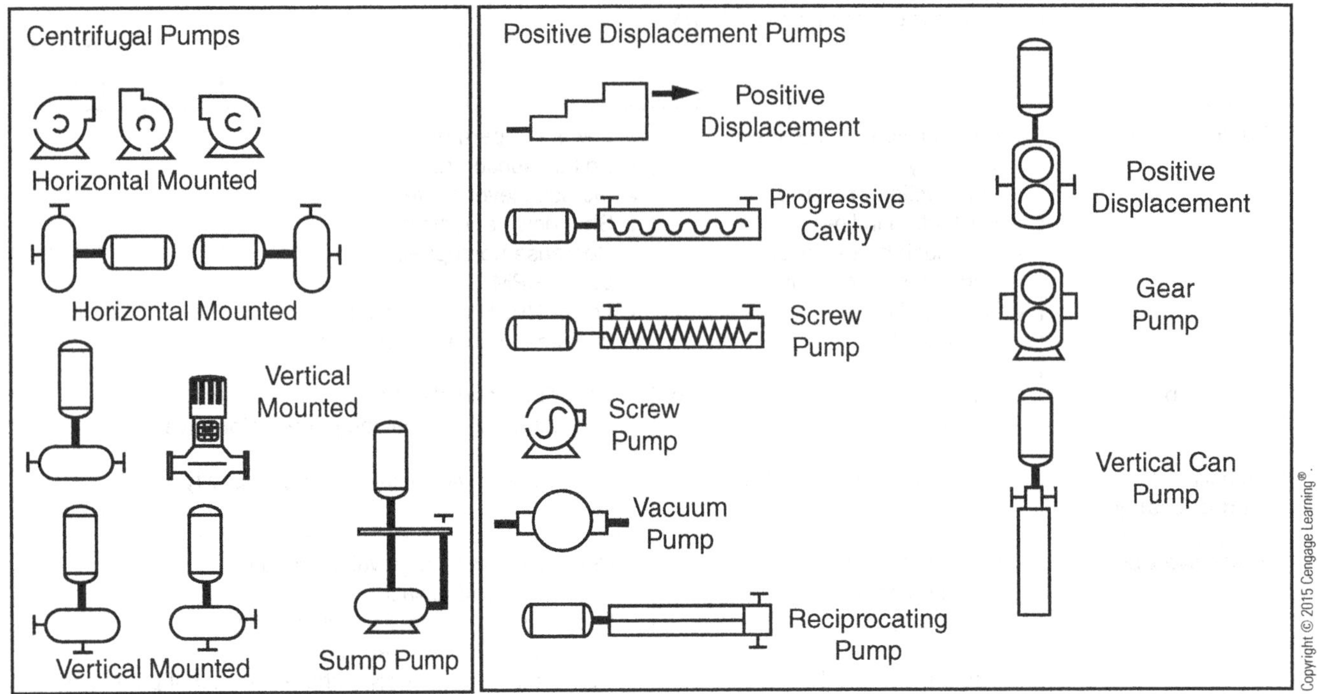

Figure 4.29 *Pump Symbols*

on each stroke or rotation. PD pumps can be classified as rotary or reciprocating. Rotary pumps displace fluids with screws, gears, vanes, or lobes. Reciprocating pumps move fluids by drawing them into a chamber on the intake stroke and positively displacing them with a piston, plunger, or diaphragm on the discharge stroke.

Internal slip is the percentage of fluid that leaks or slips past the internal clearances of a pump over a given time. Centrifugal pumps can operate with 100% slip over short periods of time (discharge valve closed). Positive displacement pumps are not designed to tolerate slip (discharge valve closed).

In a centrifugal pump, liquids must be pushed (not sucked or pulled) into the impellers. Suction head is a term used to describe the pressure required to force liquid into a pump. A centrifugal pump must be primed or liquid full before the pump can be engaged. Net positive suction head (NPSH) usually is calculated in feet of liquid.

Cavitation is defined as the formation and collapse of gas pockets around the impeller of a centrifugal pump. Cavitation occurs when suction pressure drops below required NPSH. At this point, the liquid vaporizes and forms gas pockets inside the pump casing. As these pockets form and collapse, the pump can be severely damaged.

During pump operation, a PD pump should never be blocked in on the discharge side until the pump is turned off. Equipment and personnel can be severely damaged if the discharge **pressure relief valve** fails to function.

Review Questions

1. Define NPSH, and describe how it affects industrial pump operation.
2. Briefly review the history of positive displacement pumps and centrifugal pumps.
3. Describe the primary scientific principles associated with centrifugal pump operation and identify key components.
4. Describe the operation of positive displacement pumps.
5. List the various types of rotary pumps.
6. Describe the basic components and operation of a screw pump.
7. Explain how external and internal gear pumps operate.
8. Describe the basic components and operation of sliding and flexible vane pumps.
9. Describe the basic components of a lobe pump.
10. List the various types of reciprocating pumps.
11. Explain the operation of diaphragm pumps.
12. Describe the operation and design of piston pumps.
13. Describe a plunger pump.
14. List the typical startup and shutdown procedures for a positive-displacement pump.
15. List the typical startup and shutdown procedures for a centrifugal pump.
16. What specific problems are associated with the operation of centrifugal and reciprocating pumps?
17. Draw a simple axial pump.
18. Draw a simple pump system using a tank, pump, and connecting piping.
19. Draw the symbol for a progressive cavity pump and a screw pump.
20. Explain how variations in NPSH affect the operation of a centrifugal pump.

Compressors

Objectives

After studying this chapter, the student will be able to:

- Explain the principles of compression.
- Describe how centrifugal compressors operate.
- Describe how axial flow compressors operate.
- Identify and describe centrifugal and positive displacement compressors.
- Identify the basic components of a rotary screw compressor.
- Describe the operation and basic components of sliding vane compressors.
- Explain how lobe compressors operate.
- Describe how liquid ring compressors operate.
- Explain the scientific principles associated with reciprocating compressors.
- Identify the basic components of a compressor system.
- Start up and shut down a positive displacement compressor.
- Start up and shut down a dynamic compressor.

Key Terms

Aftercooler—a heat-exchange device designed to remove excess heat from the discharge side of a multistage compressor.

Centrifugal compressor—uses centrifugal force to accelerate gas and convert energy to pressure.

Compression ratio—the ratio of discharge pressure (psia) to suction pressure (psia). Multistage compressors use a compression ratio in the 3 to 4 range, with the same approximate compression ratio in each stage. For example, if the desired discharge pressure is 1,500 psia, a 4-stage compressor with a 3.2 compression in each stage might be used. The pressure at the discharge of each stage would be: 1st = 47 psia, 2nd = 150 psia, 3rd = 480, 4th = 1,536 psia.

Demister—a cyclone-type device used to swirl and remove moisture from a gas.

Desiccant dryer—used to remove moisture from compressor gases as they are passed over a chemical desiccant, which adsorbs the water.

Diaphragm compressor—utilizes a hydraulically pulsed diaphragm that moves or flexes to positively displace gases.

Double-acting compressor—a reciprocating compressor that compresses gas on both sides of the piston.

Dryer—removes moisture from gas.

Intercooler—a heat exchange device designed to cool compressed gas between the stages of a multistage compressor.

Lobe compressor—a rotary compressor that contains kidney bean–shaped impellers.

Oil separator—removes oil from compressed gases.

Receiver—a compressed-gas storage tank.

Stage—each cylinder in a compressor; specifically, the area where gas is compressed.

Thermal shock—a form of stress resulting in metal fatigue when large temperature differences exist between a piece of equipment and the fluid in it.

Compressor Applications and Classification

The compression of gases and vapors in the process industry is very important. Compressors are used in a variety of applications. In a modern plastics facility, compressors are used to transfer granular powders and small plastic pellets from place to place. In natural gas plants, compressors are used to establish feed gas process pressures. Compressors also provide clean, dry air for instruments and control devices. In a refinery or

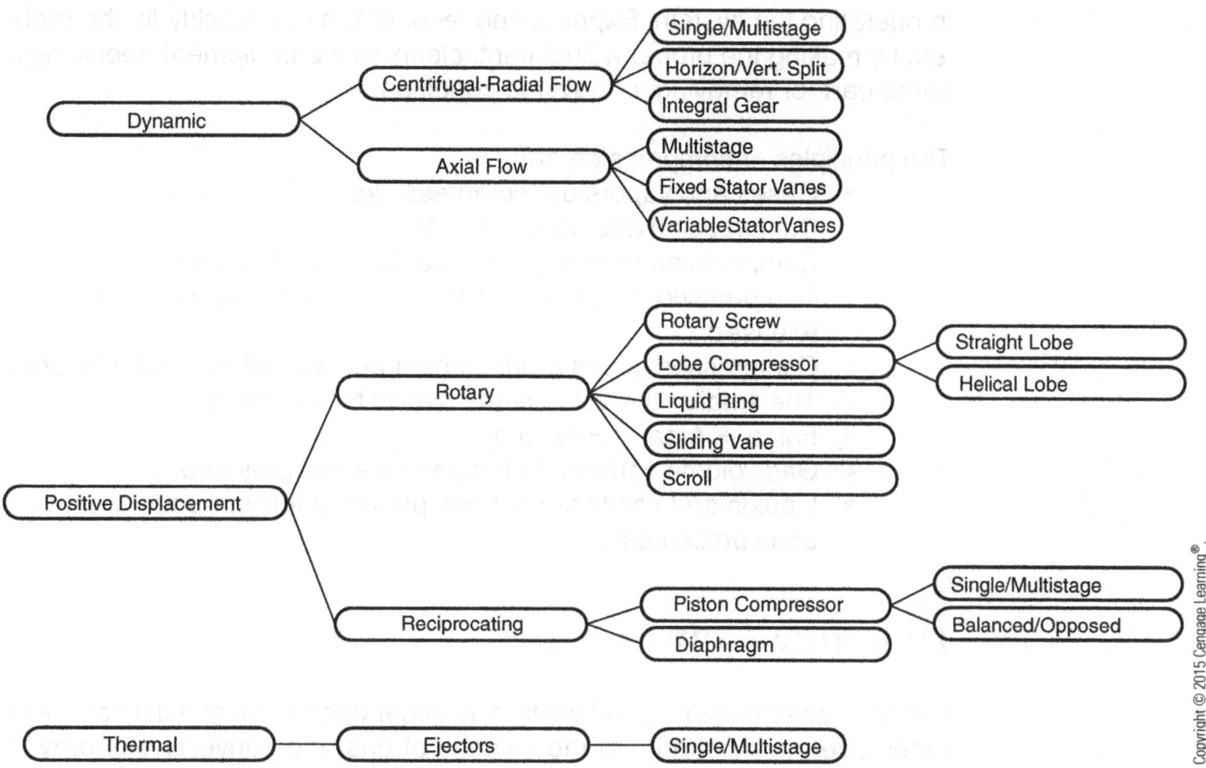

Figure 5.1 *Compressor Family Tree*

chemical plant, compressors are used to compress gases such as light hydrocarbons, nitrogen, hydrogen, carbon dioxide, and chlorine. These gases are sent to headers, from which they are distributed to a variety of applications.

There are three basic designs for compressors (Figure 5.1): dynamic, positive displacement, and thermal. Dynamic compressors include centrifugal (radial flow) and axial (straight-line) flow compressors. Positive displacement compressors include rotary and reciprocating compressors. Dynamic compressors accelerate airflow by drawing air in axially and spinning it outward (**centrifugal compressors**) or in a straight line (axial flow compressors). Positive displacement compressors compress gas into a smaller volume and discharge it at higher pressures. Thermal compressors use ejectors to direct high-velocity gas or steam into the process stream, entraining the gas, and then converting the velocity into pressure in a diffuser assembly. This chapter focuses primarily on dynamic and positive displacement compressors.

A compressor is part of a much larger system. The system's resistance to flow typically dictates compressor performance. Minor problems are occasionally experienced with compressor systems. These troubles are usually the result of dirt, adjustment problems, liquid in the system, or inexperience in operating the system. Experienced technicians can quickly fix the problem by making the proper adjustment, cleaning the equipment, replacing a minor part, or removing an adverse condition.

The principles of compression are:

- Gases and vapors are compressible.
- Compression decreases volume.
- Compression moves gas molecules close together.
- Compressed gases will resume their original shape when released.
- Compressed gases produce heat because of molecular friction.
- The smaller the volume, the higher the pressure.
- Force ÷ Area = Pressure.
- Gas volume varies with temperature and pressure.
- Liquids and solids are not compressible (except under tremendous pressures).

Dynamic Compressors

Dynamic compressors are classified as either centrifugal or axial flow. Both types operate by changing the velocity of gas and converting energy to pressure.

Centrifugal Compressors

During operation, gas enters a centrifugal compressor at the suction inlet and is accelerated radially by moving impellers (Figures 5.2 and 5.3). Centrifugal compressors have one moving element, the driveshaft and

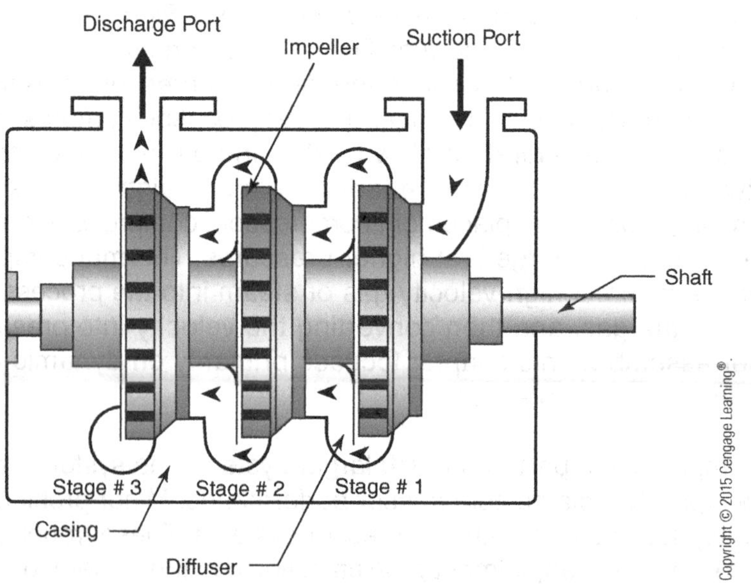

Figure 5.2 *Multistage Centrifugal Compressor*

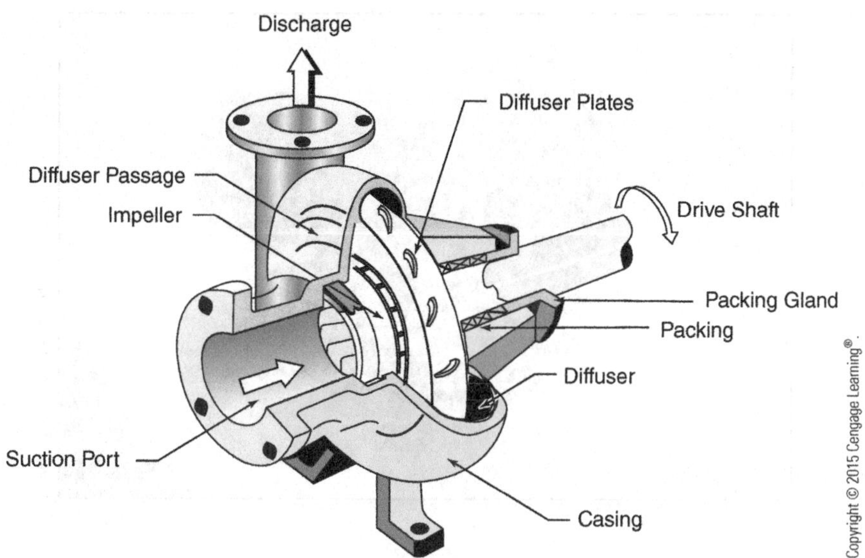

Figure 5.3 *Centrifugal Compressor*

impeller. In a centrifugal compressor, the impeller discharges into a circular, narrow chamber called the *diffuser*. This narrow opening completely surrounds the impellers. As back-pressure builds in the impeller, gas velocity is accelerated through the diffuser assembly and into a circular volute. As high-velocity gas moves through the diffuser and into the volute, kinetic energy is converted into pressure as gas speed slows in the ever-widening volute before exiting the discharge port.

Because compressor performance is linked to the compressibility of the gas it is moving, centrifugal compressors are more sensitive to density and fluid characteristics than are reciprocating compressors. Most centrifugal compressors are designed to operate at speeds in excess of 3,000 RPM. Recent advances in technology have resulted in the development of a centrifugal compressor that runs at speeds in excess of 40,000 RPM.

Centrifugal compressors can be single-**stage** or multistage. Single-stage compressors (Figure 5.4) compress the gas once, whereas multistage compressors deliver the discharge of one stage to the suction of another stage. Single-stage centrifugal compressors are designed for high gas flow rates and low discharge pressures; multistage compressors are designed for high gas flow rates and high discharge pressures. Centrifugal compressors are also used for transferring wet product gases that typically damage positive displacement compressors.

Compression ratio is defined as the ratio of discharge pressure (psia) to suction pressure (psia). Frequently, the desired discharge pressure is very high, over 100 times that of the inlet pressure. When a gas is compressed, the temperature of the gas increases. If a gas was compressed in one stage to a pressure 100 times that of the inlet pressure, the gas

Figure 5.4 *Single-Stage Centrifugal Compressor*

temperature would be extremely high. Multistage compressors, with cooling between stages, are used to develop high pressures to allow for the heat of compression. The compression ratio normally runs in the 3 to 4 range, with the same approximate compression ratio in each stage. For example, if the desired discharge pressure is 1,500 psia, a 4-stage compressor with a 3.2 compression in each stage might be used. The pressure at the discharge of each stage would be: 1st = 47 psia, 2nd = 150.5 psia, 3rd = 481.7, 4th = 1,541.4 psia. The simple calculation used to calculate the pressure increase on each stage is:

Stage One	14.7 psia	×	3.2	=	47.04 psia
Stage Two	47.04 psia	×	3.2	=	150.528 psia
Stage Three	150.528 psia	×	3.2	=	481.689 psia
Stage Four	481.689 psia	×	3.2	=	1541.407 psia

The basic components of a centrifugal compressor are shown in Figure 5.3. The part of the impeller vane that comes into contact with gas first is called the *suction vane tip*. The part of the impeller vane that comes into contact with the gas last is called the *discharge vane*. The driver is an electric motor or turbine.

The basic types of impellers used on centrifugal compressors are the open backward-bladed impeller, open radial-bladed impeller, and closed backward-bladed impeller. Figure 5.5 illustrates various impeller designs.

Centrifugal compressors are considered to be the workhorses of the chemical-processing industry. They are chosen more often than other types for new installations because they have a very low initial installation cost, low operation and maintenance cost, simple new piping installations, interchangeable drivers, large volume capacity per unit of plot area, and

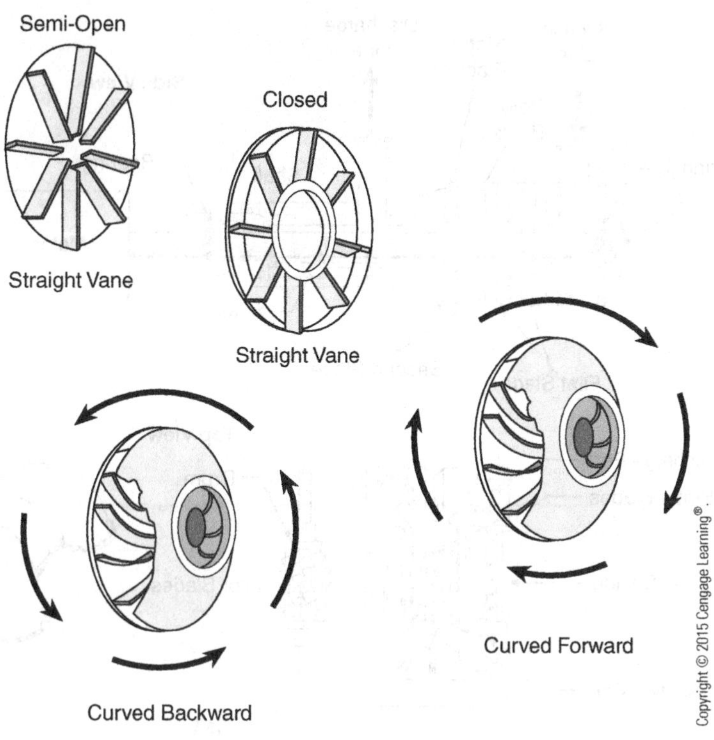

Figure 5.5
Impeller Types

long service life. In addition, they can deliver much higher flow rates than positive displacement compressors.

Axial Flow Compressors

In the industrial environment, axial compressors are the compressor of choice for jobs where the highest flows and pressures are required. Unlike centrifugal compressors, axial compressors do not use centrifugal force to increase gas velocity. An axial flow compressor is composed of a rotor that has rows of fanlike blades (Figure 5.6). Airflow is moved axially along the shaft. Rotating blades attached to a shaft push gases over stationary blades called stators. The stators are mounted on or attached to the casing. As the rotating blades increase the gas velocity, the stator blades slow it down. As the gas slows, kinetic energy is released in the form of pressure. Gas velocity increases as it moves from stage to stage until it reaches the discharge scroll. Multistage axial compressors can generate very high flow rates and discharge pressures.

As a general rule, an axial compressor requires twice as many stages as a centrifugal compressor to perform the same operation; however, axials are 8% to 10% more efficient. Axial compressors are limited to approximately 16 stages because of temperature and equipment stress. Axial flow compressors are often used in series flow with centrifugal compressors because they are capable of operating at greater capacities. The primary application of axial compressors involves the transfer of clean gases such as air. The internal components of an axial flow compressor are extremely sensitive to corrosion, pitting, and deposits.

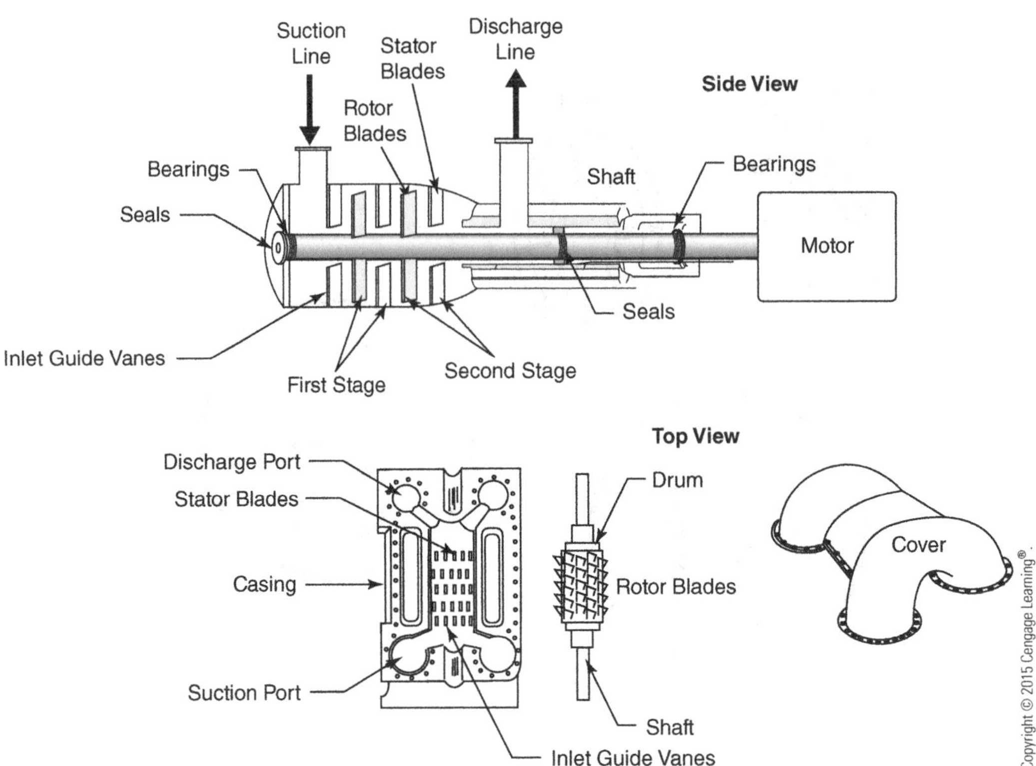

Figure 5.6 *Axial Flow Compressor*

The stator blades in an axial compressor can be fixed, individually adjustable, or continually variable. Individually adjustable stator blades can be adjusted from outside the casing. Continually variable blades are adjusted by a drive ring linked to a driveshaft that is automatically actuated by a power cylinder.

In contrast to a centrifugal compressor, axial compressors accelerate and compress gas in a horizontal, straight-through motion, without the turbulent changes in direction characterized by centrifugal compressors. Pound for pound, axial compressors are lighter, more efficient, and smaller than centrifugals. A 23,000-hp axial produces as efficiently as a 25,500-hp centrifugal. Even so, axial flow compressors are not as common as reciprocating and centrifugal compressors. One main use of axial compressors is in gas turbine applications.

Blowers and Fans

Blowers and fans are simple devices typically classified as compressors. The two basic designs are axial flow and centrifugal flow. Most blowers and fans are single-stage devices designed to perform a specific function. Single-stage, centrifugal blowers are used for low-pressure air systems, refrigeration units, leaf blowing, ventilation systems, or laboratory hoods.

Fans can be used to direct airflow into or out of industrial equipment such as cooling towers, flares, boilers, furnaces, HVAC (heating, ventilating, and air conditioning) systems, or air-cooled heat exchangers, or they can be used for ventilation of confined spaces.

Fans can be classified as centrifugal, propeller, tube-axial, or vane-axial. Centrifugal fans are designed to move gases over a wide range of conditions. Propeller fans consist of a propeller and a motor mounted on a ring. This fan is primarily designed to operate over a wide range of volumes at low pressures and to move air from one enclosed area into another. Tube-axial fans are mounted directly in the pipe cylinder and are designed to move air or gas at medium pressures. The vane-axial fan resembles the tube-axial fan. The motor and fan are mounted directly in the tube. A series of vanes help direct flow over a wide range of volumes and pressures. Each of these four fans can be direct drive or belt driven.

Positive-Displacement Compressors

Positive-displacement compressors operate by trapping a specific amount of gas and forcing it into a smaller volume. They are classified as either rotary or reciprocating. Rotary compressors are further classified as rotary screw, sliding vane, lobe, or liquid ring. Reciprocating compressors are classified as piston or diaphragm.

Positive-displacement compressors remove a set volume of gas for every rotation or stroke of the primary transfer elements. In process systems where fluid density and suction pressures vary, positive displacement devices provide steady service. Rotary compressors can deliver pressures between 100 and 130 psia. Reciprocating compressors discharge pressures that range from 0 to 30,000 psig.

Rotary Compressors

Rotary compressors take their name from the rotating motion of the transfer element. A good case could be made that centrifugal compressors are rotary. Centrifugal compressors do rotate, but they do not positively displace or compress the gas. In contrast, the rotating elements of a rotary compressor displace a fixed volume of fluid inside a durable casing on each rotation.

Rotary Screw Compressors

The rotary screw compressor is commonly used in industry. This device closely resembles the lobe compressor and operates with two helical rotors that rotate toward each other, causing the teeth to mesh (Figure 5.7). As the left rotor turns clockwise, the right rotor rotates counterclockwise,

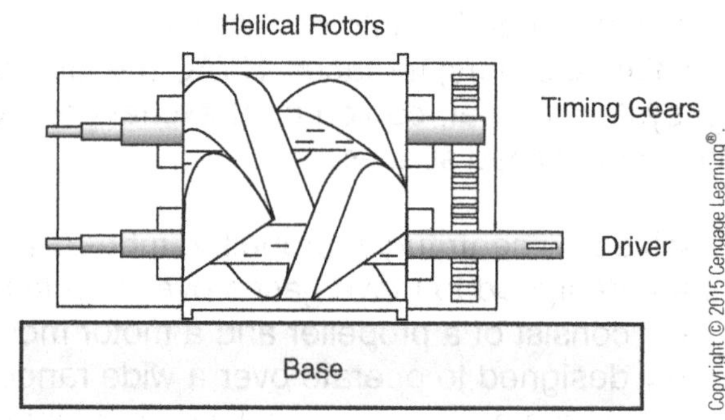

Figure 5.7 *Rotary Screw Compressor*

forcing gas to become trapped in the central cavity. Rotary screw compressors are designed with an inlet suction line and an outlet discharge port. The two rotors are attached to a driveshaft, timing gears, and a driver that provides the energy to operate.

Flow enters the device and is moved axially toward the discharge port. The majority of compression takes place very close to the compressor outlet. The moving elements of the rotary screw compressor do not touch each other or the inner wall. A set of timing gears allows the power rotor to turn the alternate rotor. Because of this design, the rotating elements do not require lubrication, making them a perfect choice for dry gas service. Because of the small tolerances that exist between the moving elements, some internal slip occurs during operation.

Rotary screw compressors operate at speeds between 1,750 and 3,600 RPM and have capacity ratings above 12,000 cfm (cubic feet per minute) on the inlet volume and discharge pressures between 3 and 20 psig. Some rotary screw units can operate between 60 and 100 psig. Another feature associated with the rotary screw compressor is its ability to be used as a vacuum device. This system is designed to handle 500 to 10,000 cfm on the suction side and to pull a vacuum between 5 and 25 inches of mercury.

Sliding Vane Compressors

The sliding vane compressor uses a slightly off-center rotor with sliding vanes to compress gases. The major components of a sliding vane compressor are shown in Figure 5.8. The gas inlet port is positioned so that gas flows into the vanes when they are fully extended and form the largest pocket. As the vanes turn toward the discharge port, the gases are compressed.

The body of the compressor is fabricated from cast iron or steel. A set of cooling water jackets is fabricated into the initial design and tested for tightness. The rotor and shaft are made of high-strength alloy steel. The rotor is precision made with slots around the entire rotor. The sliding vanes are composed of asbestos-phenolic resin, metal, or high-temperature, durable metal. Sliding vane compressors require lubrication between the vane and contact surface. Lubricating oil is injected into the suction side of the

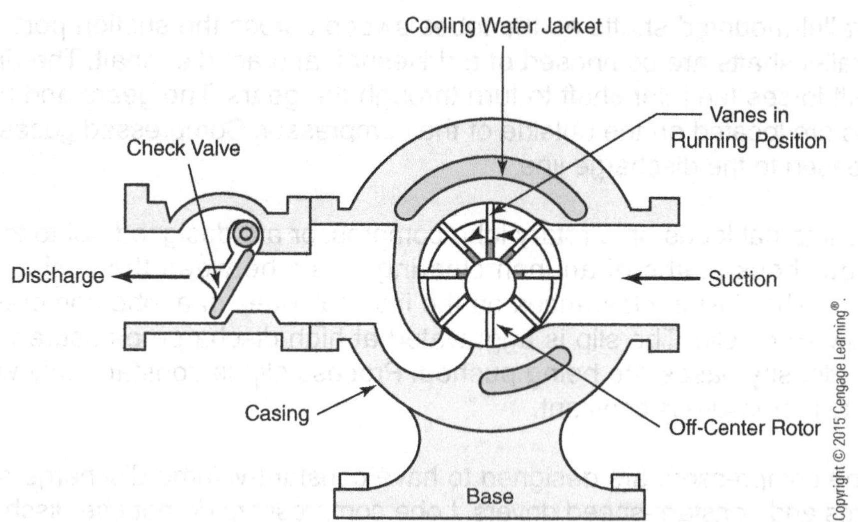

Figure 5.8
Sliding Vane Compressor

compressor. This procedure helps prevent internal slip and provides a positive seal. Sliding compressors are typically nonpulsing systems.

As gas enters the sliding vane compressor, it is captured in vanes and swept around the casing, filling the chamber. As the vanes rotate toward the discharge, the vane length shortens because of the rotor's eccentric position with the shaft, and volume is decreased. As volume decreases, pressure increases until maximum compression is achieved. At this point, the gas is discharged out of the compressor. This type of compressor does not use suction or discharge valves because it is designed to discharge against system pressure.

Lobe Compressors

Lobe compressors are characterized by the two kidney bean–shaped impellers used to trap and transfer gases (Figure 5.9). The close clearances between the casing and impellers are maintained by a set of timing gears. During operation, the two impellers move in opposite directions on

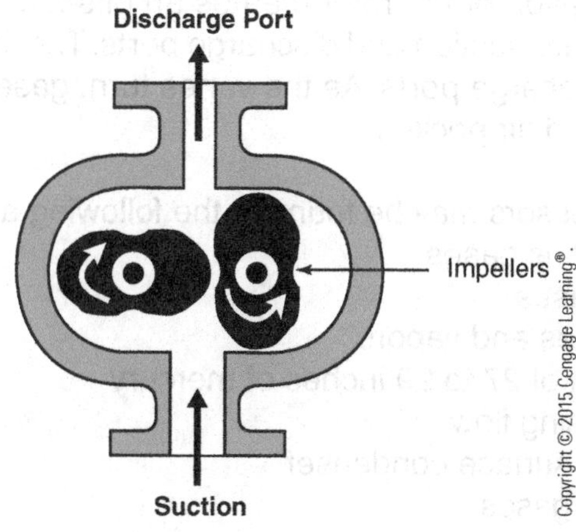

Figure 5.9
Lobe Compressor

parallel-mounted shafts as the lobes sweep across the suction port. The parallel shafts are composed of a driveshaft and an idler shaft. The driveshaft forces the idler shaft to turn through the gears. The gears and bearings are located on the outside of the compressor. Compressed gases are released to the discharge line.

The internal lobes on a rotary lobe compressor are designed not to touch. A few thousandths of an inch clearing exists between the casing and lobes. The design clearances on the internal lobes of a lobe compressor allow some slip. The slip is aggravated at high discharge pressure when low-density gases are being pushed. Process slip is constant only when system pressure is constant.

Lobe compressors are designed to have constant-volume discharge pressures and constant-speed drivers. Lobe compressors do not use discharge or suction valves because they are not designed to operate at a specific pressure. Discharge pressures are determined by the system's process pressure.

Lobe compressors can be used in wet and dry gas service. The rotation of the lobes may be up or down; that is, the discharge port can be at the top or at the bottom of the unit. In dry service, the upward rotation is preferred. In wet service, the downward rotation is recommended so any condensed liquids can escape. Lobe compressors can be used as compressors or vacuum pumps.

Liquid Ring Compressors

A very unusual compressor design is the liquid ring compressor. It combines the centrifugal action of the liquid with a positive displacement, rotary action. A liquid ring compressor has one moving transfer element and a casing that is filled with makeup water or seal liquid (Figure 5.10). As the rotor turns, the fluid is centrifugally forced to the outer wall of the elliptical casing. An air pocket is formed in the center of the casing. As the liquid ring compressor rotates, a small percentage of the liquid escapes out the discharge port. Makeup water or seal liquid is admitted into the compressor during operation. The liquid medium helps cool the compressed gases. The off-center position of the rotor creates an offset in the air pocket. Located on the rotor are suction and discharge ports. The inlet ports are much larger than the discharge ports. As the vanes turn, gases are compressed in the volute-shaped air pocket.

Liquid ring compressors may be found in the following applications:

- Hazardous gases
- Toxic gases
- Hot gases and vapors
- Vacuum of 27 to 29 inches of mercury
- Nonpulsing flow
- Jet and surface condenser
- Oil-free gases

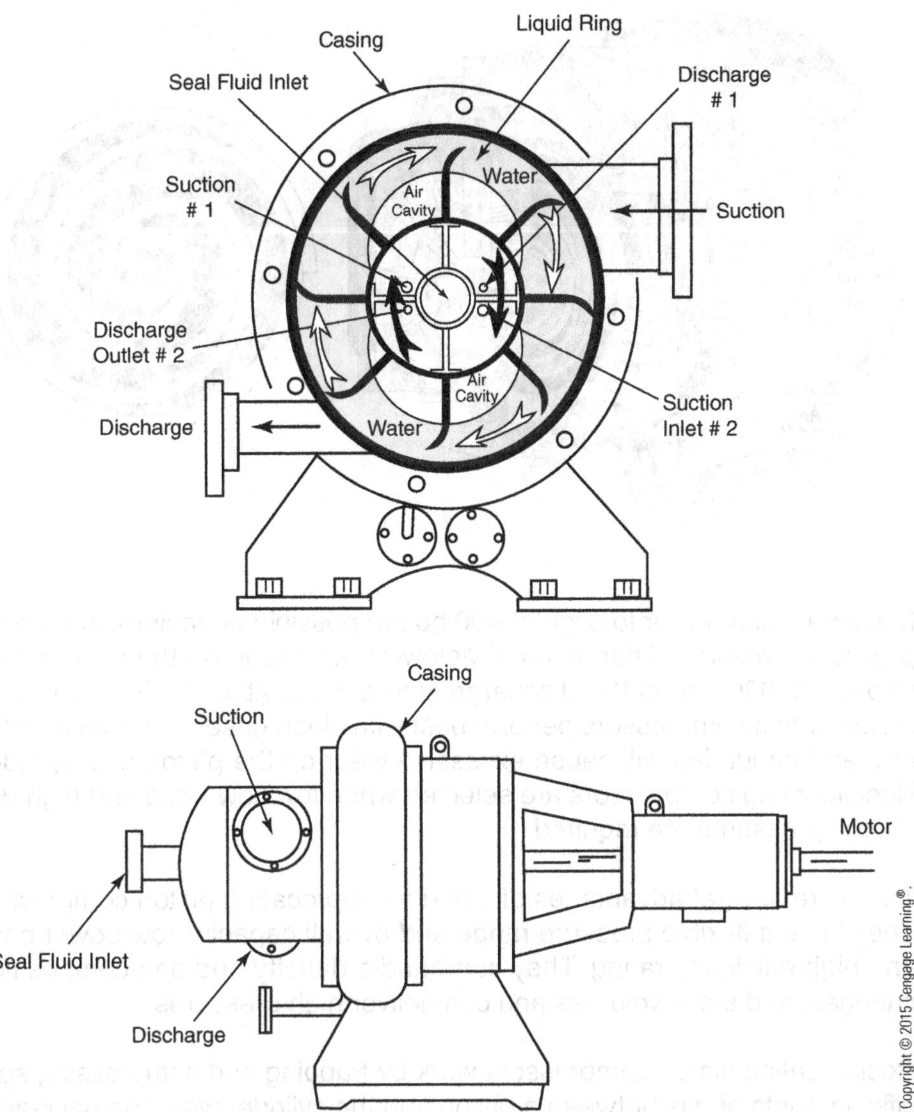

Figure 5.10
Liquid Ring Compressor

Scroll Compressors

A scroll compressor (see Figure 5.11) has two interleaved spiral vanes designed to compress fluids into ever-decreasing volumes. Scroll compressors run quietly and smoothly at lower volumes, trapping fluid between the scrolls. In most cases, one scroll is fixed and one orbits eccentrically without rotating.

Reciprocating Piston Compressors

Their distinctive back-and-forth motion characterizes reciprocating compressors. Reciprocating compressors are classified as either piston or diaphragm. Diaphragm compressors utilize a hydraulically pulsed diaphragm that moves or flexes to positively displace gases. We discuss only the piston type because it is the most popular design. Equipment ratings from

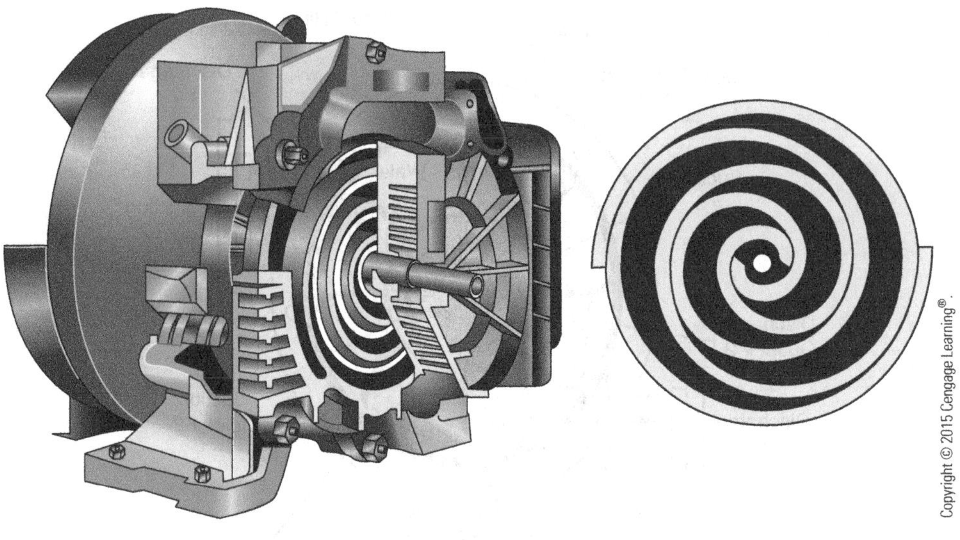

Figure 5.11
Scroll Compressor

fractional horsepower to over 20,000 hp are possible on reciprocating compressors. Pressure differences of below-atmospheric on the suction side to over 30,000 psi on the discharge side are possible. During operation, reciprocating compressors perform best with clean gases. Entrained water, dirt, and impurities will cause excessive wear on the piston and cylinder. Reciprocating compressors are selected when low flow rates and high discharge pressures are required.

There are several advantages of using a reciprocating piston compressor. They have a flexible pressure range and overall capacity, low power cost, and high efficiency rating. They can handle density and gas composition changes, and small volumes and can deliver high pressures.

Reciprocating piston compressors work by trapping and compressing specific amounts of gas between a piston and the cylinder wall. The back-and-forth motion incorporated by a reciprocating compressor pulls gas in on the suction or intake stroke and discharges it on the other. Spring-loaded suction and discharge valves work automatically as the piston moves up and down in the cylinder chamber. The basic parts of a reciprocating piston compressor are shown in Figure 5.12.

Reciprocating piston compressor design varies from model to model. These variations usually occur in the total number of cylinders and in the arrangement of the suction and discharge lines. Most piston compressors have one to four cylinders. Each cylinder has its own piston, rings, and automatic valves. Common crankshafts can be shared with multiple connecting rods. The same cylinder can be equipped with multiple suction and discharge valves in **double-acting compressors**.

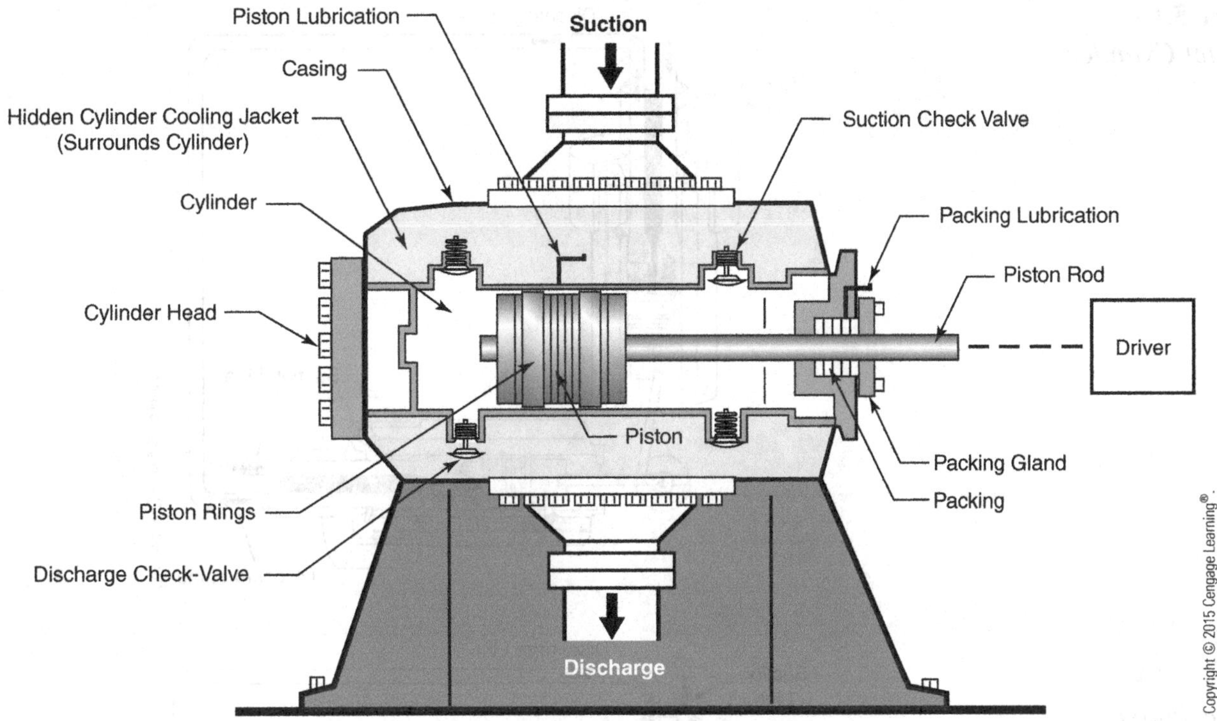

Figure 5.12 *Reciprocating Piston Compressor (Double-Acting)*

Cylinder Design

A cylinder's material is typically selected on the basis of corrosion resistance, thermal shock resistance, pressure rating, and mechanical shock resistance. (**Thermal shock** is a form of stress resulting in metal fatigue caused by large temperature differences between a piece of equipment and the fluid in it.) Common materials used to fabricate cylinders are cast iron (up to 1,200 psig), nodular iron (1,500 psig), cast steel (1,200–2,500 psig), and forged steel (over 2,500 psig).

Single-Acting Parallel Arrangement

In a parallel cylinder arrangement (Figure 5.13), two cylinders are lined up. They have separate intakes and a common discharge line. This type of operation doubles the flow rate while keeping the pressure constant.

Single-Acting Multistage Compressors

A multistage piston compressor has two or more cylinders. The first cylinder usually is referred to as the first stage. Two-stage multistage compressors (Figure 5.14) discharge from the first cylinder into the suction line of the smaller second cylinder. In a multistage compressor, flow rates usually are low and overall pressure is high.

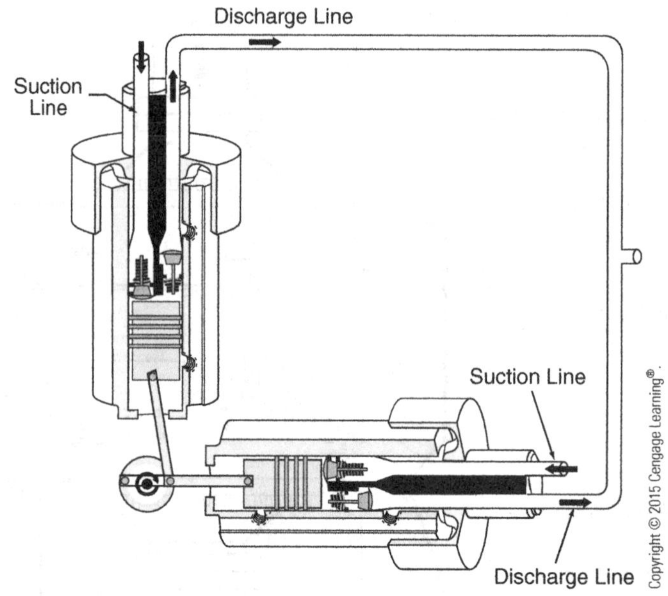

Figure 5.13
Parallel Cylinder

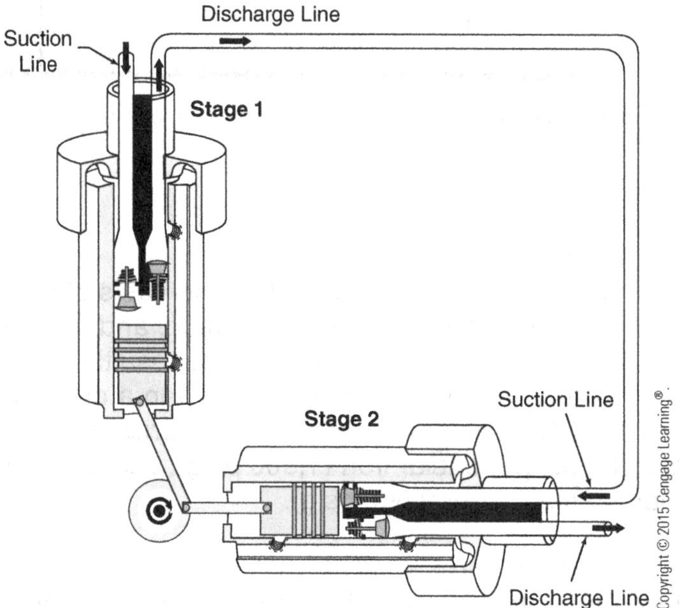

Figure 5.14
Multicylinder

Double-Acting Compressors

Double-acting compressors (Figure 5.15) use a common cylinder and piston to discharge and take in gases on each side of the cylinder. Double-acting compressor cylinders are equipped with two spring-loaded intake valves and two spring-loaded discharge valves. For example, when the piston moves to the right, the chamber to the left of the piston fills with gas, while the chamber to the right of the piston discharges. The exact opposite occurs on the reverse stroke. This technology doubles the efficiency of the compressor.

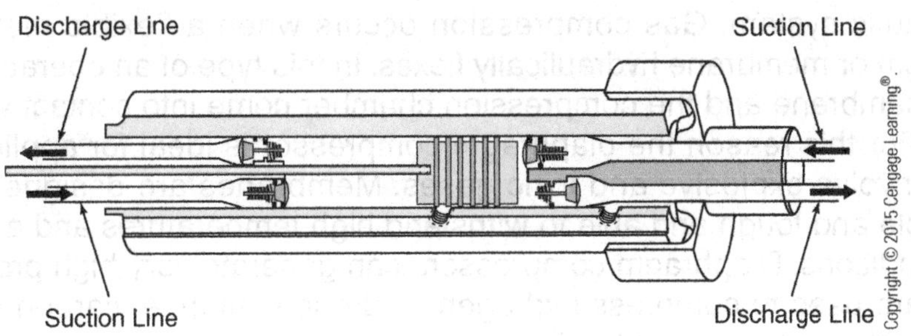

Figure 5.15
Double-Acting

Compressor Layout

The layout of a compressor can be determined easily by looking at the position of the cylinders. V- and L-shaped layouts are found frequently (see Figure 5.14).

Pulsation Control

A common problem found with reciprocating compressors is pulsation. This inherent problem occurs because the suction and discharge valves open and close during each cycle. This problem can be controlled by using a surge drum, pulsation dampener (Figure 5.16), or volume bottle. These devices provide smooth gas flow, reduce vibration, and prevent overloading or underloading the compressor.

Diaphragm Compressors

Diaphragm compressors utilize a hydraulically pulsed diaphragm that moves or flexes to positively displace gases. This type of compressor is closely related to a reciprocating compressor. This type of compressor is a combination of several systems: a gas compression system and a

Figure 5.16
Pulsation Dampener

hydraulic system. Gas compression occurs when a flexible metal diaphragm or membrane hydraulically flexes. In this type of an operation only the membrane and the compression chamber come into contact with the gas. For this reason the diaphragm compressor is ideal for applications that involve explosive and toxic gases. Membranes are designed to be durable and tough and able to withstand high temperatures and a variety of conditions. Diaphragm compressors can generate very high pressures and are used to compress hydrogen, hydrogen chloride, carbon monoxide, and compressed natural gas. Diaphragm compressors come in one, two, three, or more stages. Each stage requires the use of one diaphragm. Figure 5.17 illustrates the basic components of a diaphragm compressor.

The basic components of a diaphragm compressor include:

- Inlet and outlet gas check valves
- Hydraulic fluid check valve
- Diaphragm or membrane
- Hydraulic injection pump
- Hydraulic piston
- Sight glass
- Piston rods
- Crankshaft
- Crankcase frame

Figure 5.17 *Diaphragm Compressor*

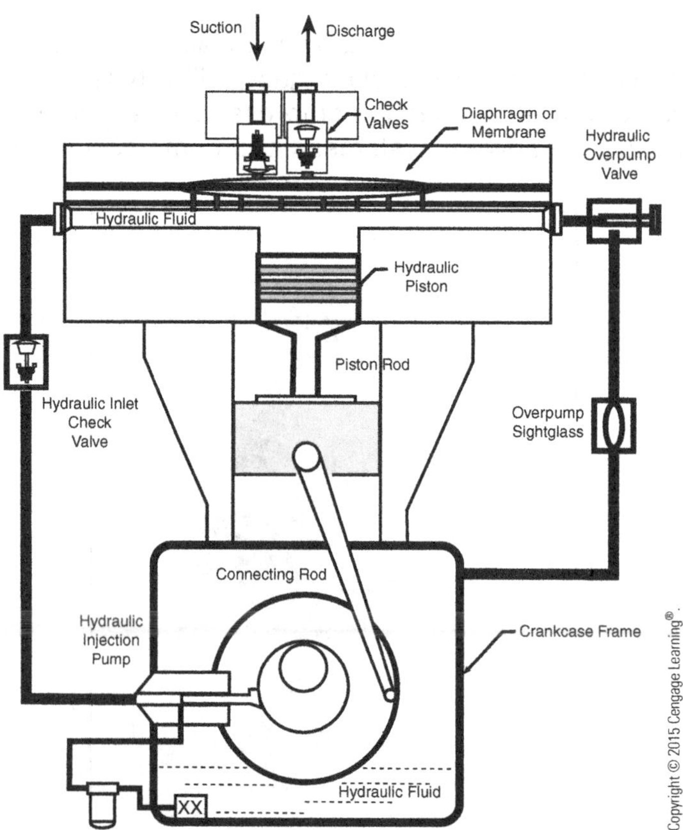

Supporting Equipment in a Compressor System

When compressors are used in a process system, a wide assortment of supporting equipment is required (Figure 5.18). Some of this equipment—such as filters, drivers, and seals—has been described elsewhere. Here we will describe **intercooler** and **aftercooler** heat exchangers, safety valves, silencers, **demisters**, and **dryers**.

Intercooler and Aftercooler Heat Exchangers

The compression of gases creates heat in a compressor. Shell and tube heat exchangers (intercoolers and aftercoolers) have been added to the design to control high temperatures. As gas is discharged out of the first stage of a compressor, the intercooler lowers the temperature. This cooled gas is directed into the suction line of the second-stage compressor. As this gas is compressed (creating more heat), it is discharged into the after-cooler before going to the **receiver**.

Safety Valves

Safety valves and pressure relief valves are used to relieve excess pressure that could damage operating equipment and hurt operating personnel.

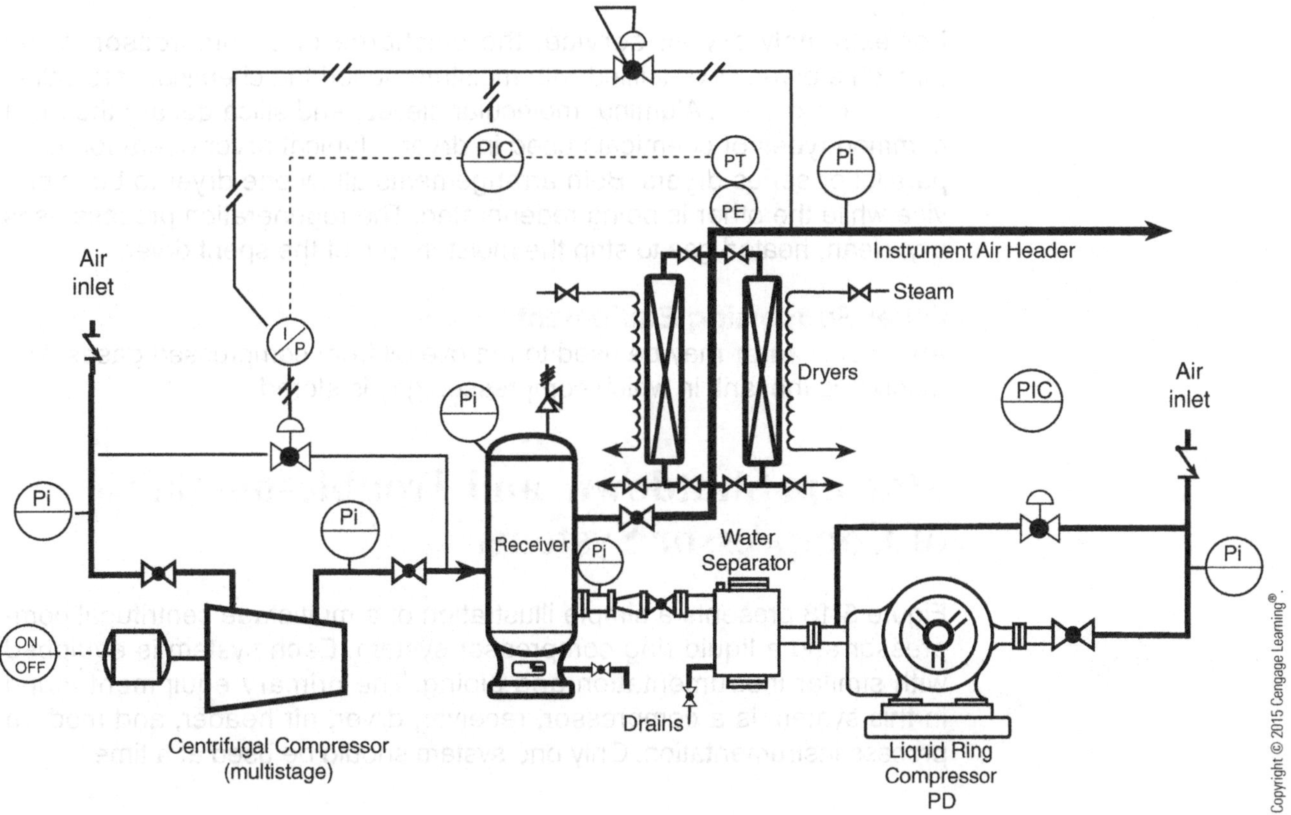

Figure 5.18 *Compressor System*

These valves are sized to handle specific flow rates and should not be replaced without ensuring that the replacement valves meet engineering specifications. Some safety valves automatically reset; other types must be removed and mechanically reset.

Silencers

Most compressors exceed OSHA (Occupational Safety and Health Administration) standards for noise pollution. Silencers are used to muffle some of the damaging noise produced by compressors. To be effective, mufflers should be mounted on the inlet and outlet of a compressor. However, hearing protection should always be worn when one is working on or near a compressor.

Demister

A demister is a device designed to remove liquid droplets from gas. Demisters function along the same lines as a cyclone. As gas enters the top of the demister, it is swirled around the outer perimeter of the canister. Because liquid is heavier than air, centrifugal force brings it into contact with the outer wall. The heavier component falls to the bottom of the demister and is removed. Clean gas escapes out the discharge line on the top of the demister. Many positive displacement compressors would be damaged if wet gases were introduced.

Dryer

For extremely dry air service, the discharge of a compressor is run through a dryer. Dryers filled with moisture-adsorbing chemicals are called **desiccant dryers**. Alumina, molecular sieves, and silica gel are the most common types of chemicals used in dryers. Typical dryer operation uses parallel or series dryers. Both arrangements allow one dryer to be in service while the other is being regenerated. The regeneration process uses dry, clean, heated gas to strip the moisture out of the spent dryer.

Other Supporting Equipment

An **oil separator** may be used to remove oil from compressed gases. The receiver is the tank in which compressed gas is stored.

Startup, Shutdown, and Troubleshooting of Compressor Systems

Figure 5.18 presents a simple illustration of a multistage centrifugal compressor and a liquid ring compressor system. Each system is equipped with similar instrumentation and piping. The primary equipment found in this system is a compressor, receiver, dryer, air header, and modern process instrumentation. Only one system should be used at a time.

Tables 5.1 and 5.2 illustrate typical startup and shutdown procedures. Tables 5.3 and 5.4 illustrate troubleshooting charts for identifying and solving compressor problems.

Table 5.1 *Starting Up and Shutting Down a Dynamic Compressor*

Procedure	Comments
1. Perform valve lineups on compressor and associated equipment.	Ensure that each valve is in the proper position before startup.
2. Check oil levels and bearing-cooling water systems.	
3. Check compressor controls for correct positioning.	
4. Press start button on panel.	Allow compressor to warm up unloaded.
5. Check for abnormal conditions on compressor.	Check temperature, pressure, noise, and excessive vibrations.
6. Open suction line to compressor.	Opening the suction line should load up the compressor as gas flow is established.
7. Monitor equipment until process fills lines.	

Table 5.2 *Starting Up and Shutting Down a Positive-Displacement Compressor*

Procedure	Comments
1. Perform valve lineup on compressor and associated equipment.	Ensure that each valve is in proper position before startup.
2. Line up cooling water to exchangers.	
3. Check oil levels and bearing-cooling water systems.	
4. Check compressor controls for correct positioning.	
5. Reset manual unloader valve to unloaded.	
6. Press start button on panel.	Allow compressor to warm up.
7. Check for abnormal conditions on compressor.	Check temperature, pressure, noise, and excessive vibrations.

Table 5.3 Troubleshooting a Centrifugal Compressor

Problem	Possible Cause
Excessive vibration	• Misaligned shaft • Damaged coupling • Damaged rotor • Bearing or seal damage
Discharge pressure low	• Leak in piping • Low suction pressure • System demand exceeding design limits • Compressor not up to speed
Lube oil pressure low	• Oil level low • Incorrect pressure setting • Lube oil pump failure • Dirty filter or strainer
High temperature on bearing oil	• Restricted flow • Oil needs to be changed • Bearing failure • Water in lube oil • Fouling in oil coolers
Driveshaft misalignment	• Foundation shift • Loose bolts on foundation • Piping strain • Grouting washed out
Water in oil system	• Ruptured tube in heat exchanger • Condensation in oil reservoir • Rain water • Steam tracing leak

Compressor Symbols

Each type of compressor can be represented by a symbol. Figure 5.19 illustrates compressor symbols.

Summary

Compressors differ from pumps in that pumps move liquids and compressors move gases. Compressors are used to transfer granular powders and small plastic pellets from place to place; to establish feed gas process pressures; to provide clean, dry air for instruments and control devices; and to compress gases such as light hydrocarbons, nitrogen, hydrogen, carbon dioxide, and chlorine.

Gases and vapors are compressible. Compression decreases volume by moving gas molecules close together. Compressed gases will resume their

Table 5.4 *Troubleshooting a Reciprocating Compressor*

Problem	Possible Cause
Compressor will not start	• MCC or switchgear problem • Low oil pressure shut down • No power to motor • Permissives or interlock not satisfied
Noise in the cylinder	• Loose piston • Worn piston rings • Piston striking cylinder head • Valve improperly seated
High discharge temperature	• Fouled intercooler • High inlet temperature • Leaking piston rings • Wrong lube oil flow rate
Safety valve lifting	• Leaking suction valve on final stage • Worn out rings on final stage • Discharge line restricted • Safety valve needs to be replaced
Packing leaking	• Packing needs to be replaced • Scored piston rod • Dirt in packing • Pressure increase
Frame knocks	• Cold oil • Low oil pressure • Wrong oil in system • Loose or worn crankpin

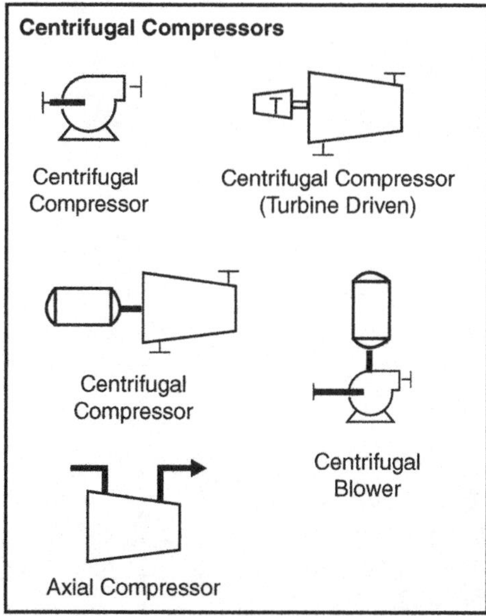

Positive Displacement Compressors

Reciprocating Compressor

Rotary Compressor

Rotory Compressor & Silencers

Liquid Ring Vacuum

Rotary Screw Compressor

Positive Displacement Blower

Reciprocating Compressor

Figure 5.19 *Compressor Symbols*

original shape when released. Compressed gases produce heat because of molecular friction. The higher the pressure, the smaller the volume. Gas volume varies with temperature and pressure.

Compressors usually can be classified into two groups: positive displacement and dynamic. Positive-displacement compressors operate by trapping a specific amount of gas and forcing it into a smaller volume. These designs include rotary (rotary screw, sliding vane, lobe, and liquid ring) and reciprocating (piston and cylinder and diaphragm).

The most common type of compressors, reciprocating compressors work by trapping and compressing specific amounts of gas between a piston and cylinder wall. The back-and-forth motion incorporated by a reciprocating compressor pulls gas in on the suction or intake stroke and discharges it on the other.

Dynamic compressors operate by accelerating the gas and converting the energy to pressure. They can deliver much higher flow rates than positive displacement compressors.

Review Questions

1. List two purposes for which industrial manufacturers use compressors.
2. List four gases that compressors are used to compress.
3. List the two types of compressors.
4. List the three rotary compressors.
5. List two reciprocating compressors.
6. Compare and contrast the operation of an axial compressor with a multistage centrifugal compressor.
7. Describe a rotary screw compressor.
8. Describe a sliding vane compressor.
9. Describe a lobe compressor.
10. Describe a liquid ring compressor.
11. Describe parallel arrangement in a reciprocating compressor.
12. Describe a multistage arrangement in a reciprocating compressor.
13. Describe a double-acting arrangement in a reciprocating compressor.
14. List the basic components of a centrifugal compressor.
15. List the main differences between a single-stage and a multistage compressor.
16. What are the components of a typical compressor system?
17. Contrast intercooler and aftercooler heat exchangers.
18. Describe how to start up a dynamic compressor.
19. Describe how to start up a positive-displacement compressor.
20. List four types of compressor fans.
21. Define thermal shock.
22. Describe the term "compression ratio" and explain how to use it.
23. Explain the operation of a pulsation dampener.
24. List the major components of a sliding vane compressor.
25. Explain the primary differences between operating a positive displacement compressor system and a centrifugal compressor system.

Turbines and Motors

Objectives

After studying this chapter, the student will be able to:

- Identify the different kinds of turbines.
- Review the history of steam turbines.
- Describe the operating principles and components of a steam turbine.
- Describe the types of steam turbines.
- List the startup procedures for a steam turbine.
- Identify the basic components of a gas turbine.
- Describe how a gas turbine operates.
- Identify the basic components of an electric motor.
- Describe how an electric motor operates.

Key Terms

Condensate—moisture produced when steam, hot gases, or vapors condense (change to the liquid state).

Exhaust valve—valve used to block steam turbine outlet steam.

Gas turbine—a device that uses high-pressure gases to turn a series of turbine wheels to provide rotational energy to turn an axle or shaft.

Governor—speed-control device that adjusts the governor valve.

Governor valve—an automatic valve that controls steam turbine speed by regulating the amount of steam admitted.

Heat soaking—a turbine warm-up procedure designed to remove condensate and warm the internal parts; includes slow rolling the turbine at low speeds between 200 and 500 RPM.

Hunting—occurs when a steam turbine's speed fluctuates while the controller searches for the correct operating speed.

Hydraulic turbine—a device that uses high-pressure liquids to turn a turbine wheel attached to a pump generator.

Impulse turbine—a steam turbine with a blading design that causes rotation of the blades and shaft when high-velocity steam from an external source pushes on it.

Journal bearing—see *Radial bearings.*

Labyrinth seal—a shaft seal designed to stop steam flow in a steam turbine; consists of a series of ridges and intricate paths.

Nozzle—a device designed to restrict flow and convert pressure into velocity.

Overspeed trip—a safety device used to shut down a steam turbine when it exceeds its rotational speed limit by closing the turbine trip valve.

Reactive turbine—a steam turbine with a fixed nozzle and an internal steam source.

Sentinel valve—a spring-loaded automatic relief valve that makes a high-pitched noise when turbine speed approaches the design maximum.

Slow roll—controlling turbine speed at low (200–500) RPM.

Steam chest—area where steam enters a steam turbine.

Steam strainer—a mechanical device that removes impurities from steam.

Trip valve—a fast-closing steam inlet valve operated by an overspeed trip lever.

Wind turbine—commonly referred to as windmill; uses air pressure to pump water, grind grain, and operate small generators.

Kinds of Turbines

Turbines are classified according to their principle of operation and the type of fluid that turns them. The four main types of turbines are steam, gas, hydraulic, and wind. In steam turbines (impulse movement), the rotor turns in response to the force (velocity) of a gas. In **hydraulic turbines** (reaction movement), the rotor turns in response to the pressure of a liquid. **Gas turbines** use high-pressure gases, and **wind turbines** (windmills) use air pressure. Steam and gas turbines are the two types most commonly used in industry.

Steam turbines are typically classified as condensing, noncondensing, reaction, or impulse. In condensing turbines, exhaust steam flows to surface condensers. Condensing turbines operate at vacuum pressure. In noncondensing turbines, exhaust steam is utilized in low-pressure steam applications. In **reaction turbines**, steam is discharged from a **nozzle** mounted on the rotor. Movement is a reactive response to the release of steam from an internal source. In **impulse turbines**, steam from an external source acts on the rotor to create movement. Most plants use this design. Each of these designs can have one or more stages. Table 6.1 lists the parameters of turbines.

Table 6.1 *Turbine Sizes*

Size	Stages	Horsepower	Speed (RPM)
Small	1	0.7–250	1,000
Medium	1 or 2	5–4,000	2,000–15,000
Large	3 or more	5,000–50,000	2,000–20,000

History of Steam Turbines

In 200 BCE, Archimedes described a device used by ancient Egyptians. This device consisted of a hollow globe, mounted on bearings, with a series of nozzles on the side. Water is poured into the globe and heated to boiling. As the water boils, it is converted to steam and escapes through the nozzles, rotating the globe. The globe rotates because of the effects of the escaping steam. The rotation of the globe demonstrates Newton's third law of motion: For every action there is an opposite and equal reaction. Modern steam turbines operate under the same principle.

In 1629, the Italian engineer Giovanni Branca designed the first impulse steam turbine. Impulse movement involves a type of turbine blading design that causes rotation of the blades and shaft when high-velocity steam from an external source pushes on the blades. Branca's innovative design directed high-velocity steam against the blades. The impulse design is used in present-day steam turbines. Technical improvements to the steam turbine design have enhanced efficiency and power.

Steam turbines have replaced the once-popular steam engine because they weigh less and occupy less space. Steam turbines operate with little or no vibration because they have little, if any, back-and-forth motion. Turbines are very efficient when run at high speed under a heavy load. Turbine efficiency drops drastically when the equipment is slowed down.

Operating Principles of Steam Turbines

The primary operating principle of a turbine is to convert steam energy into mechanical energy that can be used to drive rotating equipment. A steam turbine is a device (driver) that converts kinetic energy (steam energy of movement) to mechanical energy. Steam turbines have a specially designed rotor that rotates as steam strikes it. This rotation is used to operate a variety of shaft-driven equipment. Turbines are used primarily as drivers for pumps, compressors, ocean vessels, turbo-electric locomotives, naval vessels, and electric power generation.

As high-pressure steam enters a turbine, it passes through a device called a *nozzle*. Nozzles restrict the flow and increase the velocity of the steam. The nozzle directs this high-velocity steam against the blades of a paddlewheel, causing it to rotate. As the steam passes through alternate sets of fixed and revolving blades, it constantly expands as it moves along. The rotating paddlewheel is attached to a shaft, and the blading and shaft together make up the rotor. Impulse or reaction movement occurs as the steam strikes the rotor, converting the steam energy into mechanical energy. The amount of steam energy needed to perform useful work depends on the pressure range through which the steam expands.

The steam used to operate a steam turbine is produced in a boiler. Boilers produce steam that can enter a turbine at temperatures as high as 538°C (average 1,000 to 1,050°F) and pressures as high as 3,500 psi inlet and 200 psi outlet. (Steam turbines can also run under a vacuum.) High-pressure steam is admitted slowly into a turbine to warm it up and remove **condensate** (moisture produced by condensation).

Steam turbines are used to drive the electric generators in modern power plants. A multistage steam turbine is considered to be one of the world's most powerful engines. Modern turbine technology includes 50 or more stages linked along a horizontal shaft. Each stage consists of a set of moving and stationary blades. The curved blades of each stage are designed so that the spaces between the blades act as nozzles and increase steam velocity. As the steam zigzags between the stationary and moving blades,

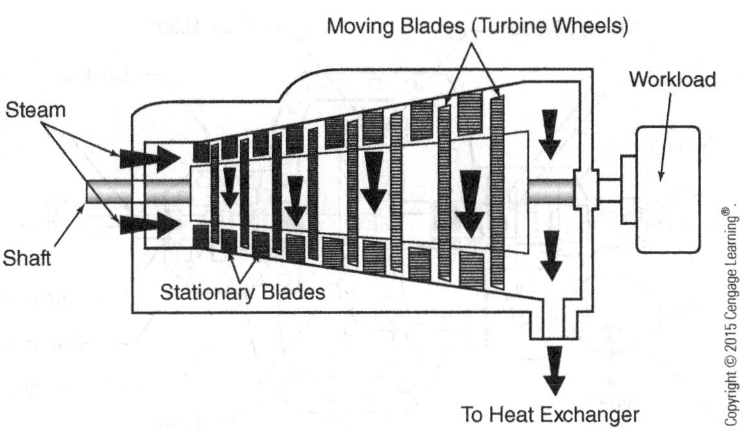

Figure 6.1
Steam Turbine

it begins to expand as much as 1,000 times its original volume. Modern turbine design increases the size of each stage, giving the turbine a conical shape (Figure 6.1).

Impulse and **reactive steam turbines** operate under similar principles. Impulse turbines have a blading design that causes rotation of the blade-and-shaft assembly, or rotor, when high-velocity steam pushes on the blades. The kinetic steam source is external. Reactive movement occurs when steam escapes from a fixed nozzle attached to the rotor, propelling the rotor. The kinetic steam source is internal.

Both impulse and reaction turbines can be either condensing or noncondensing turbines. Condensing turbines exhaust steam into a heat exchanger called a surface condenser that cools and condenses the steam. The condensate is sent to the boiler, where it is converted back to steam. Condensing-type turbines are the most efficient type because they extract the maximum amount of energy from the steam.

Noncondensing, or extraction-type, turbines are multistage turbines designed to take high-pressure steam, use it in the turbine, and then extract a portion of the steam for other use. As high-pressure, high-velocity steam passes over the turbine wheel, the steam expands. This expansion enables the turbine to divert low-pressure steam to other units. Some multistage turbines can induce steam into back stages to increase delivered horsepower. They are called *induction-type turbines*.

Basic Components of a Steam Turbine

The parts of a steam turbine may be thought of as being in four groupings: rotor, fixed parts, governing mechanism, and lubrication system (Figure 6.2). A steam turbine may have miscellaneous other parts for adjustments and safety.

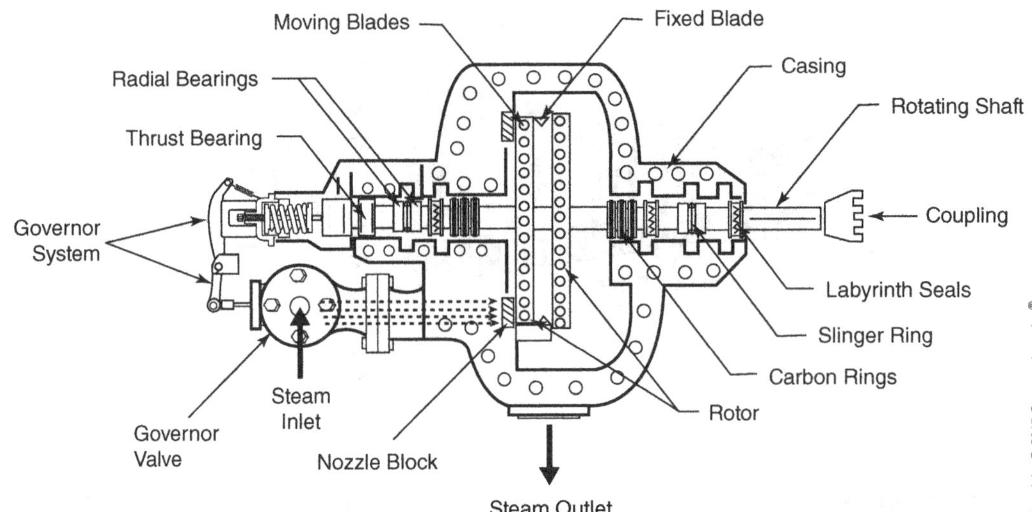

Figure 6.2 *Steam Turbine Components*

Rotor

Steam turbines have a set of rotating blades and a row of fixed "half moon" blades. The wheel-shaped rotating blades sandwich the fixed blades. Operators commonly refer to the assembly consisting of the shaft and the rotating blades as the rotor (Figure 6.3). A visual inspection of the rotor reveals that the rotating blades are firmly attached to the shaft. The rotor is statically and dynamically balanced to ensure smooth operation. Turbine blading uses a progressive cavity-type (see the Rotary Pumps section in Chapter 4) design to move steam through the rotating and fixed blades.

Moving blades are made of durable stainless steel that has been rolled and drawn. The blades are securely fastened by a series of dovetail

Figure 6.3
Steam Turbine Rotor

grooves. The correct spacing is maintained by soft-iron packing pieces. The shrouded outer ends of the blades prevent vibration and capture steam in the blade path.

During operation, steam enters the inner chamber of the turbine, striking the blades with the full force of the high-velocity steam. The blades rotate in response to the steam pressure.

Fixed Parts

The principal stationary parts in a steam turbine are the fixed blades; throttle valve; steamtight casing; steam chest; nozzle; and bearings, rings, and seals.

Fixed Blades

The fixed blades (Figure 6.4) are made of durable stainless steel that has been rolled and drawn. The fixed blades are a half-moon-shaped ring located in the lower section of the turbine, sandwiched between the moving blades. When fixed and rotating blades are aligned in the correct position, steam passages are formed across the wheel of the turbine.

Casing

The casing is composed of a base and covering made of carbon steel or turbine iron. The base and the covering are designed to form steamtight joints. Gaskets typically are not needed when reinforced flanges are used.

Steam Chest

The **steam chest** houses the governor valve, overspeed trip, and **steam strainer** (a mechanical device that removes impurities from steam). It is composed of carbon steel or iron and is bolted to the lower casing.

Figure 6.4
Fixed Blades

Nozzle

The nozzles and nozzle block constitute a precision instrument fabricated from a solid block of high-tensile carbon silicon steel that directs high-velocity steam against the rotor. Nozzle blocks are bolted to the steam chest. The nozzle has overlapping exits that allow the steam jets to converge before being directed against the buckets of the rotor.

Bearings

Bearings (Figure 6.5) provide radial and axial support for the shaft of a steam turbine. Radial bearings (also called **journal bearings**) are designed to keep the rotor of a steam turbine from moving from side to side or up and down. Oil supply passages are built into the radial housing, or a slinger ring lubricates the bearings. As the shaft rotates, the lubrication forms a thin film between the shaft and the bearing that allows the system to float. This type of bearing typically is located beside the thrust bearing on one end of the turbine and by the shaft seal on the other.

Thrust (axial) bearings are designed to control axial, or back-and-forth, movement along the shaft and to hold the rotor in correct alignment with stationary parts. Thrust bearings are located outside the steam chest next to the shaft seals. During operation, the thrust bearing draws oil up under the rounded edges of the pivot shoes. As hydraulic pressure builds, the shoes tip slightly, allowing more oil to be drawn under the collar. At normal speed, the shoes have a slight tilt, which is sufficient to accommodate an oil wedge. This oil wedge separates the shoes from the collar.

Seals

Carbon-ring and labyrinth shaft seals are located at the end of each casing along the rotor. These devices are used to minimize the outward leakage of steam under pressure and the inward leakage of air. Carbon rings prevent leakage between the rotor shaft and the casing. The stainless steel

Figure 6.5
Bearings, Seals, and Slinger Rings

spring-backed ring gland is mounted in a corrosion-resistant sleeve that is kept from rotating by a rod passing through the lower housing section. **Labyrinth seals** consist of a series of ridges and intricate paths designed to stop flow.

Governing Mechanism

The **governor** (Figure 6.6) is designed to automatically regulate the speed of the turbine. A shaft-type governor is located internally and mounted to the shaft. Centrifugal force causes the weights to rotate. The weights are constrained by a spring. As centrifugal force increases, the weights move farther from the central shaft, compressing the spring. The rotating spindle is attached to a fixed sleeve, which controls a fulcrum lever. The lever is used to position the governor valve.

The **governor valve** is a corrosion-resistant valve used to regulate steam flow into a steam turbine. An inlet block valve, the throttle valve, outside the steam chest lets steam enter the turbine's steam chest. The governor valve is located inside the steam chest. The governor valve controls the amount of steam flow to the nozzle and is a critical part of the overall turbine control system. Multistage turbines have multiple governor valves arranged in a rack.

There are two types of governors: electronic, hydraulic speed control and mechanical, hydraulic speed control. A typical governing system may be

Figure 6.6
Governor System

divided into three parts: a speed-sensitive element, a motion sensor amplifier to the control valve, and a control valve.

Lubrication System

The lubricating oil has five functions. It lubricates bearings and gears. It cools the lubricated parts. It transfers frictional heat. It acts as the hydraulic medium for the governor. It acts as the hydraulic medium for the actuation of the governor valves and safety devices.

Other Parts

Steam turbines have a hand speed changer capable of making 10% speed range adjustments. A **sentinel valve**—a spring-loaded automatic relief valve—makes a high-pitched noise when turbine speed approaches the design maximum. An **overspeed trip** is a safety device used to shut down the turbine when it exceeds 115% rotational design limits. A weighted spring attached to the governor hub is used to close the governor and stop valves. An **exhaust valve** blocks turbine outlet steam. This valve must never be closed when the steam inlet valve is open because the turbine casing could become overpressurized. A butterfly **trip valve**, which is a heavy-duty valve, is attached to the steam chest. It works independently or in conjunction with the governor mechanism. It is held in place by a powerful trip linkage.

Steam Turbine Problems

Steam turbines are susceptible to two types of problem: vibration and **hunting**. Vibration-sensing equipment is used to monitor turbine performance. Probes are used by process technicians to check inboard and outboard radial and thrust bearings. Excessive vibration could indicate failing bearings or internal problems.

There are two types of vibration: radial vibration and axial movement. Radial vibration increases when the turbine surges or after a cold start when the machinery passes through critical speed. Surging also can affect axial movement in the rotor. Higher bearing temperatures will accompany any of these problems. Operators should be aware that axial movement is the most important consideration when monitoring and troubleshooting vibration problems.

Steam turbines are designed to be operated at a controlled speed. If the steam turbine is operated above or below the normal operating set point, a problem known as *hunting* may occur. Hunting takes place when a turbine's speed fluctuates while the controller searches for the correct operating speed. Another problem that causes the steam turbine to speed up and slow down is mechanical linkage binding. As discussed earlier, the governor system is responsible for the speed the turbine operates. Occasionally, the mechanical linkage between the governor and governor valve will bind. This binding causes the steam turbine to hunt. Sticking or binding problems can be resolved by lubricating the linkage on the steam turbine.

Inlet steam pressure problems can also cause a steam turbine to hunt. The chemical processing industry attempts to maintain constant pressures on utility steam heads. Occasionally, boilers go down, or larger demands are made on the system. A fluctuation in the suction steam pressure affects the turbine. On hot days, the system may work perfectly, but cold weather might create problems for your turbine. Some steam turbines are equipped with hand valves that can be opened to admit more steam into the system or closed to reduce flow. Hand valves typically are opened during high-load operation and closed during low-load operation.

Start Up a Steam Turbine

To start up a steam turbine, follow the operating procedure in Table 6.2.

Table 6.2 *Starting Up a Steam Turbine*

Procedure	Comments
1. Visually inspect turbine for damage. Reset overspeed trip. Check lube oil level for turbine governor. Ensure that rotating equipment covers are in place.	If turbine is damaged, contact machinists for closer inspection.
2. Drain condensate from turbine. Slowly back-feed low-pressure steam into turbine exhaust.	This will slowly heat up the turbine and minimize thermal shock.
3. Check lubrication system's constant-level oiler. Check cooling system. Check seal system where the shaft enters the casing.	Cooling water circulates through the bearing jackets continuously.
4. Slowly open the throttle valve and bring the turbine speed up 10% to 20% normal operating speed.	This step is called heat soaking and includes slow rolling (controlling the speed at 200–500 RPM).
5. Bring turbine up to operating speed by opening the throttle valve, listening for line-out sound, and checking governor linkage.	If you open the throttle valve quickly, the governor will not have time to get control and the turbine may surge above limits.

Gas Turbines

A gas turbine is a device that uses high-pressure combustion gases to turn a series of turbine wheels to provide rotational energy to turn an axle or a shaft. Gas turbines are used to operate electric generators, ships, and racing cars and are a primary component of jet aircraft engines. There are three primary parts of a gas turbine system: an axial compressor, a combustion chamber, and a gas turbine (Figure 6.7). The gas turbine system mixes compressed air with fuel in a combustion chamber. A spark plug ignites the mixture, which is directed into the suction side of the gas turbine. The hot combustion gases rush into the gas turbine, causing the turbine wheels to turn. Hot exhaust gases are discharged from the body of the gas turbine. The air compressor and the gas turbine are mounted to the same axle, which is connected to the workload.

When the air compressor and the combustion chamber are used in combination, the device is frequently referred to as a gas generator. During operation, a fraction of the power generated by the turbine is used to run the compressor. When the air compressor pulls air into the system, the pressure increases. When the compressed air mixes with the fuel and is ignited, the higher pressure allows the mixture to burn better. The fuel used to operate a gas turbine is natural gas or oil. The hot combustion gases produced by the gas or oil are used in the same way a steam turbine uses steam to turn the rotor. The air for combustion is generally filtered through a bag-house arrangement to remove airborne contaminants, which would deposit on turbine components.

The basic components of a gas turbine system fall under four primary areas: the compressor, combustion chamber, gas turbine, and workload (Figure 6.8). Each of these areas has a number of critical components, and they all are linked by a common axle.

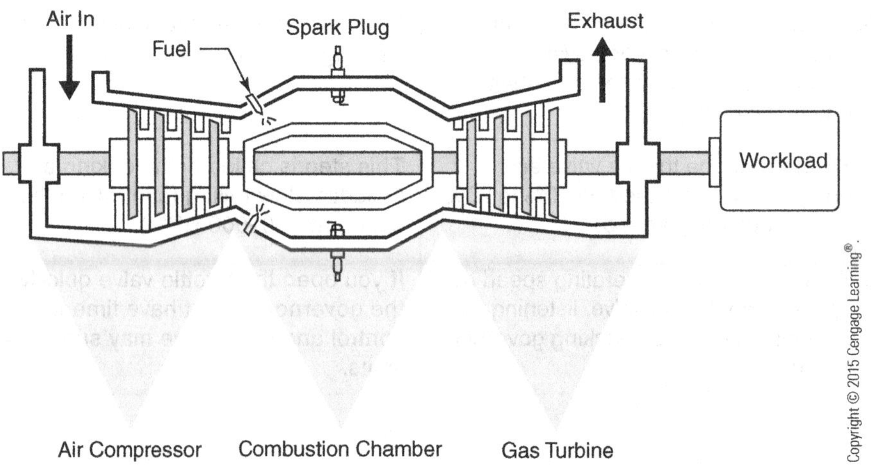

Figure 6.7
Gas Turbine System

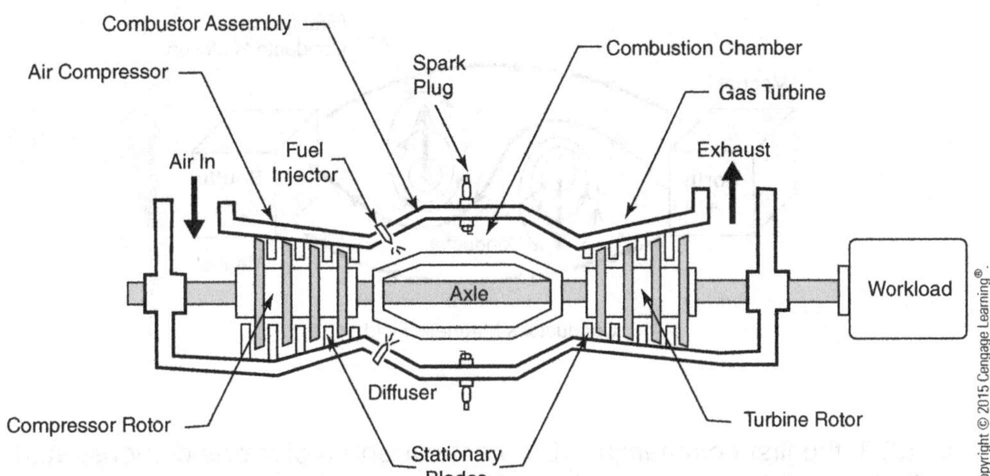

Figure 6.8
Gas Turbine Components

The compressor consists of the compressor rotor assembly, stator blades, rotor blades, compressor case assembly, air inlet filter assembly, bearings and seals, and compressor diffuser assembly. The combustion chamber consists of a fuel injector, combustor housing assembly, gas fuel manifold, bleed air valve, and igniter. The gas turbine consists of a gas producer rotor assembly, power rotor assembly, moving turbine wheels and stationary blades, nozzle case and assembly, turbine exhaust diffuser, and exhaust collector. The workload consists of the driven shaft and the driven device.

Each part of the gas turbine system is an integral part of the whole unit. Axial flow compressors have replaced most other compressor designs because of the large volume they can handle. The combustion chamber combines two feed components to produce a continuous, high-pressure flow into the turbine. The gas turbine has a number of stages that increase in size to accommodate the expanding hot gases that jet through the moving turbine wheels and stationary blades.

Electric Motors

The history of electric motors began in 1820, when a Danish physicist, Hans Christian Oersted, discovered that a wire generates a magnetic field when an electric current is passed through it. The discovery of the principle of electromagnetism opened up a new scientific field. In 1825, an Englishman, William Sturgeon, wrapped a wire around an iron bar and generated a much stronger magnetic field. As scientists learned about this new discovery, a number of improvements were made to the electromagnet. By 1831, an English chemist named Michael Faraday had discovered that when a conductor is passed through a magnetic field, electric current flows through the conductor. This discovery led to the invention of the electric generator.

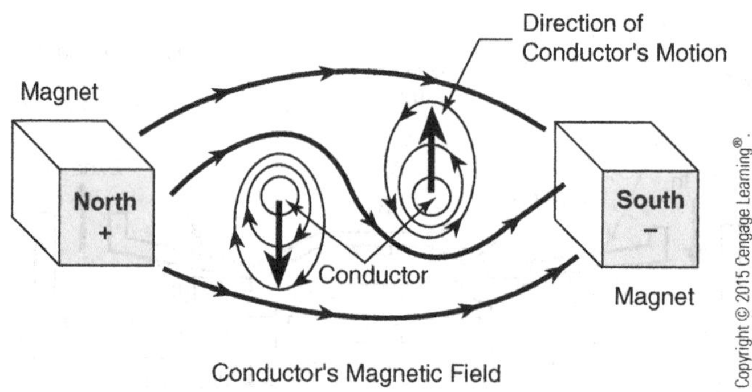

Figure 6.9 *How an Electric Motor Works*

In 1873, the first commercial DC (direct current) motor was demonstrated. A Serbian engineer named Nikola Tesla invented the first AC (alternating current) motor in 1888.

Electric motors are used to operate pumps, generators, compressors, fans, blowers, and other equipment. Electric motors can be classified as either DC or AC. The operation of both types of electric motor is based on three scientific principles: Electric current creates a magnetic field; unlike magnetic poles attract and like magnetic poles repel each other; and current direction determines the magnetic poles. An electric motor consists of a stationary magnet (stator) and a rotating conductor (rotor). A permanent magnetic field is formed by the lines of force between the poles of the stationary magnet (Figure 6.9).

In a DC motor, a conductor (armature) is located between the north and south poles of a stationary magnet (field structure) (Figure 6.10).

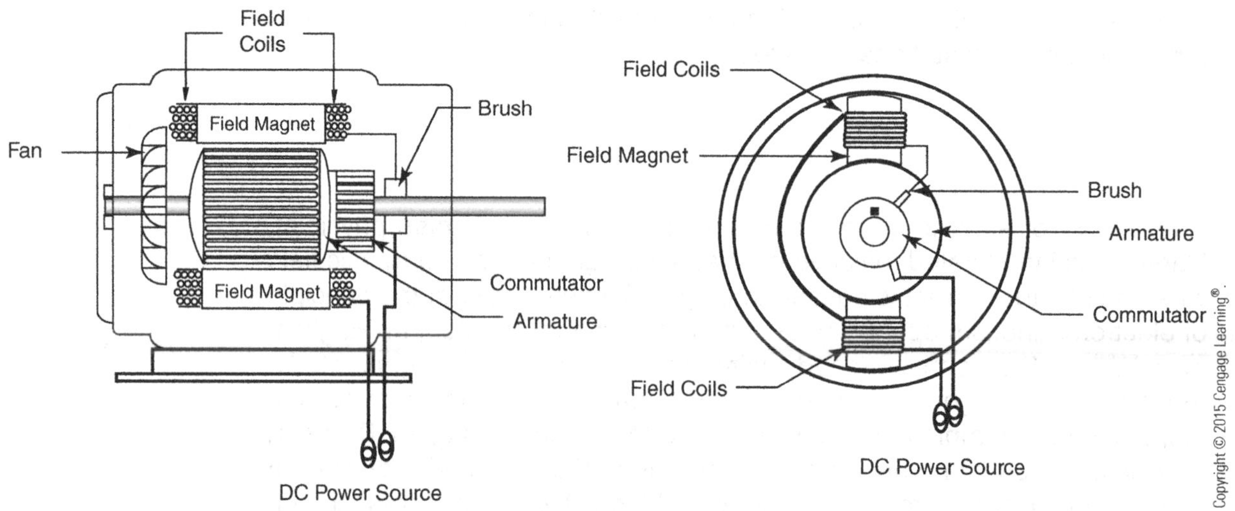

Figure 6.10 *DC Motor*

A commutator is used to reverse the direction of current, helping transmit current between the power source and armature. The armature is a cylindrical device attached to a driveshaft that is designed to become an electromagnet when current is passed through it. During operation, the armature rotates through the magnetic fields. At set intervals, the rotation cuts across the magnetic fields, reversing current flow. This process occurs on each half rotation of the armature. The field structure provides a magnetic field for the armature to move through. Magnetic fields are composed of lines of force. In DC motors, the terms field structure, field magnet, and field coils may be used to represent the stationary magnet. When electricity passes through the conductor in a DC motor, the conductor becomes an electromagnet and generates another magnetic field inside the original lines of force. As the twin fields increase in intensity, they strengthen each other and push against the conductor. The current and these strong magnetic fields determine the direction of rotation in a DC motor.

The rotor in an AC motor is a slotted iron core (Figure 6.11). Copper bars are fitted into the slots. Two thick copper rings hold the bars in place. Unlike a DC motor, electric current in an AC motor is not run directly to the rotor. Alternating current flows into the stator, producing a rotating magnetic field. The stator artificially creates an electric current in the rotor, which generates the second magnetic field. When the two fields interact, the rotor turns. A typical AC motor is composed of a stator, field coils, field magnet, rotor, shaft, bearings and seals, conduit box, frame, fan, shroud, and AC power source.

There are three basic designs for DC motors: series, shunt, and compound. The difference between designs is due to the circuit arrangement at the field structure and armature. Series motors are connected in series at the armature and the field magnets. Electric current flows through the field magnet and into the armature. Series motors are characterized by their ability to start up quickly. Motor speed is affected by the size of the load. Shunt motors are connected in parallel as current flows through the armature and magnet. Wire wrapped around the field magnet provides the resistance and determines the strength of the current and magnetic field. Shunt motors are characterized by their ability to run at an even speed regardless of load. Compound motors are connected to the armature with two magnets, one in series and the other in parallel. Compound motors have the characteristics of both the series and shunt motors.

Turbine and Motor Symbols

Turbines and motors can be represented by symbols. Figure 6.12 illustrates steam turbine and motor symbols.

Figure 6.11
AC Motor

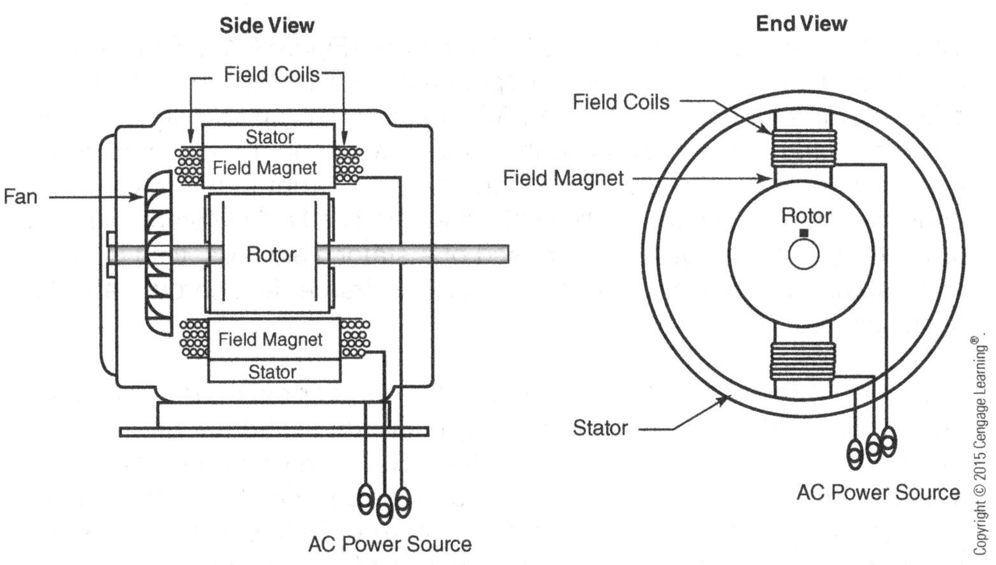

Figure 6.12
Symbols for Steam Turbines and Motors

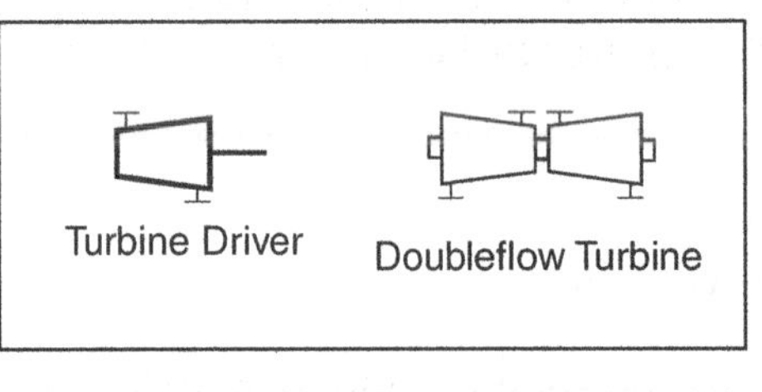

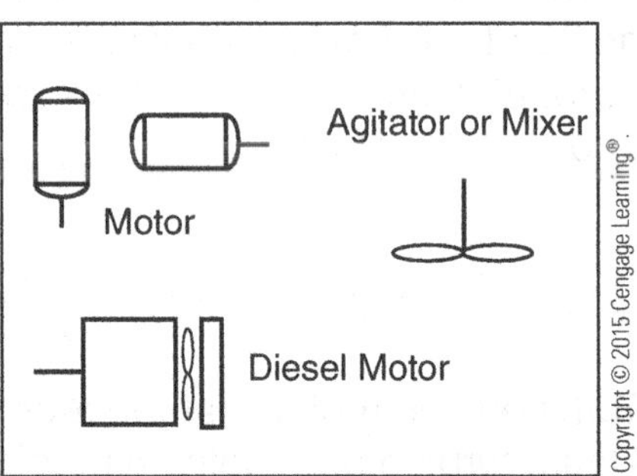

Summary

Turbines convert kinetic energy (energy of movement) to mechanical energy that is used to impart rotation to shaft-driven equipment. Turbines can use the kinetic energy of liquid (hydraulic turbines), wind (windmills), steam, or gas.

As steam enters a steam turbine, it is directed into a nozzle. The nozzle directs the steam onto the blading, which is attached to the shaft. The blading and shaft make up the rotor. Movement occurs as the steam strikes the rotor, converting the steam energy into mechanical energy.

Steam turbines typically are classified into the following categories: condensing (exhaust steam is condensed), noncondensing (exhaust steam is used in low-pressure steam applications), reaction (steam is discharged from a nozzle mounted on the rotor, and movement is a reactive response to the release of steam), impulse (steam acts upon the rotor to create movement), single stage, and multistage. Most plants use the impulse design.

A gas turbine uses high-pressure combustion to turn a series of turbine wheels to provide rotational energy to a driven device in the same way a steam turbine uses steam to turn the rotor. The gas turbine system mixes compressed air with fuel in a combustion chamber. A spark plug ignites the mixture, which is directed into the suction side of the gas turbine. The hot combustion gases rush into the gas turbine, causing the turbine wheels to turn.

Electric motors are used to operate pumps, generators, compressors, fans, blowers, and other equipment. The operation of an electric motor is based upon three scientific principles: Electric current creates a magnetic field; magnetic poles attract and repel each other; and current direction determines the magnetic poles. An electric motor consists of a stationary magnet (stator) and a moving conductor (rotor). A permanent magnetic field is formed by the lines of force between the poles of the magnet.

Review Questions

1. What is a governor valve?
2. Describe heat soaking and slow roll.
3. What does *hunting* mean?
4. Describe the main difference between reaction and impulse steam turbines.
5. What is a labyrinth seal?
6. Describe how the nozzle works on a steam turbine.
7. What is an *overspeed trip*?
8. Explain how a gas turbine operates.
9. Describe the application of safety relief valves to steam turbines.
10. Explain how an electric motor operates.
11. Describe why a throttle valve is important to steam turbine operation.
12. What is the function of a steam strainer?
13. List the components of a steam turbine.
14. What is the primary function of radial and axial bearings?
15. Explain the operation and function of a steam turbine steam chest.
16. Describe the purpose and operation of a governor system.
17. Why is vibration monitoring in a steam turbine important?
18. Describe how a steam turbine is started up.

Heat Exchangers

Objectives

After studying this chapter, the student will be able to:

- Describe the basic principles of fluid flow inside a heat exchanger.
- Explain the methods of heat transfer that apply to heat exchangers.
- Compare the operation of finned and plain tubes.
- List the basic parts of a hairpin (double-pipe) heat exchanger.
- Describe a shell-and-tube, fixed head, single-pass heat exchanger.
- Describe a shell-and-tube, fixed head, multipass heat exchanger.
- Describe a U-tube heat exchanger.
- Describe the operating principles of a kettle and thermosyphon reboiler.
- Describe the types of heat exchangers used on a distillation tower.
- Draw a simple heat exchanger system.
- Describe the basic components and operation of a plate-and-frame heat exchanger.
- Identify the basic components of an air-cooled heat exchanger.
- Explain the operation and design of a spiral heat exchanger.
- Respond to problems associated with heat exchangers.

Key Terms

Baffles—evenly spaced partitions in a shell-and-tube heat exchanger that support the tubes, prevent vibration, control fluid velocity and direction, increase turbulent flow, and reduce hot spots.

Channel head—a device mounted on the inlet side of a shell-and-tube heat exchanger that is used to channel tube-side flow in a multipass heat exchanger.

Condenser—a shell-and-tube heat exchanger used to cool and condense hot vapors.

Conduction—the means of heat transfer through a solid, nonporous material resulting from molecular vibration. Conduction can also occur between closely packed molecules.

Convection—the means of heat transfer in fluids resulting from currents.

Countercurrent—see *Counterflow.*

Counterflow—refers to the movement of two flow streams in opposite directions; also called countercurrent flow.

Crossflow—refers to the movement of two flow streams perpendicular to each other.

Differential pressure—the difference between inlet and outlet pressures; represented as ΔP, or delta p.

Differential temperature—the difference between inlet and outlet temperature; represented as ΔT, or delta t.

Fixed head—a term applied to a shell-and-tube heat exchanger that has the tube sheet firmly attached to the shell.

Floating head—a term applied to a tube sheet on a heat exchanger that is not firmly attached to the shell on the return head and is designed to expand (float) inside the shell as temperature rises.

Fouling—buildup on the internal surfaces of devices such as cooling towers and heat exchangers, resulting in reduced heat transfer and plugging.

Kettle reboiler—a shell-and-tube heat exchanger with a vapor disengaging cavity, used to supply heat for separation of lighter and heavier components in a distillation system and to maintain heat balance.

Laminar flow—streamline flow that is more or less unbroken; layers of liquid flowing in a parallel path.

Multipass heat exchanger—a type of shell-and-tube heat exchanger that channels the tube-side flow across the tube bundle (heating source) more than once.

Parallel flow—refers to the movement of two flow streams in the same direction; for example, tube-side flow and shell-side flow in a heat exchanger; also called *concurrent*.

Radiant heat transfer—conveyance of heat by electromagnetic waves from a source to receivers.

Reboiler—a heat exchanger used to add heat to a liquid that was once boiling until the liquid boils again.

Sensible heat—heat that can be measured or sensed by a change in temperature.

Shell-and-tube heat exchanger—a heat exchanger that has a cylindrical shell surrounding a tube bundle.

Shell side—refers to flow around the outside of the tubes of a shell-and-tube heat exchanger. See also *Tube side*.

Thermosyphon reboiler—a type of heat exchanger that generates natural circulation as a static liquid is heated to its boiling point.

Tube sheet—a flat plate to which the ends of the tubes in a heat exchanger are fixed by rolling, welding, or both.

Tube side—refers to flow through the tubes of a shell-and-tube heat exchanger; see *Shell side*.

Turbulent flow—random movement or mixing in swirls and eddies of a fluid.

Types of Heat Exchangers

Heat transfer is an important function of many industrial processes. Heat exchangers are widely used to transfer heat from one process to another. A heat exchanger allows a hot fluid to transfer heat energy to a cooler fluid through **conduction** and **convection**. A heat exchanger provides heating or cooling to a process. A wide array of heat exchangers has been designed and manufactured for use in the chemical processing industry.

In pipe coil exchangers, pipe coils are submerged in water or sprayed with water to transfer heat. This type of operation has a low heat transfer coefficient and requires a lot of space. It is best suited for condensing vapors with low heat loads.

The double-pipe heat exchanger incorporates a tube-within-a-tube design. It can be found with plain or externally finned tubes. Double-pipe heat exchangers are typically used in series-flow operations in high-pressure applications up to 500 psig **shell side** and 5,000 psig **tube side**.

A **shell-and-tube heat exchanger** has a cylindrical shell that surrounds a tube bundle. Fluid flow through the exchanger is referred to as tube-side flow or shell-side flow. A series of **baffles** support the tubes, direct fluid flow, increase velocity, decrease tube vibration, protect tubing, and create pressure drops. Shell-and-tube heat exchangers can be classified as **fixed head**, single pass; fixed head, multipass; **floating head**, multipass; or U-tube. On a fixed head heat exchanger (Figure 7.1), **tube sheets** are attached to the shell. Fixed head heat exchangers are designed to handle temperature differentials up to 200°F (93.33°C). Thermal expansion prevents a fixed head heat exchanger from exceeding this **differential temperature**. It is best suited for **condenser** or heater operations. Floating head heat exchangers are designed for high temperature differentials

Figure 7.1
Fixed Head Heat Exchanger

above 200°F (93.33°C). During operation, one tube sheet is fixed and the other "floats" inside the shell. The floating end is not attached to the shell and is free to expand.

Reboilers are heat exchangers that are used to add heat to a liquid that was once boiling until the liquid boils again. Types commonly used in industry are **kettle reboilers** and **thermosyphon reboilers**.

Plate-and-frame heat exchangers are composed of thin, alternating metal plates that are designed for hot and cold service. Each plate has an outer gasket that seals each compartment. Plate-and-frame heat exchangers have a cold and hot fluid inlet and outlet. Cold and hot fluid headers are formed inside the plate pack, allowing access from every other plate on the hot and cold sides. This device is best suited for viscous or corrosive fluid slurries. It provides excellent high heat transfer. Plate-and-frame heat exchangers are compact and easy to clean. Operating limits of 350°F to 500°F (176.66°C to 260°C) are designed to protect the internal gasket. Because of the design specification, plate-and-frame heat exchangers are not suited for boiling and condensing. Most industrial processes use this design in liquid-liquid service.

Air-cooled heat exchangers do not require the use of a shell in operation. Process tubes are connected to an inlet and a return header box. The tubes can be finned or plain. A fan is used to push or pull outside air over the exposed tubes. Air-cooled heat exchangers are primarily used in condensing operations where a high level of heat transfer is required.

Spiral heat exchangers are characterized by a compact concentric design that generates high fluid turbulence in the process medium. As do other exchangers, the spiral heat exchanger has cold-medium inlet and outlet and a hot-medium inlet and outlet. Internal surface area provides the conductive transfer element. Spiral heat exchangers have two internal chambers.

The Tubular Exchanger Manufacturers Association (TEMA) classifies heat exchangers by a variety of design specifications including American Society of Mechanical Engineers (ASME) construction code, tolerances, and mechanical design:

- Class B, designed for general-purpose operation *(economy and compact design)*
- Class C, designed for moderate service and general-purpose operation *(economy and compact design)*
- Class R, designed for severe conditions *(safety and durability)*

Heat Transfer and Fluid Flow

The methods of heat transfer are conduction, convection, and **radiant heat transfer** (Figure 7.2). In the petrochemical, refinery, and laboratory environments, these methods need to be understood well. A combination of conduction and convection heat transfer processes can be found in all heat exchangers. The best conditions for heat transfer are large temperature differences between the products being heated and cooled (the higher the temperature difference, the greater the heat transfer), high heating or coolant flow rates, and a large cross-sectional area of the exchanger.

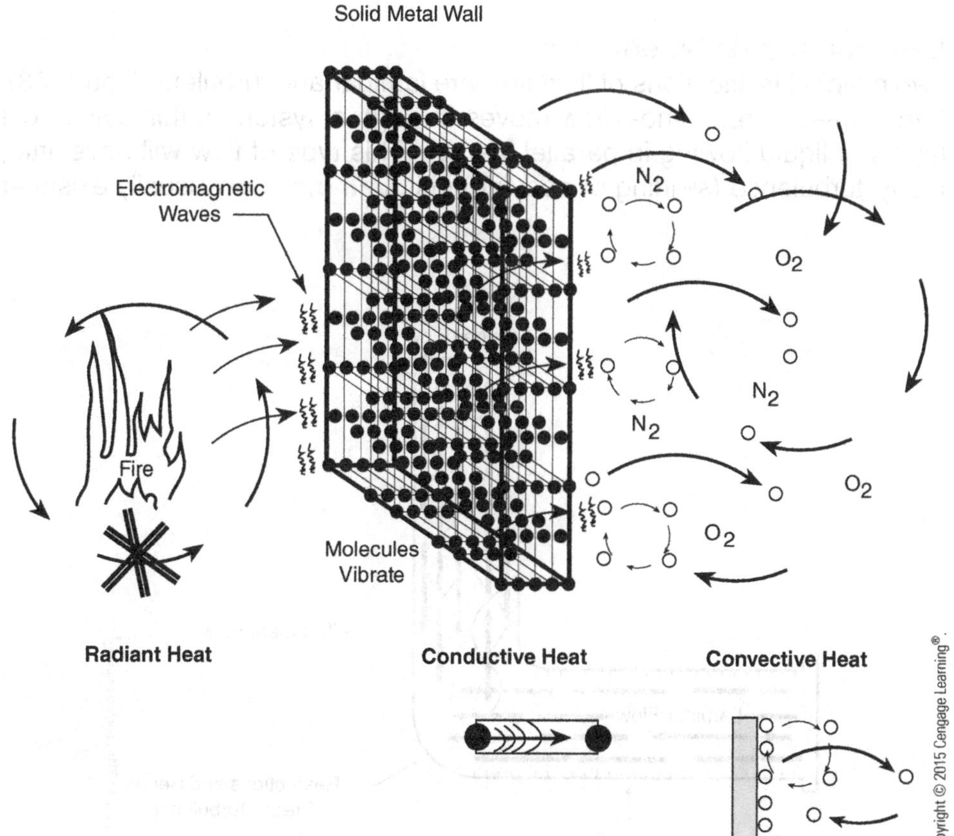

Figure 7.2
Heat Transfer

Conduction

Heat energy is transferred through solid objects such as tubes, heads, baffles, plates, fins, and shell, by conduction. This process occurs when the molecules that make up the solid matrix begin to absorb heat energy from a hotter source. Since the molecules are in a fixed matrix and cannot move, they begin to vibrate and, in so doing, transfer the energy from the hot side to the cooler side.

Convection

Convection occurs in fluids when warmer molecules move toward cooler molecules. The movement of the molecules sets up currents in the fluid that redistribute heat energy. This process will continue until the energy is distributed equally. In a heat exchanger, this process occurs in the moving fluid media as they pass by each other in the exchanger. Baffle arrangements and flow direction will determine how this convective process will occur in the various sections of the exchanger.

Radiant Heat Transfer

The best example of radiant heat is the sun's warming of the earth. The sun's heat is conveyed by electromagnetic waves. Radiant heat transfer is a line-of-sight process, so the position of the source and that of the receiver are important. Radiant heat transfer is not used in a heat exchanger.

Laminar and Turbulent Flow

Two major classifications of fluid flow are laminar and turbulent (Figure 7.3). Laminar—or streamline—flow moves through a system in thin cylindrical layers of liquid flowing in parallel fashion. This type of flow will have little, if any, turbulence (swirling or eddying) in it. **Laminar flow** usually exists at

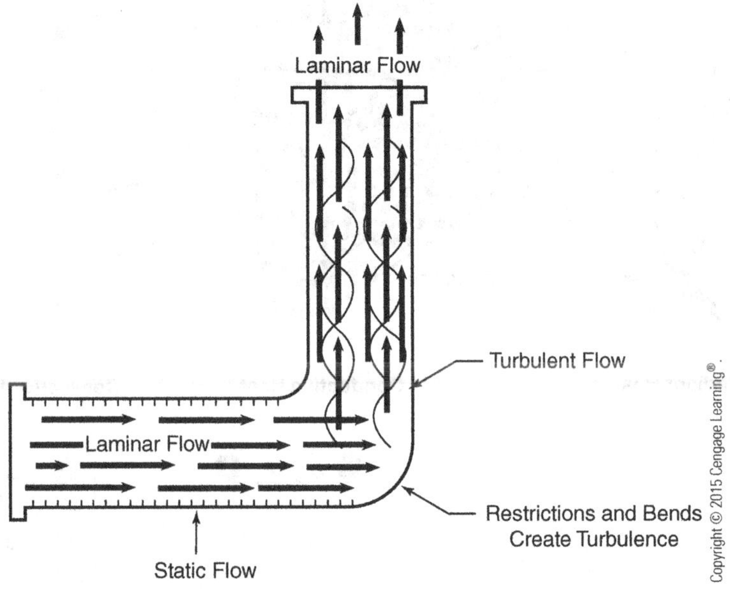

Figure 7.3 *Laminar and Turbulent Flow*

low flow rates. As flow rates increase, the laminar flow pattern changes into a turbulent flow pattern. **Turbulent flow** is the random movement or mixing of fluids. Once the turbulent flow is initiated, molecular activity speeds up until the fluid is uniformly turbulent.

Turbulent flow allows molecules of fluid to mix and absorb heat more readily than does laminar flow. Laminar flow promotes the development of static film, which acts as an insulator. Turbulent flow decreases the thickness of static film, increasing the rate of heat transfer.

Parallel and Series Flow

Heat exchangers can be connected in a variety of ways. The two most common are series and parallel (Figure 7.4). In series flow (Figure 7.5), the tube-side flow in a multipass heat exchanger is discharged into the tube-side flow of the second exchanger. This discharge route could be switched to shell side or tube side depending on how the exchanger is in service. The guiding principle is that the flow passes through one exchanger before it goes to another. In **parallel flow**, the process flow goes through multiple exchangers at the same time.

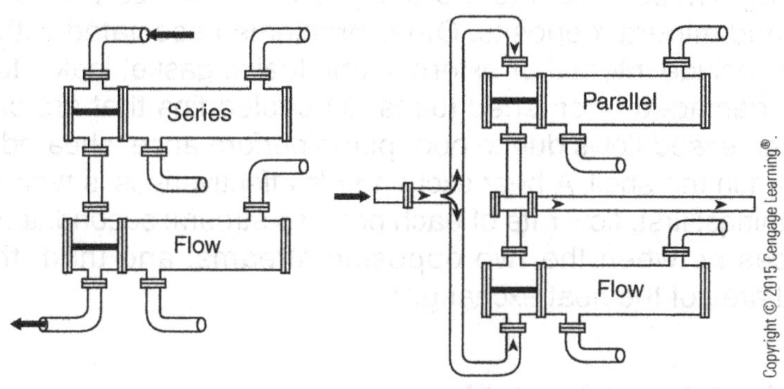

Figure 7.4 Parallel and Series Flow

Figure 7.5 Series Flow Heat Exchangers

Heat Exchanger Effectiveness and Common Problems

The design of an exchanger usually dictates how effectively it can transfer heat energy. **Fouling** is one problem that stops an exchanger's ability to transfer heat. During continual service, heat exchangers do not remain clean. Dirt, scale, and process deposits combine with heat to form restrictions inside an exchanger. These deposits on the walls of the exchanger resist the flow that tends to remove heat and stop heat conduction by insulating the inner walls. An exchanger's fouling resistance depends on the type of fluid being handled, the amount and type of suspended solids in the system, the exchanger's susceptibility to thermal decomposition, and the velocity and temperature of the fluid stream. Fouling can be reduced by increasing fluid velocity and lowering the temperature. Fouling is often tracked and identified using check-lists that collect tube inlet and outlet pressures, and shell inlet and outlet pressures. This data can be used to calculate the pressure differential or Δp. **Differential pressure** is the difference between inlet and outlet pressures, represented as ΔP, or delta p.

Corrosion and erosion are other problems found in exchangers. Chemical products, heat, fluid flow, and time tend to wear down the inner components of an exchanger. Chemical inhibitors are added to avoid corrosion and fouling. These inhibitors are designed to minimize corrosion, algae growth, and mineral deposits. Other problems associated with heat exchangers include internal or external tube leaks, gasket leaks, tube-sheet leakage, damaged or crushed tubes, air-cooled fins that are crushed or fouled, decreased flows due to poor pump performance, sheared tubes, or air trapped in the shell. A heat exchanger's effectiveness is typically linked to three things: first, flow rate of each process stream; second, temperature differences between the two opposing streams; and third, the cross-sectional area of the heat exchanger.

Double-Pipe Heat Exchanger

A simple design for heat transfer is found in a double-pipe heat exchanger. A double-pipe exchanger has a pipe inside a pipe (Figure 7.6). The outside pipe provides the shell, and the inner pipe provides the tube. The warm and cool fluids can run in the same direction (parallel flow) or in opposite directions (counterflow or **countercurrent**).

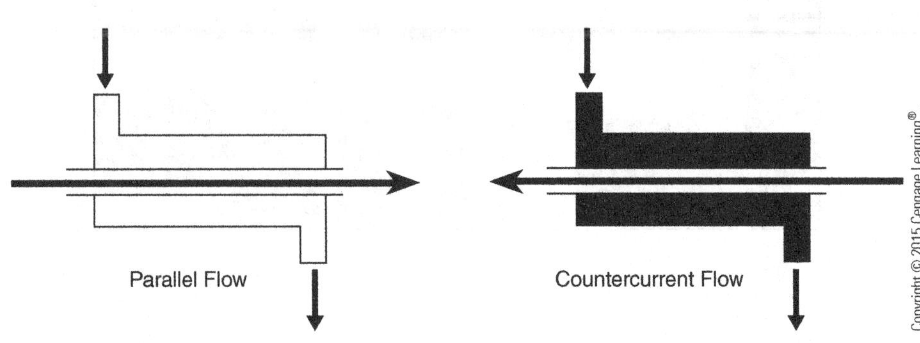

Figure 7.6
Double-Pipe Heat Exchanger

Flow direction is usually countercurrent because it is more efficient. This efficiency comes from the turbulent, against-the-grain, stripping effect of the opposing currents. Even though the two liquid streams never come into physical contact with each other, the two heat energy streams (cold and hot) do encounter each other. Energy-laced, convective currents mix within each pipe, distributing the heat.

In a parallel flow exchanger, the exit temperature of one fluid can only approach the exit temperature of the other fluid. In a countercurrent flow exchanger, the exit temperature of one fluid can approach the inlet temperature of the other fluid. Less heat will be transferred in a parallel flow exchanger because of this reduction in temperature difference. Static films produced against the piping limit heat transfer by acting like insulating barriers. The liquid close to the pipe is hot, and the liquid farthest away from the pipe is cooler. Any type of turbulent effect would tend to break up the static film and transfer heat energy by swirling it around the chamber. Parallel flow is not conducive to the creation of turbulent eddies.

One of the system limitations of double-pipe heat exchangers is the flow rate they can handle. Typically, flow rates are very low in a double-pipe heat exchanger, and low flow rates are conducive to laminar flow.

Hairpin Heat Exchangers

The chemical processing industry commonly uses hairpin heat exchangers (Figure 7.7). Hairpin exchangers use two basic modes: double-pipe and multipipe design. Hairpins are typically rated at 500 psig shell side and 5,000 psig tube side. The exchanger takes its name from its unusual hairpin shape. The double-pipe design consists of a pipe within a pipe. Fins can be added to the internal tube's external wall to increase heat transfer. The multipipe hairpin resembles a typical shell-and-tube heat exchanger, stretched and bent into a hairpin.

The hairpin design has several advantages and disadvantages. Among its advantages are its excellent capacity for thermal expansion because of

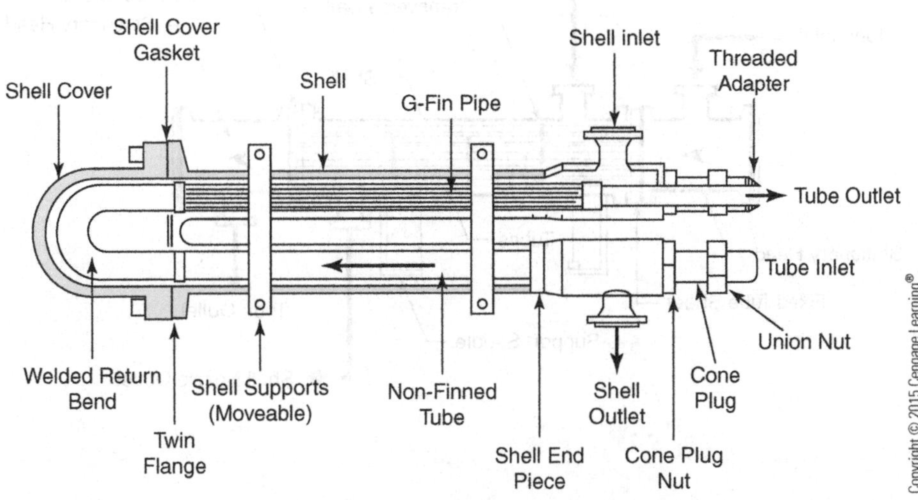

Figure 7.7 *Hairpin Heat Exchanger*

its U-tube type shape; its finned design, which works well with fluids that have a low heat transfer coefficient; and its high pressure on the tube side. In addition, it is easy to install and clean; its modular design makes it easy to add new sections; and replacement parts are inexpensive and always in supply. Among its disadvantages are the facts that it is not as cost-effective as most shell-and-tube exchangers and it requires special gaskets.

Shell-and-Tube Heat Exchangers

The shell-and-tube heat exchanger is the most common style found in industry. Shell-and-tube heat exchangers are designed to handle high flow rates in continuous operations. Tube arrangement can vary, depending on the process and the amount of heat transfer required. As the tube-side flow enters the exchanger—or "head"—flow is directed into tubes that run parallel to each other. These tubes run through a shell that has a fluid passing through it. Heat energy is transferred through the tube wall into the cooler fluid. Heat transfer occurs primarily through conduction (first) and convection (second). Figure 7.8 shows a fixed head, single-pass heat exchanger.

Fluid flow into and out of the heat exchanger is designed for specific liquid-vapor services. Liquids move from the bottom of the device to the top to remove or reduce trapped vapor in the system. Gases move from top to bottom to remove trapped or accumulated liquids. This standard applies to both tube-side and shell-side flow.

Designs and Components

Exchanger nomenclature uses the terms *front end*, *shell* or *middle section*, and *rear end* to refer to the three parts of shell-and-tube heat exchangers. The front-end design of a heat exchanger varies depending on the type of service in which it will be used. The shell has seven popular designs that are linked to the way flow moves through the shell. The rear-end section of a heat exchanger is linked to the front-end design. Industrial manufacturers are currently using over nine popular designs.

Figure 7.8
Fixed Head, Single-Pass Heat Exchanger

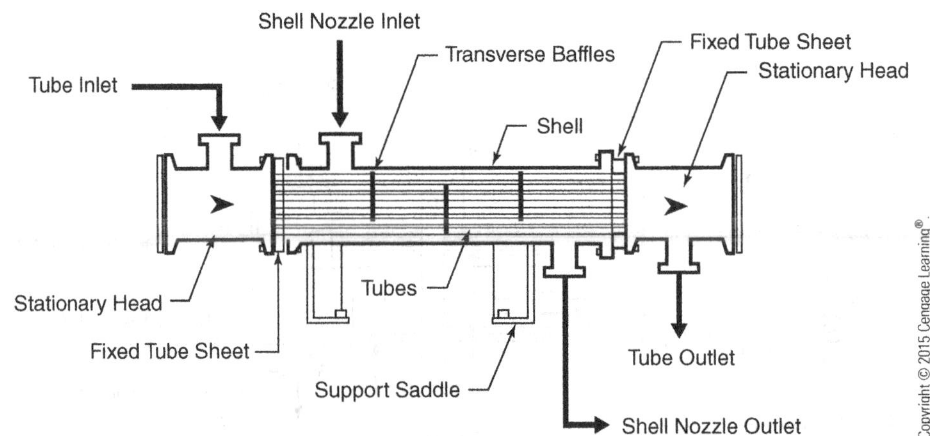

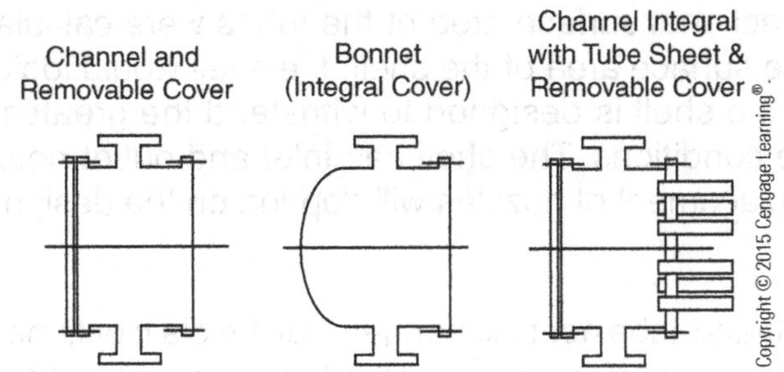

Figure 7.9
Head Designs

Head

The heads (Figure 7.9) on a shell-and-tube heat exchanger can be classified as front-end or rear-end types. The front-end head has five primary designs: (1) channel and removable cover; (2) bonnet; (3) channel integral with the tube sheet and removable cover (removable tube bundle); (4) channel integral with the tube sheet and removable cover (fixed to shell); and (5) special high-pressure closure. The rear-end (or return) header has eight possible designs: (1) fixed tube sheet with channel and removable cover; (2) fixed tube sheet with bonnet; (3) channel integral with the tube sheet and removable cover (fixed to shell); (4) outside packed floating head; (5) floating head with backing device; (6) pull-through floating device; (7) U-tube bundle; and (8) externally sealed floating tube sheet.

Shell

The shell can be classified as single pass, double pass, split flow, double-split flow, divided flow, kettle, or cross flow (Figure 7.10). The shell is designed to operate at a specific temperature and pressure, which are clearly marked on the manufacturer's code stamp plate. Process technicians can determine the type of shell flow by the positions of the inlet and outlet ports. The shell is the largest single part of the heat exchanger, but

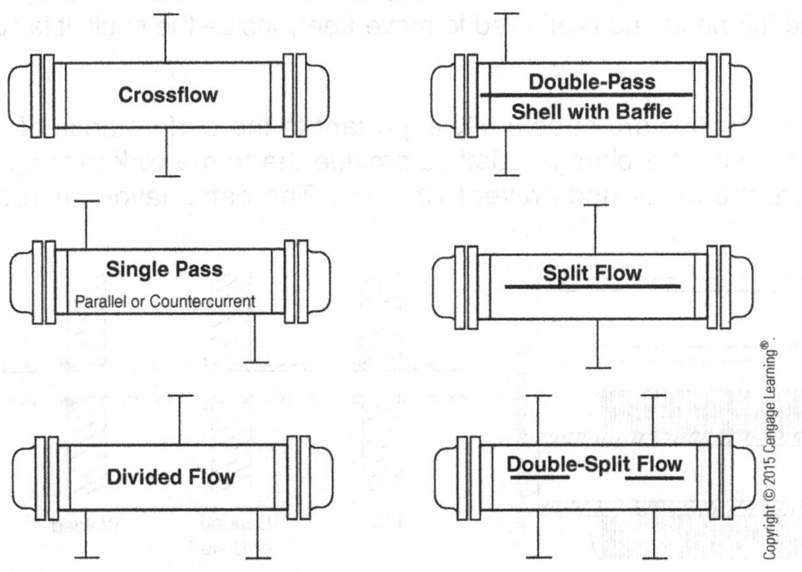

Figure 7.10
Shell Designs

if the cross-sectional surface area of the tubes were calculated and compared with the surface area of the shell, the shell would look very small. In most cases, the shell is designed to withstand the greatest temperature and pressure conditions. The shell has inlet and outlet nozzles. The total number and placement of nozzles will depend on the design.

Tubes

Tubes on shell-and-tube heat exchangers can be plain or finned (Figure 7.11). Fins provide more surface area and allow greater heat transfer to take place. Fins can be located externally or internally. Although plain tubes are more commonly used in fabrication, the enhanced features of the finned tube are starting to make an impact on new design engineers. Tube materials include brass, carbon, carbon steel, copper, cupronickel, glass, stainless steel, specialty alloys, Monel, nickel, and tantalum.

Tube Sheet

Tube sheets are often described as fixed or floating, single or double. A tube sheet is a flat plate to which the ends of the tubes in a heat exchanger are fixed by rolling, welding, or both. Tube sheets have carefully drilled holes designed to admit the end of a tube and secure it to the plate. Double tube sheets are used to prevent tube-side leakage of highly corrosive fluids. The space between the plates provides a void where these hazardous materials can be safely removed from the process stream. Tube sheet connections are identified as plain, rolled, beaded or belled, flared, or welded (Figure 7.12). Some connections are both rolled and welded. A duplex tube (tube-inside-a-tube) can be beaded or belled, plain or flared. During operation, the tubes will expand. This expansion creates a problem within a fixed head design. Engineering specifications take into account thermal tube expansion. The term *fixed tube sheet* applies to the way the tube sheet is located in the inlet or return head. If the tube sheet is welded or bolted to the shell, it is *fixed*. If the tube sheet is independently secured to the tub head and is allowed to move freely inside the shell, it is *floating*.

Baffles

Internal baffles are structurally important to the performance of a shell-and-tube heat exchanger. Baffles provide the framework to support and secure the tubes and prevent vibration. The baffle layout increases or

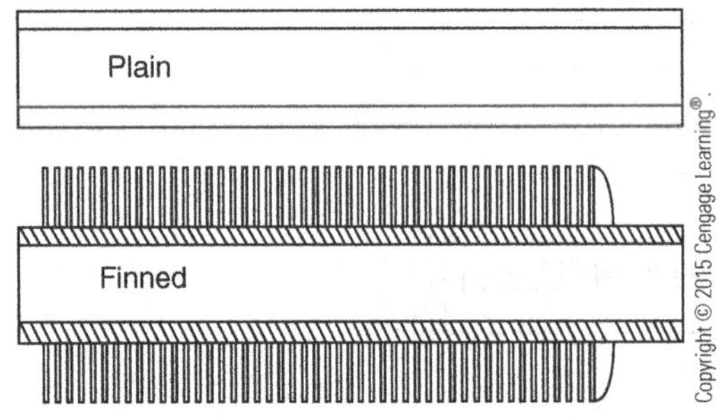

Figure 7.11 *Plain and Finned Tubes*

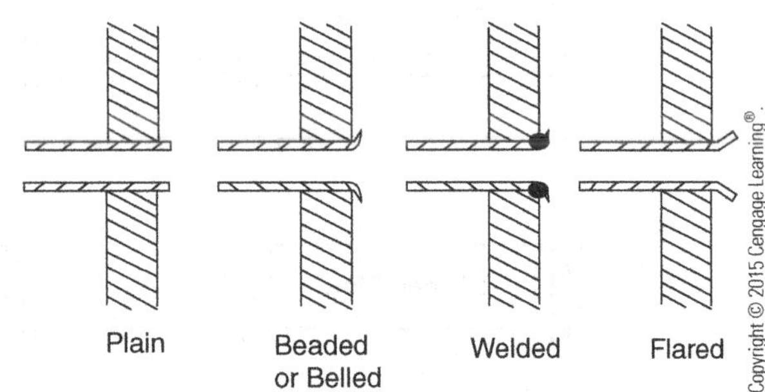

Figure 7.12 *Tube Sheet Connections*

decreases fluid and directs flow at specific points. Tube-side baffles, or pass partitions, are built into the heads to direct tube-side flow. Tube-side baffles may be cast or welded in place. Single-pass exchangers do not need a baffle in the inlet or return head. Multipass exchangers requiring two passes will have a single baffle in the inlet **channel head**. A variety of baffle arrangements are available. Cost goes up with each pass. Additional passes are often needed to provide adequate fluid velocities to prevent fouling (internal buildup of material) and to control heat transfer.

Segmental baffles (Figure 7.13) are often used in horizontal shell-and-tube heat exchangers. The holes in the baffle are drilled to fit the size of the tube. Without support, tubes will vibrate under pressure. Each segmental baffle supports half of the tubes. Baffles are evenly spaced and alternated from one side to the other to support the tube bundle and direct fluid flow. Segmental baffles may be horizontal or vertical cut. The choice of which arrangement to use is based on the required service. For example, a vertical arrangement is typically used in horizontal exchangers used as condensers, reboilers, or vaporizers. Systems transferring large quantities of suspended solids may also use this design. The vertical design allows liquid and solids to flow around baffles.

Horizontal baffles are used in vapor-phase or all-liquid-phase operations. This type of arrangement is not used where entrained gases are trapped in the liquid unless V-notches are cut in the bottom of the baffle. Horizontal baffles are used in clean service with notches at the bottom to allow liquid drainage on removal from service.

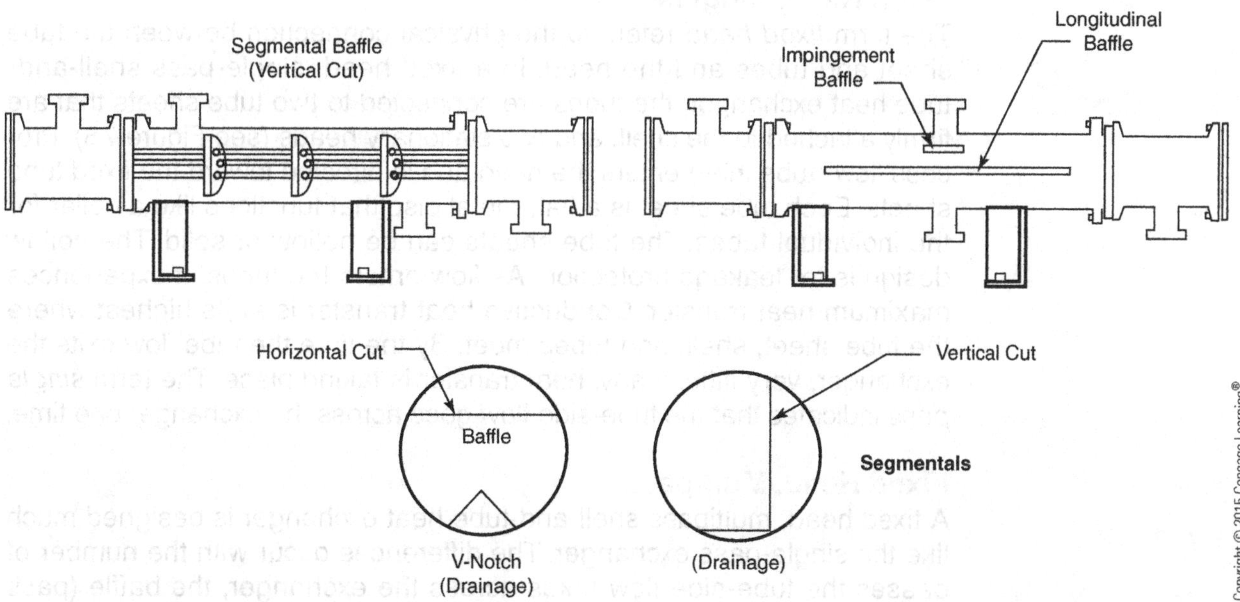

Figure 7.13 *Baffle Arrangements*

Impingement baffles are used to protect tubing from direct fluid impact. In some systems, high-pressure steam is admitted into the shell side. An impingement baffle, placed over the tubes, will deflect the steam as it enters the exchanger, thereby preventing cutting, pitting, and erosion problems in the tubes.

Longitudinal baffles are used inside the shell to split or divide the flow, increase velocity, and provide superior heat transfer capabilities. This type of baffle can be welded in place, slid into a slot, or situated with special packing. Longitudinal baffles do not extend the entire length of the exchanger because at some point the fluid must flow around it.

Tie Rods

Tie rods and concentric tube spacers keep the baffles in place and evenly spaced. Each hole in the baffle plates is 1/64 inch larger than the tube's outside diameter. Tube vibrations on the leading edge of the baffle will eventually damage the tube. Tie rods hold the baffles in place and prevent vibration and excessive tube movement.

Nozzles and Accessory Parts

Shell-and-tube inlet and outlet nozzles are sized for pressure drop and velocity considerations. Nozzle connections frequently have thermowells (a chamber that houses temperature-sensing devices) and pressure indicator connections. Safety and relief valves are located in required areas around the exchanger. Product drains are used to empty the sections between baffles during maintenance. Vents are located on the upper side of the shell to remove gases and vapors. Block valves and control valves are located in the piping entering and leaving the exchanger.

Fixed Head, Single Pass

The term *fixed head* refers to the physical connection between the tube sheet and tubes and the head. In a fixed head, single-pass shell-and-tube heat exchanger, the tubes are connected to two tube sheets that are firmly attached to the shell, and two stationary heads (see Figure 7.8). Process flow (tube inlet) enters the head and is directed toward the fixed tube sheets. Each tube sheet is a flat, metal disc that functions like a collar for the individual tubes. The tube sheets can be hollow or solid. The hollow design is for leakage protection. As flow enters the tubes, it experiences maximum heat transfer. Conductive heat transfer is at its highest where the tube sheet, shell, and tubes meet. By the time the tube flow exits the exchanger, very little, if any, heat transfer is taking place. The term *single pass* indicates that the tube-side flow goes across the exchanger one time.

Fixed Head, Multipass

A fixed head, multipass shell-and-tube heat exchanger is designed much like the single-pass exchanger. The differences occur with the number of passes the tube-side flow takes across the exchanger, the baffle (pass

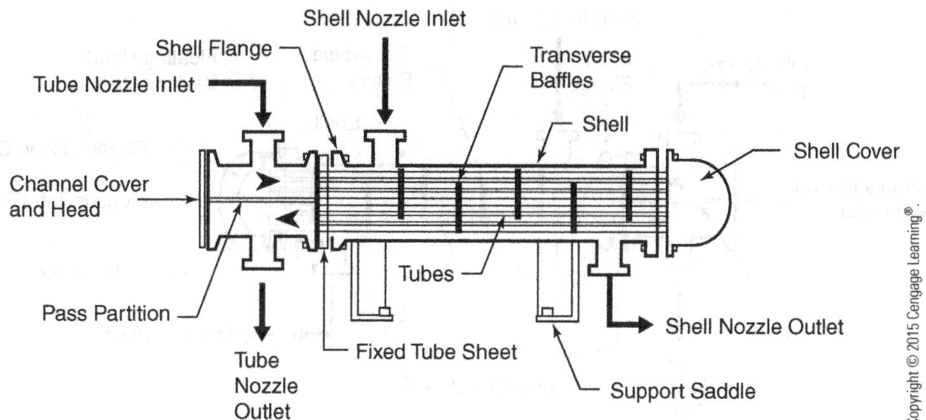

Figure 7.14 *Fixed Head, Multipass Heat Exchanger*

partition) added to the channel head, and the lack of a tube-side outlet on the discharge head (Figure 7.14).

In a fixed head, **multipass heat exchanger**, flow enters the channel head and is directed into the tubes. A baffle installed in the head limits access to a portion of tubes on the tube sheet. As fluid flows through the exchanger, heat is transferred into or out of the fluid. After completing the first pass, process flow is directed back into another portion of tubes. This second pass across the exchanger allows additional heat transfer to occur.

As tube-side flow moves through the exchanger, it encounters a variety of flow variations from the shell side. Since heat transfer in a shell-and-tube exchanger occurs primarily through conduction and convection, the hotter fluid will influence the cooler. At various points, the tube and shell flows run parallel—that is, as **counterflow**, which is also called **crossflow**. Baffle arrangement influences the directions heat transfer and fluid flow take. The basic heat transfer relation is $Q = UA$, where Q is the heat duty, in Btu per hour; U is the overall heat transfer coefficient, in Btu per hour per square foot of surface; and A is the area available for heat transfer, in square feet. Counterflow operation provides more heat transfer than parallel flow.

Floating Head

In a floating head, multipass shell-and-tube heat exchanger, one side of the tube bundle is fixed to the channel head, the other side is unsecured, or floating (Figure 7.15). Flow enters the channel head and is directed into the tubes that are attached to a common, fixed, tube sheet. As flow moves from left to right, it makes one pass before it crosses right to left for the second pass. A network of baffles is established on the tube bundle to enhance heat transfer. Adding fins to the tubes can further enhance heat transfer. An impingement baffle (pass partition) is located between the tubes and shell inlet. This redirects the flow and keeps the tubes from being damaged. This

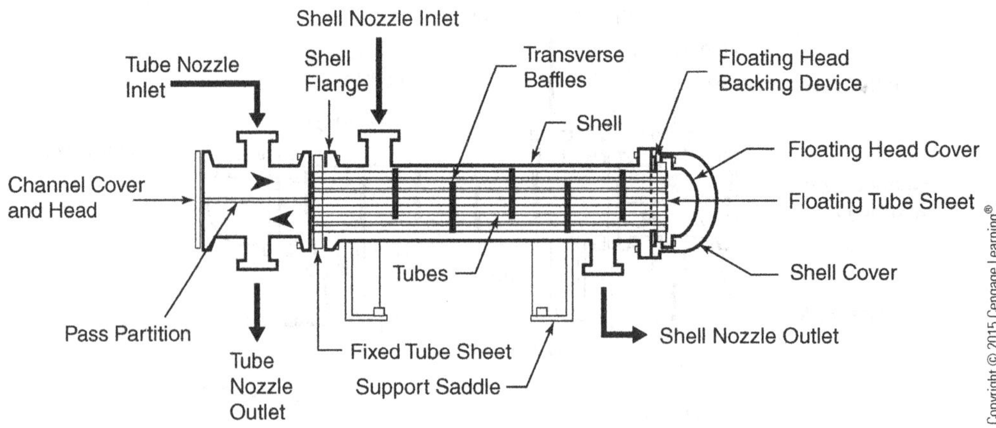

Figure 7.15 *Floating Head, Multipass Heat Exchanger*

type of heat exchange produces the highest heat transfer efficiency. Floating head exchangers, with their high cross-sectional areas (fins), are designed for high temperature differentials and high flow rates.

U-Tube

A fixed tube sheet on one end that is typically bolted to the shell characterizes a U-tube exchanger (Figure 7.16). The tube sheet connects a series of tubes bent in a U-shape (Figure 7.17). The ends of the tubes are secured to the tube sheet. This design limits the total number of tubes that can be used when compared with a fixed head. A channel head directs tube flow across the body of the exchanger twice. U-tube exchangers are specially designed for large temperature differentials. The U-shaped design allows the head to float and accommodate the thermal expansion of the tubes. Each complete U-tube has a single fundamental frequency as flow passes over it. Segmental baffles placed at equal distances provide the support and framework that bond the tubular bundle into a single unit. Longitudinal baffles may be used to direct fluid flow.

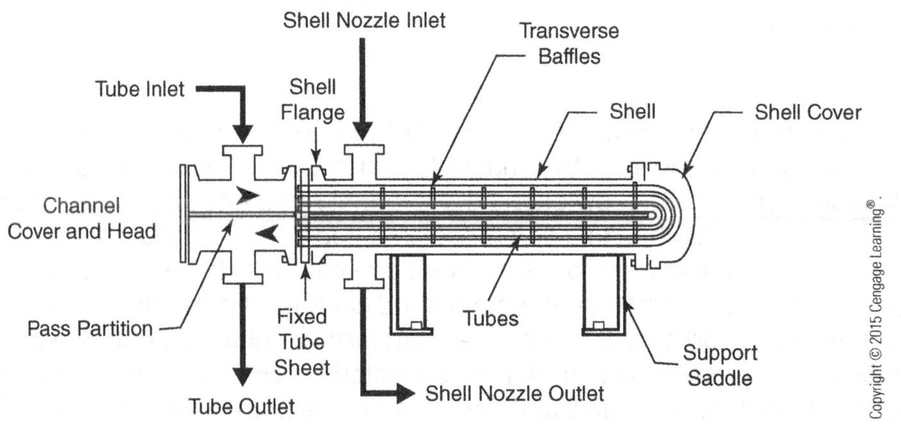

Figure 7.16 *U-Tube Heat Exchanger*

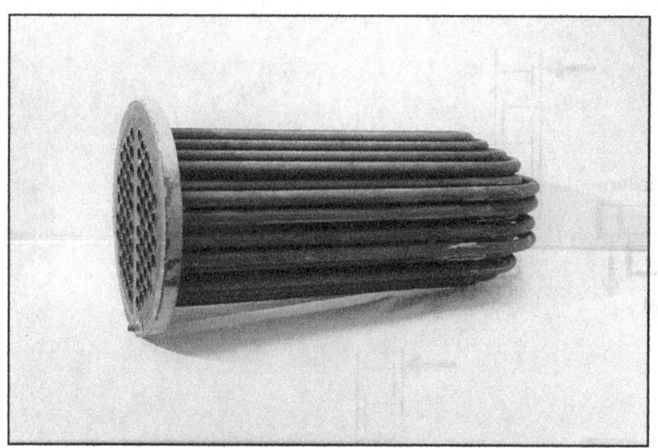

Figure 7.17 *U-Tube*

Reboilers

Reboilers are used to add heat to a liquid that was once boiling until the liquid boils again. Reboilers are closely associated with the operation of a distillation column. Typical reboiler arrangements include five basic patterns: flooded-tube kettle reboiler, natural circulation, forced circulation, vertical thermosyphon, and horizontal thermosyphon (Figure 7.18). These types of devices are classified by how they produce fluid flow. If a mechanical device, such as a pump, is used, the reboiler is referred to as a forced circulation reboiler. Circulation that does not require a pump is classified as natural circulation.

Kettle Reboiler

Kettle reboilers are shell-and-tube heat exchangers designed to produce a two-phase, vapor-liquid mixture that can be returned to a distillation column (Figure 7.19). Kettle reboilers have a removable tube bundle that uses steam or a high-temperature process medium to boil the fluid. A large vapor cavity above the heated process medium allows vapors to concentrate. Liquid that does not vaporize flows over a weir and into the liquid outlet. Hot vapors are sent back to the distillation column through the reboiler's vapor outlet ports. This process controls the level in the bottom of the distillation column, maintains product purity, strips smaller hydrocarbons from larger ones, and helps maintain the critical energy balance on the column.

Kettle reboilers operate with liquid levels from 2 inches above and 2 inches below the upper tubes. Engineering designs typically allow 10 inches to 12 inches of vapor space above the tube bundle. Vapor velocity exiting the reboiler must be low enough to prevent liquid entrainment. Bottom product spills over the weir that fixes the liquid level on the tube bundle.

An important concept with a distillation column is energy or heat balance. Reboilers are used to restore this balance by adding additional heat for the

Figure 7.18
Reboiler Arrangements

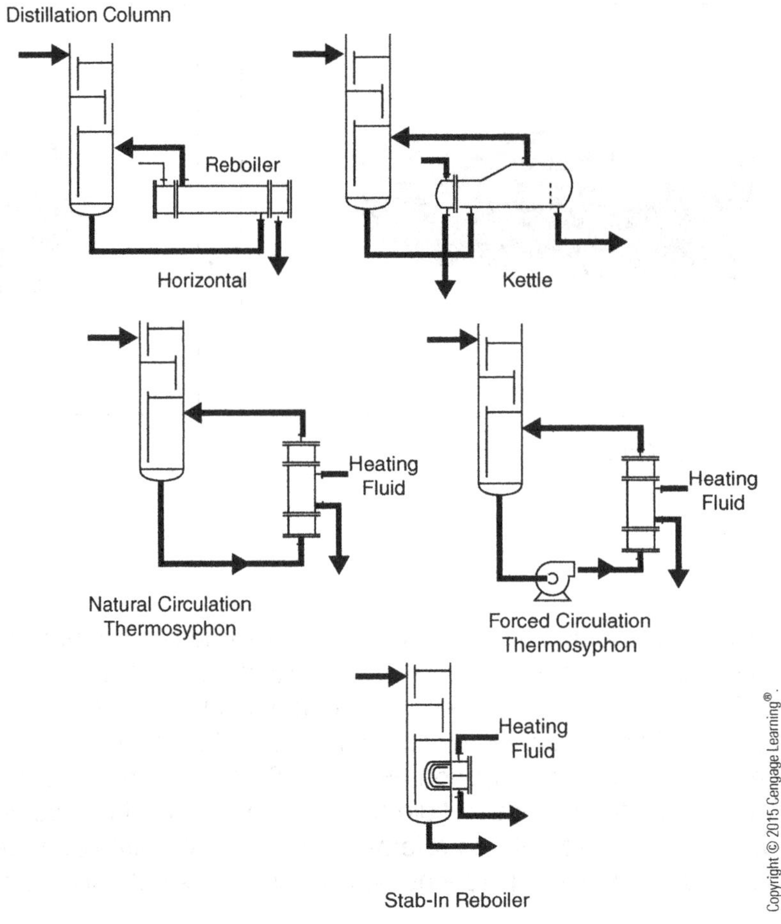

Figure 7.19
Kettle Reboiler

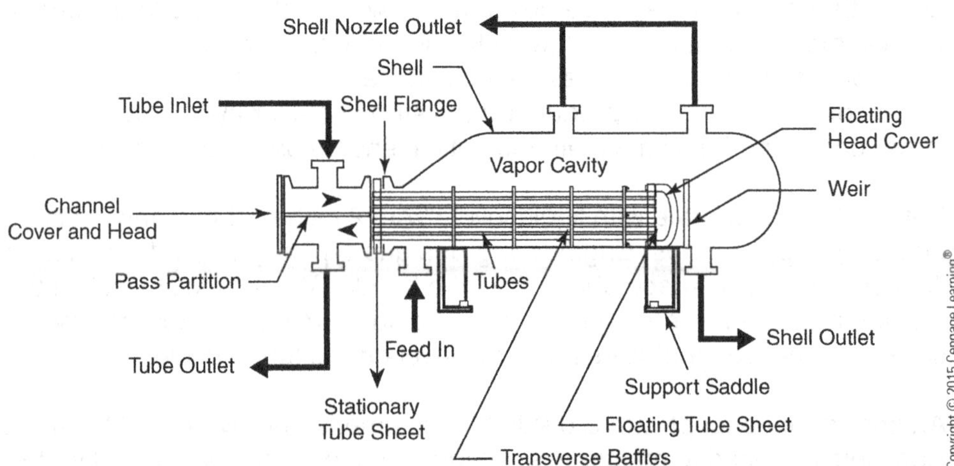

separation processes. Bottom products typically contain the heavier components from the tower. Reboilers take suction off of the bottom products and pump them through their system. Column temperatures are controlled at established set-points.

Product flow enters the bottom shell side of a reboiler. As flow enters the reboiler, it comes into contact with the tube bundle. The tubes have steam or hot fluids flowing through them. As the bottom product comes into contact with the tubes, a portion of the liquid is flashed off (vaporized) and captured in the dome-shaped vapor cavity at the top of the reboiler shell. This vapor is sent back to the tower for further separation. A weir contains the unflashed portion of the liquid in a reboiler. Excess flow goes over the weir and is recirculated through the system. Kettle reboilers are easy to control because circulation and two-phase flow rates are not considerations.

Vertical and Horizontal Thermosyphon Reboilers

A thermosyphon reboiler is a fixed head, single-pass heat exchanger connected to the side of a distillation column. Thermosyphon heat exchangers can be mounted vertically or horizontally. The critical design factor is providing sufficient liquid head in the column to support vapor or liquid flowback to the column. Natural circulation occurs because of the differences in density between the hotter liquid in the reboiler and the liquid in the distillation tower. One side of the exchanger is used for heating, usually with steam or hot oil; the other side takes suction off the column. When steam is used as the heated medium in a vertical exchanger, it enters from the top shell inlet and flows downward to the shell outlet, to allow for the removal of condensate. The lower tube inlet of the exchanger usually takes suction at a point low enough on the column to provide a liquid level to the exchanger. A pump is not connected to the column and exchanger unless a forced circulation system is required. This system uses buoyancy forces to flash off and pull in liquid. Newton's third law of motion, which states that for every action there is an equal and opposite reaction, is a basic operating principle of thermosyphon reboilers. As liquids and vapor circulate back to the column, the inlet line provides fresh liquid to support the circulation.

Stab-In Reboiler

The stab-in reboiler is mounted directly into the base of the distillation column. Steam or hot oil is used as the heating medium. Heat energy is transferred directly into the process medium. The lower section on a distillation column is specially designed to allow the bottom product to boil. This lower section maintains a liquid seal as hot vapors move up the column and heavy liquids collect in the bottom.

Hot Oil Jacket Reboiler

Some reboilers have specially designed hot oil jackets surrounding the bottom of the column. In this type of service, hot oil enters the outer shell and provides heat to the bottom product primarily through conduction and convection.

The outer jacket functions like a heat exchanger as hot fluid circulates through the shell. This type of system is used on smaller distillation systems.

Plate-and-Frame Heat Exchangers

Plate-and-frame heat exchangers are high heat transfer and high pressure drop devices. They consist of a series of gasketed plates, sandwiched together by two end plates and compression bolts (Figures 7.20 and 7.21). The channels between the plates are designed to create pressure drop

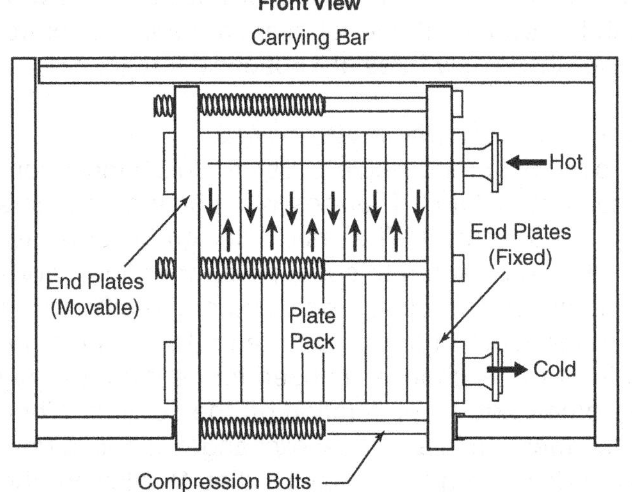

Figure 7.20 *Plate-and-Frame Heat Exchanger*

Figure 7.21 *Plate-and-Frame Assembly*

and turbulent flow so high heat transfer coefficients can be achieved. The openings on the plate exchanger are located typically on one of the fixed-end covers.

As hot fluid enters the hot inlet port on the fixed-end cover, it is directed into alternating plate sections by a common discharge header. The header runs the entire length of the upper plates. As cold fluid enters the countercurrent cold inlet port on the fixed-end cover, it is directed into alternating plate sections. Cold fluid moves up the plates while hot fluid drops down across the plates. The thin plates separate the hot and cold liquids, preventing leakage.

Fluid flow passes across the plates one time before entering the collection header. The plates are designed with an alternating series of chambers. Heat energy is transferred through the walls of the plates by conduction and into the liquid by convection. The hot and cold inlet lines run the entire length of the plate heater and function like a distribution header. The hot and cold collection headers run parallel and on the opposite side of the plates from each other. The hot fluid header that passes through the gasketed plate heat exchanger is located at the top. This arrangement accounts for the pressure drop and turbulent flow as fluid drops over the plates and into the collection header. Cold fluid enters the bottom of the gasketed plate heat exchanger and travels countercurrent to the hot fluid. The cold fluid collection header is located in the upper section of the exchanger.

Plate-and-frame heat exchangers have several advantages and disadvantages. They are easy to disassemble and clean and distribute heat evenly so there are no hot spots. Plates can easily be added or removed. Other advantages of plate-and-frame heat exchangers are their low fluid resistance time, low fouling, and high heat transfer coefficient. In addition, if gaskets leak, they leak to the outside, and gaskets are easy to replace. The plates prevent cross-contamination of products. Plate-and-frame heat exchangers provide high turbulence and a large pressure drop and are small compared with shell-and-tube heat exchangers.

Disadvantages of plate-and-frame heat exchangers are that they have high-pressure and high-temperature limitations. Gaskets are easily damaged and may not be compatible with process fluids.

Spiral Heat Exchangers

Spiral heat exchangers are characterized by a compact concentric design that generates high fluid turbulence in the process medium (Figure 7.22). This type of heat exchanger comes in two basic types: (1) spiral flow on both sides and (2) spiral flow–crossflow. Type 1 spiral exchangers are used in liquid-liquid, condenser, and gas cooler service. Fluid flow into the exchanger is designed for full counterflow operation. The horizontal axial installation provides excellent self-cleaning of suspended solids.

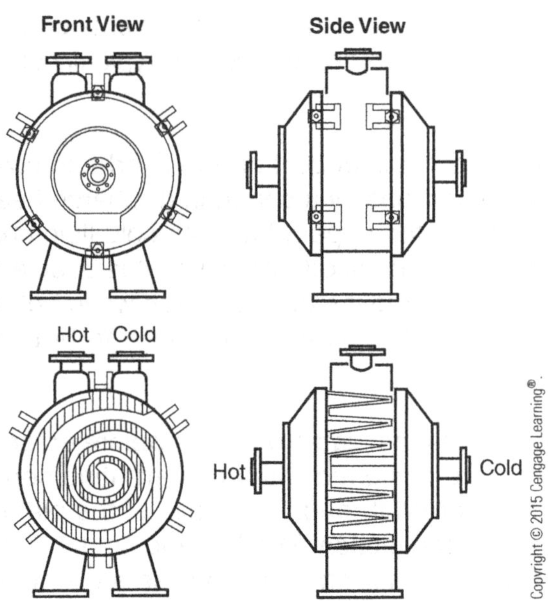

Figure 7.22 *Spiral Heat Exchanger*

Type 2 spiral heat exchangers are designed for use as condensers, gas coolers, heaters, and reboilers. The vertical installation makes it an excellent choice for combining high liquid velocity and low pressure drop on the vapor-mixture side. Type 2 spirals can be used in liquid-liquid systems where high flow rates on one side are offset by low flow rates on the other.

Air-Cooled Heat Exchangers

A different approach to heat transfer occurs in the fin fan or air-cooled heat exchanger. Air-cooled heat exchangers provide a structured matrix of plain or finned tubes connected to an inlet and return header (Figure 7.23). Air is used as the outside medium to transfer heat away from the tubes. Fans are used in a variety of arrangements to apply forced convection for heat

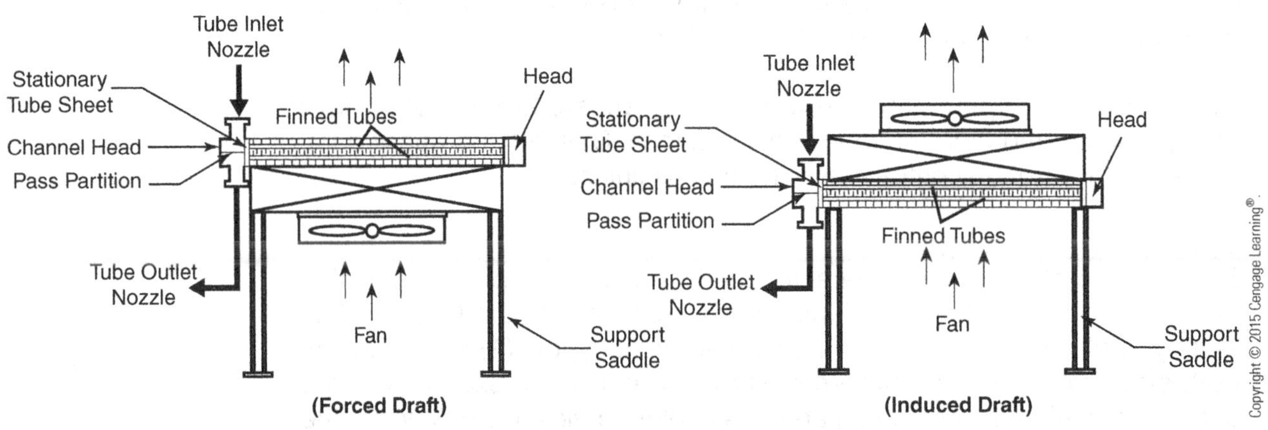

Figure 7.23 *Air-Cooled Heat Exchanger*

transfer coefficients. Fans can be mounted above or below the tubes in forced-draft or induced-draft arrangements. Tubes can be installed vertically or horizontally.

The headers on an air-cooled heat exchanger can be classified as cast box, welded box, cover plate, or manifold. Cast box and welded box types have plugs on the end plate for each tube. This design provides access for cleaning individual tubes, plugging them if a leak is found, and rerolling to tighten tube joints. Cover plate designs provide easy access to all of the tubes. A gasket is used between the cover plate and head. The manifold type is designed for high-pressure applications.

Mechanical fans use a variety of drivers. Common drivers found in service with air-cooled heat exchangers include electric motor and reduction gears, steam turbine or gas engine, belt drives, and hydraulic motors. The fan blades are composed of aluminum or plastic. Aluminum blades are designed to operate in temperatures up to 300°F (148.88°C), whereas plastic blades are limited to air temperatures between 160°F and 180°F (71.11°C, 82.22°C).

Air-cooled heat exchangers can be found in service on air compressors, in recirculation systems, and in condensing operations. This type of heat transfer device provides a 40°F (4.44°C) temperature differential between the ambient air and the exiting process fluid.

Air-cooled heat exchangers have none of the problems associated with water such as fouling or corrosion. They are simple to construct and cheaper to maintain than water-cooled exchangers. They have low operating costs and superior high temperature removal (above 200°F or 93.33°C).

Their disadvantages are that they are limited to liquid or condensing service and have a high outlet fluid temperature and high initial cost of equipment. In addition, they are susceptible to fire or explosion in cases of loss of containment.

Heat Exchangers and Systems

A heat exchanger system includes two or more heat exchangers working in series or parallel to raise or lower the temperature of a process stream. Heat exchanger systems may also include cooling towers, furnaces, distillation columns, reactors, hot oil or steam systems, pipes, pumps, valves, and complex process instruments.

Heat Transfer System

Heat exchangers are commonly used to transfer heat energy between two separate flows. In Figure 7.24 two heat exchangers are shown that heat the feed before it enters a distillation column. Feed enters the shell side of

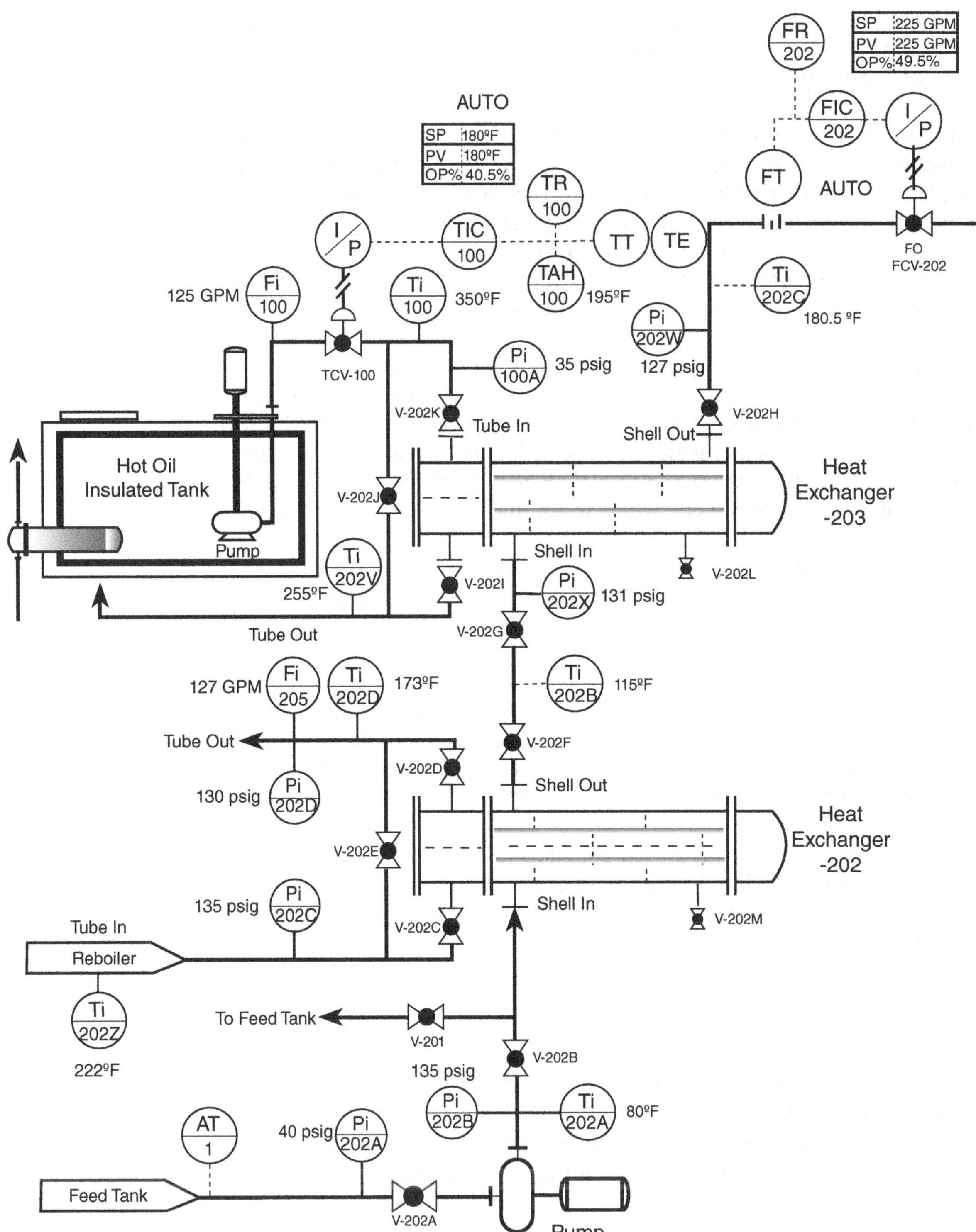

Figure 7.24 *Heat Exchanger System*

the first exchanger at 80°F (26.66°C) and exits the shell at 115°F (46.11°C). Exchanger 202 has a longitudinal baffle running through the center of the shell. This partition forces the feed through a series of lower baffles to pass across the body of the heat exchanger one time before entering the upper section of the shell and moving back across the body of the exchanger and through another series of baffles before exiting through the shell outlet.

The tube inlet on exchanger 202 has a feed temperature of 222°F (105.55°C) as it enters the channel head and passes through the lower tube sheet and into the tubes. As the feed flows through the tubes, it transfers heat energy into the cooler shell product. Heat transfer is primarily through conduction and convection. During the heat transfer process, the temperature on the reboiler feeds drops from 222°F to 173°F (105.55°C to 78.33°C). The differential temperature (Δt) is 49 and the difference in the tube inlet pressure at 135 psig and the tube outlet pressure at 130 psig is (Δp) 5. Process technicians carefully monitor these differences over extended run times.

Heat exchanger 203 also has a tube inlet and a tube outlet as well as a shell inlet and outlet. Pressure and temperature are carefully monitored and tracked on checklists and statistical process control charts. On the tube side, a hot oil system is used to transfer heat energy to the shell feed. The flow rate through the shell side is controlled at 225 GPM (gallons per minute). During operation the following variables are very important:

Ex-202

- Shell inflow rate 225 GPM @ 80°F (26.66°C) @ 135 psig
- Shell outflow rate 225 GPM @ 115°F (46.11°C) @ 131 psig
- Shell $\Delta p = 4$
- Tube inflow rate 127 GPM @ 222°F (105.55°C) @ 135 psig
- Tube outflow rate 127 GPM @ 173°F (78.33°C) @ 130 psig
- Tube $\Delta t = 49$
- Pump $\Delta p = 95$; suction = 40 psig; discharge = 135 psig

Ex-203

- Shell inflow rate 225 GPM @ 115°F (46.11°C) @ 131 psig
- Shell outflow rate 225 GPM @ 180°F (82.22°C) @ 127 psig
- Shell $\Delta p = 4$
- Tube inflow rate 125 GPM @ 350°F (176.66°C) @ 35 psig
- Tube outflow rate 125 GPM @ 255°F (123.88°C).
- Tube $\Delta t = 95$

A heat transfer system can be very complicated with modern process control instrumentation. Since a heat exchanger can explode like a bomb, proper training and care are needed during operation as well as startup and shutdown. Figure 7.24 illustrates the various components found in a heat exchanger system. A flow control loop is found on the shell outlet of Ex-203 and fails in the open position. This will prevent the feedstock from overheating and potentially rupturing the heat exchanger. Correct line-ups on heat exchangers are essential, and process technicians should carefully review the standard operating procedure. Table 7.1 illustrates a simple approach to starting up a heat exchanger system.

Table 7.1 *Startup Procedure for Heat Exchanger System*

Procedure	Notes
Prior to startup, the feed tank has been blended, sampled, and prepared for introduction into the heat transfer system for processing.	
1. Notify your supervisor and all concerned about startup.	*Downstream process technicians, I & E, engineering, maintenance, etc.*
2. Ensure all safety hazards are secured.	*Trash, locks, old permits, etc.*
3. Check equipment and instrumentation.	*Stroke control valves 0–100%, check line-ups, etc.*
4. Sample feedstock.	
5. Line up feed tank to heat exchangers.	*Check each valve line-up*
6. Set FIC 202 to 50 GPM and put in AUTO.	
7. Open V-202B, V-202F, V-202G, V-202H.	*Carefully monitor instrumentation.*
8. Close V-201.	*The feed pump has been circulating the feed and is now entirely lined up to the process. @ 50 GPM.*
9. Open V-202C on the lower tube inlet to EX-202 and open V-202D on the upper tube outlet. Ensure V-202E bypass is closed. Call tank farm technician and ensure all valve line-ups to off-specification tank are made from EX-202 tube outlet to storage.	
10. Line up the hot oil system to EX-203. Open V-202K upper tube inlet and V-202i lower tube outlet. Ensure V-202J bypass is closed.	
11. Set TIC 100 to 180°F and put in AUTO.	
12. Start hot oil pump.	
13. Increase FIC-202 to 100 GPM and let heat transfer system line out.	
14. Increase FIC-202 to 150 GPM and let heat transfer system line out.	
15. Ensure flow from reboiler to EX-202 to tank farm is moving and pressure instrument readings are normal.	

16. Increase FIC-202 to 225 GPM and let heat transfer system line out.	
17. Visually inspect all equipment, piping, and instrumentation and take initial run readings.	
18. Catch sample from EX-202 at V-202M sample port using all safety procedures and requirements.	
19. Catch sample from EX-203 at V-202L sample port using all safety procedures and requirements.	
20. Notify all involved of any operational changes to the heat transfer system.	

Cooling Towers

Heat exchangers and cooling towers often team up to form industrial cooling systems. The system consists of a cooling tower, heat exchanger, and pump (Figure 7.25). During operation, cooling water is pumped into the shell side of a heat exchanger and returned (much hotter) to the top of the cooling tower. As the hot water goes into the top of the cooling tower, it enters a water distribution header, from which it is sprayed over the internal components (fill) of the tower. As the water falls on the splash bars, cooler air contacts the water. This process removes 10 to 20% of the **sensible heat** (heat that can be measured by a change in temperature). Another 80 to 90% of the heat energy is removed through evaporation. The cooled water collects in a basin at the foot of the cooling tower, where a recirculation pump sends it back to the heat exchanger.

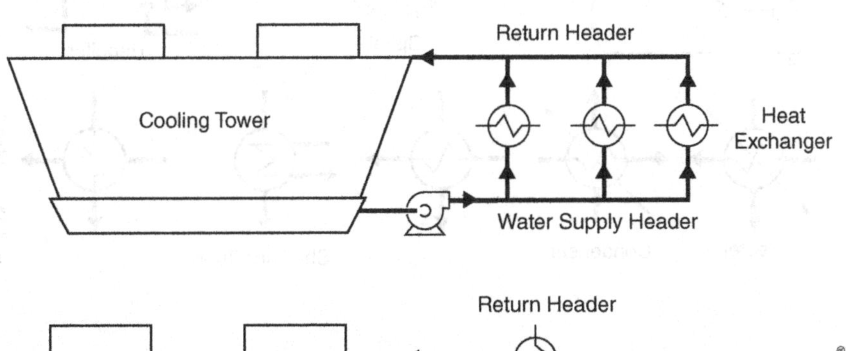

Return Header
Cooling Tower
Water Supply Header

Figure 7.25
Cooling Tower System

Distillation Columns

Distillation columns use heat exchangers to preheat feedstock (heat exchanger), condense hot vapors (condenser), or add heat to the tower and crack (separate) heavier components (kettle and thermosyphon reboilers). Condensers are typically shell-and-tube heat exchangers used to condense hot vapors into liquid. A condenser can be found at the top of most distillation columns.

Heat Exchanger Symbols

Each type of heat exchanger can be represented by a symbol. Figure 7.26 illustrates heat exchanger symbols.

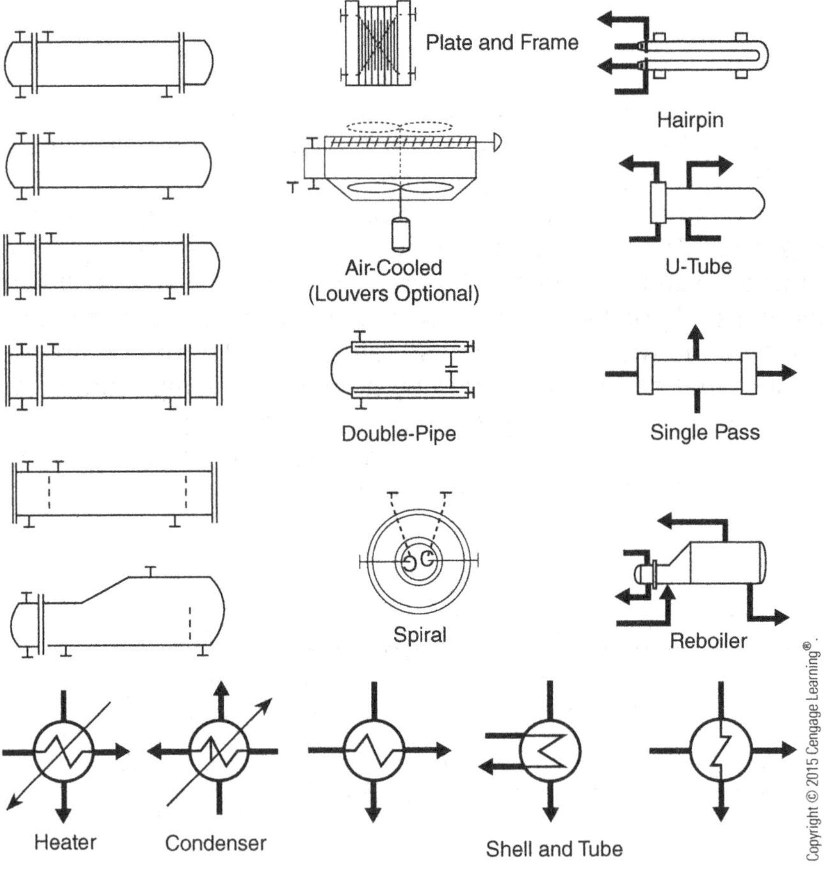

Figure 7.26 *Heat Exchanger Symbols*

Summary

A heat exchanger allows a hot fluid to transfer heat energy in the form of heat to a cooler fluid without the two fluids physically coming into contact with each other. Heat exchangers can be categorized as pipe coil, double pipe, shell and tube, reboiler, plate and frame, air cooled, and spiral. The shell-and-tube heat exchanger is the most common in the process industry. Shell-and-tube heat exchangers are designed to handle high flow rates in continuous operations. Reboilers are used to maintain heat balance in distillation columns. Kettle reboilers and thermosyphon reboilers are the types most often used. Condensers are typically shell and tube heat exchangers often used in distillation columns to condense hot vapors into liquid.

The three methods of heat transfer are conduction, convection, and radiation. Conduction and convection are used in heat exchangers, but radiation is not used. Heat transfer occurs best when large temperature differences exist between the products, flow rates are high, and cross-sectional area is large.

Laminar—that is, streamline—flow moves through a system in layers of liquid flowing in parallel. Turbulent flow is the random movement or mixing of fluids. Once turbulent flow is initiated, molecular activity speeds up until the fluid is uniformly turbulent. Laminar flow is not conducive to heat transfer.

Review Questions

1. List the methods of heat transfer. Discuss the interrelationships that exist between them and how each applies to heat exchangers. Which is (are) the most critical and why?
2. Draw and label a hairpin heat exchanger.
3. Draw and label a shell-and-tube heat exchanger.
4. Draw and label a kettle reboiler. Explain what is happening inside the device.
5. Describe the effect of laminar and turbulent flow in and out of a heat exchanger system.
6. List five types of heat exchangers.
7. What is meant by the term *floating head*?
8. How are heaters, condensers, and reboilers related to distillation systems?
9. Contrast parallel and series flow through a heat exchanger.
10. Draw an air-cooled heat exchanger.
11. Explain the purpose of using finned tubes in heat exchangers.
12. Describe the operation of a spiral heat exchanger.
13. Draw and label a plate-and-frame heat exchanger.
14. Heat exchangers and cooling towers often team up to form industrial cooling systems. Please explain this statement.
15. Does steam flow from top to bottom or from bottom to top inside a heat exchanger? Explain your answer.
16. Explain why a kettle reboiler would be preferred over a thermosyphon reboiler in a distillation system.
17. Describe the application and operation of a vertically mounted thermosyphon reboiler. Please draw and show all flows in and out and location of exchanger.
18. Describe natural circulation, forced circulation, and flooded-tube circulation in vertical and horizontal thermosyphon and kettle reboilers.
19. Describe the engineering designs associated with vapor space above the tube bundle in a kettle reboiler and liquid levels above and below the tubes.
20. Describe the principle of "fouling" and pressure differential between the tube inlets and outlets.

Cooling Towers

Objectives

After studying this chapter, the student will be able to:

- List and describe the basic components of a cooling tower.
- Describe the principles of heat transfer in a cooling tower system.
- Describe the relationship between heat exchangers and a cooling tower.
- Explain how an atmospheric cooling tower operates.
- Explain how a natural-draft cooling tower operates.
- Explain how a forced-draft cooling tower operates.
- Explain how an induced-draft cooling tower operates.
- Illustrate crossflow and counterflow in a cooling tower.
- Describe a water-cooling system.
- Describe the characteristics of water that cause problems with water-cooling systems.

Key Terms

Air intake louvers—slats located at the bottom or sides of a cooling tower to direct airflow.

Approach to tower—the temperature difference between the water leaving a cooling tower and the wet-bulb temperature of the air entering the tower.

Basin—concrete storage compartment or catch basin located at the bottom of the cooling tower.

Basin heaters—are designed to keep cooling system water from freezing during the winter months.

Biocides and algaecides—prevent biological growths from interfering with water circulation.

Blowdown—a process of controlling the level of suspended solids in a cooling tower by removing a certain amount of water from the basin and replacing it with makeup water.

Capacity—the amount of water a cooling tower can cool.

Cell—the smallest subdivision of a cooling tower that can function as an independent unit. Some cooling tower systems have multiple cells.

Cooling range—the temperature difference between the hot and cold water in a cooling tower.

Cooling towers—are evaporative coolers specifically designed to cool water or other mediums to the ambient wet-bulb air temperature.

Cooling tower types—are typically classified as induced (requires fan), forced (requires fan), atmospheric, and natural draft.

Drift eliminators—devices used in a cooling tower to keep water from blowing out.

Drift loss—entrained water lost from a cooling tower in the exiting air; also called windage loss.

Dry-bulb temperature (DBT)—the air temperature as measured without taking relative humidity into account.

Evaporate—to turn to vapor; evaporation removes heat energy from hot water.

Fill—plastic or wood surfaces that direct airflow and provide for contact of water and air in a cooling tower. See also *Splash bar*.

Forced-draft—type of mechanical-draft cooling tower that uses fans to push air into the tower.

Induced-draft—type of mechanical-draft cooling tower that uses fans to pull air out of the tower.

Leaching—is the loss of wood preservative chemicals in the supporting structure of the cooling tower as water washes or flows over the exposed components.

Parts per million—one PPM equals 1 pound to every 1,000,000 pounds. Typically associated with suspended solids in the basin or product stream.

Plenum—the open area sandwiched between the fill in the center of an induced-draft cooling tower.

Psychrometry—is the study of cooling by evaporation.

Relative humidity—a measurement of how much water air has absorbed at a given temperature.

Scale—the result of suspended solids adhering to internal surfaces of equipment in the form of deposits.

Splash bar—a device used in a cooling tower to direct the flow of falling water and increase surface area for air–water contact.

Total Dissolved Solids (TDS)—the dissolved minerals, such as magnesium and calcium, found in water, which is typically treated with sulfuric acid.

Water distribution header—a pipe that evenly disperses hot water over the fill of a cooling tower.

Water distribution system—typically consists of a deep pan with holes equipped with nozzles that distribute the water across the fill using gravity. Some systems utilize a pipe and spray nozzle design.

Wet-bulb temperature (WBT)—the air temperature as measured by a thermometer that takes into account the relative humidity.

Windage or drift loss—small water droplets that are carried out of the cooling tower by flowing air.

Cooling Tower Applications and Theory of Operation

Cooling towers (Figure 8.1) are heat transfer devices designed to cool water for reuse. They cool hot water by bringing it into direct contact with air, using countercurrent or crossflow patterns. A cooling tower contains wood or plastic slats, called **fill**, that direct airflow and the flow of water falling from the top of the tower. The downward-flowing water coats the fill and forms a film, thereby increasing the surface area for

Figure 8.1
Cooling Tower

contact between the cool air and hot water. Hot water transfers heat to the cooler air it contacts in the tower. This process results in both sensible heat loss and evaporation. Sensible heat is heat that can be measured or felt. When water changes to vapor, the vapor takes heat energy with it, leaving behind the cooler liquid. **Evaporation**, which accounts for 80 to 90% of the heat loss, is the most critical factor in cooling tower efficiency. It is affected by **relative humidity** (the amount of water in a given quantity of air at a given temperature), temperature, and wind velocity. Other factors that affect cooling tower efficiency are tower design, water contamination, and equipment problems. Cooling towers can be described as psychrometry devices. **Psychrometry** is the study of cooling by evaporation.

Temperatures in a cooling tower are closely controlled. The temperature difference (ΔT) between the inlet air temperature (wet bulb) and the outlet water temperature is referred to as the **approach to tower**. The temperature difference between the hot and cold water is referred to as the **cooling range**. Cooling tower **capacity** is defined as the amount of water a cooling tower can cool.

There are two ways to measure temperature: **dry-bulb temperature (DBT)** and **wet-bulb temperature (WBT)**. Wet-bulb temperature takes into account the relative humidity, whereas dry-bulb temperature does not. Wet-bulb temperatures usually are lower than dry-bulb temperatures. The wet-bulb temperature, perhaps the single most important factor in cooling tower performance, can be described in several ways:

- The lowest theoretical temperature to which water can be cooled in the tower.
- The temperature of the air saturated with water (also referred to as the dewpoint of air).
- A theoretical temperature that cannot be reached, only approached.

Basic Components of a Cooling Tower

Most modern cooling towers are built with treated wood, cedar, cypress, redwood, or plastic because these materials are resistant to the negative effects of water. As hot process water returns to the tower, it enters the **water distribution header**, which is a pipe located at the top of most towers (Figure 8.2). From here, water is sprayed or allowed to fall down into the tower over the **splash bars** and fill. The splash bars direct the downward flow of water and increase the surface area available for air–water contact. The material inside a tower that directs the flow of water and air is called the fill. Fill can be arranged in patterns that produce either counterflow or crossflow. Pumps suction water from the water **basin** and discharge it into the cooling water supply header. The supply header distributes water

Figure 8.2 *Water Distribution System*

to process exchangers, where it absorbs heat and returns to the top of the cooling tower through the cooling water return header.

Most towers develop significant draft or air movement due to design or air density differences. **Drift eliminators** prevent water from being blown or sucked out of the tower. This type of water loss is called **drift loss** or **windage loss**. Makeup water is added to replace water that has been lost by evaporation or **blowdown**.

Induced-draft cooling towers use fans to pull air out of the system. **Forced-draft** cooling towers use fans to push air into the system. Some cooling towers have **air intake louvers**, slats located on the side of the tower to direct airflow. These louvers can be fixed or movable depending on the tower design. A hyperbolic (i.e., chimney) tower has a stack above the fill and water distribution system.

Cooling Tower Classification

Cooling towers are classified by how they produce airflow and the direction the airflow takes in relation to the downward flow of water. Airflow is produced atmospherically, naturally, or mechanically.

Atmospheric and natural-draft cooling towers do not use fans to produce airflow. In an atmospheric cooling tower, wind velocity activates the heat transfer process. In a natural-draft cooling tower, temperature differences inside and outside the tower change the density of the air. Water vapor and air rise naturally up the chimney. Atmospheric towers initially have a crossflow airflow, whereas natural-draft cooling towers can be arranged with crossflow or counterflow fill patterns.

Induced-draft and forced-draft cooling towers use fans to move air into and out of their systems. Although mechanical fans use energy, heat transfer rates are much higher when these devices are used. Forced-draft cooling towers primarily use counterflow arrangements. Induced-draft cooling towers primarily use crossflow, but counterflow designs are available.

The world's tallest cooling tower is Germany's Niederaussem Power Station, standing 200 meters (656 ft.). Wet cooling towers operate using the scientific principle of evaporation. Examples of wet cooling towers include (1) induced draft crossflow, (2) forced draft counterflow, (3) atmospheric, and (4) hyperbolic or chimney. Dry cooling towers operate using the scientific principle of convection by heat transmission through pipes or tubes or surfaces that separate the working fluid from the ambient air. Hybrid wet *and* dry cooling towers can be found in operation at many systems in the chemical processing industry.

Atmospheric Cooling Tower

In atmospheric towers (Figure 8.3), wind moves air into and out of the tower. Airflow rates are determined by wind velocity. The tower is designed so that winds blow in horizontally, so the air moves in a crossflow direction. Cool air enters the tower through the louvered sides and passes across the downward-flowing hot water. As the air becomes heated by contact with the hot water, it travels up because hot air rises. This air is moving in a counterflow direction, opposite the direction of the falling water.

The top of an atmospheric tower has drift eliminators to stop water loss when wind velocity surges. The location of an atmospheric tower is important because wind velocities of 4.5 to 6.5 MPH are required in order for it to function properly. These towers are designed for water leaving the tower to be 4°F or 5°F (−15.55°C) lower than the wet-bulb temperature of entering air.

Atmospheric cooling towers have a 30 to 55% efficiency rating for cooling water. Because this system does not require a fan, it is very cost-effective; however, its efficiency can fluctuate greatly because it depends on an uncontrollable factor, wind velocity. Heat transfer drops significantly with the loss of airflow.

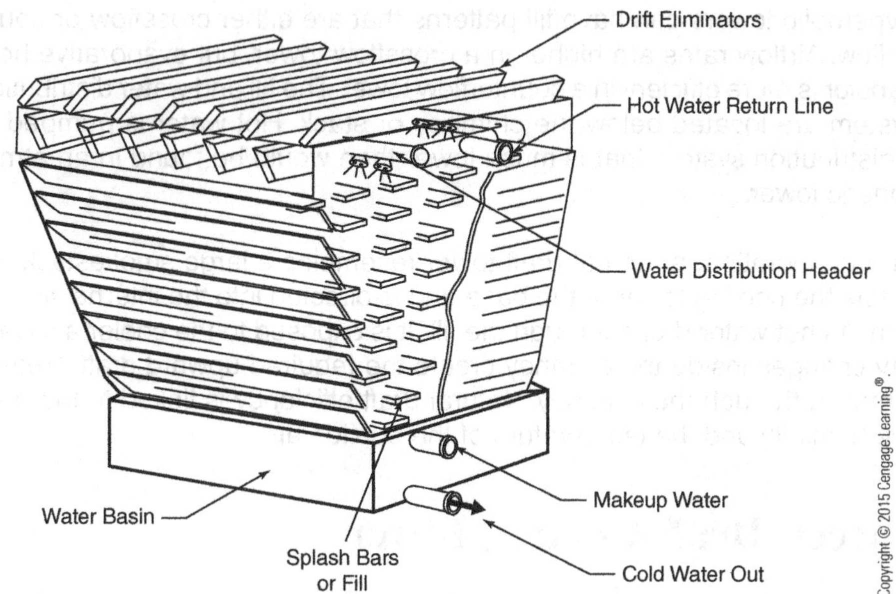

Figure 8.3 *Atmospheric Cooling Tower*

Natural-Draft Cooling Tower

Hyperbolic, or chimney, towers are natural-draft towers that have a large stack, or chimney (Figure 8.4). They usually are associated with power plant operation. Commercial towers are typically around 310 feet high with a lower diameter of 210 feet and a throat around 120 feet that gradually widens to 134 feet at the top. They are designed for flows in excess of 500,000 GPM (gallons per minute). Airflow is produced by temperature-induced density differences inside and outside the stack.

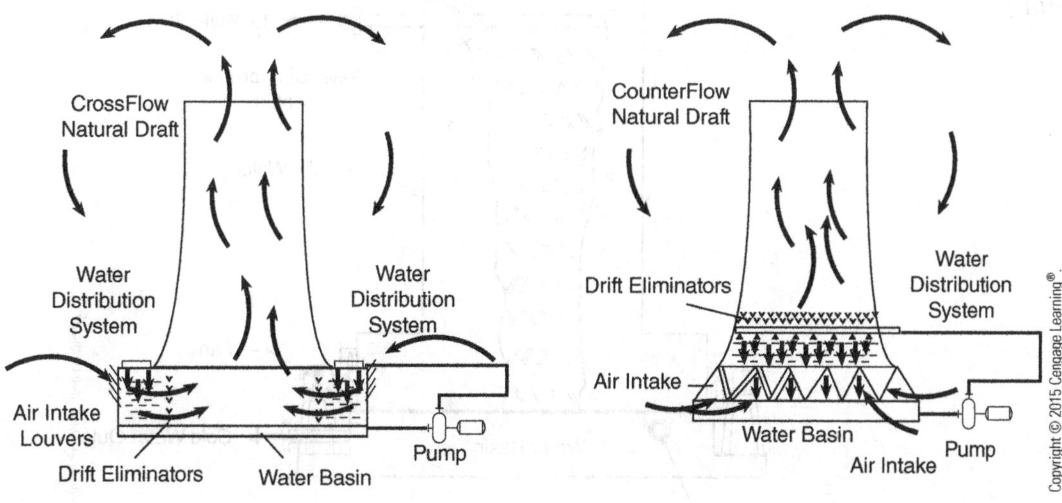

Figure 8.4 *Natural-Draft Cooling Tower*

Hyperbolic towers can have fill patterns that are either crossflow or counterflow. Airflow rates are higher in a crossflow tower, but evaporative heat transfer is more efficient in a counterflow tower. The fill and water distribution system are located below the chimney or stack. Hot water is pumped to a distribution system that is much lower than would be found in an atmospheric tower.

During operation, a natural-draft tower resembles a large smokestack. Air enters the cooling tower at the base and is directed into the internal fill pattern. As hot water drops through the fill, it is exposed to the cooler air. Density changes inside the chimney create the required upward draft. Heat is removed through the chimney. Natural-draft efficiency is linked to the relative humidity and the temperature of the outside air.

Forced-Draft Cooling Tower

Forced-draft cooling towers force air in mechanically by the use of fans (Figure 8.5) on the lower side of the tower. Forced-draft towers usually have solid sides without louvers. The fans push in 100% of the process air. Flow direction is counterflow; the fans push air upward against the downward flow of water. Air in forced-draft towers can have high velocities, but the exiting air slows so much that it is recirculated back into the tower, cutting efficiency by 20%. Forced-draft cooling towers have much higher heat transfer rates than atmospheric and natural-draft cooling towers, and they are significantly lower in height. This type of cooling tower is less efficient than induced-draft towers because some hot air is recirculated.

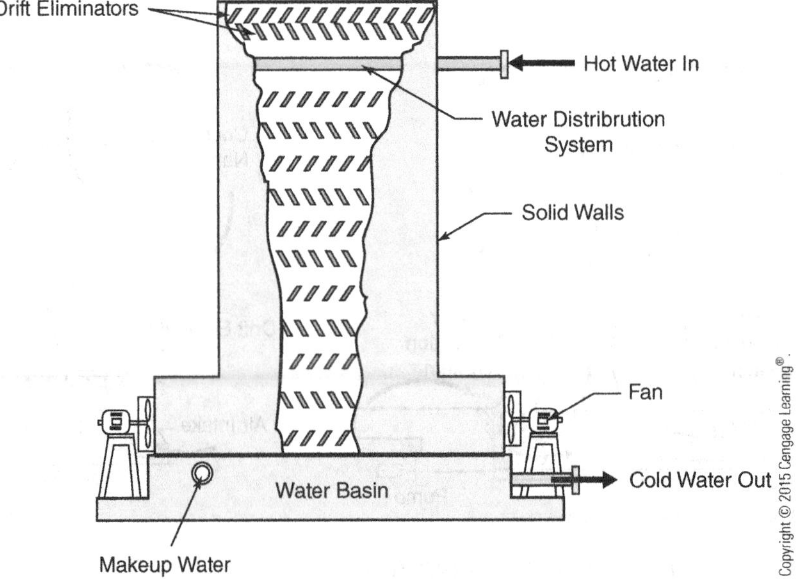

Figure 8.5
Forced-Draft Cooling Tower

Induced-Draft Cooling Tower

An induced-draft tower is another kind of cooling tower that produces airflow mechanically (Figure 8.6). It differs from the forced-draft cooling tower in that it pulls air out of the tower rather than forcing it in. Airflow in an induced-draft tower is slower than in a forced-draft tower, but heat transfer through evaporation is more efficient. The tower fan, located on top of the tower (Figures 8.7 and 8.8), produces discharge rates strong enough to lift the hot air above the tower, so hot air is not recirculated into the tower. Induced-draft towers can circulate airflow horizontally (crossflow) or vertically (counterflow). During crossflow operation, drift eliminators are located in the center of the tower to reduce water loss. Counterflow operations force air vertically across a solid area of fill, and drift eliminators are located above the fill and water distribution headers.

Figure 8.6
Induced-Draft Cooling Tower

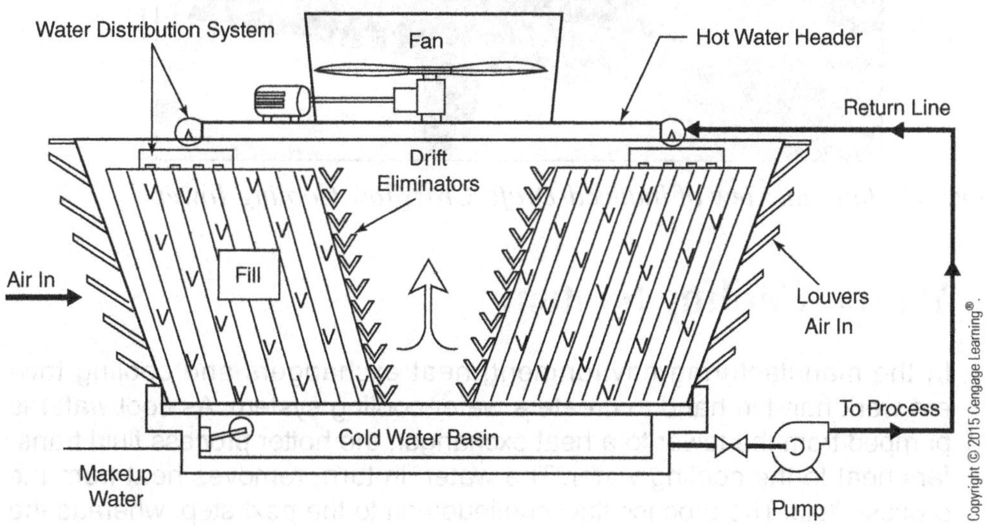

Figure 8.7
Induced-Draft, Crossflow Cooling Tower

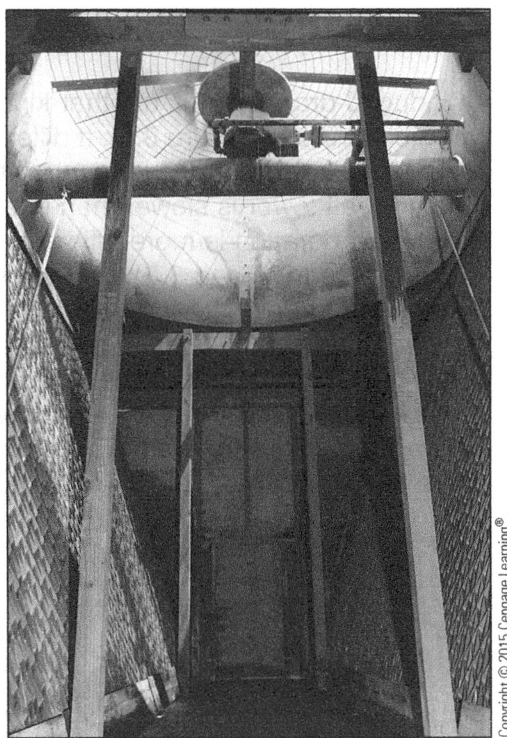

Figure 8.8 *Internal View of Induced-Draft, Crossflow Cooling Tower*

Water-Cooling System

In the manufacturing environment, heat exchangers and cooling towers work hand in hand to create a water-cooling system. As cool water is pumped from the tower to a heat exchanger, the hotter process fluid transfers heat to the cooling water. The water, in turn, removes heat from the process fluid. The process flow continues on to the next step, whereas the hot water is returned to the tower to be cooled. A centrifugal pump sends

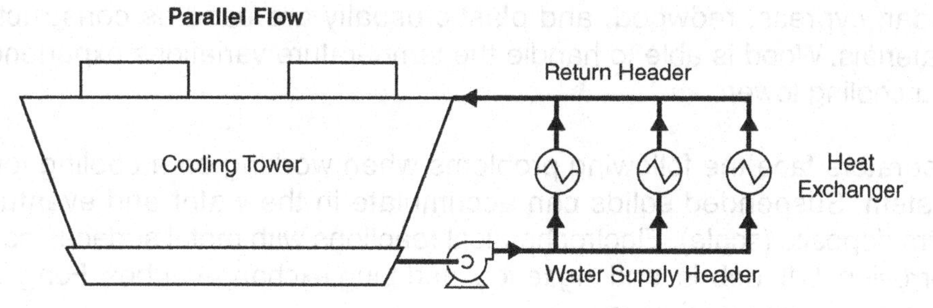

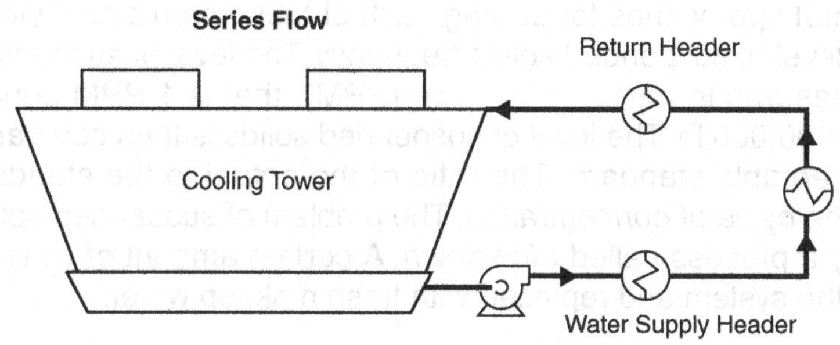

Figure 8.9
Series and Parallel Flow Heat Exchangers

the hot water to the top of the tower, where it enters the water distribution header. The hot water is distributed evenly throughout the tower, where much of the heat it had gained is given up in evaporation. The cooled liquid is recirculated back to the heat exchanger and the process loop continues.

Heat exchangers can be connected to cooling towers in a variety of ways. The two most common are parallel and series (Figure 8.9). In parallel flow, the process flow goes through multiple exchangers at the same time. In series flow, the flow passes through one exchanger before it goes to another.

The Trouble with Water

Over time, water dissolves everything it touches. The Grand Canyon is strong evidence that water can remove large and small obstacles. Hot water dissolves solids faster than cold water.

Hot water dissolves a little bit of everything it comes into contact with. By the time the hot water returns to the tower, it is full of suspended solids. Hot water also has a tendency to become corrosive and to form deposits. When hot water enters the cooling tower laced with suspended solids, it undergoes evaporation. This process removes water and leaves the solids. The remaining fluid concentrates in the basin. Over time, this concentration builds up to levels that must be controlled.

The materials used to construct a cooling tower must be durable and capable of withstanding wide temperature differences. Treated wood, cedar, cypress, redwood, and plastic usually are used as construction materials. Wood is able to handle the temperature variations experienced in a cooling tower.

Operators face the following problems when working on a cooling tower system. Suspended solids can accumulate in the water and eventually form deposits (scale). Electrochemical reactions with metal surfaces cause corrosion. Silt, debris, and algae foul and plug exchanger tubes. Fungi and bacteria cause wood decay.

There are several approaches for solving each of these problems. Operators check the level of suspended solids frequently. The level of suspended particles is measured in **parts per million** (PPM), that is 1 PPM equals 1 lb. to every 1,000,000 lb. The level of suspended solids is then compared against an acceptable standard. The ratio of the actual to the standard level is called the *cycle of concentration*. The problem of suspended solids is controlled by a process called *blowdown*. A certain amount of water is removed from the system and replaced with fresh makeup water.

Scale-forming solids can be removed with softening agents, or suspension of solids can be prevented by adding chemicals. Another approach is to precipitate the scale so it can be removed (to precipitate is to get particles to fall from suspension). Corrosion can be minimized by the addition of chemical inhibitors, which form a film that protects metal. Fouling can be controlled by filtering devices, alone or with dispersants that prevent suspended solids from accumulating. Biocides (such as chlorine or bromine) can be used to prevent wood decay. Another problem found in cooling tower operation is leaching. **Leaching** is the loss of wood preservative chemicals in the supporting structure of the cooling tower as water washes or flows over the exposed components.

All of these problems and solutions require monitoring. Operators are responsible for testing and checking:

- pH of water
- Total dissolved solids (TDS)
- Inhibitor concentration
- Chlorine or bromine concentration
- Precipitant concentrations
- Tower equipment checklists
- Filters and screens
- Wet-bulb temperature and humidity

Cooling Tower System

Cooling tower 302 is classified as an induced draft (or draw-through), crossflow, single-cell device that is primarily designed to control the temperature on condenser Ex-204. A cooling tower is often referred to as a heat rejection

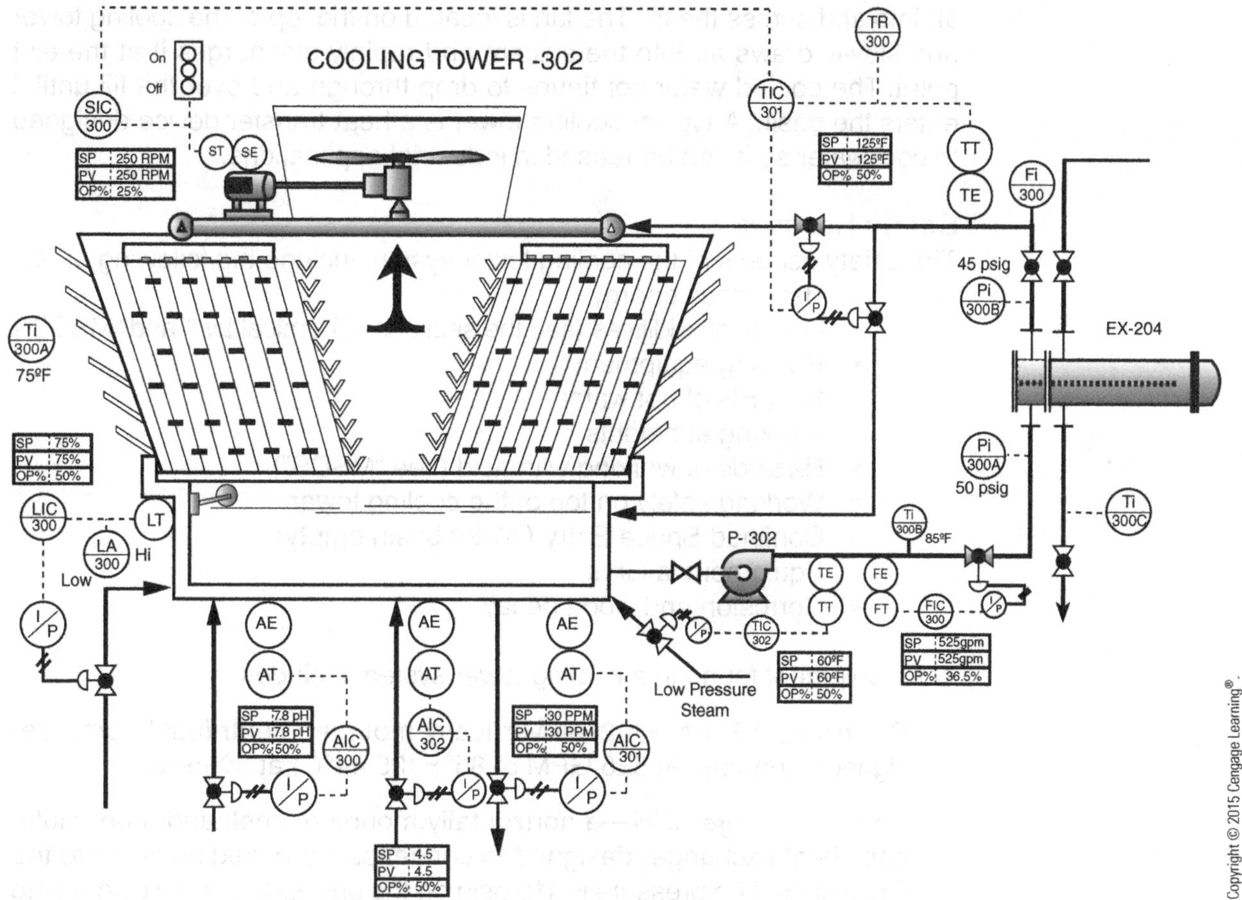

Figure 8.8 *Cooling Tower System: CTW-302*

device designed to extract excess heat from the returning water and expel it into the atmosphere. Figure 8.10 shows what the cooling tower system looks like. This type of heat transfer relies on the principle of evaporation. When this process occurs, heat from the water is absorbed by the air stream that raises the relative humidity to near 100%. These heated currents are quickly dissipated by the wind. Cooling tower 302 is an evaporative heat rejection device that can significantly reduce water temperatures.

Cooling Tower 302

CTW-302 is an enclosed structure with a system of air louvers designed to direct airflow across the fill. As warm water enters the top of the cooling tower, the water distribution system carefully sprays or directs fluid flow over a labyrinth-like honeycomb, splashboards, or fill. The purpose of the fill is to allow the hot water to spread out over the surface of the boards. The fill provides a vastly expanded surface area interface that enhances air liquid contact. As evaporation occurs, the air becomes saturated with water and is carried out of the cooling tower system. CTW-302 uses a fan to draw air into and across the fill. The fan is located on the top of the cooling tower and slowly draws air into the system and rapidly discharges it at the exit point. The cooled water continues to drop through and over the fill until it enters the basin. A typical cooling tower is a heat transfer device designed to cool water so it can be reused in industrial applications.

Safety Hazards

The safety aspects of the cooling tower system include the following areas:

- Hazardous energy
- Chemical additives (see the sections "Chemical List" and "MSDS")
- Rotating equipment
- Hazards of hot water
- Working at heights
- Hazards of working with acid (see "MSDS")
- Working safely on top of the cooling tower
- Confined Space Entry (Water basin empty)
- Equipment failures
- Corrosion and wood decay

The equipment found in a cooling tower system includes:

Centrifugal Pump-302—a vertically mounted centrifugal pump designed to operate at 525 GPM at 85°F (29.44°C) at **50 psig.**

Heat Exchanger-204—a horizontally mounted, shell-and-tube, multipass heat exchanger designed to condense vaporized butane into the liquid state. The pressure is 115 psig on the shellside and 50 psig on the tubeside. The tube side contains cooling water from the cooling tower system. The shell side contains vapor or liquid butane at 115 psig. The heat exchanger is comprised of stainless steel and is rated at 225 psig at 250°F (121.11°C) on the tubes and 225 psig at 300°F (148.88°C) on the shell. The inlet pressure on the tube side is run at 50 psig and 45 psig on the tube outlet. The tube delta pressure is 5 psig.

Cooling Tower Fan—is mounted on the top of the water distribution pipe in the center of the cooling tower. A local controller and on/off switch are located near the motor. The motor is designed to run at 250 RPM and can be adjusted depending upon need.

Plenum—is the open area directly under the fan. In an induced draft, crossflow cooling tower, the velocity of the air creates a partial vacuum under the fan as it expels the vapor-enriched air.

Water Distribution System—consists of a 12" deep pan, 4' × 20', with holes equipped with stationary spiral nozzles that distribute the water across the fill. A pan with holes and nozzles is located on the east and west side of the cooling tower.

Water Basin—is a concrete reinforced structure designed to store water and provide a foundation upon which the rest of the cooling tower can be supported. The basin is designed to collect the water as it flows across the horizontal fill and downward. The water basin provides suction for the water pump and must be able to resist chemical attack. Technicians working on the cooling tower basin must place careful attention to water pH, temperature, parts per million, and biological problems. **Basin heaters** come in a variety of designs: low pressure steam, electric, or hot oil. The primary purpose of a basin heater is to keep the cooling tower clear of ice during the cold winter months.

Louvers—are evenly spaced and designed to direct airflow across the downward flow of water.

Fill—provides a vastly expanded surface area interface that enhances air–liquid contact. As evaporation occurs, the air becomes saturated with water and is carried out of the cooling tower system. The fill is comprised of pressure-treated materials: wooden slates, run through a plastic support structure. The fill is designed to provide plenty of surface area so good air–liquid contact is ensured. Evaporation takes place inside the fill area transferring heat energy to the moist area, increasing the relative humidity in this area to 100%, and moving this heated plume into the atmosphere. The plume or fog is virtually harmless except that it can reduce visibility when it touches the ground. Evaporation accounts for 80 to 90% of the heat transfer in a cooling tower. This convective process takes place as the hot water cascades down the fill or splash boards. As the hot water spreads out across the fill, air flows over the area. Figure 8.11 illustrates how the rising hot water vapor or

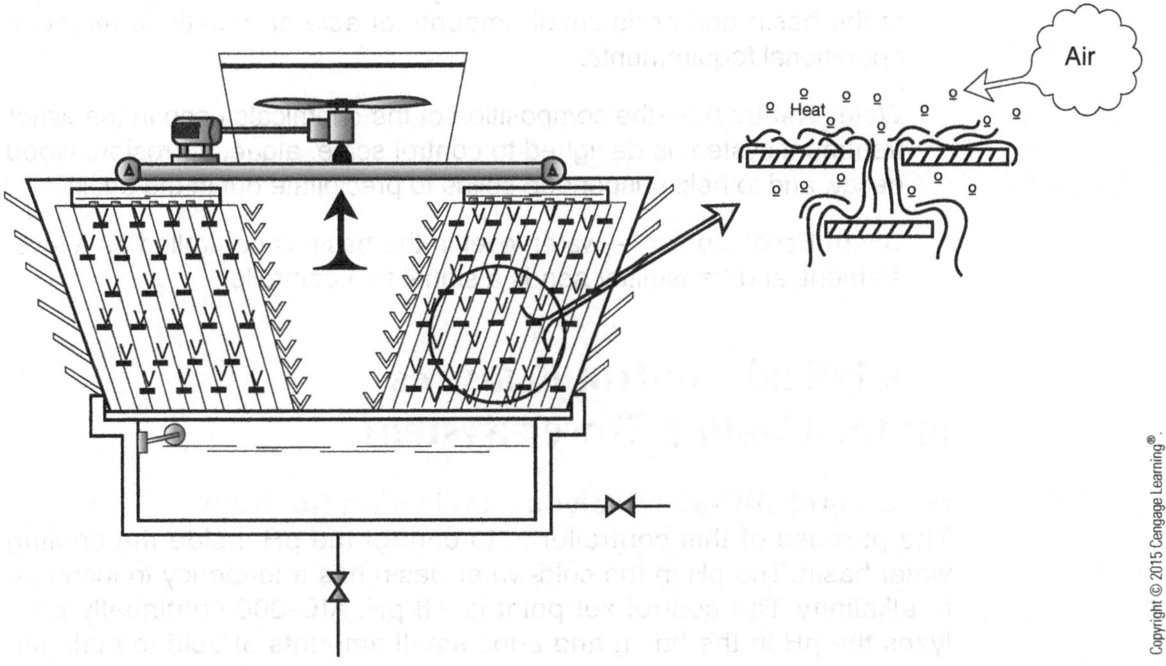

Figure 8.11 *Splash Boards or Fill*

heat energy is carried out of the cooling tower and the cooler water drops down into the water basin.

Drift Eliminators—block and prevent water loss from the system. As hot, moisture-rich vapor flows across the fill, it is lighter than the outside air and accelerates as it enters the plenum. This warm air increases in velocity as it is propelled high above the cooling tower by the fan.

Blowdown—Draw-off or blowdown is primarily used to control the buildup or concentration of minerals in the recirculation water. The blowdown system is designed to control the level of suspended solids in the water basin. High levels of suspended solids will cause fouling. Blowdown is closely related to the term *concentration cycles*. Blowdown refers to automatically removing 7 to 10% of the water in the water basin and replacing it with fresh water. In most cases, the water makeup system runs continuously due to evaporative losses and drift losses, and draw-off.

Concentration Cycles—in the cooling tower typically can range from 3 to 7. The cooling tower is typically set at 6 concentration cycles before it blows down the system. A cycle describes one pass from the cooling tower to the process and back to the cooling tower. Cycles of concentration display the accumulation of dissolved minerals in the recirculation system.

Cell—is the smallest subdivision of a cooling tower that can function as an independent unit.

pH Control—An automatic control system continually analyzes the pH in the basin and adds small amounts of acid or caustic to maintain operational requirements.

Water Treatment—the composition of the chemicals used in the water treatment system is designed to control scale, algae, corrosion, wood decay, and to help suspended solids to precipitate out in the basin.

Basin Level Control—water level in the basin is controlled as a level element, and transmitter sends a signal to a controller.

Analytical Control Features on the Cooling Tower System

pH Control: AIC-300 Analytical Indicating Controller

The purpose of this controller is to control the pH inside the cooling water basin. The pH in the cold-water basin has a tendency to increase in alkalinity. The control set point is 7.8 pH. AIC-300 continually analyzes the pH in the basin and adds small amounts of acid to maintain operational requirements. The cooling tower specification range is 7.6 pH to 8.4 pH on water in the recirculation system. Figure 8.12 shows a pH scale.

Blowdown Control: AIC-301 Analytical Indicating Controller

The purpose of this controller is to monitor and control the levels of suspended solids. Using either draw-off or blowdown, this process is primarily used to control the buildup or concentration of minerals in the recirculation water. The blowdown system is designed to control the level of suspended solids in the water basin. High levels of suspended solids will cause fouling. Blowdown is closely related to the term *concentration cycles*. Blowdown refers to automatically removing 7 to 10% of the water in the water basin and replacing it with fresh water. In most cases, the water makeup system runs continuously due to evaporative losses, drift losses, and draw-off.

The concentration cycles in CTW-302 range is set at 6 concentration cycles before it blows down the system. A cycle describes one pass from the cooling tower to the process and back to the cooling tower. Cycles of concentration display the accumulation of dissolved minerals in the recirculation system.

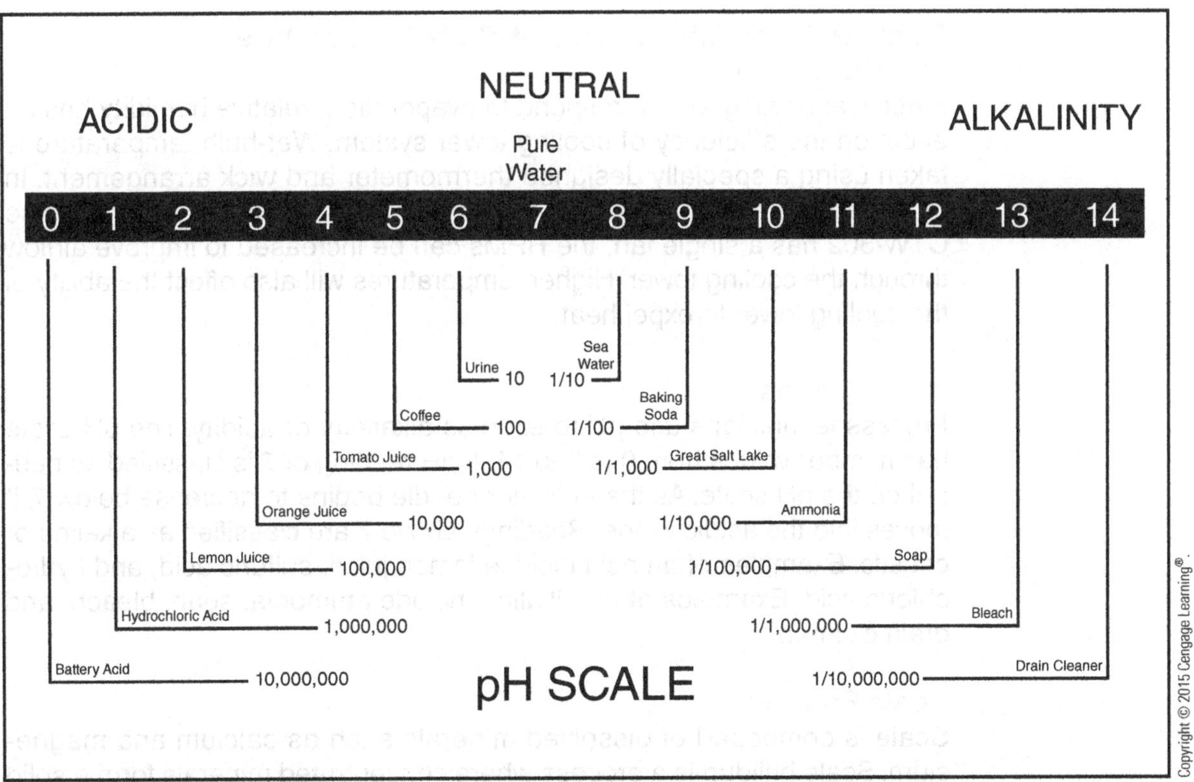

Figure 8.12 *pH Scale*

Suspended Solid Chemical Control: AIC-302 Analytical Indicating Controller

The purpose of this controller is to control scale, algae, corrosion, wood decay, and to help suspended solids to precipitate out in the basin. AIC-302 monitors conditions in the basin and maintains unit specifications by adding liquid chemical treatment. Liquid treatment is set at 4.5.

Fan Speed & Temperature Control: SIC-300 Speed Indicating Controller

The speed on the fan is set at 250 RPM and can be adjusted by the technician. An on/off switch is also located near the motor.

Common Cooling Tower Problems and Solutions

Cooling towers can be shut down easily; however, unless a serious problem occurs or equipment repair and turnaround are scheduled, the tower is kept in continuous operation. There are some common problems and concerns for which the cooling tower system must be monitored.

Cooling Tower Efficiency: Wet-Bulb Temperature and Humidity Cooling

Since wet cooling towers respond to evaporation, relative humidity has an effect on the efficiency of cooling tower system. Wet-bulb temperature is taken using a specially designed thermometer and wick arrangement. In multi-cell operations, additional fans are typically applied. However, since CTW-302 has a single fan, the RPMs can be increased to improve airflow through the cooling tower. Higher temperatures will also affect the ability of the cooling tower to expel heat.

pH Problems

Process technicians use pH to express alkalinity or acidity. The pH scale has number values from 0 to 7 to 14. A pH reading of 7 is classified as neutral on the pH scale. As the indicator needle begins to decrease below 7, it moves into the acidic range. Readings above 7 are classified as alkaline or caustic. Examples of an acid include lemon juice, sulfuric acid, and hydrochloric acid. Examples of an alkaline include ammonia, soap, bleach, and drain cleaner.

Scale Formation

Scale is composed of dissolved minerals such as calcium and magnesium. Scale buildup is a process where concentrated minerals form a solid coating on the inside of piping and tubes, reducing fluid flow and heat

Simple Startup Procedure for Cooling Tower System

Procedure	Notes
1. Notify your supervisor and all concerned about Startup.	*Downstream process technicians, I & E, engineering, maintenance, etc.*
2. Ensure all safety hazards are secured.	*Trash, locks, old permits, etc.*
3. Check equipment and instrumentation.	*Stroke control valves 0–100%, check line-ups, etc.*
4. Establish level in water basin and set LIC-300 @ 75% and place system in AUTO.	
5. Sample water in basin and send to lab.	
6. Line up Pump 302 from basin to pump to exchanger 204 to cooling tower 302 water distribution system.	
7. Line up EX-204.	
8. Start pump 302 and monitor Pi-300A (50 50psig) and Pi-300B (45 psig).	
9. Set FIC-300 TO 525 RPM and place in AUTO.	
10. Start Fan-300 and set SIC-300 to 1250 RPM.	
11. Set AIC-300 to 7.8 pH and place in AUTO.	
12. Set AIC-301 to 30 PPM and put in AUTO.	
13. Set AIC-302 to 4.5 GPH and place controller in AUTO.	
14. Set TIC-301 to 125°F and put in AUTO.	
15. Set TIC-302 (low pressure steam) to 60°F and put in AUTO.	
16. Record Ti-300A wet bulb temperature (WBT).	
17. Record Ti-300B.	
18. Calculate the approach to the tower.	
19. Carefully monitor all conditions and collect make-up water sample, basin samples, and P-302 discharge.	
20. Verify all process variables with standard operating procedure.	

transfer. Cooling tower water that has a high pH tends to enhance scale formation. Sulfuric acid is used to decrease the pH in the catch basin.

Total Dissolved Solids (TDS)

Water naturally contains magnesium and calcium in the form of dissolved solids. As water evaporates out of the cooling tower, dissolved minerals are left behind and, over time, concentrate and form scale. Dissolved solids break down the components in wood fibers that provide support for the cooling tower's internal structure. Sulfuric acid can minimize problems caused by dissolved solids. Blowdown on CTW-302 is timed so chemical injection has time to blend into the recirculation system. A chemical additive is added in a section of the cooling tower where turbulent flow enhances mixing and where it is not drawn into pump 302's suction before adequate blending has occurred.

Microorganisms: Chlorine and Bromine Concentrations

Cooling towers provide the perfect breeding ground for microorganisms since they are exposed to warm water and sunlight. Microorganisms form slime that can build up and restrict water flow and conductive heat transfer. Biocides such as chlorine and bromine can control the growth of these organisms. Unfortunately, chlorine is an extremely hazardous material and must be handled with caution. Bromine is a little safer to handle and use. CTW-302 utilizes a special blend of biocides to keep the cooling tower operating smoothly. Other chemical inhibitors are added to this blend and need to be thoroughly mixed with basin water.

Dissolved Gases

Cooling tower water may have dissolved gases like oxygen, hydrogen sulfide, and carbon dioxide that chemically react with iron. These gases will enhance corrosion, destroy metal surfaces, damage valuable equipment, and deteriorate metal pipes, brackets, and bolts used to secure wood products together. High quantities of dissolved gases will make the catch basin water acidic.

Ruptured Heat Exchanger Tubes

If the tube ruptures on Ex-204, the butane, pentane, and catalyst mix will pour into the recirculation water since it is at 100 psig and the recirculation water is at 45 to 50 psig. The pH of the cooling tower water would also become very alkaline and set off the high pH alarm.

Fan 300 Failure

CTW-302 operates the fan when temperatures and relative humidity reach predetermined levels. During the winter months, the fan is rarely operated and the hot water bypass system is frequently used. A spare motor, gearbox, and fan blade arrangement are located in the maintenance shop.

Several days, however, are typically required for large projects. It is possible to operate the cooling tower in a natural draft condition.

Pump 302 Failure

If pump 302 fails, the distillation system will immediately go into alarm condition. The pump backup system will need to be placed on-line immediately. P-302 is a vertically mounted centrifugal pump with a series of small screens installed on the suction side of the pump. These screens keep the liquid entering the pump clean and prevent internal damage. NPSH (net positive suction head) and NPDH (net positive discharge head) must be kept between specific specifications in order for the recirculation system to operate properly.

Instrument Problems

Instruments on the cooling tower include three analytical control loops, one level control loop, two temperature control loops, and one speed control loop on the fan. A variety of simple instruments are mounted on the cooling tower system.

Suspended Solids, Ex-204 Tube Fouling

As air circulates through the cooling tower, it carries solids in the form of small dust particles. These solids are captured by the downward flow of water and collected in the catch basin where they form sludge. Sludge is removed through the blowdown system. Suspended solids can build up inside the tubes of a heat exchanger and cause fouling. *Fouling* is a term used to describe restriction or plugging.

Blowdown and Cycles of Concentration Problems

Concentration cycles in cooling towers range from three to seven and are dependent upon the quality of the makeup water. Circulating water in CTW-302 is filtered and treated with **biocides and algaecides** to prevent microorganism growth and a control system for pH adjustment.

Broken or Collapsed Fill

High winds, ice buildup, corrosion, microorganisms, and other variables can cause structural damage to the cooling tower. While a cooling tower is a simple design, it needs all of its parts to operate efficiently. Fill or splashboards can be replaced and put back in place. Bolts and brackets that have been corroded need to be repaired.

Water Distribution System Problems

The water distribution system has a variety of components including a series of valves, pans, water distributors, and pan covers that can be damaged. Pipe leaks are not uncommon and can quickly lead to larger problems. If the water distributors become dislodged from the pan, uneven amounts of water can flow over the fill before it has time to contact air flowing through the tower.

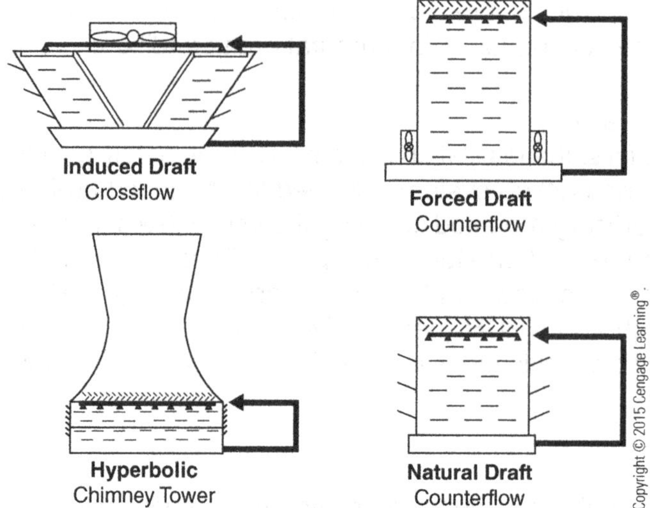

Figure 8.13 *Cooling Tower Symbols*

Cooling Tower Symbols

Each type of cooling tower can be represented by a symbol. Figure 8.13 illustrates cooling tower symbols.

Summary

A cooling tower is a heat transfer device used by industry to cool hot water for reuse in process systems by evaporation and sensible heat loss. Hot water transfers heat to air as it passes in the tower. Evaporation accounts for 80 to 90% of the heat transfer in a cooling tower. Cooling tower efficiency is affected by relative humidity, temperature, wind velocity, tower design, water contamination, and equipment problems.

Cooling towers are classified by how they produce airflow and the direction the airflow takes in relation to the downward flow of water. Atmospheric towers use wind to move air into and out of the tower. Their efficiency depends on wind velocity. Hyperbolic, or chimney, towers are natural-draft towers that produce airflow by temperature-induced density differences. Forced-draft cooling towers produce airflow mechanically through the use of fans located on the lower side of the tower. The fans push air into the tower. Induced-draft towers have a fan on top of the tower. The fan pulls hot air out of the top of the tower.

Water-cooling systems use heat exchangers and cooling towers in combination. Cool water is pumped from the tower to a heat exchanger, where hot process fluid transfers heat to the cool water. The hot water is returned to the tower to be cooled.

Water-cooling systems can be affected by corrosion, scale formation, fouling, and wood decay. The materials used to construct a cooling tower system must be durable and capable of withstanding wide temperature differences. They must be resistant to the bad effects of water, attack from treating chemicals, and biological agents. Treated wood, cedar, cypress, redwood, and plastics are used as cooling tower construction materials.

Review Questions

1. Draw a diagram of a cooling tower and three heat exchangers in parallel service.
2. Draw a diagram of a cooling tower and three heat exchangers in series service.
3. Draw a simple atmospheric cooling tower. List and describe seven components.
4. Draw a simple hyperbolic cooling tower. List the major components. What is the other name for this type of tower?
5. Draw a simple forced-draft cooling tower. List the components.
6. Draw a simple induced-draft cooling tower with crossflow. List the components.
7. Draw a simple induced-draft cooling tower with counterflow. List the components.
8. Evaporative cooling accounts for what percentage of the cooling effect in a cooling tower?
9. List five factors that affect the efficiency of a cooling tower.
10. What is the measurement of how much water the air has absorbed at a given temperature?
11. On what does the rate of airflow in a natural-draft cooling tower depend?
12. Hyperbolic towers are designed primarily for service in what industry?
13. Which tower gets higher airflow: counterflow or crossflow?
14. Which tower is more efficient: counterflow or crossflow?
15. Does air move faster through a forced-draft or an induced-draft cooling tower?
16. Which tower is most efficient: forced-draft or induced-draft? Why?
17. List four objectives of water treatment.
18. Describe pH control on a cooling tower system.
19. Describe the purpose of blowdown and illustrate how it works.
20. Define *approach to tower*.

Boilers

Objectives

After studying this chapter, the student will be able to:

- Describe the basics of boiler operation.
- Describe a fire-tube boiler.
- Describe the main components of a water-tube boiler and explain how it operates.
- List some boiler operating problems.
- Distinguish between superheated and desuperheated steam.
- Describe the primary responsibilities of a boiler technician.
- Describe an inverted bucket steam trap.
- Describe a float steam trap.
- Describe the bellows thermostatic steam trap.

Key Terms

Bellows trap—a thermostatic steam trap that operates by opening or closing a bellows as the temperature changes; this movement opens and closes a valve.

Boiler load—plant demand for steam.

Burner—used to evenly distribute air and fuel vapors over an ignition source and into a boiler firebox.

Damper—a device used to regulate airflow.

Desuperheating—a process applied to remove heat from superheated steam.

Downcomers—the inlet tubes from the upper to lower drum of a water-tube boiler; these tubes contain hot water.

Economizer—a section of a fired boiler used to heat feedwater before it enters the steam drums.

Fire-tube boiler—a type of boiler that passes hot gases through tubes to heat and vaporize water.

Flame impingement—frequent or sustained contact between flames and tubes in fire-tube boilers and furnaces.

Float steam trap—a steam trap that operates with a float that opens a valve as the condensate level rises.

Inverted bucket steam trap—a mechanical steam trap that operates with an inverted bucket inside a casing; effective on condensate and noncondensing vapors.

Mud drum—the lower drum of a water-tube boiler.

Risers—the tubes from the lower drum to the upper drum of a water-tube boiler; these tubes contain steam and water.

Spuds—gas-filled sections in a boiler-fuel gas burner.

Steam-generating drum—a large upper drum partially filled with feedwater. This drum is the central component of a boiler. It is connected to the lower mud drum by the downcomer and riser tubes and receives steam from the steam-generating tubes.

Steam trap—a device used to separate condensate from steam and return it to the boiler to be converted to steam.

Superheated steam—steam that is heated to a higher temperature.

Thermostatic steam trap—a type of steam trap that is controlled by temperature changes.

Water hammer—a condition in a boiler in which slugs of condensate (water) flowing with steam damage equipment.

Water-tube boiler—a type of boiler that passes water-filled tubes through a heated firebox.

Boiler Applications and Basic Operation

Steam generators or, as they are commonly called, *boilers*, are used by industrial manufacturers to produce steam. Steam is used to operate steam turbines, distillation systems, and reaction systems. They can be used for such processes as laminating, vulcanizing, extrusion, firefighting, and flare systems; and to provide cooling or heating to process equipment.

Boilers use a combination of radiant, conductive, and convective heat transfer methods to change water to steam. A simple boiler consists of a heat source, water-containing drum, water inlet, and steam outlet (Figure 9.1). As heat is added to the drum, the temperature increases until the water boils. As the steam rises, it is captured in a line and sent on for further processing. Factors that affect boiler operation are density differences for internal circulation, pressure, temperature, and water level.

Fire-Tube Boilers

A more complicated boiler is the **fire-tube boiler**, which resembles a modified shell-and-tube heat exchanger. This type of boiler is composed of a shell and a series of tubes designed to transfer heat from the fire-tubes and into boiler feedwater. Combustion gases exit through a chamber similar to an exchanger head and pass safely out of the boiler. The water level in the boiler shell is maintained above the tubes to protect them from overheating. The term *fire-tube* denotes that the heat source is from within the tubes. A fire-tube boiler (Figures 9.2 and 9.3) consists of a boiler shell with feed inlet and outlet connections, fire-tubes, a combustion tube, **burner**, feedwater inlet, steam outlet, combustion gas exhaust port, and tube sheets.

Water-Tube Boilers

The most common type of large commercial boiler is a **water-tube boiler** (Figure 9.4). A water-tube boiler consists of an upper and lower drum

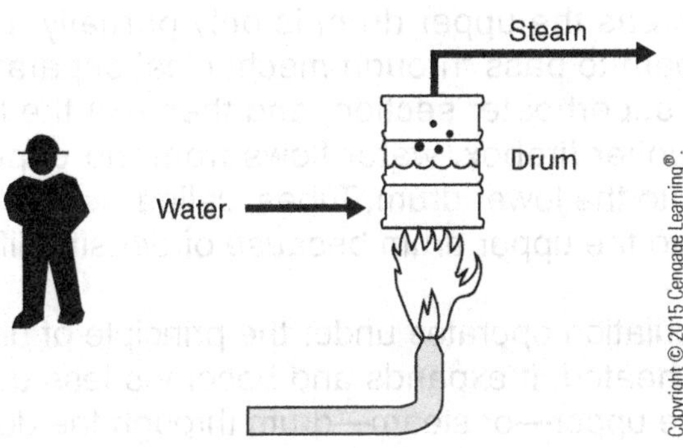

Figure 9.1
Simple Boiler

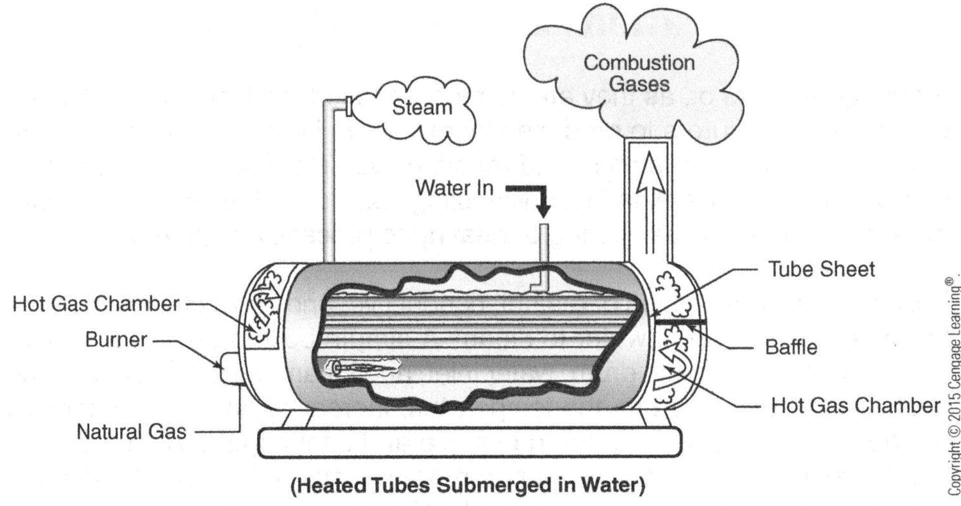

Figure 9.2
Fire-Tube Boiler Operation

Figure 9.3
Fire-Tube Boiler

connected by tubes. The lower drum and water-tubes are filled completely with water, whereas the upper drum is only partially full. This arrangement allows steam to pass through mechanical separators in the upper drum, flow to a superheater section, and then exit the boiler. As heat is applied to the boiler firebox, water flows from the upper drum through **downcomers** into the lower drum. Tubes, called **risers**, cause water and steam to flow into the upper drum because of density differences.

Boiler water circulation operates under the principle of differential density. When a fluid is heated, it expands and becomes less dense. Cooler water flows from the upper—or steam—drum through the downcomers to the

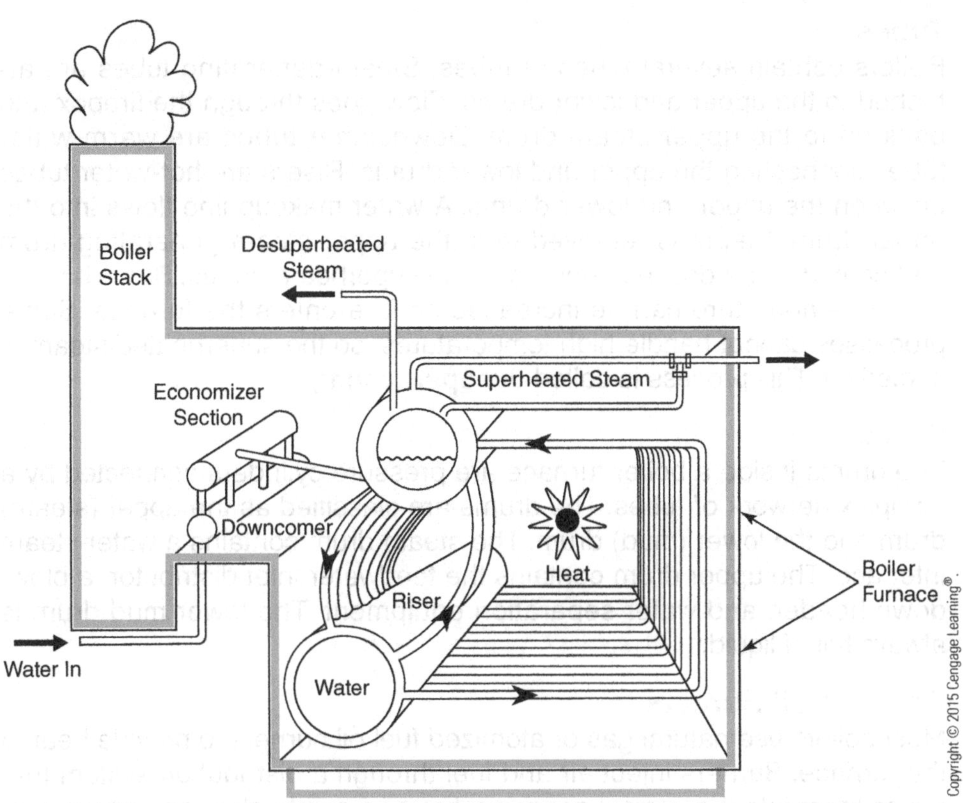

Figure 9.4
Water-Tube Boiler

mud drum (the lower drum) and then rises as some steam is generated. Circulation continues, and makeup water is added to the upper drum to replace the steam that is generated.

Water circulation continues in a water-tube boiler because steam bubbles in the lower drum move up the riser tubes and cause water density to decrease. The cooler water in the downcomer flows into the mud drum. The riser and steam-generating tubes are physically located near the burners. Steam moves up the riser and steam-generating tubes and into the upper **steam-generating drum**. Steam generation causes pressure to rise. When the target pressure is achieved, the boiler is "placed on the line." Pressure is maintained by adding makeup water and continuously applying heat.

Main Components

Furnace

The water-tube boiler firebox (that is, the furnace) is designed to reduce the loss of heat and enhance the heat energy being applied to the boiler's internal components. Boiler furnaces have a refractory lining, burners, convection-type section, radiant section, fans, oxygen control, stack, **damper**, and many other components associated with fired heaters.

Tubes

Boilers contain several types of tubes. Steam-generating tubes are attached to the upper and lower drums. Flow goes through the firebox and back up to the upper steam drum. Downcomer tubes are warm-water tubes connecting the upper and lower drums. Risers are hot-water tubes between the upper and lower drums. A water makeup line flows into the upper drum. Steam is removed from the upper steam-generating drum and heated to the desired temperature in superheater tubes. **Superheated steam** temperature can be increased as it re-enters the furnace. Some processes cannot handle high temperatures, so the superheated steam is cooled off. This process is called **desuperheating**.

Drums

The drums inside a boiler furnace are pressure cylinders connected by a complex network of tubes. The drums are classified as the upper (steam) drum and the lower (mud) drum. The steam drum contains a water-steam interface. The upper drum contains the feedwater inlet distributor, a blowdown header, and water separation equipment. The lower mud drum is always full of liquid.

Gas and Oil Burners

Most boilers use natural gas or atomized fuel oil burners to provide heat to the furnace. Burners inject air and fuel through a distribution system that mixes them into the correct concentrations so combustion can occur easily. Some large boilers, primarily in electrical generating plants, burn coal. The key components of the combustion apparatus (Figure 9.5) include the following:

- Dampers that regulate air into the burner
- Air ducts with fixed blades that create a swirling effect as air enters the furnace
- Components called **spuds** that distribute fuel gas
- An igniter that works like a spark plug to ignite the flammable mixture

Flame detection instruments shut off fuel gas if the flame goes out; and factory mutual valves (FM valves) shut off fuel gas when potentially dangerous

Figure 9.5
Natural Gas Burner

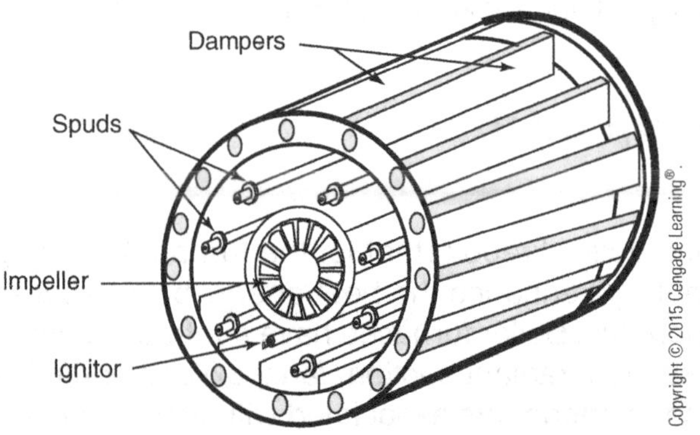

situations arise, such as low drum level and flame failure. Most plant boilers use forced-draft fans to supply combustion air.

Economizer Section

The **economizer** section (see Figure 9.4) is used to increase boiler efficiency by preheating the water as it enters the system. This section is a series of headers and tubes located between the firebox and the stack. Temperatures are typically lower in the economizer section than in the rest of the system, but the hot flue gases moving out of the firebox and into the stack still have enough heat to offset energy costs. The economizer section in a boiler is very similar to the convection section in a fired heater system. Both operate under the energy-saving concept of using the hot flue gases before they are lost out the stack.

Boiler Functions

When a boiler is being started up, the following process occurs. The furnace, which contains cool water in drums and tubes, starts to heat up. When the burners are lit, hot combustion gases begin to flow over the generating tubes, riser tubes, downcomer tubes, and drums. Radiant, convective, and conductive heat transfer begin to take place. Hot gases flow out of the firebox, into the economizer section, and out the stack. Water temperature increases at programmed rates. Pressure begins to increase. Steam may initially be vented to the atmosphere. As the temperature of the water inside the generating and riser tubes increases, the density of the water decreases and initial circulation is established. Bubbles begin to form and rise in the water, increasing circulation and pressure (Figure 9.6). This circulation rate can easily reach 2 million pounds per hour. At this point, approximately 65,000 pounds per hour of steam is being produced.

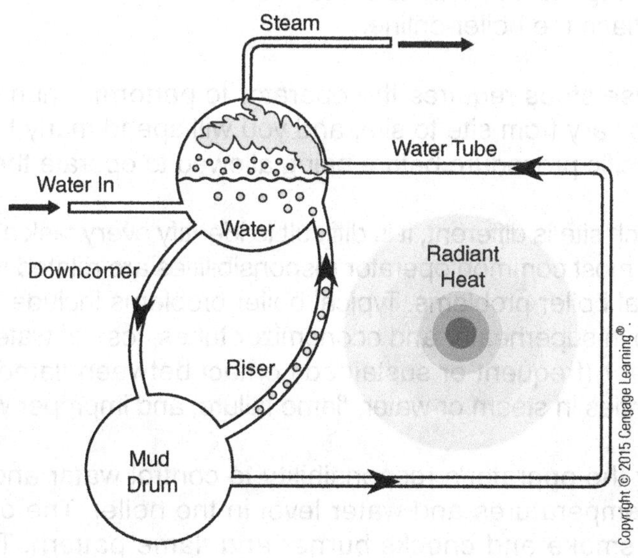

Figure 9.6 *Steam and Water Drum Circulation*

Each time the water passes through the tubes, it picks up more heat energy. When the pressure increases to slightly above the system pressure, steam will flow through the nonreturn valve into the system. **Boiler load** is a term used to describe the plant's demand for steam.

Steam

As long as steam and water are in contact with each other, the steam is saturated. This saturated condition means that for every temperature of water, a corresponding pressure of steam exists. The pressure on the water sets the temperature as long as the steam and water are in contact. Basic boiler design removes the steam from the upper steam water drum and heats it up at essentially the same pressure. This process is referred to as *superheating*.

Some plant processes cannot tolerate high temperatures. The process of cooling the superheated steam is referred to as *desuperheating*. During the desuperheating process, part of the superheated steam is returned to the steam drum. The cooler liquid in the steam drum removes heat from the superheated stream and allows it to be used in specific plant processes.

Boiler Operation

Starting up a boiler requires the following steps:

1. Fill the steam drum with water to the normal level.
2. Start the fan.
3. Purge the furnace.
4. Check furnace for percentage of flammables.
5. Light the burners.
6. Bring the boiler up to pressure.
7. Place the boiler online.

Each of these steps requires the operator to perform a number of tasks. These tasks vary from site to site, and you will spend many hours training for your specific procedure before being allowed to operate the boiler.

Because each site is different, it is difficult to identify every task a boiler operator has. The most common operator responsibilities are related to the prevention of typical boiler problems. Typical boiler problems include tube rupture, soot buildup in superheater and economizer tubes, loss of water flow, **flame impingement** (frequent or sustained contact between flame and tubes), scale, impurities in steam or water, flame failure, and improper water level.

It is usually the operator's responsibility to control water and steam flow rates and temperatures and water level in the boiler. The operator also checks for smoke and checks burner and flame pattern. The operator maintains good housekeeping and unit logs and checks fuel pressure and temperature and oxygen level. Finally, the operator monitors the pressure of the firebox and drum; the temperatures in the firebox, stack, superheater, and desuperheater temperatures; and ensures fan operation.

Steam Systems

Steam is used in a variety of applications in industrial manufacturing environments. There is a considerable cost incurred in the treatment and production of steam, so steam reclamation is an important and common feature at most companies that use steam in their processes.

As steam flows from the boiler to the plant, it begins to cool. As it cools, condensate is formed. Condensate can cause many of serious problems as it flows with the steam. Slugs of water can damage equipment and lead to a condition known as **water hammer**. Devices known as **steam traps** are used to remove condensate. Steam traps are grouped into two categories: mechanical and thermostatic. Mechanical steam traps include inverted buckets and floats. Thermostatic traps include bellows-type traps. A steam system that includes a steam trap is shown in Figure 9.7.

Inverted Bucket Steam Trap

The **inverted bucket steam trap** (Figure 9.8) is a simple mechanical device used to remove condensate from steam and return it to a condensate header. The condensate header runs back to the boiler, where the

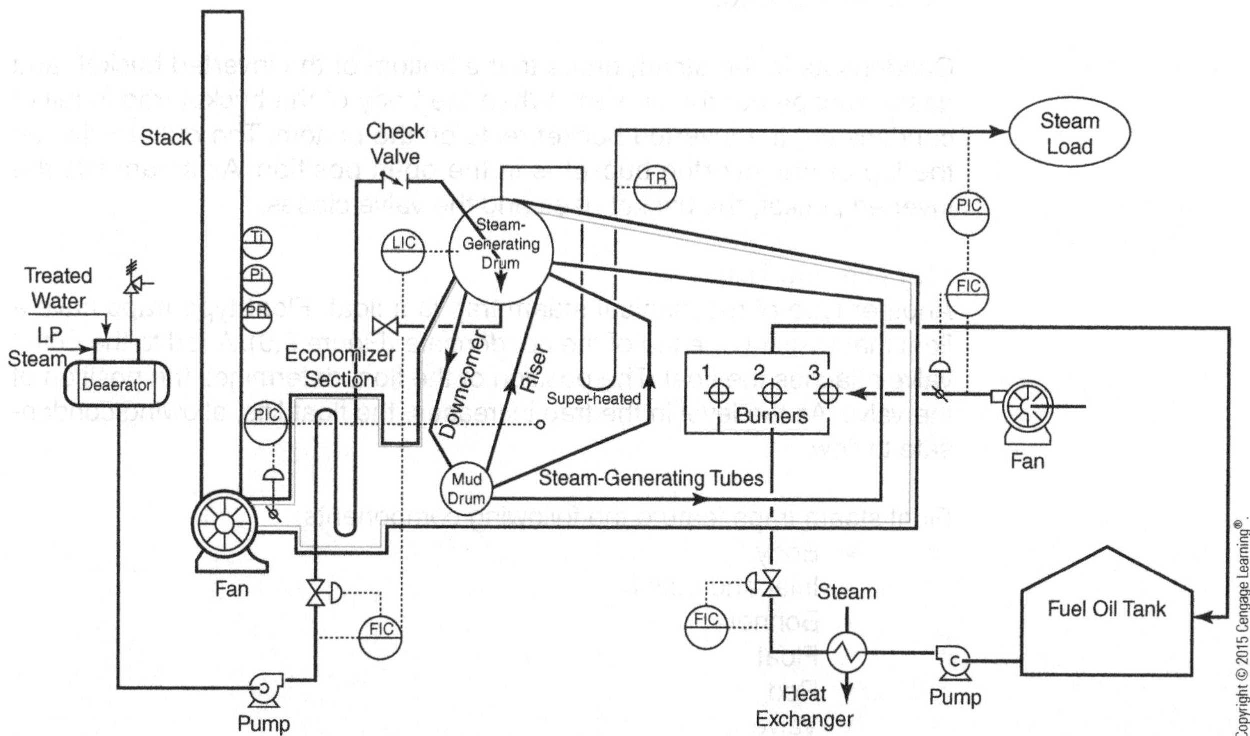

Figure 9.7 *Steam System*

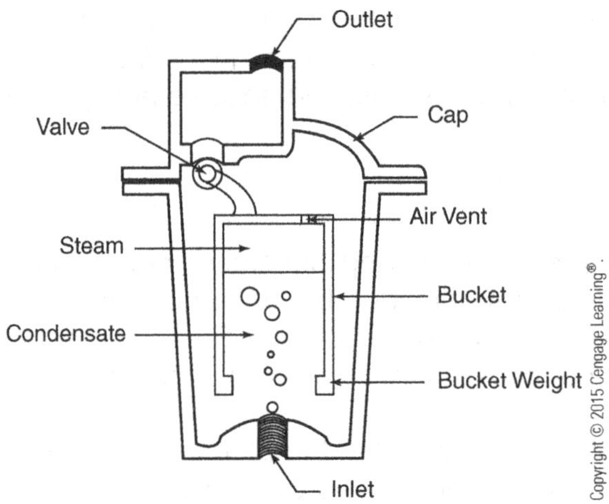

Figure 9.8
Inverted Bucket Steam Trap

clean condensate is converted to steam. Inverted bucket traps can handle condensate, air, and other noncondensable gases such as nitrogen and oxygen.

During operation, the steam enters the bottom of the trap via the inlet and fills the inverted bucket. An air vent is located on the top of the bucket. Gases escape through this hole and into the outlet line. The outlet valve is also located on the top of the inverted bucket. The position of the bucket determines whether the valve is open or shut. When the bucket is in the lower position, the valve is open. When the bucket is in the upper position, the valve is closed.

Condensate in the steam drops to the bottom of the inverted bucket, and gases escape out the air vent. When the body of the bucket trap is full of condensate, the inverted bucket rests on the bottom. The outlet valve on the top of the inverted bucket is in the open position. As steam fills the inverted bucket, the bucket rises and the valve closes.

Float Steam Trap

Another type of mechanical steam trap is a float. Float-type traps have a float that rests on the top of the condensate (Figure 9.9). A rod to the outlet valve attaches the float. The position of the float determines the position of the valve. As the level in the trap increases, the float lifts, allowing condensate to flow.

Float steam traps feature the following components:

- Body
- Inlet and outlet
- Bonnet
- Float
- Rod
- Valve

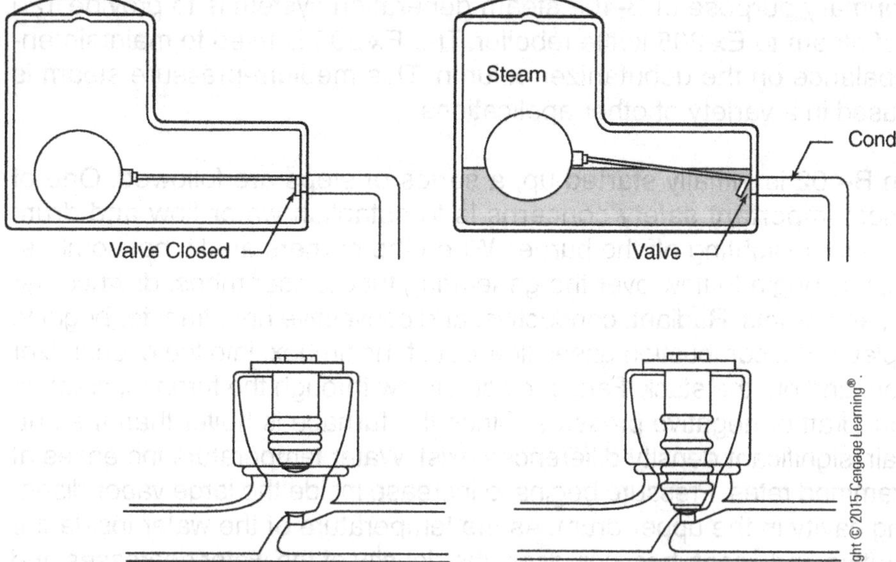

Figure 9.9
Float Steam Trap

Figure 9.10
Thermostatic Steam Trap

Float traps are not designed to handle noncondensable gases. Noncondensable gases can keep the float trap from operating properly. This condition is referred to as being *air-bound*.

Bellows Thermostatic Steam Trap

One of the most popular steam traps is the **thermostatic steam trap**. Thermostatic steam traps are cheaper and selected more frequently than any other. This type of trap responds to the temperature differences between condensate and steam. A common thermostatic trap design is the **bellows trap** (Figure 9.10).

During operation, steam enters the bottom of the trap and comes into contact with the bellows. Condensate causes the bellows to contract and open. Steam causes the bellows to expand and close. Bellows traps can handle condensate and noncondensable gases.

Steam Generation System

Steam-generating systems are very large and very complex. Modern control instrumentation makes the operation and control of this type of system much easier. There are a number of hazards associated with the boiling water and the steam produced. High-pressure steam directed in a narrow beam can cut a broom stick in half. High-pressure steam can also provide rotational energy to a steam turbine. Instrument systems are only as useful as the technicians are that work with them. Alarms that are ignored or bypassed, control loops that are left in manual, or process problems that are ignored can lead to serious consequences.

The primary purpose of B-402 steam generation system is to provide 120 psig of steam to Ex-205 kettle reboiler. The Ex-205 is used to maintain energy balance on the debutanizer column. This medium-pressure steam is also used in a variety of other applications.

When B-402 is initially started up, a series of steps are followed. One of the most important safety concerns is to establish water flow and drum levels prior to lighting off the burner. When the burners are lit, hot combustion gases begin to flow over the generating tubes, riser tubes, downcomer tubes, and drums. Radiant, conductive, and convective heat transfer begin to take place. Hot combustion gases flow out of the firebox, into the economizer section, and out the stack. Fans provide airflow through the furnace, creating a slight draft or negative pressure. Since the furnace is hotter than the outside air, significant density differences exist. Water temperature increases at programmed rates. Pressure begins to increase inside the large vapor disengaging cavity in the upper drum. As the temperature of the water inside the generating and riser tubes increases, the density of the water decreases and initial circulation is established. Bubbles begin to form and rise in the water, increasing circulation and pressure. Each time the water passes through the tubes, it picks up more heat energy. When the pressure increases to slightly above the system pressure set point, steam will flow to the header.

Inside the upper steam-generating drum of B-402, steam and water come into physical contact, saturating the steam. This saturated condition means that for every temperature of water, a corresponding pressure of steam exists. The pressure on the water sets the temperature as long as the steam and water are in contact. Basic boiler design removes the steam from the upper steam water drum and superheats it at essentially the same pressure. B-402 is designed to operate at 120 psig. However, some operating facilities require low-pressure steam. This is when desuperheating is used. During the desuperheating process, part of the superheated steam is routed through the boiling liquid in the steam drum, cooling it down to a lower pressure. The boiling water is cooler than the 120 psig steam and reduces the pressure to around 60 psig.

A number of hazards are associated with the operation of a boiler system. Some of these hazards include:

- Hazards associated with high-temperature steam, “burns”
- Hazards associated with using natural gas
- Hazards associated with leaks
- Instrument failures
- Hazards associated with *confined space entry*
- Opening and blinding
- Isolation of *hazardous energy* permit labeled “Lock-out, Tag-out”
- Routine work on equipment and facilities
- Hazards associated with lighting burners
- Exceeding boiler temperatures or pressures

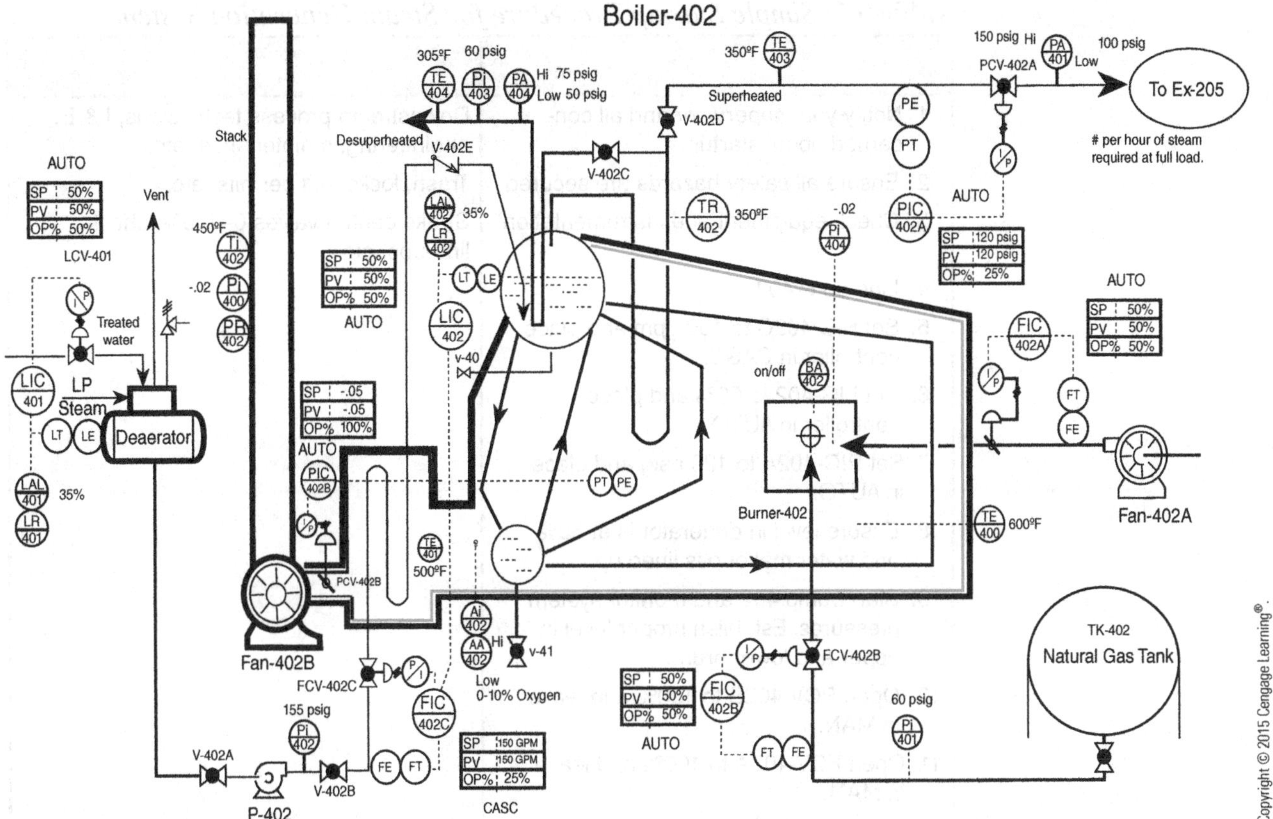

Figure 9.11 *Steam Generation System: Boiler B-402*

- Hazards associated with using water treatment chemicals
- Error with valve line-up resulting in explosion or fire

While a large list of potential hazards exist beyond the above list, it indicates that careful training is required for all new technicians assigned to utilities. Figure 9.11 illustrates a steam generation system.

Steam System Symbols

Steam system devices can be represented as symbols. Figure 9.12 shows steam system symbols. Table 9.1 describes a simple approach to starting up a steam generation system.

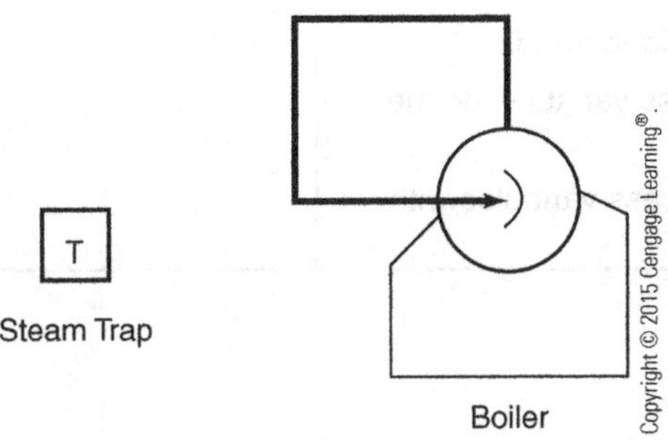

Figure 9.12 *Steam System Symbols*

Table 9.1 *Simple Startup Procedure for Steam Generation System*

Procedure	Notes
1. Notify your supervisor and all concerned about startup.	Downstream process technicians, I & E, engineering, maintenance, etc.
2. Ensure all safety hazards are secured.	Trash, locks, old permits, etc.
3. Check equipment and instrumentation.	Stroke control valves 0–100%, check line-ups, etc.
4. Line up P-402	
5. Set FIC-402C to 150 gpm and place controller in CASC.	
6. Set LIC-402 to 50% and place controller in AUTO.	
7. Set PIC-402A to 120 psig and place in AUTO.	
8. Ensure level in deaerator is at 50% and water makeup is lined up.	
9. Start pump 402 and monitor system pressures. Establish proper level in upper and lower drum.	
10. Open PCV-402B to 100% and leave in MAN.	
11. Open FCV-402A to 100% and leave in MAN.	
12. Ensure boiler has been purged with steam for 10 minutes.	
13. Monitor process pressures and level in TK-402.	
14. Ensure pilot light on burner 1 is active.	
15. Set FIC 402A to 50% and place controller in AUTO.	
16. Set PIC 402B to -.05 and place controller in AUTO.	
17. Set FIC 402B to 50% and place in AUTO.	
18. Start fans 402A and 402B.	
19. Light burner and allow boiler to come up to pressure and temperature.	
20. Monitor oxygen level and ensure it stays between 0 and 10%.	
21. Monitor TR-402 video trends.	
22. Monitor all process variables on the stack.	
23. Cross-check process variables with SPEC sheet.	

Table 9.2 *Specification Sheet And Checklist*

Level			
1. LIC-402	50%	AUTO	Upper steam-generating drum level
2. LIC-401	50%	AUTO	Deaerator level
3. LA-402	35%	Low	Upper steam-generating drum level
4. LR-402		Trend	Upper steam-generating drum level
5. LA-401	35%	Low	Deaerator level
6. LR-401		Trend	Deaerator level
Flow			
7. FIC-402A	50%	AUTO	Air to furnace
8. FIC-402B	50%	AUTO	Natural gas feed
9. FIC-402C	150 GPM	CASC	Makeup water
Analytical			
10. Ai-402		AUTO	Stack discharge
11. AA-402	0–10%	hi/lo	Combustion gases
12. BA-402	On/Off		Burner
Pressure			
13. PIC-402A	120 psig	AUTO	Steam header
14. PIC-402B	-.05 in water	AUTO	Damper
15. Pi-402	155 psig	gauge	P-402 discharge
16. Pi-400	-.02 in water	gauge	Stack temperature
17. Pi-401	60 psig	gauge	Natural gas supply pressure
18. Pi-403	60 psig	gauge	Desuperheated steam
19. Pi-404	-.02 in water	gauge	Fire box
20. PA-404	75/50 psig	hi/lo	Desuperheated steam pressure
21. PA-401	150/100 psig	hi/lo	Steam header
22. PR-402		Trend	Stack
Temperature			
23. TR-402	350°F	Trend	Steam header
24. Ti-402	450°F	gauge	Upper stack temperature
25. TE-400	600°F		Radiant section
26. TE-401	500°F		Economizer section
27. TE-403	350°F		Steam header
28. TE-404	305°F		Desuperheated steam

Summary

Boilers—steam generators—are devices that produce steam. They use a combination of radiant, conductive, and convective heat transfer methods to change water to steam.

Factors that affect boiler operation are density differences for internal circulation, pressure, temperature, and water level.

A fire-tube boiler resembles a shell-and-tube heat exchanger in that it has a series of tubes enclosed in a shell. The tubes are heated by hot combustion gases and are submerged in water. Heat is transferred from the hot tubes to the liquid through conduction and convection.

The most common type of large commercial boiler is the water-tube boiler, which consists of a furnace that contains an upper and lower drum connected by tubes. Circulation through the system depends on density differences in the water in the various tubes. This type of boiler produces superheated and desuperheated steam.

Steam systems designed to reclaim steam use steam traps to remove condensate. Steam traps are grouped into two categories: mechanical (inverted bucket steam trap and float steam trap) and thermostatic. Thermostatic steam traps are cheaper and selected more frequently than any other. They respond to the temperature differences between condensate and steam. A common thermostatic trap design is the bellows trap.

Review Questions

1. What is the name of the section in a water-tube boiler that preheats the water?
2. What is a spud?
3. Contrast a water-tube boiler and a fire-tube boiler.
4. Contrast a downcomer tube with a generating, or riser, tube.
5. Identify the key components of a water-tube boiler, and describe the water circulation in the boiler.
6. Contrast superheated steam, desuperheated steam, and saturated steam.
7. List five operations in which steam is used.
8. List six types of tubes found in a water-tube boiler.
9. Contrast the upper and lower drum in a water-tube boiler.
10. List the key components of a natural gas burner.
11. What are the seven major things an operator does when starting up a boiler?
12. List three operating problems found in a boiler.
13. What is the purpose of a steam trap?
14. Name the two classes of steam traps.
15. Name and describe two types of mechanical steam traps.
16. Name and describe a type of thermostatic steam trap.
17. What term is used for a condition in which slugs of water cause damage to equipment?
18. Describe hazards associated with boiler operation.
19. Define *placed on the line*.
20. Define *boiler load*.

Furnaces

Objectives

After studying this chapter, the student will be able to:

- Describe the various types of direct fired heaters.
- Explain the operation of an indirect fired heater.
- Apply the principles of heat transfer to fired heater operation.
- Describe the basic components of a furnace.
- Describe the different types of furnaces.
- Describe common solutions to furnace problems.

Key Terms

A-frame furnace—a furnace that has an A-frame-type exterior structure.

Air preheater—heats air before it enters a furnace at the burners.

Air registers—located at the burner of a furnace, these devices adjust secondary airflow.

Arch—a neck-like structure that narrows as it extends between the convection section and stack of a furnace.

Box furnace—a square or rectangular furnace with both a radiant and convection section.

Bridgewall—sloping section inside a furnace that transitions between the radiant section and convection section; or the section of refractory that separates fireboxes and burners.

Broken burner tiles—are located directly around the burner and are designed to protect the burner from damage. The furnace rarely needs to be shut down to replace a broken tile unless it is affecting the flame pattern.

Broken supports and guides—tend to fall to the furnace floor. Missing supports or guides will result in tubes sagging or bowing.

Burner alarms—immediately notify technicians when a burner goes out.

Cabin furnace—a cabin-shaped, aboveground furnace that transfers heat primarily through radiant and convective processes.

Charge—the process flow in a furnace.

Coking—formation of carbon deposits in the tubes of a furnace.

Color chart of steel tubes—shows 10 tube color variations associated with temperature.

Convection section—the upper area of a furnace in which heat transfer is primarily through convection.

Convection tubes—tubes located above the shock bank of a furnace or away from the radiant section where heat transfer is through convection. The first pass of tubes directly above the radiant section is referred to as the *shock bank*.

Cylindrical furnace—a cylindrical, vertical furnace, primarily designed to transfer radiant heat to a process stream.

Draft—negative pressure of air and gas at different elevations in a furnace.

Feed composition—the composition of the fuel entering a furnace, which must remain uniform or furnace operation will be affected.

Firebox—the area in a furnace that contains the burners and open flames; the area of radiant heat transfer.

Flameout—extinguishing of a burner flame during furnace operation.

Flame impingement—direct flame impingement occurs when the visible flame hits the tubes. Flame impingement can be classified as periodic or sustained.

Flashback—intermittent ignition of gas vapors, which then burn back in the burner; can be caused by fuel composition change.

Fuel pressure control loop—includes a pressure sensor and transmitter, controller, transducer and control valve designed to maintain a constant fuel pressure to the furnace burners.

Fuel pressure control—a pressure control loop located on the natural gas fuel line to the furnace that is designed to maintain constant pressure to the furnace burners.

Furnace flow control—a critical feature in furnace operation, temperature, and pressure control that regulates fluid feed rates in and out of the process furnace.

Furnace hi/lo alarms—alarm warnings that warn when the process flow is off specification and prevent equipment damage and harm to the environment and human life.

Furnace pressure control—monitors furnace pressure in the bottom, middle, and top of the furnace with a pressure control loop connected to the stack damper. The middle pressure reading on the furnace is compared to a set point and adjustments are made at the damper if necessary.

Furnace temperature control—adjusts fuel flow to the burners, and, as flow exits the process furnace, monitors process conditions. The natural gas flow controller (*slave*) is cascaded to the (*master*) temperature controller. The temperature controller adjusts fuel flow to the burners.

Hazy Firebox or Smoking Stack—often occur when not enough excess air is going into the firebox or the fuel air mixing ratio is incorrect.

Header box gaskets—provide access to the terminal penetrations or bends on the convection tubes; also called header box doors.

Header box doors and gaskets—provides access to the terminal penetrations or bends on the convection tubes; also called *header box doors.* The gaskets provide a positive seal between the inside and outside of the furnace.

Hot spot—a glowing red spot on the metal or refractory inside a furnace.

Hot tubes—glow different colors when the inside or outside of the tubes foul and when there is flame impingement, reduced flow rate, and overfiring of the furnace.

Low burner turn-down—a condition that can result in hazy firebox.

Low NO_x burners—a type of gas burner, invented by John Joyce, that significantly reduces the formation of oxides of nitrogen. Low NO_x burners are 100% efficient as all heat energy released from the flame is converted to useful heat.

Oxygen analyzer—an instrument specifically designed to detect the concentration of oxygen in an air sample. Oxygen flow rates are carefully controlled through a furnace.

Peepblocks with Peepholes—refractory blocks with holes in the center provide visual access that enable operators to inspect visually the inside of the furnace.

Peepholes—holes in the side of a furnace that enable operators to inspect visually the inside of the furnace.

Plugged burner tips—flame pattern erratic, shoots out toward a tube instead of up the firebox.

Preheated air—a compressed air system that typically pushes the air through tubes located in the upper section of the furnace. This preheated air takes full advantage of energy flow passing out of the furnace stack.

Process heaters—combustion devices that transfer convective and radiant heat energy to chemicals or chemical mixtures. Process tubes pass through the convection and radiant sections as energy is transferred to them. This transferred energy allows the liquid to be utilized in a variety of chemical processes that require higher temperatures.

Radiant tubes—tubes located in a furnace firebox that receive heat primarily through radiant heat transfer; also called *radiant coils*.

Refractory—the lining of a furnace firebox that reflects heat back into the furnace.

Ruptured tubes—flames come from opening in tubes. May cause excess oxygen levels to drop and bridge wall temperatures to increase.

Sagging or Bulged tubes—occur when guides or supports break, when inside or outside of tube fouls, due to flame impingement, reduced flow rate, or over-firing furnace. Note: Diameter of tube does not change when it sags; however, it does when it bulges.

Shock bank—tubes located directly above the firebox of a furnace that receive radiant and convective heat. The shock bank is part of the convection section.

Spalled refractory—an aging refractory that has cracked or deteriorated over time; a refractory that has not cured or dried properly; or a refractory whose anchors have failed; thus resulting in the refractory breaking loose from the sides of the furnace and falling to the furnace floor. Caused by old refractory that has cracked or deteriorated over time, or refractory that has not cured or dried properly, or broken refractory anchors.

Stack—outlet on the top of a furnace through which hot combustion vapors escape from the furnace.

Soot blowers—remove soot from tubes in the convection section that consist of hollow metal rods that are inserted into the convection section and incorporate a series of timers that admit nitrogen in quick bursts.

Terminal penetrations—provide 180° turns or pipe bends in the convection section as the pipes scroll from one side of the furnace to the other.

Vibrating tubes—tend to jump or move back and forth. Typically occurs in tubes outside the furnace. Vibrating tubes are often caused by two-phase slug-type flow inside the tubes. May be stopped by changing flow rates.

Furnace Applications and Theory of Operation

A furnace—that is, a fired heater—is a device used to heat up chemicals or chemical mixtures. Fired heaters transfer heat generated by the combustion of natural gas, ethane, propane, or fuel oil. Furnaces consist essentially of a battery of pipes or tubes that pass through a **firebox**. These tubes run along the inside walls and roof of a furnace. The heat released by the burners is transferred through the tubes and into the process fluid. The fluid remains in the furnace just long enough to reach operating conditions before exiting and being pumped to the processing unit.

Furnaces are used in crude processing, cracking, olefins production, and many other processes. Furnaces heat up raw materials so that they can produce products such as gasoline, oil, kerosene, chemicals, plastic, and rubber. The chemical-processing industry uses a variety of fired heater designs. These elaborate furnace systems can be complicated and equipped with the latest technology.

Heat Transfer

The primary means of heat transfer in a fired heater are radiant heat transfer and convection (see Figure 7.2); however, heat must pass through the walls by conduction to be absorbed by the flowing fluid. In the fired furnace, the flame on the burner is the radiant heat source. Radiant heat transfer takes place primarily in the firebox. Tubes located in the firebox are referred to as *radiant coils* or *tubes*. The tubes transfer heat to the fluid by conduction. In a fired furnace, radiant heat is emitted from the combustion of natural gas or light oil. As the radiant heat travels from the bottom of the furnace, contacting the tubes or passing in the furnace, and then continues to the top, heat is transferred to the surrounding air. This process initiates the convective heat transfer process that causes the lighter air and hot combustion gases to rise above the radiant heat source. The top of the furnace is referred to as the **convection section** because most of the heat it receives is by convection.

Combustion

Combustion is a rapid chemical reaction that occurs when proper amounts of fuel and oxygen (O_2) come into contact with an ignition source and release heat and light. Furnaces use this principle to provide heat. Complete combustion occurs when reactants are ignited in the correct proportions. Incomplete combustion occurs in a fired furnace when not enough oxygen exists to completely convert all of the fuel to water and carbon dioxide.

Many furnaces use natural gas or methane (CH_4) as fuel for the burners. Methane (CH_4) reacts with O_2 to form carbon dioxide (CO_2) and water (H_2O):

$$CH_4 + 2O_2 \rightarrow CO_2 + 2H_2O$$

Incomplete combustion may result in the production of carbon monoxide. The chemical processing industry also uses ethane, propane, and light oils for fuel. Figure 10.1 illustrates the basic components of the fire triangle or fire tetrahedron. Another common combustion reaction with oxygen is $C_3H_8 + 5O_2 \rightarrow 3CO_2 + 4H_2O$. Propane and oxygen form similar products to methane and oxygen.

Fuel Heat Value

Different fuels release different amounts of heat energy as they are burned. The heat energy released, referred to as the heat value, is measured in British thermal units per cubic foot. The British thermal unit (Btu) is a measurement of heat energy. One Btu is the amount of heat required to raise the temperature of one pound of water one degree Fahrenheit. Hydrogen has the lowest fuel heat value (274 Btu/foot^3), whereas natural gas, or

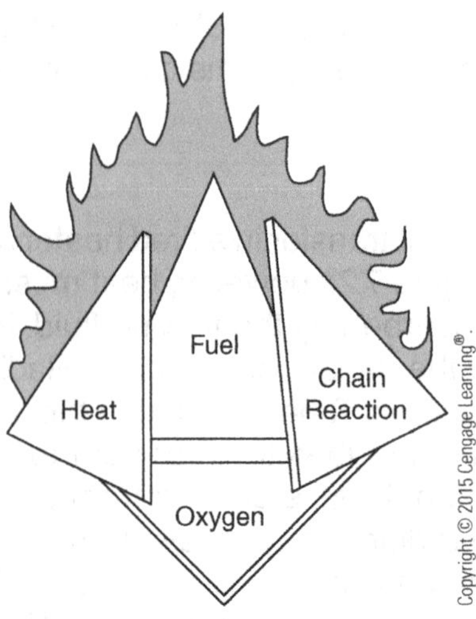

Figure 10.1
Fire Triangle or Tetrahedron

methane, has a heat value of 909 Btu/foot^3. Charts are available that list the heating values of fuels used in furnaces. It is important to realize that the more Btus a fuel gives off, the more oxygen is required for combustion.

Basic Components of a Furnace

Fired heaters come in a variety of shapes and sizes. They have different tube arrangements and feed inlets and burn different types of fuels and have different burner designs. All furnaces do, however, have certain things in common: firebox, radiant tubes or coils, convection tubes, damper and **stack**, **refractory** lining, burners and **air registers**, fuel system, instruments, and induced- or forced-draft fans.

Firebox and Refractory Layer

The section in a furnace that contains the burners and open flames is called the *firebox*. The firebox is lined with a refractory layer, a brick lining that reflects heat back into the furnace. The refractory brick is classified as firebrick or insulating brick, both of which are specially designed to withstand and reflect heat. Firebrick has a density range of 131 to 191 lb./foot^3 and maximum temperature ranges between 2,500°F (1371.1°C) and 3,300°F (1815.55°C). Insulating firebrick has much lower densities, 27.3 to 78.7 lb./foot^3, and maximum temperature ranges between 1,600°F (871.1°C) and 3,250°F (1787.77°C).

The refractory bricks are attached to stainless steel rods that are attached to 3- to 6-inch ceramic fiber insulation bat. The insulation bat and metal shell of the furnace touch. The insulation barrier between the furnace shell and brick prevents heat loss. The upper convection section and the **arch** section (the neck that narrows as it runs between the convection section and the stack) are usually insulated with heavy or light high-temperature cement (castable) or firebrick. Castable peep blocks contain **peepholes** that allow for visual inspection. Castables have a temperature range between 1,600°F (871.1°C) and 3,300°F (1815.5°C).

Typical heat loss from a furnace is between 2 and 3% of the total heat release. Since the insulation is porous, a protective coating may be applied to the inside of the steel shell to protect it from corrosive materials such as sulfur oxides.

Temperatures inside the firebox range from 1,600 to 2,000°F (871 to 1,093°C). Furnace pressures usually run below atmospheric pressure in the range of 0.4 to 0.6 inches of H_2O **draft** (negative pressure) at the bottom of the furnace.

When the east pass (tube) enters the firebox, it receives radiant heat directly from the burners. The west pass (tube) enters the opposite side of the firebox and also receives heat from the burners. A **bridgewall** may separate the two passes in the furnace. As the **charge** (i.e., process flow) leaves the furnace, the passes (tubes) enter a common header and are pumped to the processing unit.

Radiant and Convection Tubes

The tubes located along the walls of the firebox are called the **radiant tubes** or **coils**. Radiant tubes receive direct heat from the burners. These tubes operate at high temperatures and are constructed of high-alloy steels. Radiant tubes may be mounted parallel or perpendicular to the furnace wall. Radiant heat transfer accounts for 60 to 70% of the total heat energy picked up by the charge in the furnace. A **color chart of steel tubes** shows 10 tube color variations associated with temperature. Process technicians visually inspect these tubes and compare them to the color chart.

Convection tubes are located in the roof of the furnace and are not in direct contact with burner flames. Hot gases transfer heat through the metal tubes and into the charge. Convection tubes usually are horizontal and are equipped with fins to increase efficiency. Convective heat transfer to the process charge accounts for about 30 to 40% of the total heat energy picked up in the furnace. This area is often referred to as the **convection section**. It is best described as the upper area of a furnace in which heat transfer is primarily through convection. Feed is introduced into the furnace through these tubes and exits out the radiant tubes. Tubes in this area are referred to as *convection tubes* and can be accessed through the header box doors at the terminal penetrations where the return bends or rolled headers are located. Rolled headers typically have removable plugs for maintenance and tube inspection.

Air preheaters are often used to heat air before it enters a furnace at the burners. Air tubes are typically located in the stack or convection section that allow outside air to be brought in by a compressor or blower and gradually warmed up before mixing with fuel at the burner.

Soot Blower

Soot blowers are devices found in the convection section of **process heaters**. Soot blowing is required when the efficiency of the convection section decreases. This can be calculated by looking at the temperature change from the crossover piping and at the convection section discharge. Soot blowers utilize a transfer media such as nitrogen, water, air, or steam to remove deposits from the tubes. Air movement in the convection section is slower because of the finned tubes and close proximity of each pass. The initial blast of hot combustion gases tends to accumulate deposits here, specifically along the **shock bank**.

There are several different types of soot blowers: wall blowers and finned tube blowers. Furnace wall blowers have a very short lance with a nozzle at the tip. The lance has holes drilled into it at intervals so that when it is turned on, it rotates and cleans the deposits from the wall in a circular pattern. Soot blowing continues until a preset timer goes off.

Stack Damper

Combustion gases leave the furnace through the stack and are dispersed into the atmosphere at a height to ensure against any immediate deleterious effect such as carbon monoxide poisoning. As the hot air rises in the stack, it entrains combustion by-products and carries them out of the stack. This natural draft creates a lower pressure inside the furnace that is essential to good operation. Draft is defined as the difference between atmospheric pressure and the lower pressure inside the fired heater.

A damper in the stack permits adjustment of stack drafts. The stack damper is typically set to give pressures from 0.05 to 0.15 inches of H_2O (vacuum) draft. At 0.05 H_2O, approximately 350,000 lb./hour of gas flow can be obtained.

Some dampers resemble huge butterfly valves and require only one-quarter turn to be 100% open or closed. Other dampers resemble ordinary window blinds. Any rise in furnace pressure keeps secondary and primary air out of the furnace. The damper can be positioned to increase or decrease airflow. The different drafts or pressures found in a furnace are illustrated in Figure 10.2. These readings are given by an inclined furnace tube gauge.

Controlling excess oxygen in the furnace is the single most important variable affecting efficiency. For heat transfer in the firebox or radiant section, the greatest efficiency is obtained when maximum furnace temperatures are achieved. Decreasing excess air in the furnace maximizes radiant heat transfer.

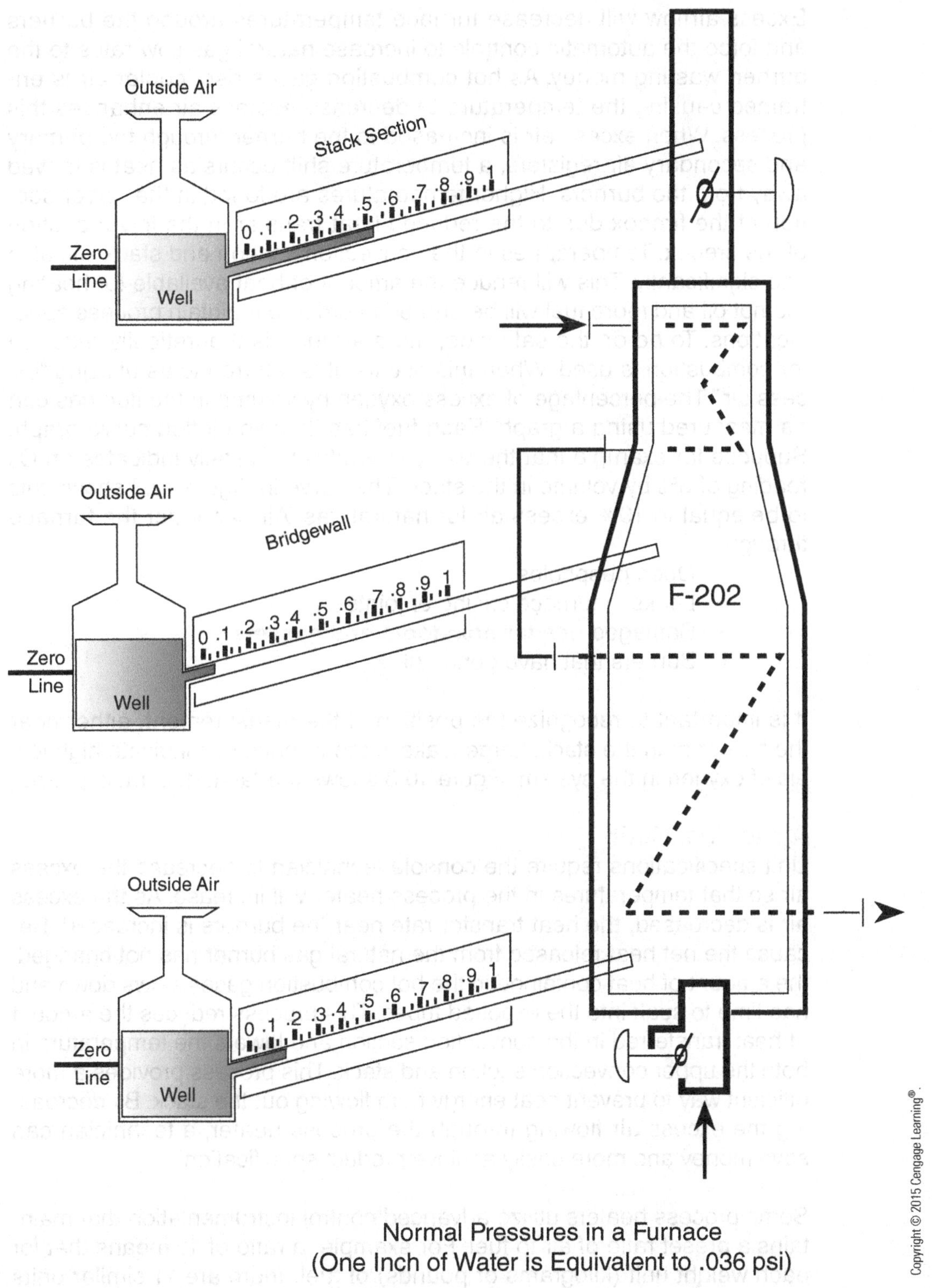

Figure 10.2 *Normal Furnace Pressures*

Excess airflow will decrease furnace temperatures around the burners and force the automatic controls to increase natural gas flow rates to the burner, wasting money. As hot combustion gases rise, cooler air is entrained causing the temperature to decrease. Excess air enhances this process. When excess air is increased to the burner through the primary and secondary air registers, a temperature shift occurs as heat is moved away from the burners. Higher temperatures are found in the upper section of the firebox due to the reduced heat transfer in the lower section of the firebox. Temperatures in the convection section and stack will also rise significantly. This will reduce the amount of heat available for heating the hot oil and more fuel will be burned in order to maintain process specifications. To be on the safe side, more air than is theoretically required for combustion is used. When this occurs, it is referred to as utilizing "excess air." The percentage of excess oxygen by volume in the flue gas can be measured using a graph. Each fuel has its own plotted curve graph. Suppose for example that the **oxygen analyzer** digitally indicates an O_2 reading of 3% by volume in the stack. The curve in Figure 10.3 shows this to be equal to 10% excess air for natural gas. Air can enter the furnace through:

- Open peepholes
- Leaks in furnace casing or joints
- Damaged **header box doors and gaskets**
- Burners that have gone out

It is important to recognize the position of the measurement, either near the burner or in the stack. Large leaks in the furnace can indicate high levels of oxygen in the system. Figure 10.3 shows the "air-to-fuel ratio" chart.

Air-to-Fuel Ratio

Unit specifications require the console technician to decrease the excess air so that temperatures in the process heater will increase. As the excess air is decreased, the heat transfer rate near the burners is increased. Because the net heat released from the natural gas burner has not changed, the amount of heat contained in the hot combustion gases slows down and has time to soak into the exposed tubes. This process reduces the amount of heat transferred in the convection section and lowers the temperature in both the upper convection section and stack. This process provides a more efficient way to prevent heat energy from flowing out the stack. By decreasing the excess air flowing through the process heater, a technician can save money and more easily achieve product specification.

Some process heaters utilize advanced control instrumentation that maintains a preset ratio of air to fuel. For example, a ratio of 11 means that for each weight unit (kilograms or pounds) of fuel, there are 11 similar units of oxygen being supplied. Higher ratios indicate that there is more excess air—a lower ratio translates to less excess air. Theoretical air can be indicated in terms referred to as *air-to-fuel ratio*. When the air is specified

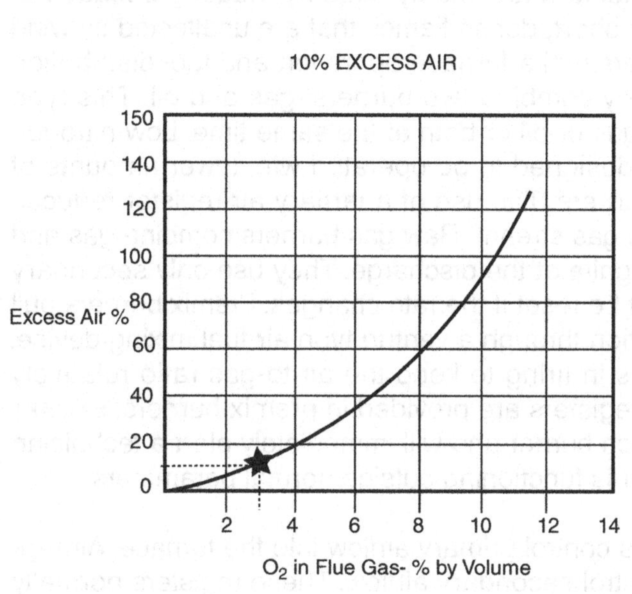

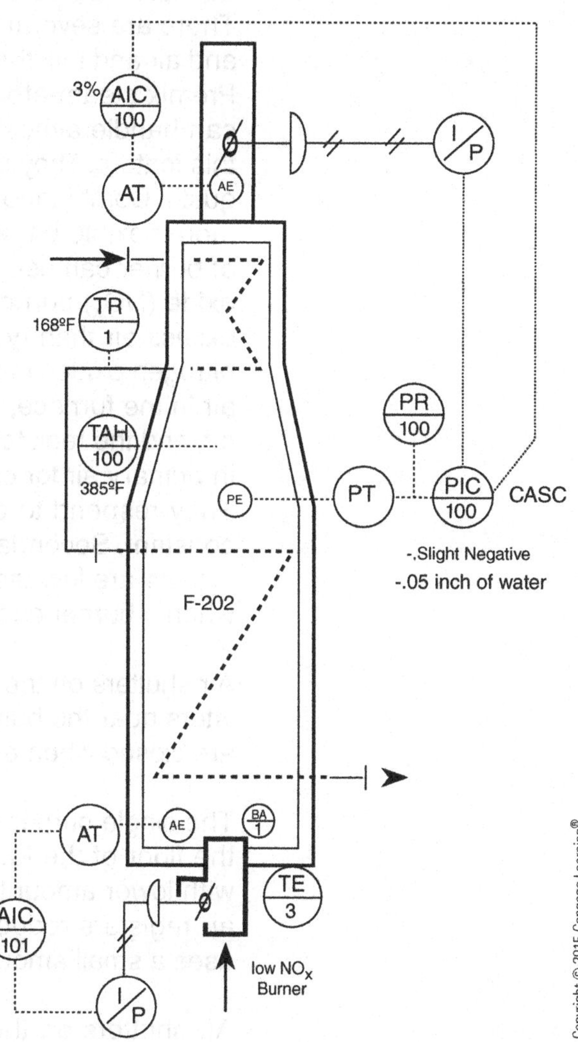

Figure 10.3 *Air-to-Fuel Ratio Chart*

in terms of air-to-fuel ratio, the amount of combustion air is calculated by adding 1 to the ratio and multiplying the results times the fuel rate. If the ratio is 11and the fuel rate is 4 pounds per minute, this can be expressed as follows:

Fuel rate = 4 lb. per minute

Air rate = 11 + 1 = 12

That is,

$4 \times 12 = 48$ pounds of flue gas produced by combustion

Burners and Air Registers

Burners can be arranged on the floor or the lower walls of the firebox. There are several types of burners. Oil burners set the proportion of fuel and air and mix them by atomizing the fuel with high-pressure steam or air. Premix steam-atomizing burners are internal-mix atomizing burners that can handle almost any fuel and are widely used by industry because of this feature. They produce short, dense flames that are unaffected by wind gusts. Combination burners make furnace operation and fuel distribution more flexible because they combine two burners: gas and oil. This type of burner can use either gas or oil or both at the same time. Low nitrogen oxide (NO_x) burners are designed to be operated with lower amounts of excess air than typical burners. The use of a tertiary air register reduces nitrogen oxides in the flue gas stream. Raw gas burners combine gas and air in the furnace, which ignite at the discharge. They use only secondary air, and the registers must be reset if the rate changes. Premix burners pull in primary air for combustion through a venturi type air fuel mixing device. They respond to changes in firing to keep the air-to-gas ratio relatively constant. Secondary air registers are provided in premix burners. **Burner alarms** are located on each burner and will immediately alert a technician when a burner goes out or is functioning outside normal parameters.

Air shutters on the burners control primary airflow into the furnace. Air registers near the burner control secondary airflow. These registers normally are closed when excess oxygen is detected in the furnace.

The single burner in this system is a low nitrogen oxide system located on the floor of the furnace. **Low NO_x burners** are designed to be operated with lower amounts of excess air than typical burners. The use of tertiary air registers reduces nitrogen oxides in the flue gas stream. The burner uses a small amount of steam to better disperse the fuel and oxygen.

Air shutters on the burners control primary airflow into the furnace. Air registers near the burner control secondary airflow. These registers normally are closed when excess oxygen is detected in the furnace. A **fuel pressure control loop** is located on the natural gas fuel line to the furnace that is designed to maintain constant pressure to the furnace burners.

The perfect mixing of air and fuel is impossible and no practical way has been found to determine when the combustion process is complete. Incomplete combustion indicates that unburned vapors will be present in the hot combustion gases. To be on the safe side, most facilities use excess air to ensure all of the fuel has been burned. The burners are designed to avoid direct contact of the flames with the tubes in the firebox. A space of 1.5 to 2 feet is considered to be a safe distance between the open flames and the radiant tubes. The burners' flame pattern should be less than 60% the height of the firebox. Figure 10.4 illustrates the basic components of a burner.

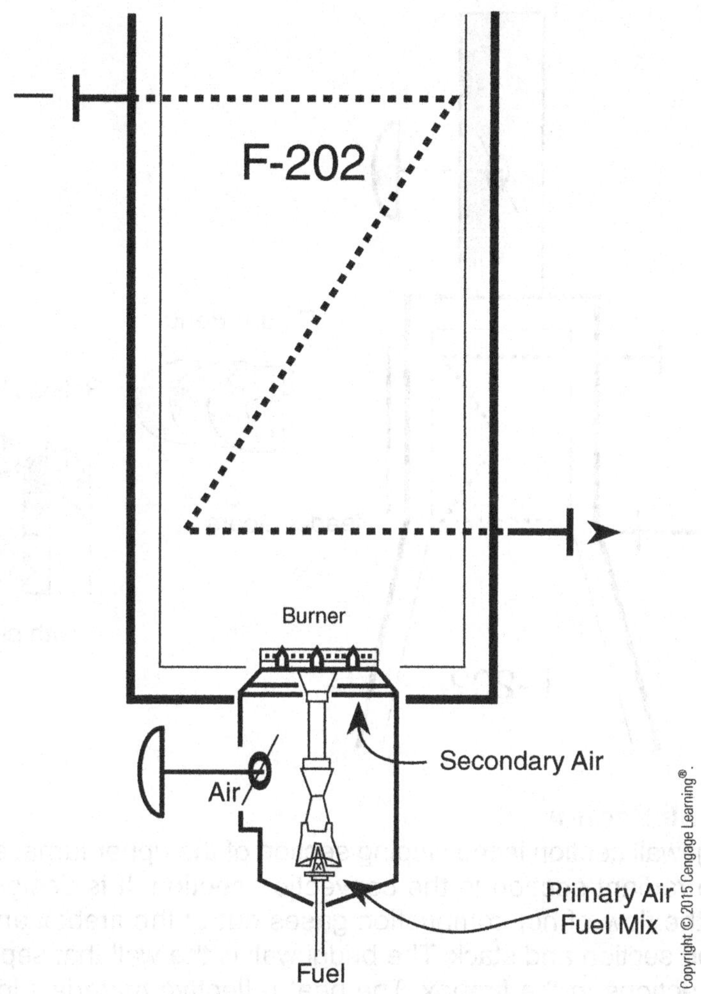

Figure 10.4
Gas Burner: F-202 Furnace

The radiant section is engineered to distribute the radiant heat energy evenly. Modern burner design consumes 100% of the fuel with a nominal excess of 10 to 15% oxygen. Excess oxygen in the furnace is carefully controlled as it enters the secondary and primary registers. This control takes place as the fuel and primary air mix at the burner and is enhanced by adjustments on the secondary air registers mounted on the outside of the burner. An oxygen monitor carefully tracks the composition of the hot combustion gases. Adjustments to the airflow rate are made at the burners and the stack damper.

The floor of the process heater has a 6 in. layer of heat-resistant castable, capped with high-temperature firebrick. Four ceramic high-temperature refractory blocks are positioned around the burner. The refractory system can withstand a wide range of high-temperature conditions. The refractory layer can be over a foot thick. The convection tubes in the upper section of the furnace have a variety of return bend designs illustrated in Figure 10.5.

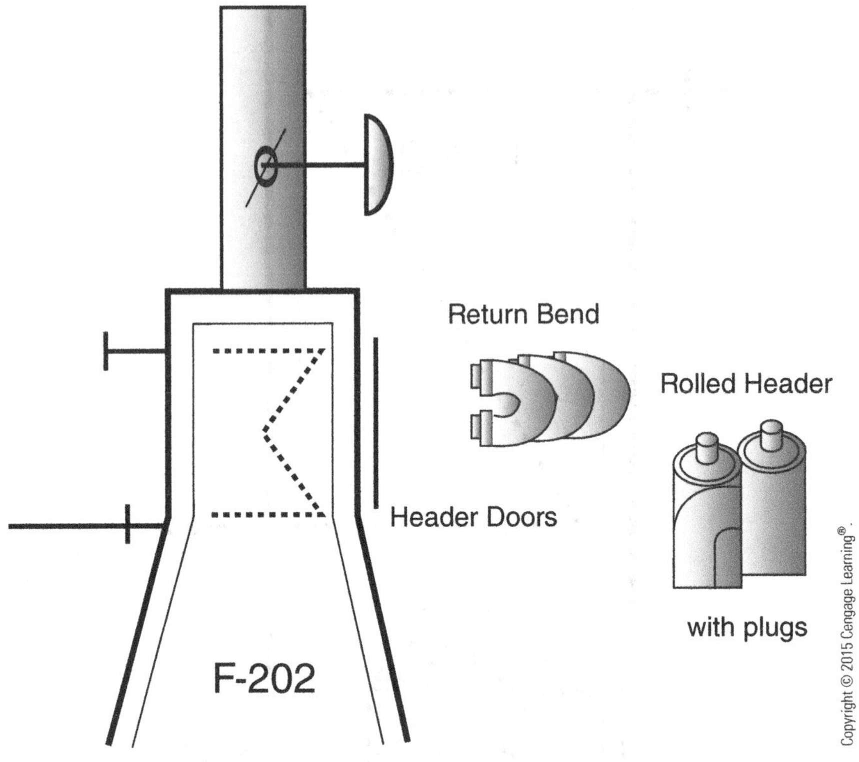

Figure 10.5 *Return Bends and Rolled Headers: F-202 Furnace*

Bridgewall Section

The bridgewall section is the sloping section of the upper furnace that connects the radiant section to the convection section. It is designed to accelerate the flow of hot combustion gases out of the firebox and into the convection section and stack. The bridgewall is the wall that separates the various sections in the firebox. The heat reflective materials in this area are designed to withstand temperatures between 1,600°F (871.1°C) and 3,300°F (1815.55°C).

Forced-Draft Process Heater

Forced-draft furnaces utilize a centrifugal blower to push **preheated air** to the burner for combustion. The preheated air is run through tube coils located above the convection section and directed to the suction of a blower (e.g., Blower 100 in Figure 10.15), which discharges under automatic control to the burner.

Fuel System

Located under or on the side of the furnace is a complex network of lines that provides fuel gas and air to the burners. The fuel is stored in a tank located a safe distance from the furnace. In an oil-burning system, atomizing steam and an oil preheating system are added to the network of pipes. Most injuries encountered in furnace operation occur during startup of the fuel burning system. **Feed composition** is best described as the composition of the fuel entering a furnace, which must remain uniform or furnace operation will be affected. The charge composition must also remain uniform or variations in process variables will occur. **Fuel pressure control** utilizes a pressure control loop located on the natural gas fuel line to the furnace that is designed to maintain constant pressure to the furnace burners. **Furnace flow control** is a critical feature in furnace operation, temperature, and pressure control that regulates fluid feed rates in and out of the process furnace. **Furnace hi/lo alarms** are alarm warnings that warn when the process flow is off specification and prevent equipment damage and harm to the environment and human life. **Furnace pressure control** monitors furnace pressure in the bottom, middle, and top of the furnace with a pressure control loop connected to the stack damper. The middle pressure reading on the furnace is compared to a set point and adjustments are made at the damper if necessary. **Furnace temperature control** adjusts fuel flow to the burners; and, as flow exits the process furnace, monitors process conditions. The natural gas flow controller (slave) is cascaded to the (master) temperature controller. The temperature controller adjusts fuel flow to the burners.

Furnace Types

Furnaces can be classified by several features: type of draft, number of fireboxes, number of passes, volume occupied by combustion gases, and shape.

Draft

Furnace draft can be natural, forced, induced, or balanced. In a natural-draft furnace (Figure 10.6), buoyancy forces induce draft as the hot air rises through the stack and creates a negative pressure inside the firebox. This pressure is lower than normal atmospheric pressure. Forced-draft furnaces (Figure 10.7) use a fan to push fresh air to burners for combustion. Forced draft is used in

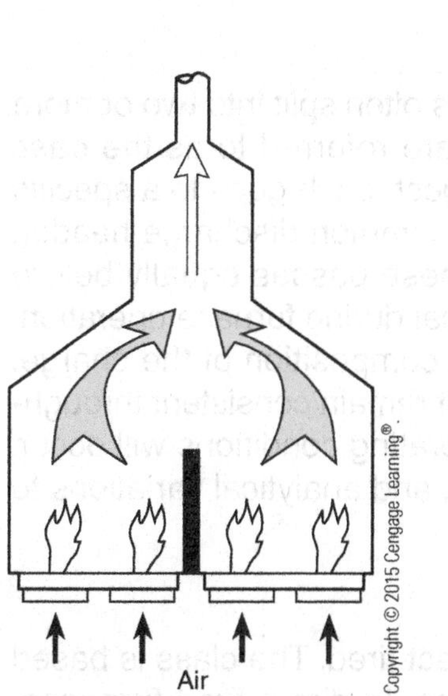

Figure 10.6 *Natural-Draft Furnace*

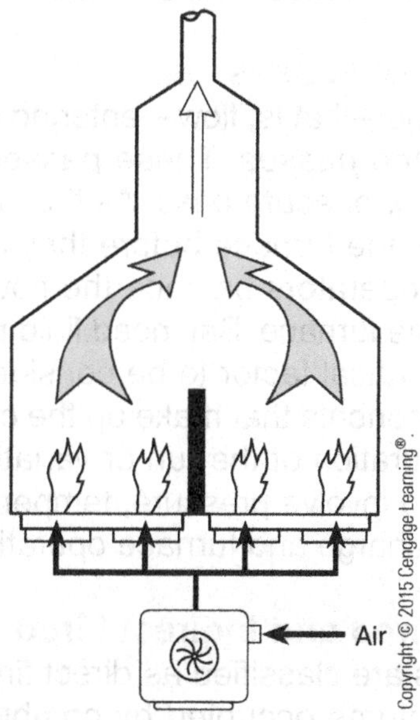

Figure 10.7 *Forced-Draft Furnace*

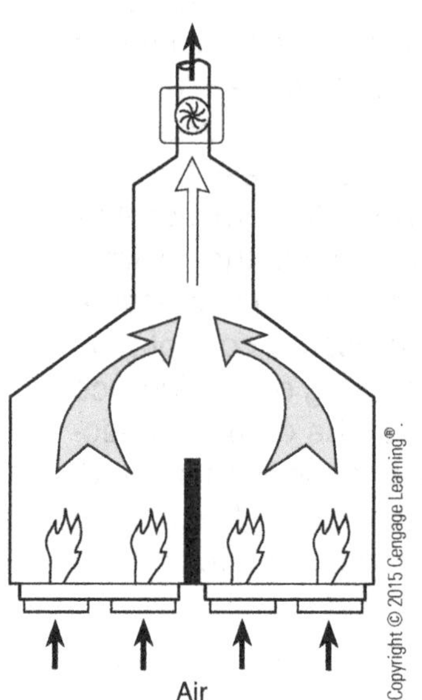

Figure 10.8 *Induced-Draft Furnace*

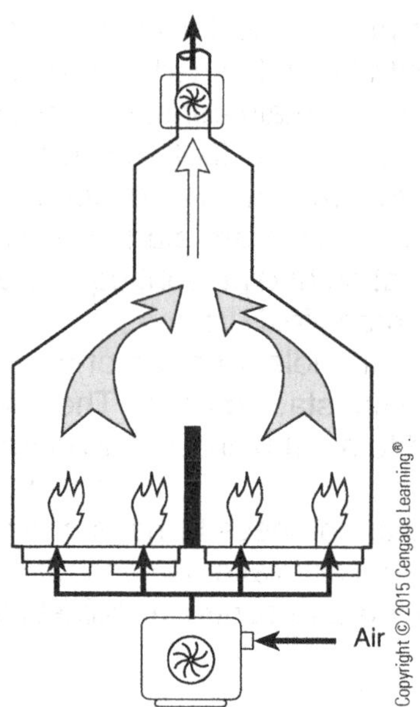

Figure 10.9 *Balanced-Draft Furnace*

furnaces that preheat the combustion air to reduce fuel requirements. In an induced-draft furnace (Figure 10.8), a fan located below the stack pulls air up through the firebox and out the stack. Balanced-draft furnaces (Figure 10.9) require two fans: one inducing flow out the stack and one providing positive pressure to the burners. Figure 10.9 shows a balanced-draft furnace.

Number of Fireboxes

A furnace can have one or two fireboxes. A double-firebox furnace has a center wall that divides two combustion chambers. Hot gases leaving the two chambers meet in a common convection section.

Number of Passes

The charge—that is, flow—entering a furnace is often split into two or more flows called *passes*. These passes usually are referred to as the *east, west, north,* or *south pass.* As the names suggest, each goes to a specific section of the furnace before they all enter a common discharge header. Furnace operators balance the flow rate of these passes equally before starting the furnace. Balanced fluid flow is critical during furnace operation. Another critical factor to be considered is the composition of the charge. The components that make up the charge must remain consistent throughout the duration of the run or variations in operating conditions will occur. This could involve pressure, temperature, flow, and analytical variations to both the charge and furnace operation.

Direct Fired and Indirect Fired

Furnaces are classified as direct fired or indirect fired. The class is based on the volume occupied by combustion gases. In direct-fired furnaces, the combustion gases typically fill the interior. Direct-fired furnaces heat process streams such as heavy hydrocarbons, glycol, water, and molten salts. Cabin, cylindrical, box, and **A-frame furnaces** are direct fired.

Fire-tube heaters are indirect fired. They contain the combustion gases in tubes that occupy a small percentage of the overall volume of the heater. The heated tubes run through a shell that contains the heated medium. A fire-tube heater resembles a multipass, shell-and-tube heat exchanger. This type of heater is composed of a shell and a series of steel tubes designed to transfer heat through the combustion chamber (tube) and into the horizontal fire tubes. Exhaust fumes exit through a chamber similar to an exchanger head and pass safely out of the boiler. The water level in the boiler shell is maintained above the tubes to protect them from overheating. The term *fire tube* comes from the way the boiler is constructed. A fire-tube heater consists of the boiler shell, fire tubes, combustion tube, burner, feedwater inlet, steam outlet, combustion gas exhaust port, and tube sheet.

Cabin Furnace

The **cabin furnace** is a very popular direct-fired heater used in the chemical-processing industry for large commercial operations. Most cabin furnaces (Figure 10.10) are located above the ground, making it possible to drain the tubes and provide easy access to the burners, which can be located on the bottom, sides, or ends. Radiant tubes may be configured in a helical or serpentine layout. The radiant section in a cabin furnace is designed to contain the flames while avoiding direct contact with the tubes.

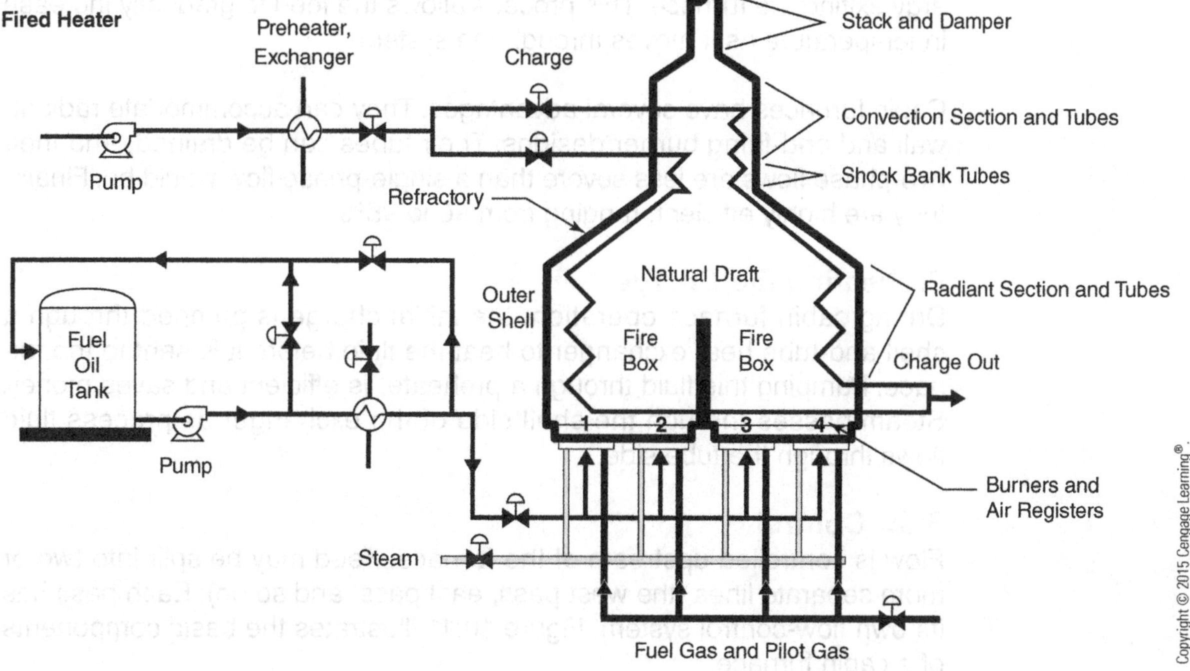

Figure 10.10 *Cabin Furnace*

A space of 1.5 to 2 feet is considered to be a safe distance between the open flames and the radiant tubes. The burners' flame pattern should be less than 60% the height of the firebox.

The radiant section is engineered to distribute the radiant heat evenly. Modern burner design consumes 100% of the fuel with a nominal excess of 10 to 15% oxygen. Excess oxygen in the furnace is carefully controlled as it enters the base of the furnace. This control takes place as the fuel and primary air mix at the burner and is enhanced by adjustments on the secondary air registers mounted on the outside of the burner. An oxygen monitor carefully tracks the composition of the hot combustion gases. Adjustments to the airflow rate are made at the burners and the stack damper.

The floor of the furnace has a 6 in. layer of heat-resistant castable, capped with high-temperature firebrick. Some cabin furnaces have a bridgewall that equally divides the firebox. A split-flow tubing arrangement exits the upper convection section as product flow drops down and into the hotter radiant section. These two separate pipe coils are also referred to as the north and south pass (or the east and west pass). The refractory system can withstand a wide range of high-temperature conditions. The refractory layer can be over a foot thick.

As hot combustion gases leave the firebox, a series of tubes—that is, the shock bank—is encountered. The shock bank in a cabin furnace receives the initial blast of the hot combustion gases. The tubes in the shock bank are the lower two or three pipe rows in the cooler convection section. The tubes in the convection section are designed to make use of the heat energy exiting the furnace. This process allows the feed to gradually increase in temperature as it moves through the system.

Cabin furnaces have several advantages. They can accommodate radiant-wall and end-firing burner designs. Their tubes can be drained, and their two-phase flows are less severe than a single-phase flow would be. Finally, they are highly efficient, ranging from 90 to 95%.

Preheating the Charge

During cabin furnace operation, the initial charge is pumped through a shell-and-tube heat exchanger to heat the fluid before it is sent to the furnace. Pumping this fluid through a preheater is efficient and saves money. Steam passes through the shell side of the exchanger as process fluid flows through the tube side.

Flow Control

Flow is controlled upstream of the furnace. Feed may be split into two or more separate lines (the west pass, east pass, and so on). Each pass has its own flow-control system. Figure 10.11 illustrates the basic components of a cabin furnace.

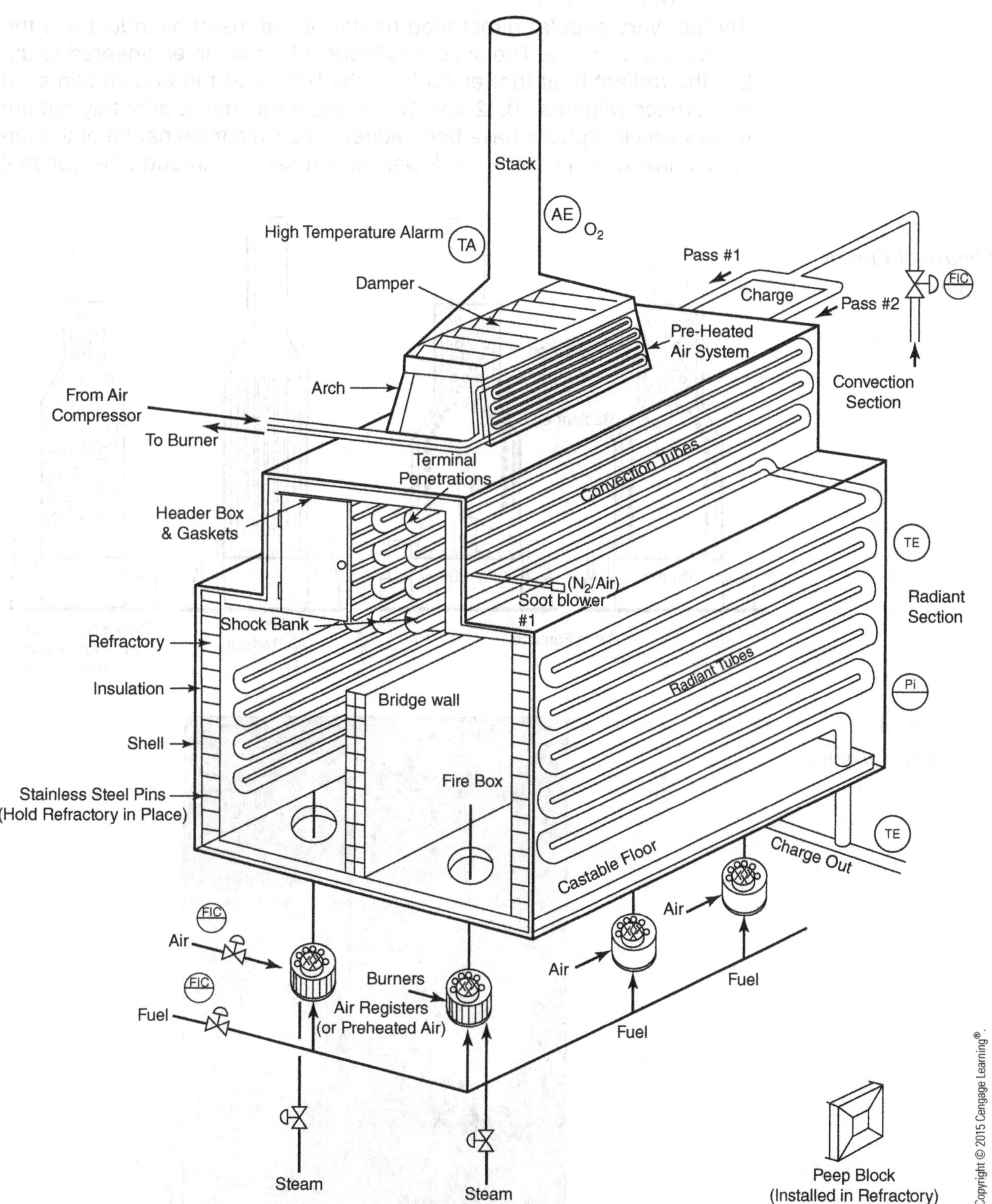

Figure 10.11 *Furnace*

Cylindrical Furnace

Another very popular direct-fired heater design used by industry is the **cylindrical furnace**. The simple cylindrical furnace is engineered to utilize the radiant heat that emits from the burner at the bottom center of the furnace (Figures 10.12 and 10.13). Heat transfer is primarily radiant unless special options have been added. The cylindrical nature of the furnace enhances draft from the lower radiant section, through the optional

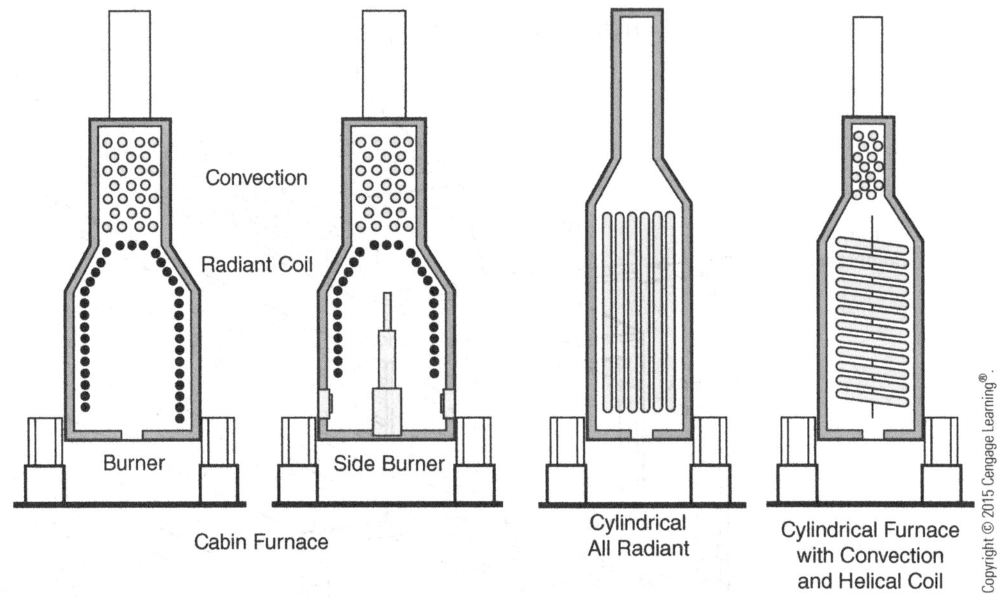

Figure 10.12 *Cylindrical Furnace Designs*

Figure 10.13 *Cylindrical Furnace*

convection section, and out the stack. The vertical tubes are typically composed of alloy steel. The net thermal efficiency is around 60% for a standard cylindrical furnace; however, if a convection section is added this rating jumps to 80%. A damper system to control airflow through the furnace is typically not used with a cylindrical design. Compared with other designs, cylindrical furnaces cost 10 to 15% less to construct, require less space and money to operate, have a higher firebox, have more parallel tube passes, and have higher flue gas velocity. A major disadvantage of cylindrical furnaces is that they have a lower efficiency than other designs because stack temperatures are higher.

Box Furnace

A **box furnace** takes its name from the square or rectangular design of the heating unit. This kind of furnace is best suited for oil firing. Burners can be on the sides or ends of the furnace (Figure 10.14). The firebox and radiant section use the roof, walls, and floor to reflect heat back into the firebox. Tubes may be arranged in a helical or serpentine layout. A bridgewall separates the radiant and convection sections. This wall may extend up from the floor or down from the ceiling. The refractory section in a box furnace is similar to that in a cabin furnace. Feed is introduced into a box furnace through the convection section and moves toward the firebox. Air tubes can be run through the convection section to warm up air that is being used for combustion. Box furnaces can use inlet and outlet fans to control draft. Induced-, forced-, natural-, and balanced-draft box furnace designs are used in industry.

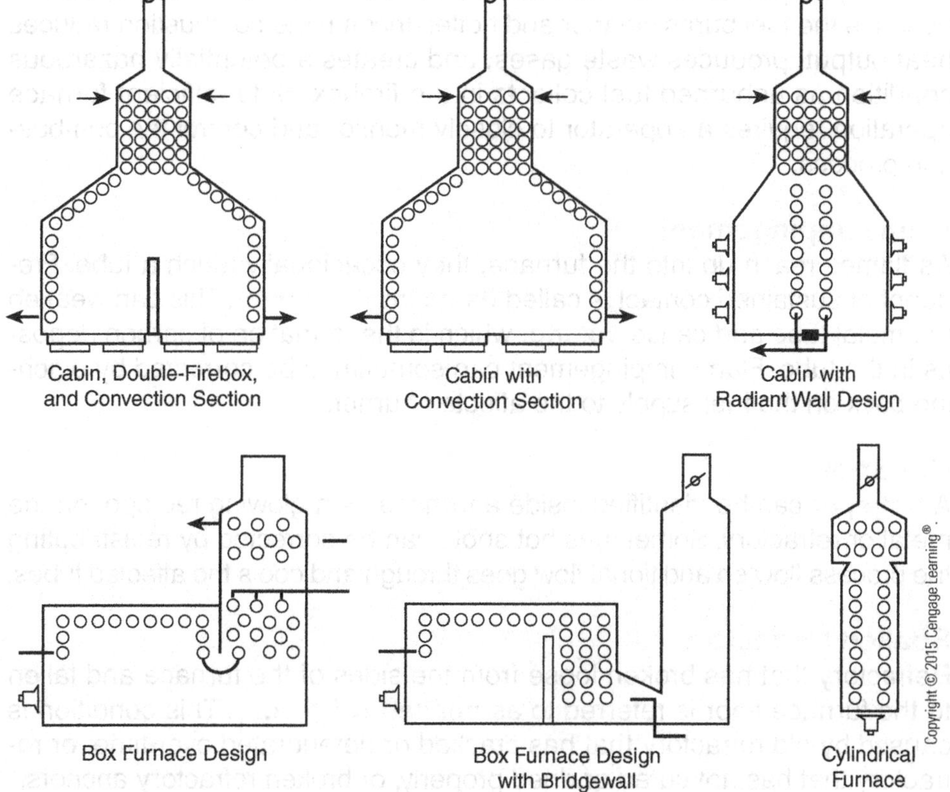

Figure 10.14
Furnace Designs

Common Furnace Problems and Solutions

Large commercial furnaces are not frequently shut down unless a serious problem occurs or equipment repair and turnaround are scheduled. There are some common problems and concerns for which equipment must be continually monitored.

Furnace Efficiency

Running furnaces efficiently is a major operating concern because two-thirds of a plant's fuel budget is needed for furnace fuel cost. Furnace efficiency is linked to environmental regulations that stipulate a clean operation.

Most furnaces use fuel gas or fuel oil. Natural gas burns cleaner and more efficiently than oil. In a natural gas furnace, oxygen and natural gas combine in the right proportions at the burner. In an oil-burning furnace, steam is used to atomize the oil and mix it with air at the burner tip.

Primary air enters the furnace through air shutters located on the burner that premix air and gas within the burner. Secondary air enters through air registers located on the burner. Air also can enter the furnace through idle burners, open peepholes, leaks in the furnace casing or casing joints, or damaged **header box gaskets**.

Ensuring proper air concentrations is critical to efficient furnace operation because the fuel burns cleaner and hotter. Incomplete combustion reduces heat output, produces waste gases, and creates a potentially hazardous condition as unburned fuel collects in the firebox. Safe, efficient furnace operation requires an operator to closely monitor and control the combustion process.

Flame Impingement

As flames reach up into the furnace, they occasionally touch a tube. Frequent or sustained contact is called **flame impingement**. This can weaken the metal tube and cause **coking**, which is the formation of carbon deposits in the tube. Flame impingement can sometimes be corrected by pinching back on the fuel supply to the affected burner.

Hot Spot

A **hot spot** can be identified inside a furnace as a glowing red spot on the metal or refractory. Sometimes hot spots can be corrected by redistributing the process flow so additional flow goes through and cools the affected tubes.

Spalled Refractory

Refractory that has broken loose from the sides of the furnace and fallen to the furnace floor is referred to as **spalled refractory**. This condition is caused by old refractory that has cracked or deteriorated over time, or refractory that has not cured or dried properly, or broken refractory anchors.

Ruptured Tubes

When flames are emitted from openings in the tube, this is an indication of **ruptured tubes**. These can cause excess oxygen levels to drop and bridgewall temperatures to increase.

Vibrating Tubes

Vibrating tubes tend to jump or move back and forth and typically take place in tubes outside of the furnace. Vibrating tubes are often caused by two-phase, slug-type flow inside the tubes. Such harmonic vibrations can be stopped by changing flow rates.

Plugged Burner Tips

Plugged burner tips tend to make the flame pattern appear erratic as it shoots up and out toward a tube instead of up the firebox.

Coke Buildup

A certain amount of coking occurs naturally over a period of time inside the radiant tube section. Shutting down the furnace and injecting superheated steam can clean up this problem.

Fuel Composition Changes

Fuels have different heat release rates that can be identified on standard charts. Fuel compositions change occasionally, and furnace efficiency also changes. Fuel composition changes can cause **flashback** (intermittent ignition of vapors that burn back in the burner). Operators must be prepared for these changes and make the correct adjustment. These adjustments include fuel flow rate, damper, charge flow, and burner alignment.

Feed Pump Failure

If the furnace feed pump fails, the furnace immediately begins to overheat, causing coking and equipment damage. If the pump does not restart, put the spare pump online, isolate the primary feed pump, and have it repaired immediately.

Flameout

Flameout occurs when the flame on a burner goes out with fuel still being pumped to the furnace. This situation puts unburned fuel into the furnace and creates a potentially dangerous situation. Flameout usually occurs when there is too much fuel being sent to the burner or when there is a loss of draft or oxygen. When this situation occurs, the furnace needs to be shut down.

Broken Burner Tiles

Broken burner tiles are located directly around the burner and are designed to protect the burner from damage. The furnace rarely needs to be shut down to replace a broken tile unless it is affecting the flame pattern.

Broken Supports and Guides

Broken supports and guides tend to fall to the furnace floor and result in sagging or bowing tubes. The falling supports and guides can also hit and damage a burner.

Hazy Firebox or Smoking Stack

Hazy fireboxes often occur when insufficient excess air is going into the firebox or the fuel–air-mixing ratio is incorrect.

Sagging or Bulged Tubes

As previously discussed, sagging tubes occur when guides or supports break. They also occur because of other factors such as when the inside or outside of a tube fouls and when there is flame impingement, reduced flow rate, or an overfiring furnace. Note that the diameter of a tube changes when it bulges.

Hot Tubes

Hot tubes glow different colors and occur when the inside or outside of the tubes foul, where there is flame impingement, reduced flow rate, or furnace overfiring.

Valve Failure

Furnace feed control valves are designed to fail in the open position. Valve failure cannot be detected until after an adjustment fails to respond. In any case, the operator will be unable to control the flow. Fortunately, bypass loops exist that allow the control valve to be isolated and repaired without shutting down the equipment. Furnace fuel gas valves are designed to fail closed.

Furnace System

Running process heaters efficiently is a major operating concern because significant funds are expended on natural gas costs. Process heaters are also closely monitored by the environmental protection agency for consistent clean operation and emissions. An advantage to using natural gas is its clean burning operation. F-202 is a clean operating system that keeps excess airflow between 8 and 10%. Ensuring proper air fuel concentrations is critical to efficient furnace operation because the fuel burns cleaner and hotter. Incomplete combustion reduces heat output, produces waste gases, and creates a potentially hazardous condition as unburned fuel collects in the firebox. Efficient, safe, process heater operation requires a technician to carefully observe and control the combustion process. Small commercial furnaces are not frequently shut down unless a serious problem occurs or equipment repair and turnaround are scheduled. There are some common problems and concerns for which equipment must be monitored. Figure 10.15 illustrates the basic components of a furnace system.

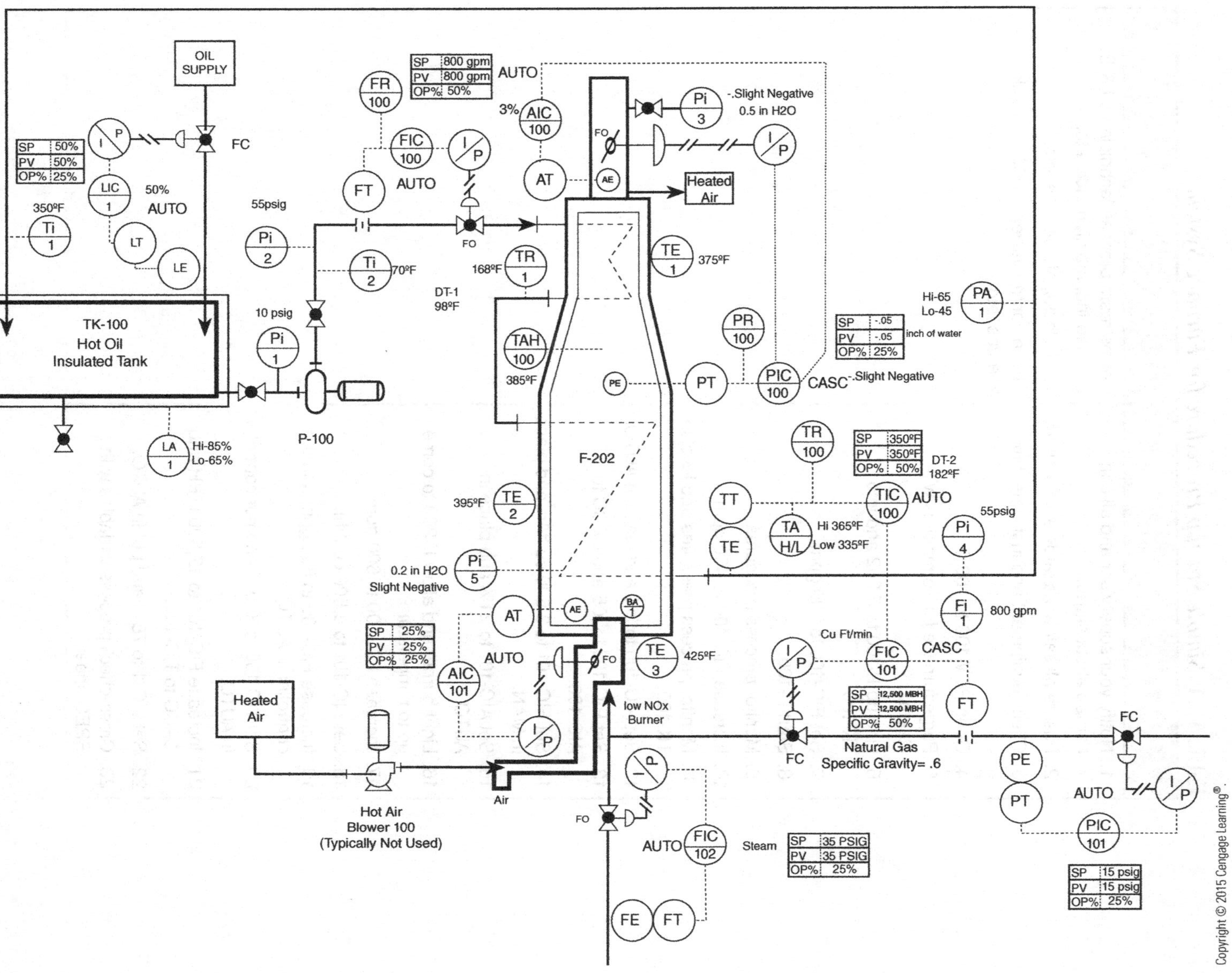

Figure 10.15 *Furnace System: F-202 Furnace*

Table 10.1 is a step-by-step approach to starting up a furnace system.

Table 10.1 *Simple Startup Procedure for Furnace System*

Procedure	Notes
1. Notify your supervisor and all concerned about startup.	Downstream process technicians, I & E, engineering, maintenance, etc.
2. Ensure all safety hazards are secured.	Trash, locks, old permits, etc.
3. Check equipment and instrumentation.	Stroke control valves 0–100%, check line-ups, etc.
4. Open PCV-100 to 100%.	
5. Purge furnace for 15 minutes with steam.	
6. Line up P-100 to F-202 and into TK-100.	
7. Set FIC-100 to 200 gpm.	
8. Start P-100.	
9. Monitor process pressures.	
10. Increase FIC-100 to 400 gpm.	
11. Monitor process pressures and level in TK-100.	
12. Set AIC-100 to 3% and place in AUTO.	
13. Set PIC-100 to .05 and CASC to AIC-100.	
14. Set FIC-101 to 3,500 cu ft/hr. and place in MAN.	
15. Set AIC-101 to 21% and place in AUTO.	
16. Light burner and allow F-202 to come up to temperature.	
17. Increase FIC-100 to 600 gpm.	
18. Set FIC-101 to 8,500 cu ft/hr.	
19. Increase FIC-100 to 800 gpm and put controller in AUTO.	
20. Set TIC 100 to 350°F and put controller in AUTO.	
21. Increase FIC-101 to 12,500 MBH and CASC to TIC-100.	
22. Set LIC-1 to 75% and put in AUTO.	
23. Cross-check process variables with SPEC sheet.	

Table 10.2 *Specification Sheet and Checklist*

Level			
1. LIC-1	75%	AUTO	Tk-100 level
2. LA-1	85%	High	Tk-100 high
3. LA-2	65%	Low	Tk-100 low
Flow			
4. FIC-100	800 GPM	AUTO	Feed to furnace
5. FIC-101	12,500 MBH	CASC	Natural gas feed
6. FIC-102	35 psig	AUTO	Steam
7. Fi-1	800 GPM		Furnace discharge
Analytical			
8. AIC-100	3%	AUTO	Stack discharge
9. AIC-101	21%	AUTO	Air to registers
10. BA-1	On/Off		Burner
Pressure			
11. PIC-100	.05 in water	CASC	Bridgewall draft
12. PIC-101	15 psig	AUTO	Natural gas feed
13. Pi-3	.5 in water		Top-stack
14. Pi-5	.2 in water		Radiant section
15. Pi-1	10 psig		P-100 suction
16. Pi-2	55 psig		P-100 discharge
17. Pi-4	55 psig		Furnace discharge pressure
18. PR-100	.05 in water		Bridgewall draft pressure
19. PA-1	Hi-65, Lo-45		Hot oil discharge
Temperature			
20. TR-1	168°F		Convection section exit temp.
21. DT-1	98°F	Δ-temp	Delta inlet/outlet convection
22. DT-2	182°F	Δ-temp	Delta inlet/outlet radiant
23. TIC-100	350°F	AUTO	Furnace exit temp
24. TAH-100	385°F		Bridgewall high temp
25. TE-1	375°F		Conv-Sect
26. TE-2	395°F		Rad-Sect
27. TE-3	425°F		Burner
28. Ti-1	350°F		At start-up
29. Ti-2	70°F		At start-up
30. TA-100 high	365°F		Hot oil discharge
31. TA-100 low	335°F		Hot oil discharge

Figure 10.16
Furnace Symbols

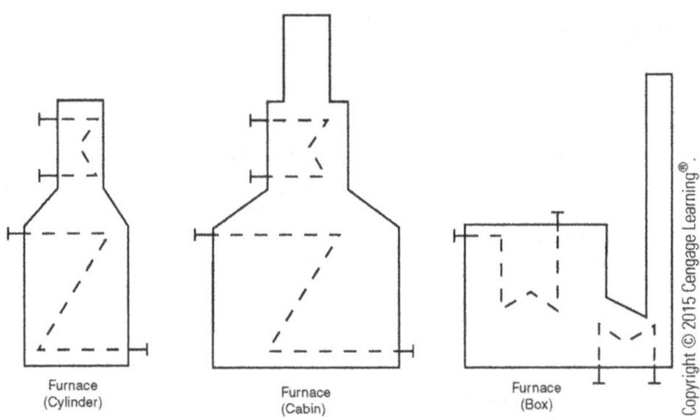

Furnace Symbols

Figure 10.16 shows the symbols for a furnace.

Summary

Fired heaters, or furnaces, are classified as direct fired or indirect fired (fire tube). In direct-fired furnaces the combustion gases occupy most of the interior space of the furnace. In fire-tube heaters the combustion gases are contained in tubes, which occupy a small percentage of the overall volume of the heater. The heated tubes run through a shell that contains the heated medium. Cabin, cylindrical, box, and A-frame furnaces are direct fired.

Combustion is a rapid chemical reaction that occurs when proper amounts of fuel and oxygen come into contact with an ignition source and release heat and light. Furnaces use this principle to provide heat to the charge. Complete combustion occurs when reactants are ignited in correct proportions. Incomplete combustion occurs when not enough oxygen exists to convert all of the fuel to water and carbon dioxide. A combination of radiant, conductive, and convective heat transfer processes occur in a furnace. The methods of heat transfer that have the greatest impact in furnaces are radiant and convective.

Direct-fired furnaces are used in many processes, including distillation, reactors, olefins, and other processes. Furnaces heat up raw materials so they can produce products such as gasoline, oil, kerosene, chemicals, plastic, and rubber. Direct-fired furnaces consist essentially of a battery of pipes or tubes that contain the process fluid and pass through a firebox. These tubes run along the inside walls and roof of a furnace. The heat released by the burners is transferred through the tubes and into the process fluid. The fluid remains in the furnace just long enough to reach operating conditions before exiting and being shipped to the downstream processing unit.

Air enters the furnace through primary air shutters on the burners and secondary air registers on the burner. Air can also enter through the stack, peepholes, cracks in the furnace, and header box gaskets. Furnaces can be classified as natural, induced, forced, and balanced draft. Burners can typically be found in the following types: oil, premix steam atomizing, combination, low NO_x, and raw gas.

Review Questions

1. What are carbon deposits in the tubes of a furnace called?
2. Describe flameout.
3. What is a hot spot?
4. What section of the furnace has the highest temperature?
5. Where are the convection tubes located in a furnace?
6. Where are the radiant tubes located in a furnace?
7. Describe flame impingement.
8. Air enters the furnace in five ways. Name them.
9. List the basic components of a furnace.
10. What is shock bank?
11. What is the charge of a furnace?
12. Describe the bridgewall section in a furnace.
13. Describe a natural-draft furnace.
14. Describe how forced-, induced-, and balanced-draft furnaces operate.
15. Define *spalled refractory*.
16. Explain the importance of proper draft control in a furnace.
17. Describe *sagging or bulged tubes*.
18. The floor of the furnace is covered with what type of material?
19. Describe the return bend designs in the convection section.
20. Excess airflow in the furnace will have what effect?

Instruments

Objectives

After studying this chapter, the student will be able to:

- Describe the basic instruments used in the process industry.
- Identify and draw standard instrument symbols.
- Describe temperature, pressure, flow, and level-measurement techniques.
- Identify the elements of a control loop.
- Describe cascaded control.
- Compare automatic and manual control.
- Explain the importance of the operating percentage on a control valve.

Key Terms

Absolute pressure (psia)—the pressure above a perfect vacuum (zero pressure).

Actuator—a device that controls the position of the flow-control element on a control valve by automatically adjusting the position of the valve stem.

Automatic control—allows a control loop to utilize all five elements and work to match the set point.

Bellows pressure element—a corrugated metal tube that contracts and expands in response to pressure changes.

Board-mounted equipment—instruments, gauges, or controllers that are mounted in a control room.

Bourdon tube—a hook-shaped, thin-walled tube that expands and contracts in response to pressure changes and is attached to a mechanical linkage that moves a pointer.

Cascade control—a term used to describe how one control loop controls or overrides the instructions of another control loop in order to achieve a desired set point.

Control loop—a collection of instruments that work together to automatically control a process. The basic elements of a control loop include primary element or sensor, transmitter, controller, transducer, and final control element. These devices are typically connected with electric lines and a pneumatic line to operate the valve.

Control valve—an automated valve used to regulate and throttle flow; typically provides the final control element of a control loop.

Controller—an instrument used to compare a process variable with a set point and initiate a change to return the process to a set point if a variance exists.

Differential pressure (DP) cell—measures the difference in pressure between two points.

Distributed control system (DCS)—a computer-based system that controls and monitors process variables.

Electric actuated valves—a valve that utilizes electricity to actuate or move the flow control device. An example of this type of valve is a solenoid.

Field-mounted equipment—instruments or controllers that are mounted near the equipment in the field.

Final control element—the device in a control loop that actually adjusts the process; typically a control valve.

Gain—the ratio of the output signal from the controller to the error signal.

Gauge pressure (psig)—the pressure above atmospheric pressure; zero is equivalent to approximately 14.7 psi at sea level.

Hydraulic actuated valves—a valve that utilizes a hydraulic actuator to position the flow control element. Internal designs include piston or vane.

Indicator gauge—an instrument used to show the value of process variables such as pressure, level, temperature, and flow.

Interlock—a device that prevents damage to equipment and personnel by stopping or preventing the start of certain equipment if a preset condition has not been met.

Manometer—a device used to measure pressure or vacuum.

Manual control—allows the controller to open the control valve and set it at a predetermined percent.

Permissive—special type of interlock that controls a set of conditions that must be satisfied before a piece of equipment can be started.

Pneumatic actuated valves—utilize air to actuate the flow control element. Internal designs may be piston, vane, or diaphragm.

Primary elements and sensors—the first element of a control loop. Primary elements and sensors come in a variety of shapes and designs depending on whether they are to be used with pressure, temperature, level, flow, or analytical control loops. An example of a temperature element is a thermocouple. A flow control primary element is a turbine meter or orifice plate. A level element is a displacer. A pressure element is a bourdon tube. Many different types of primary elements or sensor care are used in the chemical processing industry.

Process instrumentation—devices that control and monitor process variables; transmitters, controllers, transducers, primary elements, and sensors.

Programmable logic controller (PLC)—a simple, stand-alone, programmable computer that could be used to control a specific process or be networked with other PLCs to control a larger operation. PLCs are inexpensive, flexible, provide reliable control, and are easy to troubleshoot.

Resistance temperature detector (RTD)—a device used to measure temperature changes by changes in electrical resistance in a platinum or nickel wire.

Rotameter—a flow meter that allows fluid to move through a clear tube that has a ball or float in it; numbers on the side of the tube indicate flow rate.

Set point—desired value of a process variable.

Sight glass gauge—a level-measurement device consisting of a transparent tube and gauge attached to a vessel that allows an operator to see the corresponding liquid level.

Thermocouple—a temperature-measuring device composed of dissimilar metals that are connected at one end; heat applied to the connected ends causes the generation of voltage that corresponds to the temperature change, which is indicated on a temperature scale.

Thermowell—a chamber installed in vessels or piping to hold thermocouples and RTDs.

Transducer—a device used to convert one form of energy into another, typically electric to pneumatic or vice versa.

Transmitter—a device used to sense a process variable such as pressure, temperature, composition, or flow and produce a signal that is sent to a controller, recorder, or indicator.

Vacuum—any pressure below atmospheric pressure.

Vacuum pressure—pressure below zero gauge; often expressed in inches of mercury.

Basic Instruments

Automatic control is the foundation for efficient continuous flow processes. At one time, operators controlled processes manually. This type of process was "valve intensive"; it required the technician to open and close valves in piping lineups manually. Modern advances in instrumentation have made it possible for industrial manufacturers to automate their processes. To a process operator, this means that an instrument or computer can control the opening, closing, and positioning of valves; start and stop equipment; measure process variables; and respond automatically. This automation enables a single process technician to monitor and control large, complex process networks from a single control center.

Basic **process instrumentation** (Figure 11.1) includes gauges, **transmitters**, **controllers**, **transducers**, **primary elements and sensors**, computers, **control valves**, and other **final control elements**. Each instrument can be represented by a symbol discussed in the next chapter (see Figure 12.1).

Temperature Measurement

Process technicians are required to closely monitor the temperatures of process streams. When heat energy is applied to an area, molecular activity increases, and energy is transferred from molecule to molecule. As this process occurs, pressure and temperature increase in an enclosed environment, materials expand, and density changes.

Temperature is defined as the degree of hotness or coldness of an object or environment. Two commonly used scales are Fahrenheit and Celsius. Fahrenheit scales operate by using 32°F as the freezing point for water

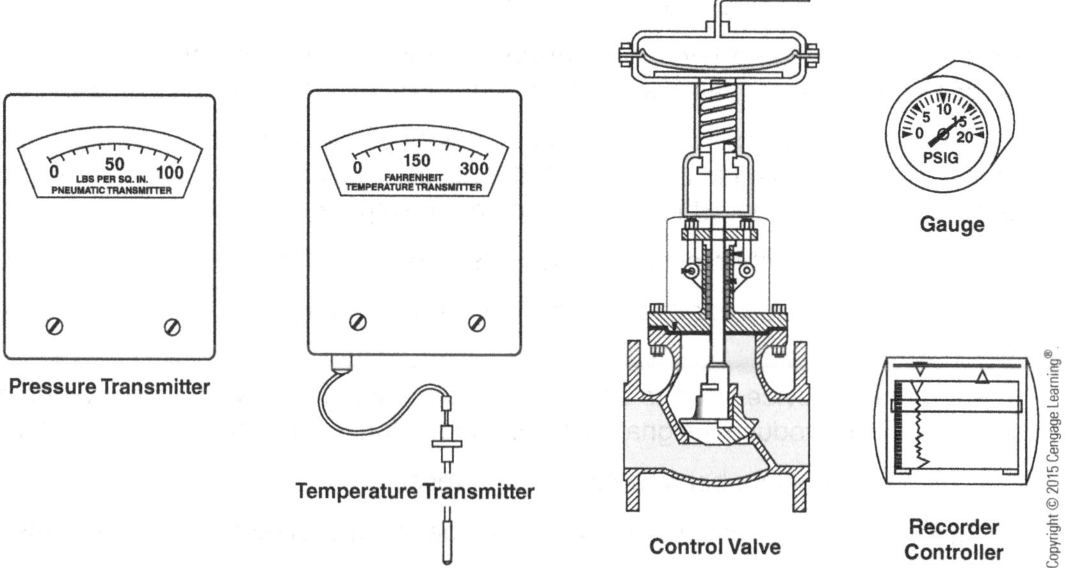

Figure 11.1 *Basic Process Instruments*

and 212°F as the boiling point of water. Celsius uses 0°C as the freezing point of water and 100°C as the boiling point of water. Process operators use Fahrenheit and Celsius thermometers to measure temperature. Local temperature **indicators** usually contain a bimetallic strip that differentially expands with increasing temperature, causing a deflection that is correlated with temperature. Bimetallic thermometers **(thermocouples)** can range from –300° to 800°F. Another familiar type of thermometer has a primary element or sensor that consists of a filled thermal bulb and capillary tubing, resistance bulb, or thermocouple. In industry, mercury is not used in thermal bulbs or capillary tubing. The most common temperature-measuring devices used in the chemical processing industry are thermoelectric.

Thermoelectric Temperature-Measuring Devices

Thermoelectric temperature-measuring devices come in two types: **resistance temperature detectors (RTDs)** and thermocouples. Both RTDs and thermocouples are held in a **thermowell**, a chamber installed in vessels or piping. An RTD is a thermoelectric temperature-measuring device composed of a small platinum or nickel wire encased in a rugged metal tube (Figure 11.2). The electrical resistance in the wire is influenced by changes in temperature. Temperature changes in an RTD are sensed by an electronic circuit and directed to a temperature indicator.

Thermocouples are composed of two different types of metal (see Figure 11.2). A thermocouple is designed to convert heat into electricity. When heat is applied to the connected ends of a thermocouple, a low-level current is generated. The higher the temperature, the greater the current generated. Electric current is detected easily by the associated electronic circuit and is converted to a corresponding temperature scale. Thermocouples come in several types; J-type and K-type thermocouples are the most common. Type K is preferred for higher temperature measurements.

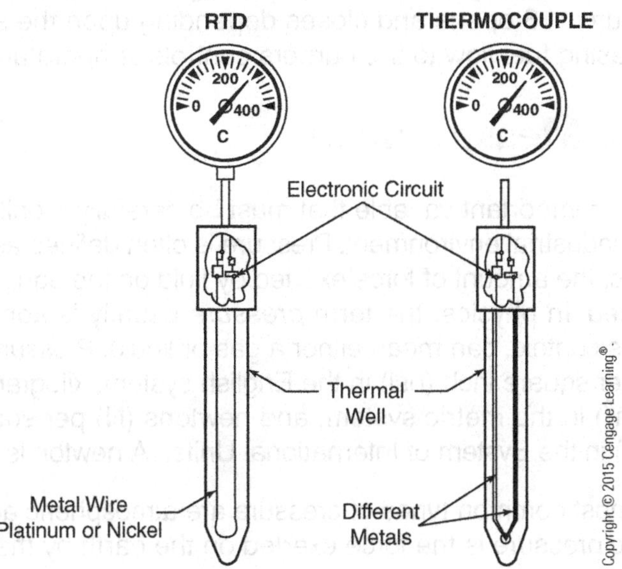

Figure 11.2
An RTD and a Thermocouple

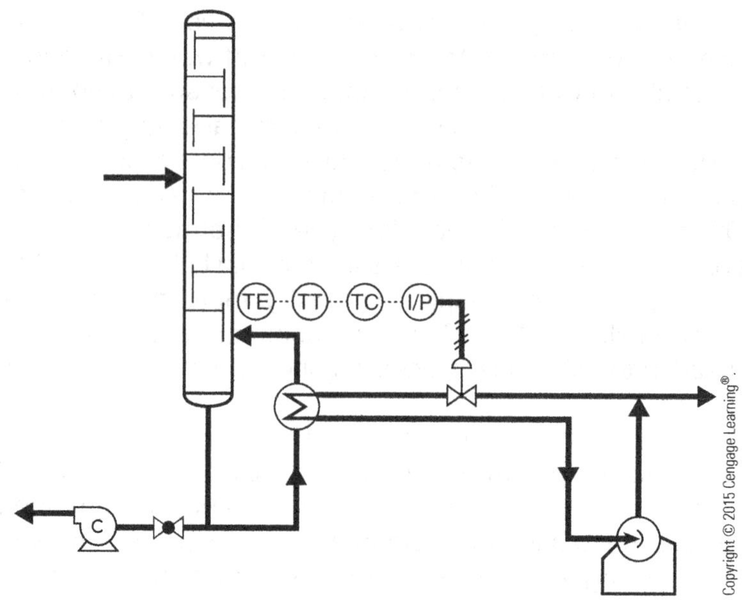

Figure 11.3
Temperature Control Loop

Temperature Control Loop

Figure 11.3 is a simple layout for a temperature control loop. In large fired heaters, a temperature measurement is taken at the furnace or from the exiting charge. The primary sensors used to detect temperature are thermocouples or RTDs, often called *temperature elements*. Temperature elements are linked to transmitters (devices that sense a process variable and produce a signal that is sent to a controller, recorder, or indicator). A 4- to 20-mA (milliamp) signal is sent to a controller that compares it with a **set point** (the desired value of a process variable). Controllers are designed to initiate a change to return a process to its set point if a variance exists. Controllers may be located in the field near the equipment or in a remote location. The controller sends an electric signal to a transducer, which is typically located near the valve to eliminate process lag. The transducer converts the electric signal to a pneumatic signal of 3 to 15 psi. The control valve in Figure 11.3 opens and closes depending upon the signal. Reducing or increasing fuel flow to the burners controls temperature.

Pressure Measurement

Pressure is an important variable that must be carefully monitored and controlled in an industrial environment. Pressure is often defined as force per unit area—that is, the amount of force exerted by fluid on the equipment in which it is contained. In physics, the term *pressure* usually is applied to a fluid, which, in this context, can mean either a gas or liquid. Pressure is measured in pounds per square inch (psi) in the English system, kilograms per square meter (kg/m^2) in the metric system, and newtons (N) per square meter, or pascals (Pa) in the System of International Units. (A newton is 1 kgm/sec^2.)

Two of the most common types of pressure are atmospheric and hydrostatic. Atmospheric pressure is the force exerted on the earth by the weight of the gases that surround it. At sea level, atmospheric pressure is about 14.7 psi (1.013 Pa). This pressure decreases with altitude because of the reduced height (and therefore weight) of gas. Hydrostatic pressure is the pressure exerted on a contained liquid and is determined by the depth of the liquid. Even a novice swimmer is acquainted with the pressure differences at the surface of the water and the bottom of the pool. This pressure difference is what causes your ears to "pop" as you swim to the bottom of a 10-foot swimming pool (hydrostatic) or drive over a high mountain range in Colorado (atmospheric).

Blaise Pascal, a French scientist, discovered in the 1650s that pressure in fluids is transmitted equally to all distances and in all directions. From this discovery, Pascal formulated what is known as Pascal's law, which states that in a fluid at rest in a closed container, a pressure change in one part is transmitted without loss to every portion of the fluid and to the container. Fluids act this way because the molecules move about freely. The distance between molecules depends on whether they are in solid, liquid, or gas states. Molecules in gases are much farther apart than they are in solids.

Also in the 1650s, an Irish scientist, Robert Boyle, developed laws that describe the relation of pressure to the volume of a gas. Boyle's law states that at constant temperature, the pressure of a gas varies inversely with its volume (Figure 11.4). The higher the gas pressure, the closer the gas molecules are and the smaller the volume they occupy. Under ordinary conditions, gas volumes decrease by half when the pressure doubles. Liquids and solids also respond to pressure increases but in much smaller proportions than gases. Liquids and solids are generally considered noncompressible.

The converse of Boyle's law is that pressure can affect temperature. It also affects level and flow rate. Pressure changes the boiling point of chemicals, their reaction rates, and the speed at which fluids flow through piping. These changes can affect product quality, so many instruments, or pressure elements, have been invented and designed to monitor and control pressure. Figure 11.5 shows some pressure elements.

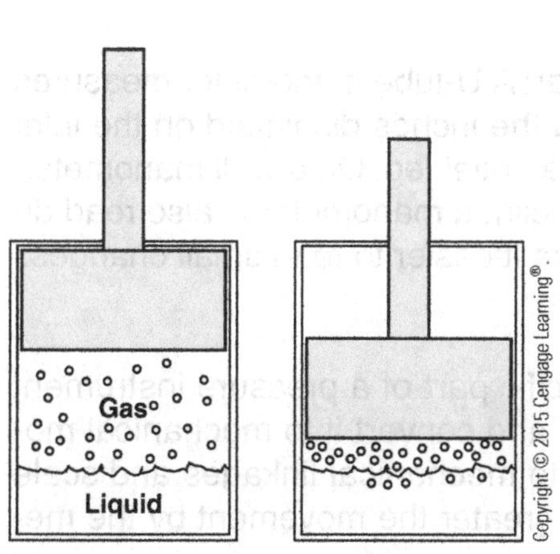

Figure 11.4 *Pressure Effects*

Figure 11.5 *Pressure Elements*

Pressure Gauges

There are three commonly used pressure scales in the manufacturing environment: gauge, absolute, and vacuum. The most commonly used pressure scale is the gauge scale, on which values are expressed in pounds per square inch gauge (psig). The gauge scale starts with atmospheric pressure (14.7 psi) as zero and moves up the pressure scale. The **absolute pressure** scale starts with a perfect vacuum as zero. Values are expressed in pounds per square inch absolute (psia). The absolute scale takes into account atmospheric pressure. To convert from **gauge pressure** to absolute, add 14.7 pounds to the psig value. To convert from absolute pressure to gauge, subtract 14.7 pounds from the psia value.

CAUTION: *Do not stand directly in front of the gauge when opening the valve that admits pressure to the gauge. Once the valve has been opened, stand directly in front of the gauge face to take the gauge reading. If you position yourself to the left or right of the gauge face, an effect known as parallax occurs. Parallax is an optical illusion that shifts the gauge face reading left or right of actual. The space between the pointer and the face of the gauge causes the parallax problem.*

A psig gauge cannot be used with system processes that operate under a vacuum. Negative pressures cause the primary elements to contract beyond design limits. If a psig gauge accidentally encounters a vacuum, the reading scale is compromised (low). Vacuum gauges overcome this problem. They are designed to operate at less than atmospheric pressure. Vacuum gauges express pressure in inches of mercury (Hg). **Vacuum** is considered to be anything below atmospheric pressure. Compound gauges can indicate both vacuum and gauge pressure readings.

Manometer

A **manometer** is a device that can be used to measure pressure or vacuum. It operates under the hydrostatic pressure principle that a column of water of a given volume always exerts a specific force. The liquid level of the water indicates the pressure.

There are three basic types of manometer: A U-tube manometer measures pressure in units of inches of water. Add the inches displaced on the inlet leg plus the inches above the zero on the outlet leg. On a well manometer, read the scale directly. The scale of an inclined manometer is also read directly, but the scale is "expanded" to make it easier to read small changes.

Primary Pressure Elements

Primary pressure elements are the specific part of a pressure instrument designed to sense changes in pressure and convert it to mechanical motion. Pressure elements are connected to mechanical linkages and scale indicators. The higher the pressure, the greater the movement by the mechanical linkage.

Bellows Pressure

A **bellows pressure element** consists of an accordion-type bellows, a spring that resists expansion of the bellows, a pressure inlet, a mechanical linkage, and a pointer (Figure 11.6). This type of device can be used to measure a variety of pressures. During operation, a bellows pressure element admits flow into the bellows. As the bellows expands, tension on the spring increases, and the mechanical linkage moves. This movement operates the pointer on the gauge. If the pressure is reduced, the spring forces the bellows back to its original position.

Bourdon Tubes

The most common type of pressure element is a **bourdon tube**. Bourdons come in a variety of shapes and designs. The most common are the C-type, helical-type, and spiral-type.

C-type bourdon tubes are named for their C-shaped, hollow pressure element (Figure 11.7). C-type bourdons are composed of a C-shaped hollow tube, a pressure inlet attached to one end of the tube, a mechanical linkage attached to the top of the bourdon tube, and a pointer. During operation, the tube expands and contracts in response to pressure changes. This process is sometimes referred to as elastic deformation. This expansion moves the mechanical linkage and pointer. Bourdons measure a wide variety of pressures, including vacuum.

Helical- and spiral-type bourdons operate the same way the C-type bourdon does. The main difference is in the actual shape of the pressure element. In an automatic control system, a spiral-type bourdon can be connected to a transmitter. As the spiral element responds to pressure changes, the transmitter sends a signal to the controller, which sends a signal to the control valve.

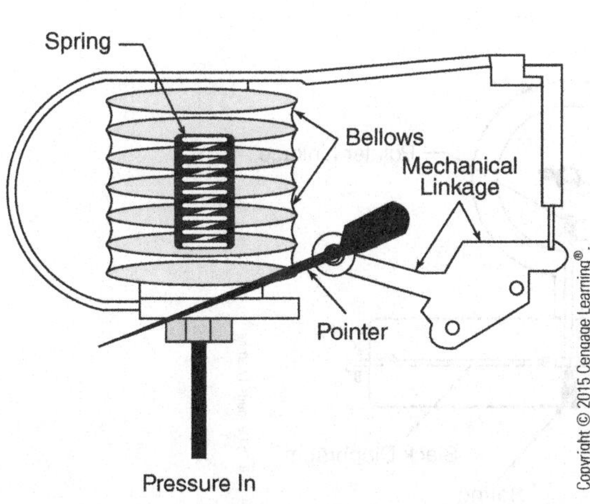

Figure 11.6 *Bellows Pressure Elements*

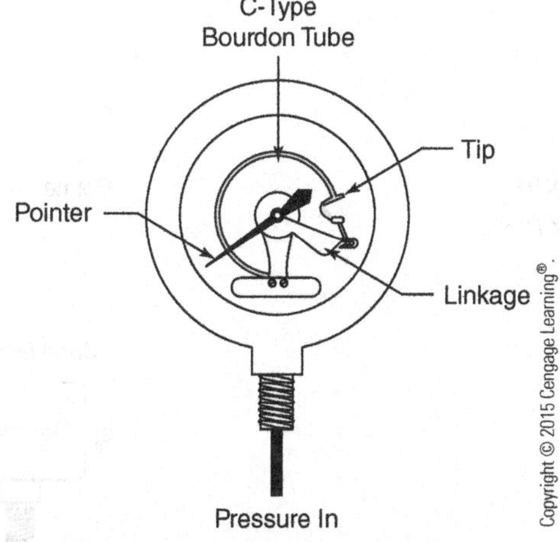

Figure 11.7 *Bourdon Tube–type Pressure Elements*

Diaphragm Capsule Pressure Element

The third type of pressure element is a diaphragm. Diaphragms come in two basic types: diaphragm capsule pressure element and slack diaphragm pressure element. The metallic diaphragm capsule pressure element consists of a metal cup covered by a flexible metal plate, a pressure inlet line to the cup, a mechanical linkage, and a pointer. Diaphragm capsule pressure elements are designed to measure small pressure changes. During operation, the dome of the cup flexes up or down. Because this movement is transferred to the pointer proportionally, a little movement on the dome can equal a lot of travel on the pointer.

The key components of a slack diaphragm pressure element (Figure 11.8) are a flexible diaphragm attached to a spring, a pressure inlet, a mechanical linkage, and a pointer. Slack diaphragm pressure elements are designed to operate under very low pressures, 0 to 0.5 psi.

Pressure Transmitter

A pressure transmitter (Figure 11.9) uses a pressure element to sense pressure and sends a signal to a controller or recorder. Pressure transmitters use all of the primary pressure elements just discussed. Linkage movement allows the transmitter to transmit a signal that is representative of the pressure to a controller or recorder. A controller opens or closes control valves depending on the signal it receives from the transmitter.

Pressure Control Loop

Pressure control loop design has the same elements as temperature, level, and flow control loops. The one area that changes consistently is the first, primary elements and sensors. Pressure control loops use devices to detect and respond to pressure changes. These primary elements are typically expansion-type devices. Figure 11.10 includes a pressure transmitter, controller, transducer, and control valve.

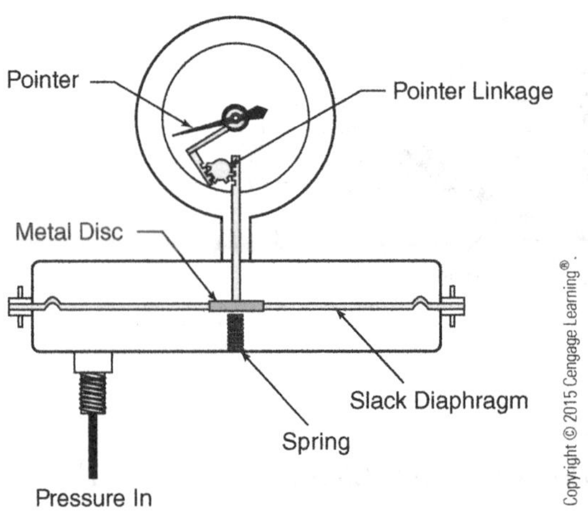

Figure 11.8
Slack Diaphragm Pressure Elements

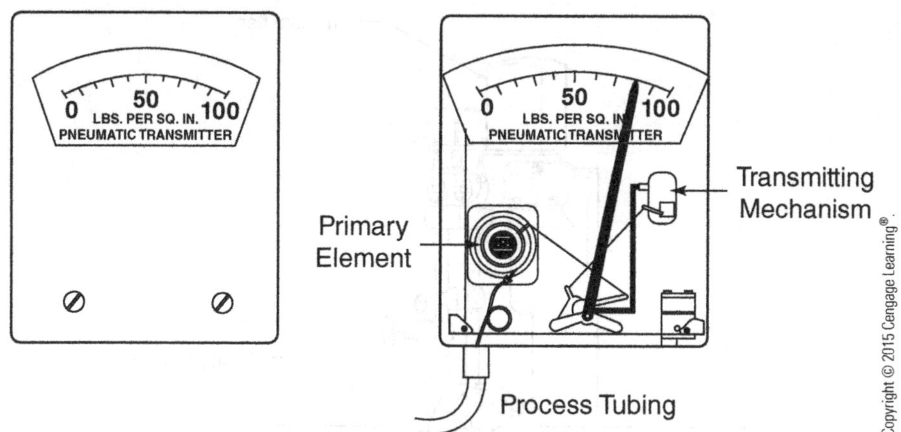

Figure 11.9
Pressure Transmitter

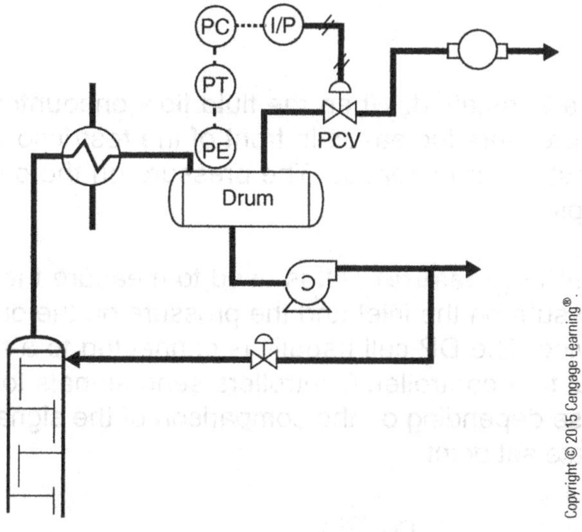

Figure 11.10
Pressure Control Loop

Fluid Flow Measurement

Fluids flow through a series of pipes, valves, pumps, and vessels. Knowing and controlling the flow rate of a particular process stream are critical to the operation of the unit. Continuous chemical reactions require precise measurements to ensure that all of the reactants (raw materials) are combined in the proper proportions to form the final products. Feed rates and product rates must be accurately controlled for economic reasons. Process flow measurements can be taken by any kind of flow meter, but flow control most often requires a flow transmitter.

Flow Transmitter

Certain types of flow meters, such as orifice plate meters (Figure 11.11) and venturi meters, use differential pressure to measure flow rate. Fluid flow through a pipe can be related to pressure differences inside the pipe when flow-restrictive devices, such as orifice plates, venturi tubes, or

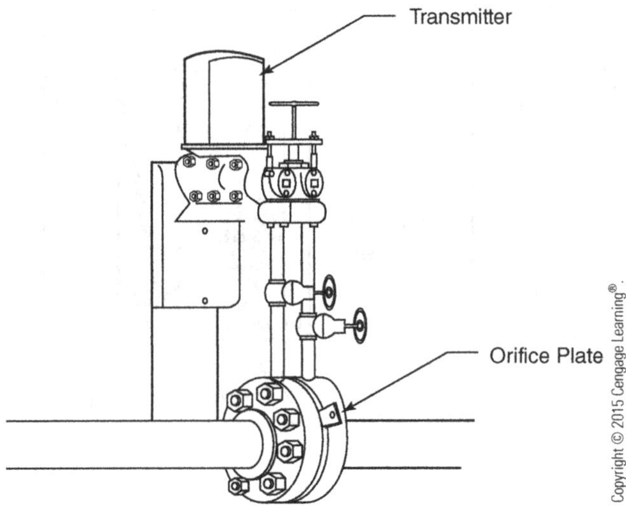

Figure 11.11
Flow Transmitter and Primary Elements

flow nozzles, are installed. When the fluid flow encounters a restriction in a pipe, the pressure increases in front of the restriction. Fluid velocity through the restriction increases. The pressure on the other side of the restriction drops.

A **DP (differential pressure) cell** is used to measure the difference between the pressure on the inlet and the pressure on the outlet side of the restrictive device. The DP cell usually is connected to a transmitter that sends a signal to a controller. Controllers send signals to control valves to open or close depending on the comparison of the signal from the field with the flow rate set point.

Positive Displacement Meters

There are two types of positive displacement meters: nutating disc and oval gear. The nutating disc meter is composed of a counter, a nutating disc (resembles a spinning top), flow inlet, and flow outlet (Figure 11.12). Nutating disc meters measure fluid flow directly by counting the rotations of the disc as fluid passes through it.

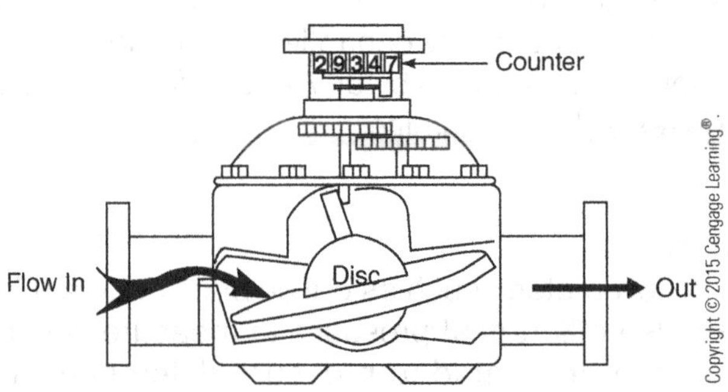

Figure 11.12
Nutating Disc Meter

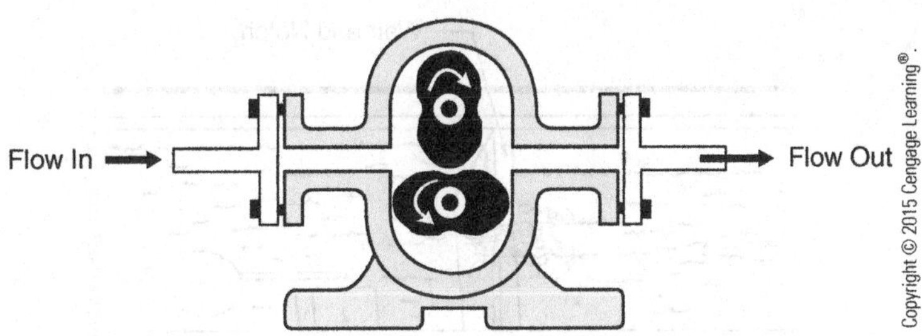

Figure 11.13
Oval Gear Meter

An oval gear meter has an internal structure that resembles that of a lobe pump (Figure 11.13). The lobe-shaped elements rotate as fluid passes through the internal chamber. The rotation of the gears is used to calculate the total flow rate.

Rotameter

Another type of flow-measuring device is a **rotameter**. A rotameter is composed of a tapered tube, scale, ball or float, and inlet and outlet (Figure 11.14). During operation, flow enters a tapered tube at the bottom of the rotameter and lifts the ball off its seat. The ball provides a constant restriction to the flow and corresponds to the flow rate on the scale that runs the length of the tube. The higher the flow rate, the higher the ball rises in the tube. Fluid flows around the ball and out the top of the rotameter.

Turbine Flow Meter

Turbine flow meters usually consist of a section of pipe with a rotor mounted in the pipe and a sensor on the outside of the pipe. As fluid enters the turbine flow meter, the turbine blades begin to rotate (Figure 11.15). The speed of the rotation is proportional to the velocity of the fluid. It is important to understand that flow velocity and flow rate are not the same. Flow velocity is the actual speed of the fluid, measured as distance per unit time (for example, feet per second). Flow rate is the total quantity of liquid that passes

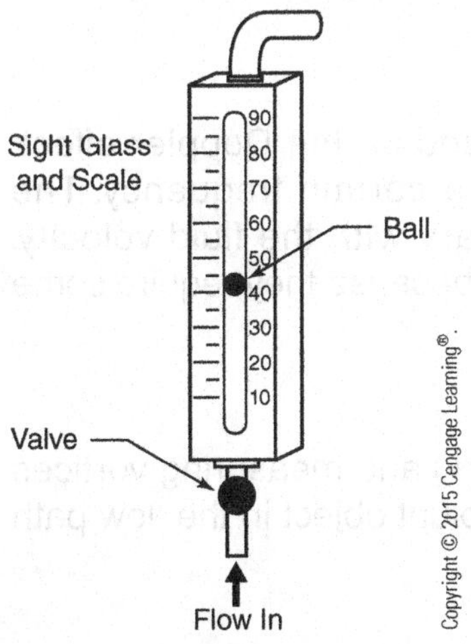

Figure 11.14 *Rotameter*

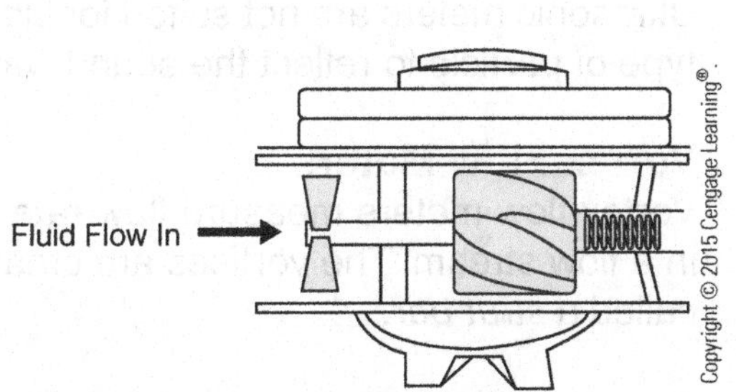

Figure 11.15 *Turbine Flow Meter*

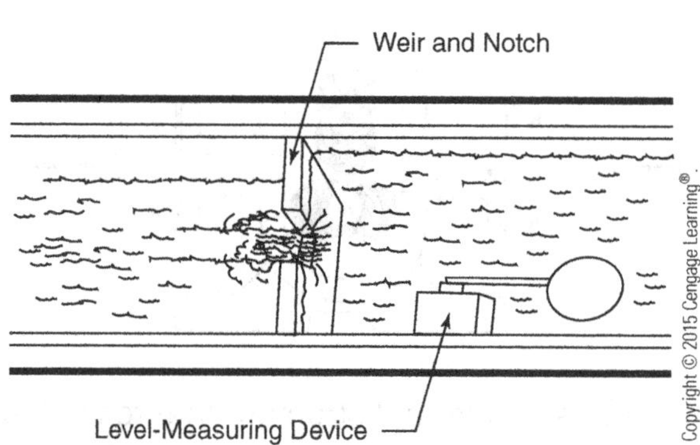

Figure 11.16
Weir and Flume

a specific point, measured as volume per unit time (for example, gallons per minute). Turbine flow meters are accurate over a wide range of flows.

Weir and Flume Flow-Measuring Devices

Weir and flume flow- and level-measuring devices are used to calculate flows in open channels (Figure 11.16). When a weir (a dam) is placed into a process stream, it creates a restriction that forces the level to build. This level is used in a calculation to identify the flow rate. There is a direct correlation between the liquid level and the flow rate; the lower the level, the slower the flow.

Flumes operate under the same principles as a weir but are used for higher flows. The flume is a narrow, sloping pass that funnels flow. At the inlet of the flume, the water level rises and is measured by a level-measuring instrument that converts the level to a flow signal.

Magnetic Flow Meters

Magnetic flow meters measure flow velocity based on the voltage created by the fluid flowing through a magnetic field. This type of meter is very effective for toxic or corrosive fluids because the fluid stream does not contact the measurement device.

Ultrasonic Flow Meters

Ultrasonic flow meters measure flow rate based on the Doppler effect. A sound wave is transmitted to the fluid at a certain frequency. The frequency of the returned sound wave will vary with the fluid velocity. Ultrasonic meters are not suited for clean fluids because they require some type of particle to reflect the sound wave.

Vortex Flow Meters

Vortex flow meters measure flow rate by creating and measuring vortices in a flow stream. The vortices are created by a blunt object in the flow path called a *strut bar*.

Thermal Flow Meters

Thermal flow meters measure flow rate based on changes in resistance due to changes in temperature. Thermistors are used to measure the change in

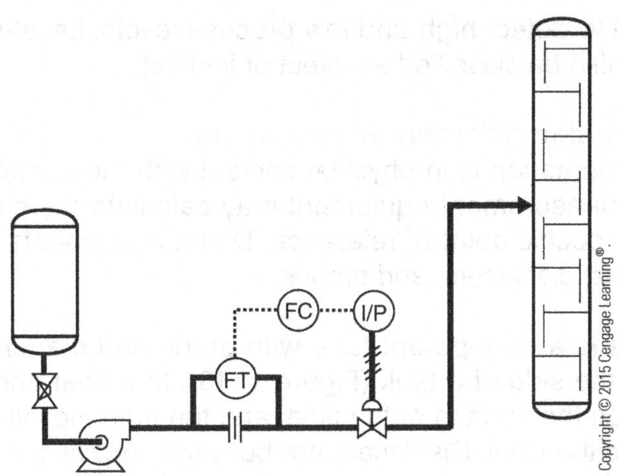

Figure 11.17
Flow Control Loop

resistance. Thermal flow meters are sensitive to the thermal conductivity of the fluid and are usually designed for a specific flowing material.

The Coriolis Meter

The Coriolis meter is a true mass flow meter that uses a vibrating U-tube to measure changes in momentum (mass flow rate). This type of meter is very effective for precise measurements of mass flow rate.

Flow Control Loop

Flow control loops (Figure 11.17) are typically designed so a measurement of the flow rate is taken first and then the flow is interrupted or controlled downstream. Flow control loops start at the primary element, which could be an orifice plate, venturi tube, flow nozzle, nutating disc, oval gear, or turbine meter. The most common primary element is the orifice plate. Orifice plates create a pressure differential that can be measured by a DP transmitter. Primary elements are typically used in conjunction with a transmitter. Although it appears that the primary element is interrupting the flow, it is not. Increased velocity across the orifice plate compensates for the restriction. The transmitted signal is sent to a controller that compares the incoming signal with the desired set point. If a change is required, the controller will send a signal to a final control element.

Level Measurement

Process technicians use fixed reference points, typically vessel taps, on which to base level measurements. The lowermost tap represents zero level, and the uppermost tap is 100%. Correct level readings and control help make modern processing possible and profitable.

Level measurements can be continuous (levels monitored continuously) or single point. In single-point measurements, readings are taken from a single point or from multiple points on a vessel. Single-point measurements are used to turn equipment (valves, pumps, compressors, motors, alarms) on or off and to detect high and low process levels. Level-measurement devices can also be classified as direct or indirect.

Direct Level-Measurement Instruments

Direct instrumentation is in physical contact with the surface of the fluid. Direct level-measurement equipment may calculate the product surface level from a specific point of reference. Direct instruments include **sight glasses**, floats, displacers, and probes.

A sight glass is a transparent tube with graduated markings (a gauge) mounted on the side of a tank (Figure 11.18). In a float and tape device, a float rests on the surface of the fluid, and the tape moves up and down, depending on the level. Displacers are buoyancy devices, or weights, that can be linked to a transmitter to control flow (Figure 11.19). Conductivity probes are high- and low-level alarms. They use electricity to complete the lower leg circuit. If liquid reaches the higher leg, the circuit is broken. This type of system, which is designed to keep the level between the high and low conductivity probes, typically is used for nonflammable material. Capacitance probes are radiation devices or load cells.

Indirect Level Measurement

Indirect instrumentation incorporates pressure changes that respond proportionally to level changes. DP cells convert pressure differences to a level indication. They measure the hydrostatic pressure difference between two points on a pressurized vessel. A continuous level detector gauge is a pressure-sensitive instrument that measures hydrostatic pressure in open vessels and converts it to a level indication. A bubbler system is a level-measurement device that forces air through a tube that is positioned in the

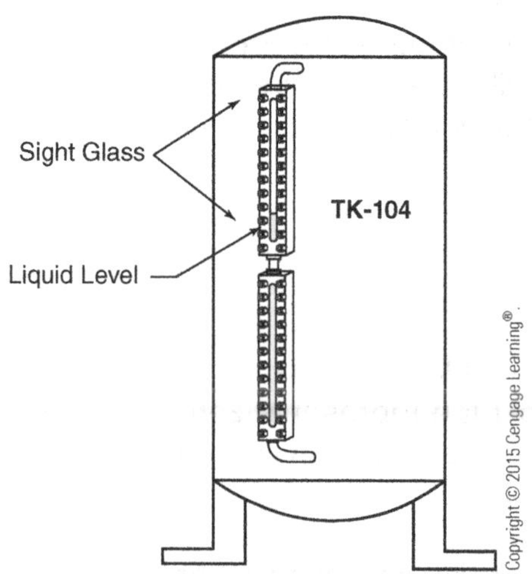

Figure 11.18 *Sight Glass*

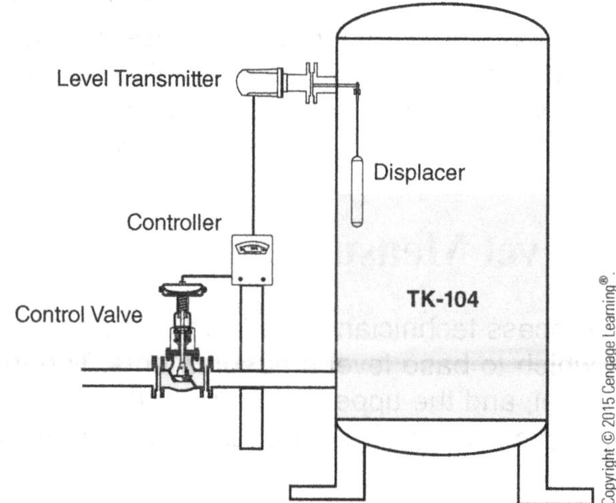

Figure 11.19 *Level Control by Displacer*

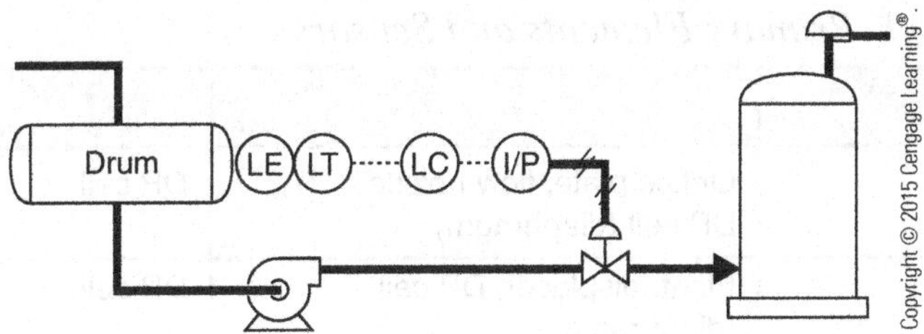

Figure 11.20
Level Control Loop

liquid. The liquid's resistance to flow registers on a pressure-sensitive level gauge that converts the pressure to a level indication. Bubbler systems are used to measure levels in open vessels.

Level Control Loop

A typical level control system uses a primary element sensing device (direct or indirect), level transmitter, controller, transducer, and control valve. Level control loops use floats, displacers, or differential pressure transmitters. Figure 11.20 uses a DP cell to detect level changes. The primary element or sensor is inside the transmitter, as indicated by the lack of a line separating LE and LT. These two devices couple up to detect and send a signal to a level controller. A transducer converts the signal and opens or closes the control valve.

Basic Elements of a Control Loop

The key component of automatic control is the control loop, the group of instruments that work together to control a process. As we have seen, these instruments typically include a transmitter coupled with a sensing device or primary element, a controller, a transducer, and a control valve. Process plants contain many control loops that are used to maintain pressure, temperature, flow, level, and composition.

The basic elements of a **control loop** are:

1. *Measurement device*—primary elements and sensors (Table 11.1).
2. *Transmitter*—a device designed to convert a measurement into a signal. This signal will be transmitted to another instrument.
3. *Controller*—a device designed to compare a signal with a set point and transmit a signal to a final control element.
4. *Transducer*—a device designed to convert an air signal to an electric signal to a pneumatic signal. Sometimes referred to as an *I* to *P* or as a converter.
5. *Final control element*—that part of a control loop (for example, control valve, damper, or governor for speed control) that actually makes the change to the process. The final control element is governed by a controller. Figure 11.21 is an example of a cascaded control loop.

Table 11.1 *Primary Elements and Sensors*

	Primary Element	Sensor
Flow	Orifice plate, flow nozzle, DP cell (diaphragm)	DP cell
Level	Float, displacer, DP cell (diaphragm)	DP cell
Pressure	Helix, spiral, bellows, bourdon tube, DP cell	DP cell
Temperature	Capillary tubing, thermal and resistance bulb	Thermocouple, RTD

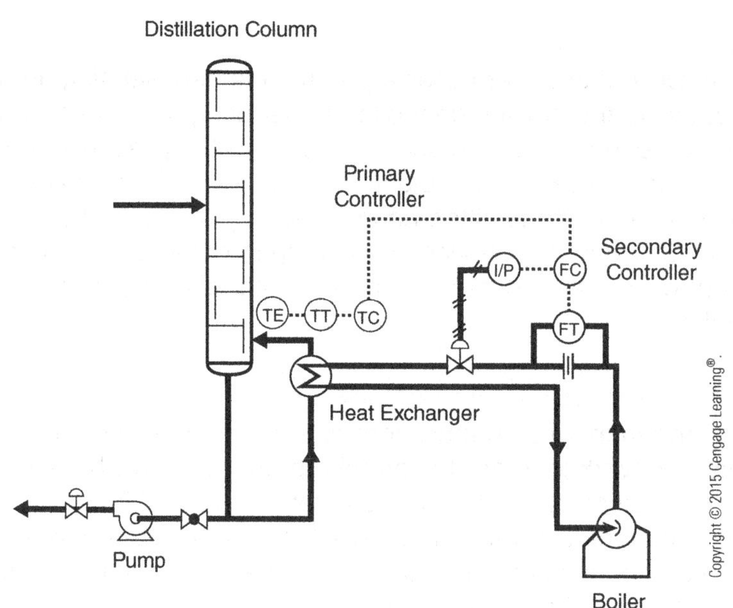

Figure 11.21 *Cascaded Control Loop (temperature)*

Transmitters

DP cell transmitters can be found in two basic designs: pneumatic and electronic. Controllers are typically mounted between 400 (closed loop) and 1,000 (open loop) feet from the transmitter. The signal from an electronic transmitter is proportional to the pressure difference between the high-pressure leg and the low-pressure leg. Standard output signals are 4 to 20 mA (milliamps), 10 to 50 mA, and 1 to 5 V (volts). The 10 to 50 mA transmitter is used because it has a higher tolerance to outside interference than does the 4 to 20 mA transmitter. Pneumatic transmitters require a 20 to 30 psig air supply in order to run the standard 3 to 15 psig output.

DP cells (Figure 11.22) function by running a high- and low-pressure tap to each side of an internal twin diaphragm capsule. Pressure changes cause

Figure 11.22 *ΔP Cell Transmitter*

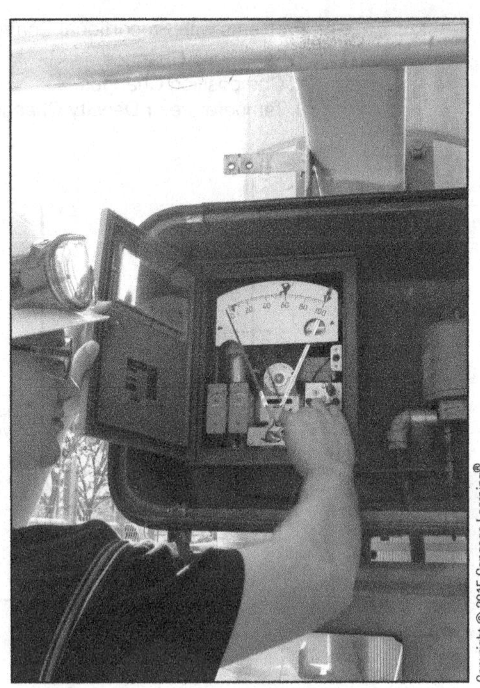

Figure 11.23 *Field-Mounted Controller*

the diaphragms to move. This process increases or decreases the signal to the controller.

Smart transmitters are another type of transmitter frequently found in the chemical processing industry. This type of transmitter is very reliable and does not need constant attention. Smart transmitters have an internal diagnostic system that warns the operator if a problem is about to occur. This type of transmitter can be used in liquid or gas service to control pressure, viscosity, temperature, flow, and level. Several advantageous features of the smart transmitter are speed, reliability, internal diagnostics, strong digital signal, and remote calibration capabilities.

Controllers and Controller Modes

The primary purpose of a controller (Figure 11.23) is to receive a signal from a transmitter, compare this signal with a set point, and adjust the process (that is, the final control element) to stay within the range of the set point. Controllers come in three basic designs: pneumatic, electronic, and digital. Electronic controllers were first introduced in the early 1960s. Before then, only pneumatic controllers were available. Pneumatic controllers require a clean air supply pressure of 20 to 30 psig. Several of the more attractive features of electronic controllers are the reduction of lag time in process changes, low installation expense, and ease of installation.

With the widespread use of the PC (personal computer), several applications were found for controller use. **Distributed control systems (DCS)**

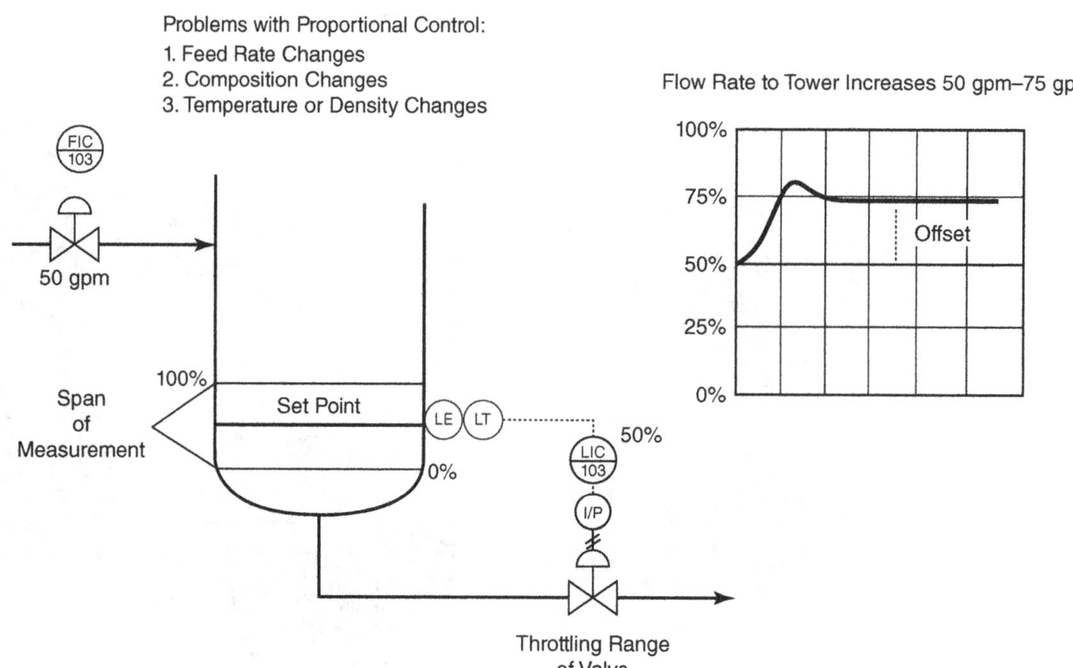

Figure 11.24 *Proportional Control*

software in computers replaced pneumatic and electronic controllers. The primary reason was the ease with which a DCS could be installed and the relatively few wires required to do it.

Most modern plants include a combination of all three systems—pneumatic, electronic, and digital. The diagram of a control loop will identify the type of controller (pneumatic, electronic, or DCS). For example, in Figure 11.24, the diagonal lines between the transducer and the valve signify pneumatic control.

Controllers can be operated in manual, automatic, or cascaded (remote) control. During plant startup, the controller is typically placed in the manual position until the process has lined out. In **manual control**, the controller sends an output to the final control element, as set by the operator. After the process is stable, the operator places the controller in automatic and allows the controller to supervise the control loop function. At this point, the controller will attempt to open and close the control valve to maintain the set point. Cascade control (Figure 11.24) is a term used to describe how a second controller will reset a controller's set point in order to achieve a desired outcome.

Proportional Band

The proportional band on a controller describes the scaling factor used to take a controller from 0% to 100% output. If the proportional band is set at 50% and the amount of lift the final control element (globe valve) has off the seat is 4 inches, the control valve will open 2 inches. Range is defined

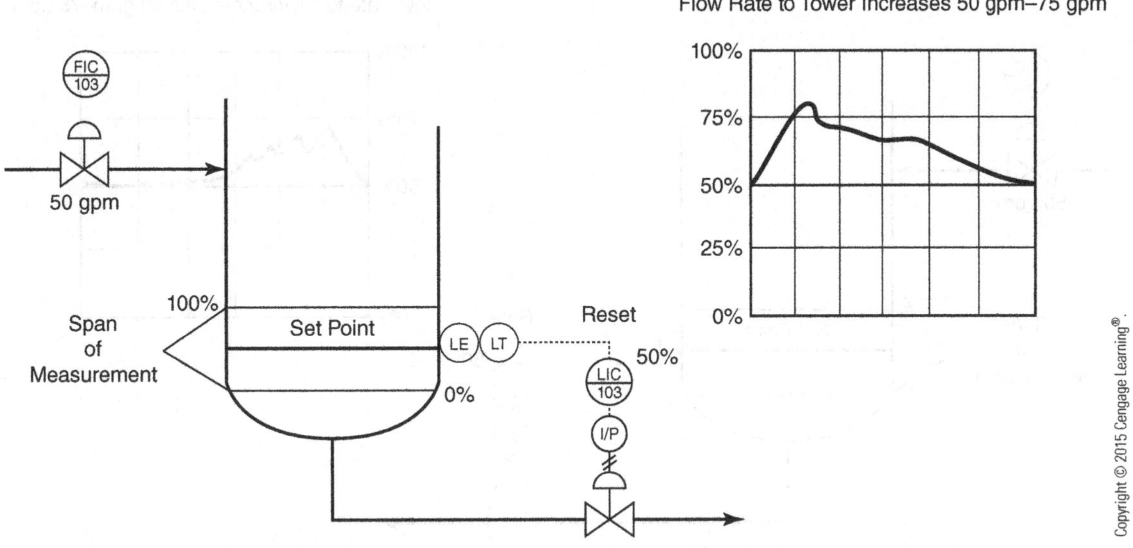

Figure 11.25 *Reset*

as the portion of the process controlled by the controller. For example, the temperature range for a controller may be limited to 80 to 140°F. Span is the difference (Δ) between the upper and lower range limits. This value is always recorded as a single number. For example, the difference between 80 and 140 is 60. (See Figure 11.25.)

Controller Modes

Process technicians use a number of controller modes. The most common types are gain, reset, and rate. Gain may be used individually or in combination with reset or rate. Gain is most frequently paired with reset. Rate is typically applied to temperature control applications. The most advanced system uses a combination of gain, reset, and rate. Proportional control is primarily used to provide gain where little or no load change typically occurs in the process. Gain plus reset is used to eliminate offset between the set point and process variables. Gain works best where large changes occur slowly. Gain plus rate is designed to correct fast-changing errors and to prevent overshooting the set point. It works best when frequent small changes are required. Gain plus reset plus rate is applied where massive rapid load changes occur. Gain, reset, and rate combine to reduce swinging between the process variable and the set point.

Gain

Most modern controllers use the term *gain* as a tuning constant. The **gain** is simply the ratio of the output signal from the controller to the error signal (difference between the process variable and set point). The gain is the reciprocal of the proportional band (PB). For example, if the PB is 100%, the gain = 1.0. If the PB is 200%, the gain is .5 (1/2). Gain is rarely used alone, because it can provide only a ratio response to a process change.

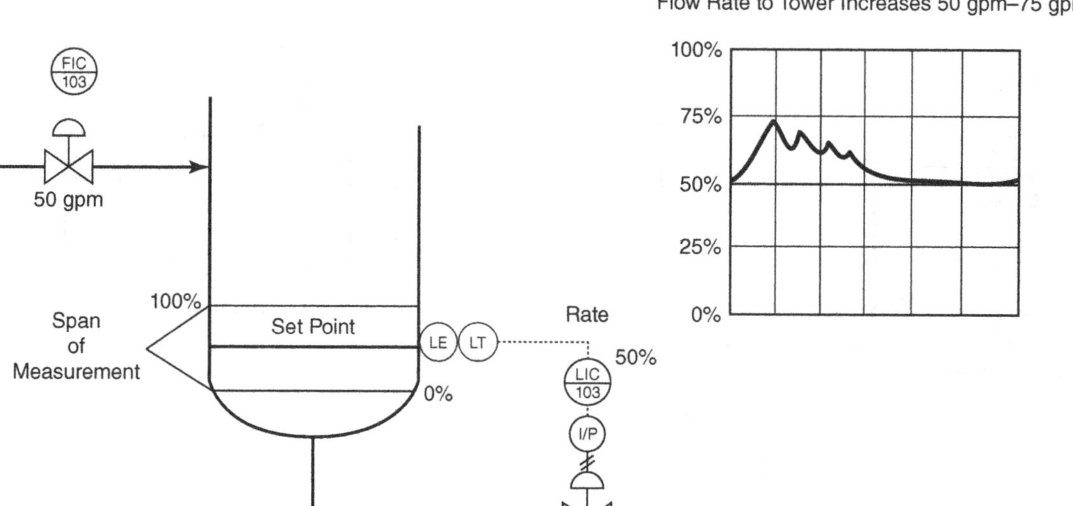

Figure 11.26 *Rate*

Reset (Integral)

Reset is a control mode that responds to a process change by calculating a slope, in repeats per minute, back toward the set point. Reset is designed to reduce the difference between the set point and process variable by adjusting the controller's output continuously until the offset is eliminated. The reset mode responds proportionally to the size of the error, the length of time that it lasts, and the reset gain setting. (See Figure 11.25.)

Rate (Derivative)

Rate mode enhances controller output by increasing the output in relationship to the changing process variable. As the process variable approaches the set point, the rate relaxes, providing a braking action that prevents overshooting the set point. The rate responds aggressively to rapid changes and passively to smaller changes in the process variable. (See Figure 11.26.)

Final Control Elements

Automatic Valves

Final control elements typically are automatic valves, but motors or other electric devices can be used. The final control element is the last link in the control loop and is the device that actually makes the change in the process. Automatic valves will open or close to regulate the process. Because they can be controlled from remote locations, they are invaluable in modern processing. An **actuator** is the device that automates a valve. The actuator controls the position of the flow control element by moving and controlling the position of the valve stem. Actuators come in three basic designs: **pneumatic**, **electric**, and **hydraulic**. Pneumatic (air operated) actuators are the most common. Pneumatic actuators convert air pressure to mechanical energy. They can be found in three designs: diaphragm, piston, and vane. The diaphragm actuator is a dome-shaped device that has a flexible diaphragm running through the center. It is typically mounted on the top of the valve. The center of the diaphragm in the dome is attached to the stem. The valve position (on or off) is held in place by a powerful spring. When air enters the dome on one side of the flexible diaphragm, it opens, closes, or throttles the valve, depending on design. The piston actuator uses an airtight cylinder and piston to move or position the stem. It is commonly found in use with automated gate valves or slide valves. It is used where a lot of stem travel is needed. Vane actuators direct air against paddles or vanes.

Electrically operated actuators convert electricity to mechanical energy. There are two types: solenoid valve and motor-driven actuator. Solenoid valves are designed for on/off service. The internal structure of a solenoid resembles a globe valve. The disc rests in the seat, stopping flow. The stem is attached to a metal core or armature that is held in place by a spring. A wire coil surrounds the upper spring and stem. When the wire coil is energized, a magnetic field is set up, causing the armature to lift and compressing the spring. The armature is held in place until the current stops. A motor-driven actuator is attached to the stem of a valve by a set of gears. Gear movement controls the position of the stem.

Hydraulically operated actuators convert liquid pressure to mechanical energy. The hydraulic actuator uses a liquid tight cylinder and piston to move or position the stem. It is commonly found in use with automated gate valves or slide valves and is used where a lot of stem travel is needed.

The following expressions are used commonly to describe actuators:

- *Air to open, spring to close.* Fails in the closed position if air system goes down. Air line is typically located on the bottom of the dome.
- *Air to close, spring to open.* Fails in the open position if air system goes down. Air line is typically located on the top of the dome.
- *Double acting, no spring.* Air lines located on both sides of the dome.
- *Fails in last position.* Air pressure to diaphragm is locked on instrument air failure.

The most common type of automatic valve is a globe valve because of its versatile on/off or throttling feature. Control loops use on/off or throttling-type valves to regulate the flow of fluid into and out of a system. In addition to the pneumatic, electric, and automatic control valves, there are also spring (or weight) operated valves. Spring-operated valves hold the flow control element in place until pressure from under the disc grows strong enough to lift the element from the seat. A check valve would be in this category.

Interlocks and Permissives

An **interlock** is a device designed to prevent damage to equipment and personnel by stopping or preventing the start of certain equipment if a preset condition has not been met. There are two types of interlocks: softwire and hardwire. Softwire interlocks are contained within the logic of a programmable computer. Hardwire interlocks are a physical arrangement. The hardwire interlock usually involves electrical relays that operate independently of the control computer. In many cases, they run side by side with the computer softwire interlocks. However, hardwire interlocks cannot be bypassed. They must be satisfied before the process they are part of can take place.

A **permissive** is a special type of interlock that controls a set of conditions that must be satisfied before a piece of equipment can be started. Permissives deal with startup items, whereas hardwire interlocks deal with shutdown items. A permissive is an interlock controlled by the DCS. This type of interlock will not necessarily shut down the equipment if one or more of its conditions are not met. It will, however, keep the equipment from starting up.

Manual and Automatic Control

During a unit startup, console operators will use manual control to initially establish a steady state flow through the system. Typically, when a new process is being brought on line, a control loop will swing high and low, as it attempts to match the set point. Experienced technicians are familiar with the operating valve percentage that is established when a unit is lined out. By placing the valve in manual control, process variations are reduced. Once the system is on line, the console technician will insert the set point on the controller, and place it in AUTO. This procedure enhances the time it takes to bring a unit up to steady state and produce a high-quality product.

Automatic control uses all five elements of the control loop in a dynamic system that measures, compares, and attempts to control a specific process. Process units may have hundreds of control loops that make it possible to control level, pressure, temperature, flow, and a variety of analytical processes. Automatic control allows a process technician to control a much larger and more complex arrangement of process systems.

Cascaded Control

Cascade control uses elements from two separate control loops. One of the controllers functions as the primary controller, and the other functions as the secondary controller. This relationship is often referred to as master–slave control. Figure 11.27 illustrates the basic elements of a cascaded system. It is important to point out that in this type of process control, controllers will communicate only with other controllers.

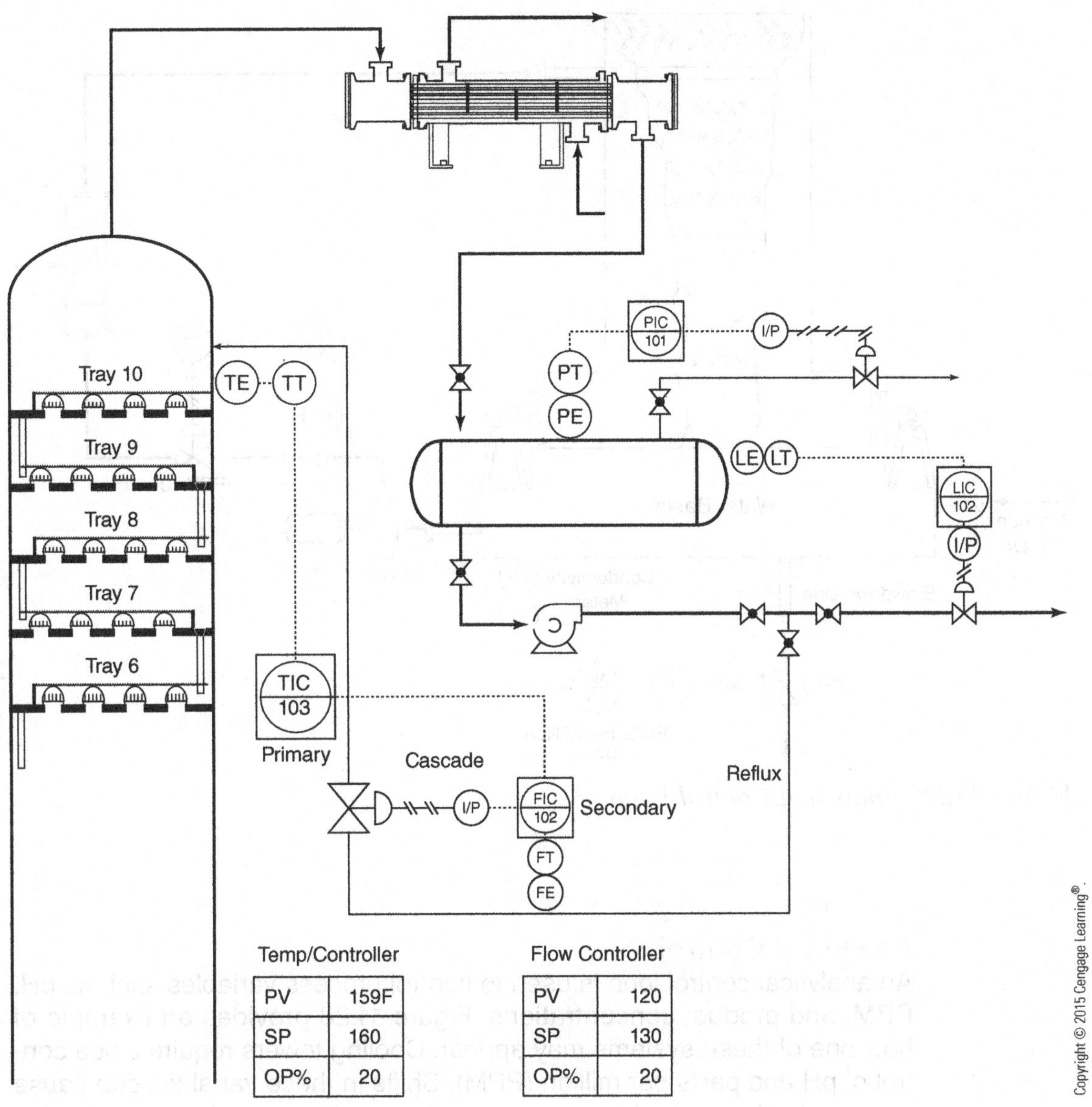

Figure 11.27 *Cascaded Control Loop*

In many process systems, it is possible to accomplish a variety of similar objectives simultaneously. For example, in a distillation system, reflux is used to control temperature and improve product purity. In most cases, a flow control loop is used to ensure steady flow; however, temperature control is considered to be the single most important factor in this system. When a temperature element is placed on the top tray and connected to a transmitter, controller, transducer, and control valve on the reflux line, a conflict appears to exist between the flow control valve and the temperature control valve. By selecting the temperature to be the primary controller, one can allow both the flow control and temperature control to work together to accomplish operational objectives.

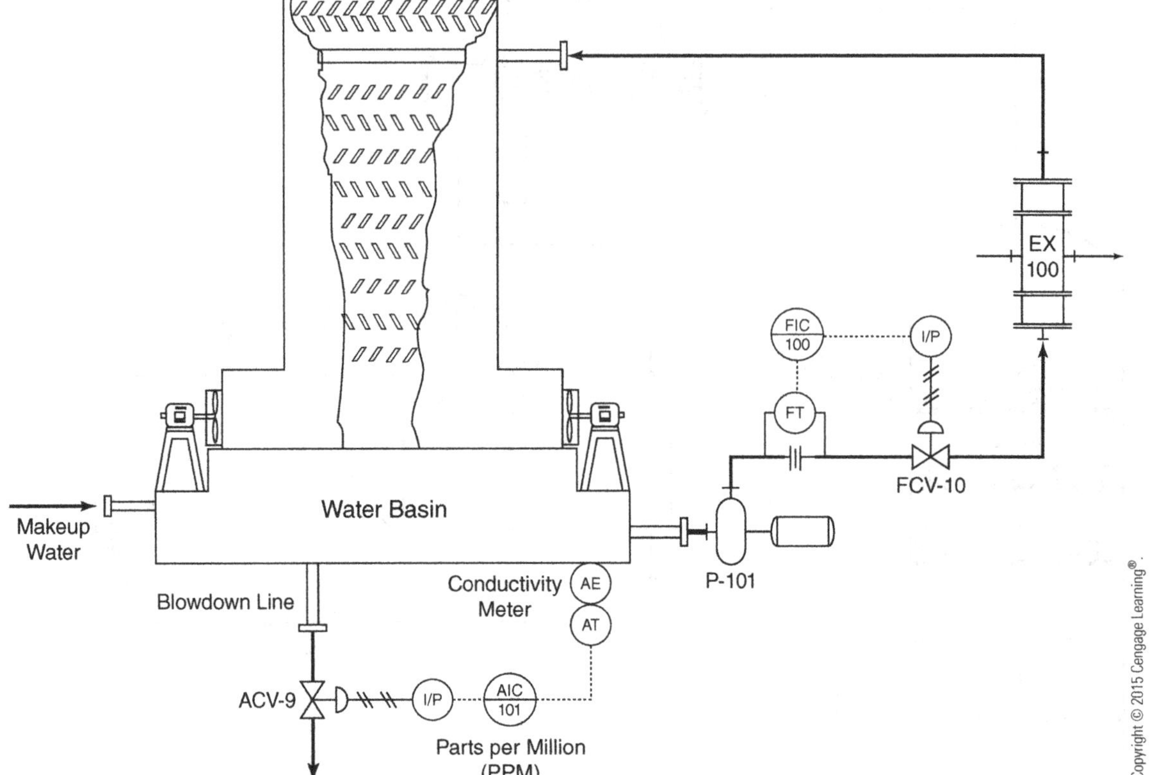

Figure 11.28 *Analytical Control Loop*

Analytical Control

An analytical control loop is used to control process variables such as pH, PPM, and product concentrations. Figure 11.28 provides an example of how one of these systems may appear. Cooling towers require close control of pH and parts per million (PPM). Shifts in these variables can cause serious problems. Control systems can be used to control these analytical variables.

Summary

Automation enhances the ability of a process operator to control large and complex process networks.

The basic instruments include gauges (pressure, level, temperature, and flow); differential pressure (DP) cells; transmitters (pressure, level, temperature, and flow); controllers (pneumatic and electronic), transducers, control valves, and primary elements and sensors (displacer and buoyancy float, thermocouples, orifice plates, etc.); and computers.

The most common type of pressure element is a bourdon tube. There are three commonly used pressure scales in the manufacturing environment: gauge scale (psig), absolute scale (psia), and vacuum scale (Hg).

Process flow measurements frequently are taken by one of the following devices: a flow transmitter, nutating disc meter, oval gear meter, rotameter, turbine flow meter, weir, and flume. Level measurements are identified by a fixed reference point (typically zero) above or below a product. Level-measurement devices can be classified as direct or indirect. Direct instrumentation is in physical contact with the surface of the fluid. Indirect instrumentation incorporates pressure changes that respond proportionally to level changes.

A control loop is a group of instruments that work together to control a process. These instruments typically include a transmitter coupled with a sensing device or primary element, a controller, a transducer, and a control valve. The controller receives a signal from a transmitter, compares this signal with a set point, and adjusts the (final control element) process to stay within the range of the set point. Controllers come in three basic designs: pneumatic, electronic, and digital. Final control elements are typically classified as automated valves, but motors or other electrical devices can be used. The final control element is the last link in the modern control loop and is the device that actually makes the change in the process.

An interlock is a device designed to prevent damage to equipment and personnel by stopping or preventing the start of certain equipment if a preset condition has not been met. A permissive is a special type of interlock that controls a set of conditions that must be satisfied before a piece of equipment can be started. Permissives deal with startup items, whereas hardwire interlocks deal with shutdown items.

Review Questions

1. List seven basic types of instruments found in the processing industry.
2. Describe the operation of a DP cell.
3. What is a rotameter used for?
4. List the basic elements of a control loop.
5. Draw a pressure control loop.
6. Draw a level control loop.
7. Draw a flow control loop.
8. Draw a temperature control loop.
9. How do the basic elements of a control loop work together?
10. Describe controllers and control modes.
11. List the primary elements and sensors used with temperature and pressure.
12. List the primary elements and sensors used with level and flow.
13. Define the term DCS.
14. Draw a cascaded control loop.
15. Describe manual and automatic control.
16. Draw an analytical control loop.
17. Explain how a process technician uses a valve's operating percentage in manual operation.
18. List the primary elements or sensors for temperature.
19. Describe a thermocouple and explain how it is used.
20. Explain how a bourdon tube works.

Process Control Diagrams

Objectives

After studying this chapter, the student will be able to:

- Review process diagram symbols.
- Describe a block flow diagram.
- Describe the use of process diagrams and the information they contain.
- Draw a process flow diagram.
- Draw and label the following control loops: level, pressure, temperature, flow, composition, and cascade.
- Describe manual, automatic, and cascade control features.
- Identify the major components and purpose of a(n) equipment location drawing, electrical drawing, elevation drawing, and foundation drawing.
- Draw a process and instrument drawing.
- Describe the various process equipment relationships.
- Describe the equipment, instruments, control loops, piping, and operational data found in process control diagrams.

Key Terms

Block flow diagram (BFD)—a set of blocks that move from left to right to show the primary flow path of a process.

Controller—an instrument used to compare a process variable with a set point and initiate a change to return the process to a set point if a variance exists.

Electrical drawings—symbols and diagrams that depict an electrical process.

Elevation drawings—a graphical representation that shows the location of process equipment in relation to existing structures and ground level.

Equipment location drawings—show the exact floor plan for location of equipment in relation to the plan's physical boundaries.

Flow diagram—see *Process flow diagram (PFD).*

Foundation drawings—concrete, wire mesh, and steel specifications that identify width, depth, and thickness of footings, support beams, and foundation.

Legends—a document used to define symbols, abbreviations, prefixes, and specialized equipment.

Piping—used to convey all kinds of fluids, liquid, or gas.

Process flow diagram (PFD)—a simplified sketch that uses symbols to identify instruments and vessels and to describe the primary flow path through a unit.

Process and instrument drawing (P&ID)—a complex drawing that uses equipment and line symbols, instruments, control loops, and electrical drawings to identify primary and secondary flow paths through the plant. P&IDs may provide operating specifications, temperatures, pressures, flows, levels, analytical variables, and mass relationship data. These documents may also include pipe sizes, equipment specifications, motor sizes, and so forth. Also called a piping and instrumentation drawing.

Introduction to Process Drawings

Process technicians or operators, research technicians, pharmaceutical technicians, laboratory technicians, environmental technicians, engineering technicians, food processing technicians, paper and pulp technicians, power generation technicians, and engineers use process drawings or schematics to describe the primary flow and equipment used in their facilities. The I & E and maintenance departments and the construction division also use these drawings and refer to them each time work is performed in a specific section. Process schematics can vary from very simple to complex. These drawings simplify complex processes and provide a window through which a technician can study **piping**, instrumentation, equipment, and locations. Process schematics include symbols, block flow diagrams, process flow diagrams, control loops, operational data, and piping and instrumentation drawings. These drawings are used by every level of the plant and are a critical document.

Block Flow Diagrams

Simple block flow diagrams can be used to show technical and non-technical people how material moves from one location to another. Figure 12.1 illustrates the basic components of a block flow diagram (BFD). **Block flow diagrams** provide a simplistic set of sequences that move from left to right and show the primary flow path of a process. In reality, these drawings are helpful to individuals who are not expected to operate the unit. The best set of teaching tools for a new trainee is a good supervisor, a trainer, a process flow diagram, and training materials that accurately reflect the plants standard operating procedures (SOPs). Technicians learn best when the expectations and qualification list are clearly spelled out. Most process units are very complex and provide challenging opportunities for apprentice technicians.

Types of Process Diagrams

Process diagrams can be broken down into three major categories: process flow diagrams (PFDs), process flow and instrumentation drawings (PFIDs), and **process and instrument drawings (P&IDs)**—officially

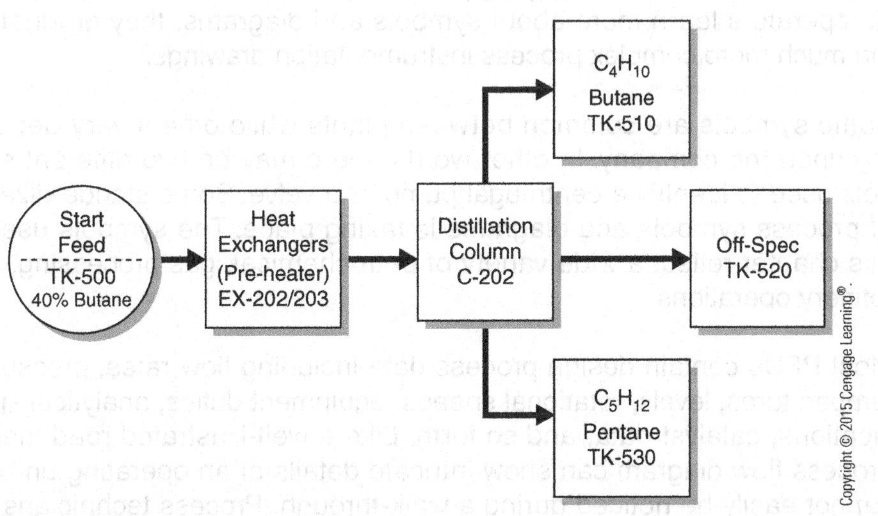

Figure 12.1
Simple Block Flow Diagram

called *piping and instrumentation drawings* by the Instrument Society of America (ISA). A **process flow diagram** is a simple illustration that uses process symbols to describe the primary flow path through a unit. A process flow diagram provides a quick snapshot of the operating system. **Flow diagrams** include all major equipment and flows. A technician can use this document to trace the primary flow of chemicals through the unit. Secondary or minor flows are not included. Complex control loops and instrumentation are not included. The flow diagram is used for visitor information and new employee training.

A PFID is a step up from the simplistic PFD to a little more complex drawing that includes some major and minor instrumentation and is very helpful during the training phase. Apprentice technicians will use a version of this document to trace lines, locate equipment, and check operating variables.

A process and instrument drawing is much more complex. The P&ID includes a graphic representation of the equipment, piping, and instrumentation. Modern process control can be clearly inserted into the drawing to provide a process technician with a complete picture of electronic and instrument systems. Process operators can look at their process and see how the engineering department has automated the unit. Pressure, temperature, flow, analytical, speed, and level control loops are all included on the unit P&ID. Cascaded control systems, indicators, and alarms are clearly visible.

PFDs, PFID's, and P&IDs are used to outline or explain the complex flows, equipment, instrumentation, and equipment layouts that exist in a process unit. New technicians are required to study a simple flow diagram of their assigned operating system. Process flow diagrams typically include the major equipment and the piping path the process takes through the unit. As operators learn more about symbols and diagrams, they graduate to the much more complex process instrumentation drawings.

Some symbols are common between plants while others vary depending upon the company. In other words, there may be two different symbols used to identify a centrifugal pump or a valve. Some standardization of process symbols and diagrams is taking place. The symbols used in this chapter reflect a wide variety of petrochemical, gas processing, and refinery operations.

Most PFDs contain design process data including flow rates, pressures, temperatures, levels, rotational speeds, equipment duties, analytical specifications, catalyst data, and so forth. Like a well-illustrated road map, a process flow diagram can show intricate details of an operating unit that cannot easily be noticed during a walk-through. Process technicians are expected to memorize the company's symbols and diagram file, interpret simple flow diagrams, and apply this knowledge within hours of starting their initial training. Technicians will graduate to complex P&IDs over the course of their training and into their work careers.

Figure 12.2 shows the basic relationships and flow paths found in a process unit. It is easier to understand a simple flow diagram if it is broken down into sections: feed, pre-heating, the process, and the final products. This simple left-to-right approach allows a technician to first identify where the process starts and where it will eventually end. The feed section includes the feed tanks, mixers, piping, and valves. In the second step, the process flow is gradually heated up for processing. This section includes heat exchangers and furnaces. In the third section, the process is included. Typical examples found in the process section could include distillation or reaction. The process area is a complex collection of equipment that works together in a system. The process is designed to produce products that will be sent to the final section.

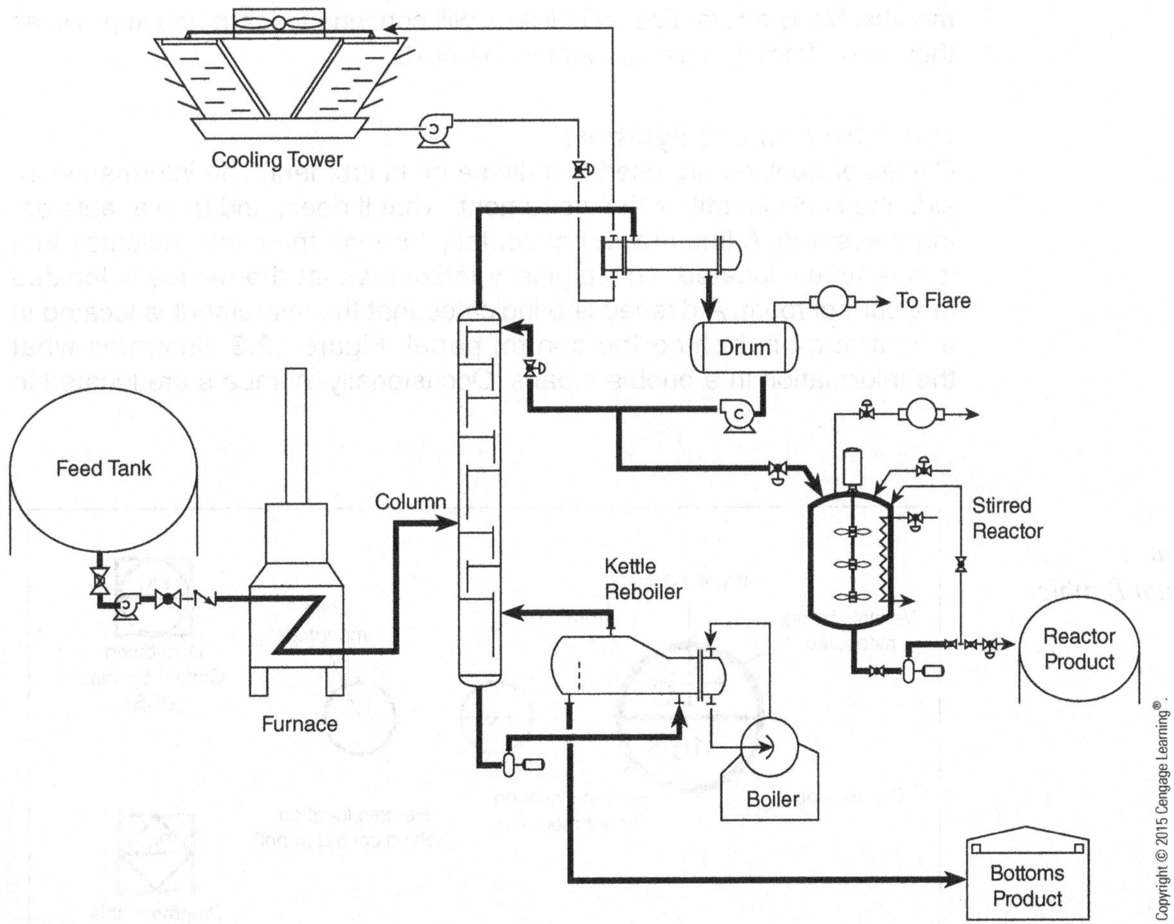

Figure 12.2 *Process Flow Diagram (PFD)*

Process Flow Diagrams

A process flow diagram is a simple illustration that uses process symbols to describe the primary flow path through a unit. A process flow diagram provides a quick snapshot of the operating unit. Flow diagrams include all primary equipment and flows. A process technician (operator), engineer, I & E, or safety or quality manager can use this document to trace the primary flow of chemicals through the unit. Secondary or minor flows are not included. Complex control loops and instrumentation are not included. The flow diagram is used for customer information, visitor information, and new employee training.

A typical plant will be composed of hundreds of smaller processes or systems. Process flow diagrams are used to illustrate these systems. New employees typically learn a series of simple flow processes as they move into more complex equipment arrangements. Given adequate time and correct instruction, it is possible to learn the most complicated process. While the concepts of life-long learning are clearly found in the chemical processing industry, it is possible to qualify on a typical job post in three to four months. Many apprentice technicians will continue to learn and improve as they move from one assignment to the next.

Basic Instrument Symbols

Circles or bubbles are used to indicate an instrument. The information inside the circle identifies the instrument, what it does, and the variable being measured. A line drawn horizontally through the circle indicates that it is remotely located. This typically indicates that the device is located in a control room. A dashed line indicates that the instrument is located in a control room behind the control panel. Figure 12.3 illustrates what the information in a bubble means. Occasionally, numbers are located in

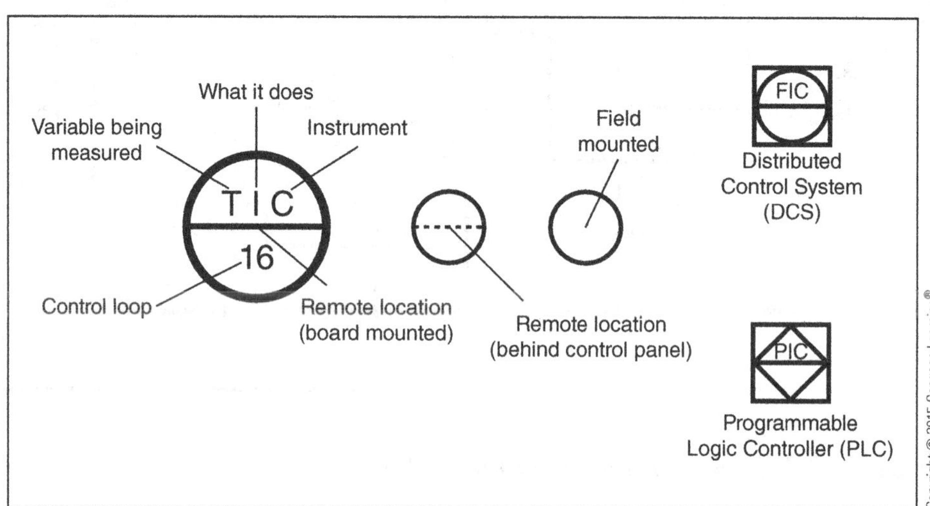

Figure 12.3 *Instrument Information Bubble*

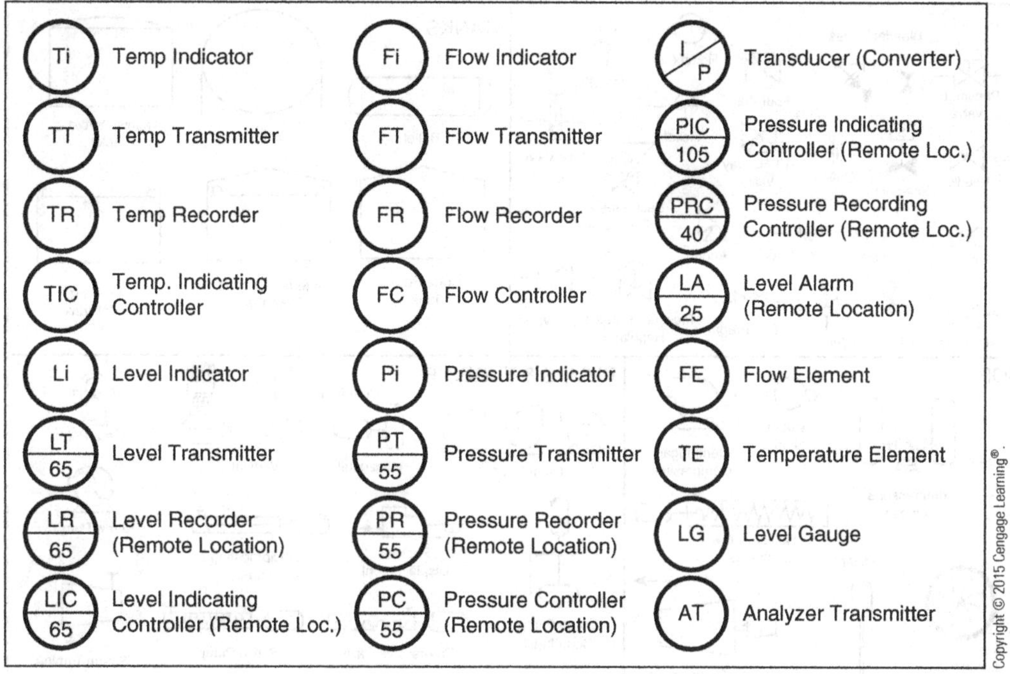

Figure 12.4 *Instrument Symbols*

the circle that may indicate the number of the control loop. This could indicate which system the device is in or any variety of special applications used by the company.

There are a large variety of instruments and instrument symbols used in the chemical processing industry. The variables that instruments measure are pressure, temperature, level, flow, and analytical variables. If the information in a circle starts with the letter "L," it is typically some type of level instrument. If the first letter in the circle is the letter "A," it is an analytical instrument. Figure 12.4 shows a list of simple instruments. These instruments will be covered in more detail later in this text.

Process technicians use P&IDs to identify all of the equipment, instruments, and piping found in their units. New technicians use these drawings during their initial training period. Knowing and recognizing these symbols is important for a new technician. The chemical processing industry has assigned a symbol for each type of valve, pump, compressor, steam turbine, heat exchanger, cooling tower, basic instrumentation, reactor, distillation column, furnace, and boiler (Figure 12.5). There are symbols to represent major and minor process lines and pneumatic, hydraulic, or electric lines, and there is a wide variety of electrical symbols.

Each plant will have a standardized file for their piping symbols. Process technicians should carefully review the piping symbols for major and minor

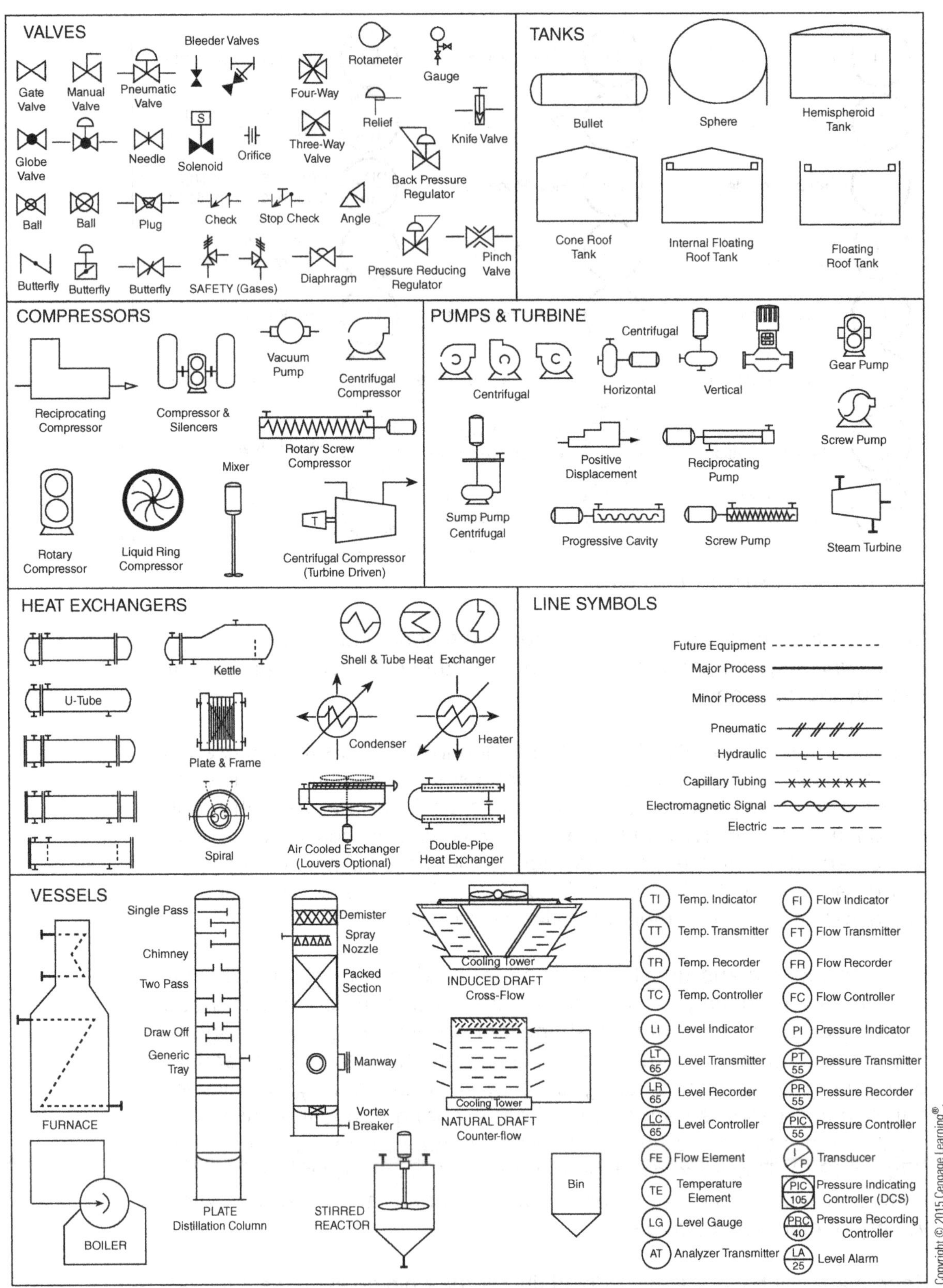

Figure 12.5 *Process and Instrument Symbols*

flows, electric, pneumatic, capillary, and hydraulic and future equipment. The major flow path through a unit illustrates the critical areas a new technician should concentrate on. A variety of other symbols are included on the piping. These include valves, strainers, flexible spool, filters, flanges, removable spool pieces, blinds, insulation, piping size, expansion joint, vent cover, in-line mixer, eductor, pulsation dampener, exhaust head, pressure rating, material codes, and steam traps. Figure 12.6 illustrates the basic line symbols used by process technicians. It should be noted that not every symbol will be covered here.

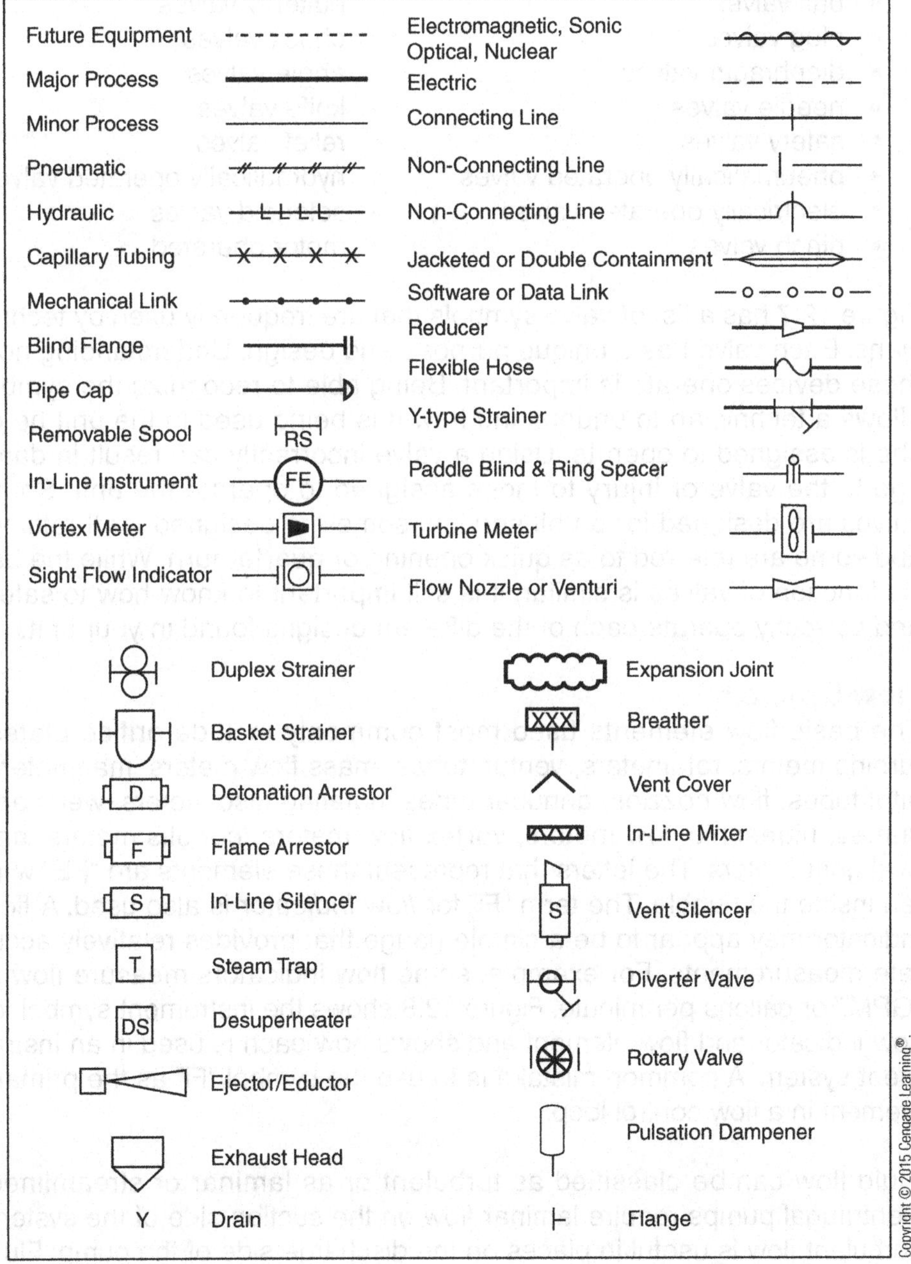

Figure 12.6
Piping and Auxiliary Symbols

Valve Symbols

In a refinery or chemical plant, there are thousands of different valves. These valves come in a variety of designs and sizes. Modern process control is linked to the operation of a device called an automated valve. An automated valve has an actuator mounted on the valve that is designed to open, close, or throttle flow through the pipe. A process system is a network of piping and valves. Valves are designed to control the flow of liquids and gases through the system. There is a corresponding symbol for each valve you will study. The most common valves are as follows:

- gate valves
- ball valves
- plug valves
- diaphragm valves
- needle valves
- safety valves
- pneumatically operated valves
- electrically operated valves
- pinch valves
- globe valves
- butterfly valves
- check valves
- angle valves
- knife valves
- relief valves
- hydraulically operated valves
- solenoid valves
- motor operated

Figure 12.7 has a list of valve symbols that are frequently used by technicians. Each valve has a unique purpose and design. Understanding how these devices operate is important. Being able to recognize the symbol allows a technician to understand how it is being used in the unit he or she is assigned to operate. Using a valve incorrectly can result in damage to the valve or injury to those assigned to operate the unit. Some valves are designed for on/off service, some are designed for throttling, and some are referred to as quick opening or quarter turn. While the basic function of valves is similar, it is still important to know how to safely and correctly operate each of the different designs found in your unit.

Flow Symbols

The basic flow elements used most commonly include orifice plates, turbine meters, rotameters, venturi tubes, mass flow meters, magmeters, pitot tubes, flow nozzles, annubar tubes, nutating disc meters, weirs and flumes, ultrasonic flow meters, vortex flow meters, coriolis meters, and oval gear meters. The letters that represent these elements are "FE" written inside the bubble. The term "Fi" for *flow indicator* is also used. A flow indicator may appear to be a simple gauge that provides relatively accurate measurements. For example, some flow indicators measure flow in "GPM," or gallons per minute. Figure 12.8 shows the instrument symbol for flow indicator and flow element and shows how each is used in an instrument system. A common mistake is to use the symbol "Fi" as the primary element in a flow control loop.

Fluid flow can be classified as turbulent or as laminar or streamlined. Centrifugal pumps require laminar flow on the suction side of the system. Turbulent flow is useful in places on the discharge side of the pump. Fluid velocity inside a pipe changes depending upon the diameter of the pipe.

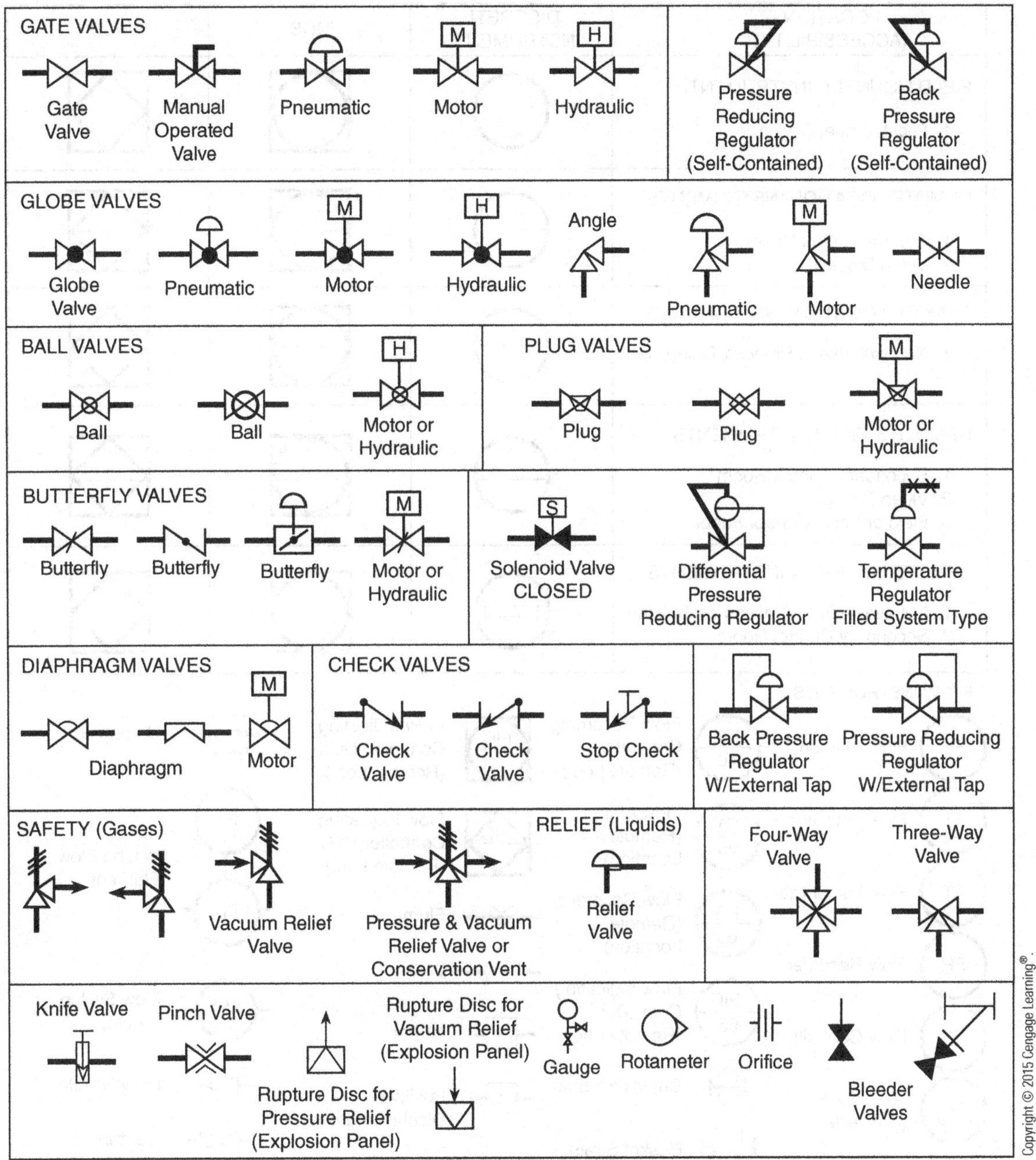

Figure 12.7 *Valve Symbols*

These changes also have a direct relationship with pressure. In pipes where the diameter suddenly gets smaller, pressure will increase at the inlet point. Fluid velocity will increase through the narrow passage, resulting in a pressure drop. Bends in the piping will cause the liquid to become turbulent. Turbulent flow tends to transfer heat energy better.

Another set of factors that will affect the velocity of a fluid is density, viscosity, and temperature increases or decreases. Many flow elements work

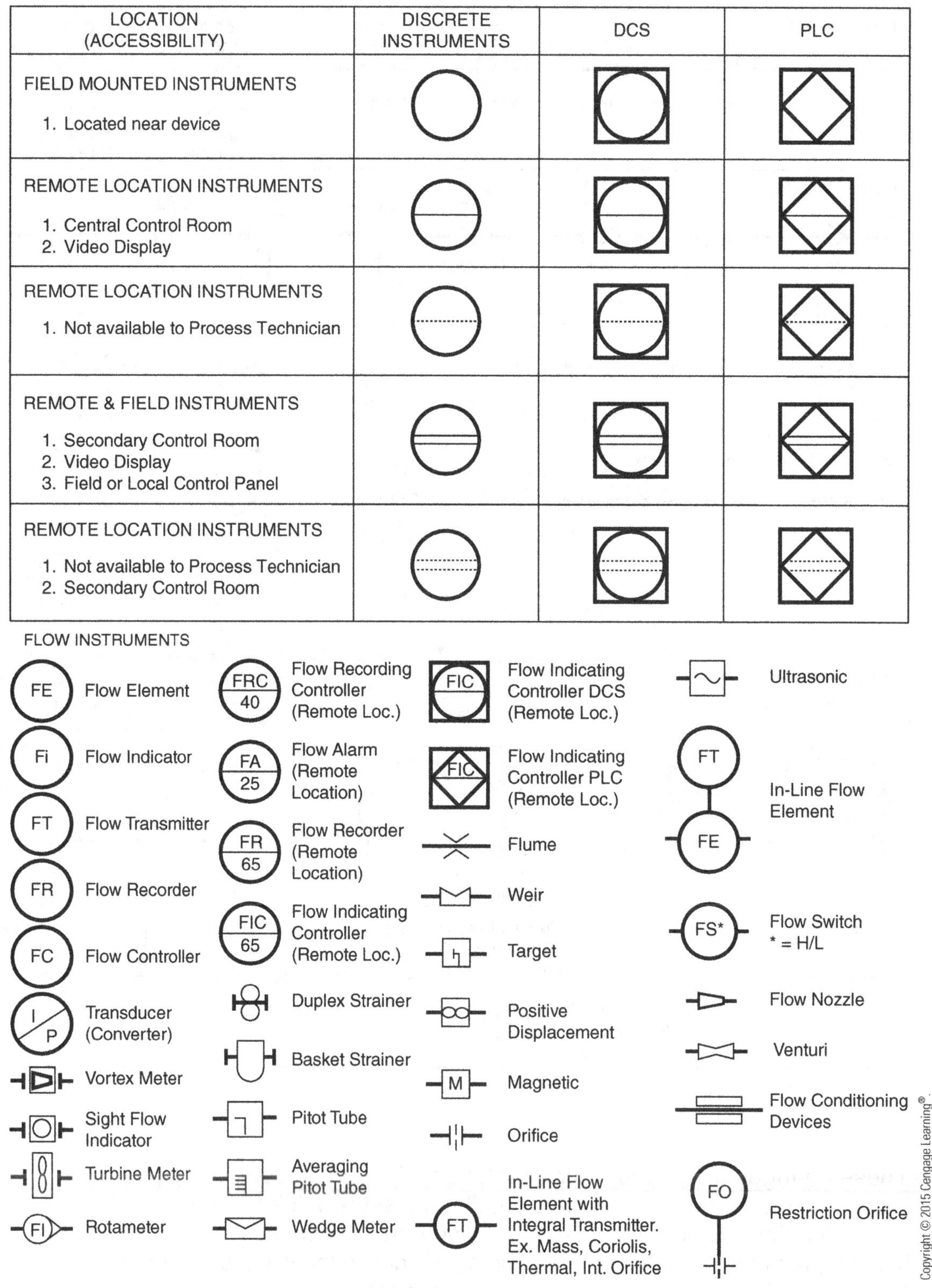

Figure 12.8 *Flow Instrument Symbols*

directly with transmitters. These flow transmitters are designed to send a signal to a **controller**. An example of this is a differential pressure transmitter that is working with an orifice plate. The transmitter is typically mounted below the line to keep the legs liquid full. High-pressure and low-pressure sides are located on each side of a pressure capsule. The orifice plate creates an artificial low-pressure situation inside the pipe that can be measured and transmitted to a controller.

Fluid flow is typically controlled with a flow control loop (FIC). In this system each of the five elements of the control loop will be flow related and labeled appropriately. It is also possible to have other flow symbols like flow recorders (FRs), high-flow alarms (HFAs), and low-flow alarms (LFAs).

Pressure Symbols

The most common pressure devices are pressure gauges, manometers, and pressure transmitters. Pressure gauges are represented on a P&ID as "Pi," or pressure indicator. A circle with "PE" is the symbol for a pressure element. A diaphragm in a DP cell is an example of a "PE." A simple equation can be used to determine pressure, P (pressure) = F (force) ÷ A (area). A pressure gauge has a variety of internal designs. Some of these designs use a bourdon tube, diaphragm capsule, bellows, diaphragm, spiral, or helical. The basic physics has the device responding to pressure changes. These responses or movements are transferred by a mechanical linkage to an indicator. For example, a bourdon tube is a hollow tube shaped like a hook. As pressure is admitted into the hollow curved tube, it attempts to straighten out. A mechanical linkage transfers this travel to an indicator that shows the pressure in pounds per square inch (psi), pounds per square inch absolute (psia), pounds per square inch vacuum (psiv), or pounds per square inch differential (psid). Figure 12.9 shows the basic instrument symbols used to show pressure.

Pressure is typically controlled with a pressure control loop (PIC). In this system, each of the five elements of the control loop will be pressure related and labeled appropriately. It is also possible to have other pressure symbols like pressure recorders (PR), high-pressure alarms (HPA), and low-pressure alarms (LPA).

Level Symbols

Level measurement is directly linked to pressure measurement. Controlling the level in a tank, vessel, reactor, or distillation column is an important concept. The common instruments used to measure level are level gauges (LGs), displacer bulbs coupled to a level transmitter, capacitance probes, a bubbler system, load cells, and differential pressure transmitters (ΔP cells). As the level in a tank increases, the pressure at the bottom of the tank increases. A simple equation, H (height of liquid above point being

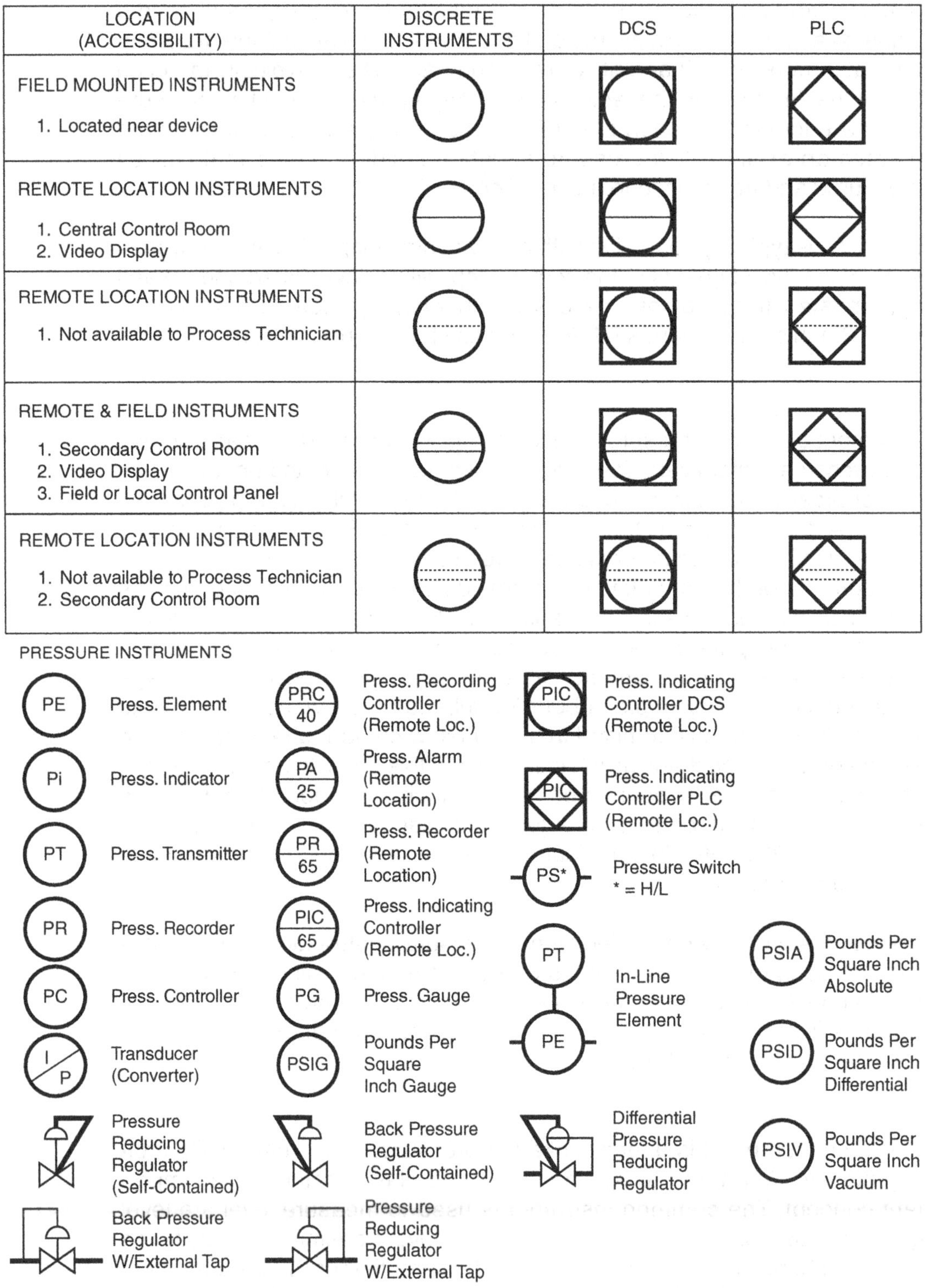

LOCATION (ACCESSIBILITY)	DISCRETE INSTRUMENTS	DCS	PLC
FIELD MOUNTED INSTRUMENTS 1. Located near device			
REMOTE LOCATION INSTRUMENTS 1. Central Control Room 2. Video Display			
REMOTE LOCATION INSTRUMENTS 1. Not available to Process Technician			
REMOTE & FIELD INSTRUMENTS 1. Secondary Control Room 2. Video Display 3. Field or Local Control Panel			
REMOTE LOCATION INSTRUMENTS 1. Not available to Process Technician 2. Secondary Control Room			

Figure 12.9 *Pressure Instrument Symbols*

calculated) × .433 × specific gravity = pounds per square inch. If the liquid is in an enclosed tank, things like vapor pressure or added gas pressure must be added to the total pressure. Figure 12.10 illustrates the basics symbols used to illustrate level. Level indicators are very common on process drawings and are represented with an instrument symbol that looks like a

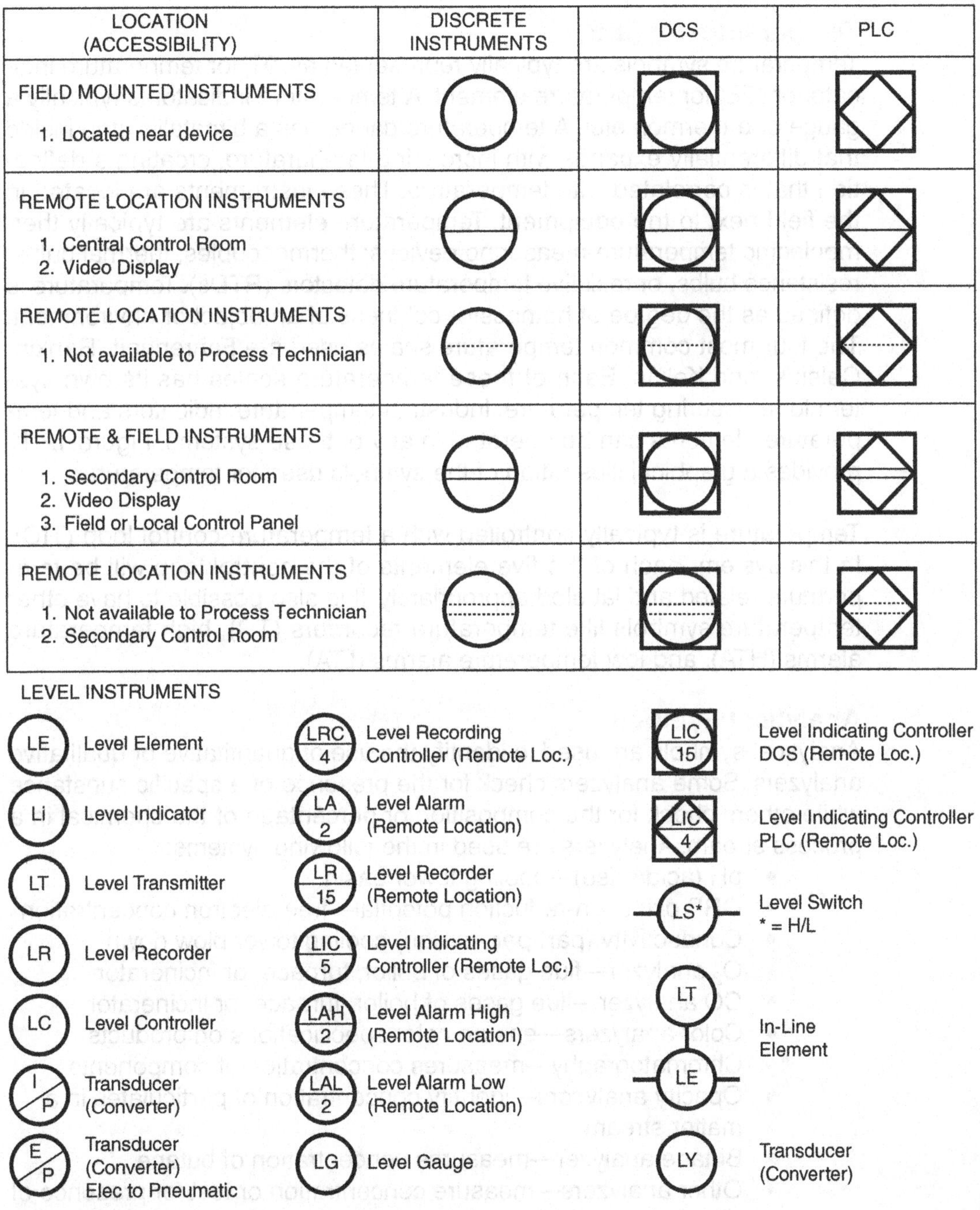

LOCATION (ACCESSIBILITY)	DISCRETE INSTRUMENTS	DCS	PLC
FIELD MOUNTED INSTRUMENTS 1. Located near device			
REMOTE LOCATION INSTRUMENTS 1. Central Control Room 2. Video Display			
REMOTE LOCATION INSTRUMENTS 1. Not available to Process Technician			
REMOTE & FIELD INSTRUMENTS 1. Secondary Control Room 2. Video Display 3. Field or Local Control Panel			
REMOTE LOCATION INSTRUMENTS 1. Not available to Process Technician 2. Secondary Control Room			

LEVEL INSTRUMENTS

LE Level Element
Li Level Indicator
LT Level Transmitter
LR Level Recorder
LC Level Controller
I/P Transducer (Converter)
E/P Transducer (Converter) Elec to Pneumatic
LRC 4 Level Recording Controller (Remote Loc.)
LA 2 Level Alarm (Remote Location)
LR 15 Level Recorder (Remote Location)
LIC 5 Level Indicating Controller (Remote Loc.)
LAH 2 Level Alarm High (Remote Location)
LAL 2 Level Alarm Low (Remote Location)
LG Level Gauge
LIC 15 Level Indicating Controller DCS (Remote Loc.)
LIC 4 Level Indicating Controller PLC (Remote Loc.)
LS* Level Switch * = H/L
LT / LE In-Line Element
LY Transducer (Converter)

Figure 12.10 *Level Instrument Symbols*

circle with "Li" written on the inside. The symbol "LE" is used to describe a level element. An example of an LE is a displacer bulb. Level is typically controlled with a level control loop (LIC). In this system, each of the five elements of the control loop will be level related and labeled appropriately. It is also possible to have other level symbols like level recorders (LR), high-level alarms (HLA), and low-level alarms (LLA).

Temperature Symbols

Temperature symbols are typically represented as "Ti" for temperature indicator or "TE" for temperature element. A temperature indicator is typically a gauge or a thermometer. A temperature gauge has a bimetallic strip inside that differentially expands with increasing temperature, creating a deflection that is correlated with temperature. These instruments are located in the field next to the equipment. Temperature elements are typically thermoelectric temperature-measuring devices: thermocouples, thermal bulbs, resistance bulbs, or resistive temperature detectors (RTDs). Temperature is defined as the degree of hotness or coldness of an object or environment. The four most common temperature scales used are Fahrenheit, Rankin, Celsius, and Kelvin. Each of these temperature scales has its own system for measuring temperature. Industrial temperature indicators and temperature elements can be operated in any of these systems. Figure 12.11 provides a graphical illustration of the symbols used for temperature.

Temperature is typically controlled with a temperature control loop (TIC). In this system, each of the five elements of the control loop will be temperature related and labeled appropriately. It is also possible to have other temperature symbols like temperature recorders (TR), high-temperature alarms (HTA), and low-temperature alarms (LTA).

Analytical Symbols

Analytical symbols are used to identify the use of quantitative or qualitative analyzers. Some analyzers check for the presence of a specific substance while others check for the composition or percentage of the chemical in a process stream. Analyzers are used in the following systems:

- pH (acid/base)—cooling tower basin
- ORP oxidation-reduction potential—free electron concentration
- Conductivity (part per -million) cooling tower blow down
- O_2 analyzer—flue gases of boiler, furnace, or incinerator
- CO analyzer—flue gases of boiler, furnace, or incinerator
- Color analyzers—ensure color specifications on products
- Chromatography—measures concentration of components
- Opacity analyzers—identify concentration of particulates in a matter stream
- Butane analyzer—measures concentration of butane
- Other analyzers—measure concentration or test for presence of a specific chemical
- Other analyzers include mass spectrometers and total carbon analyzers

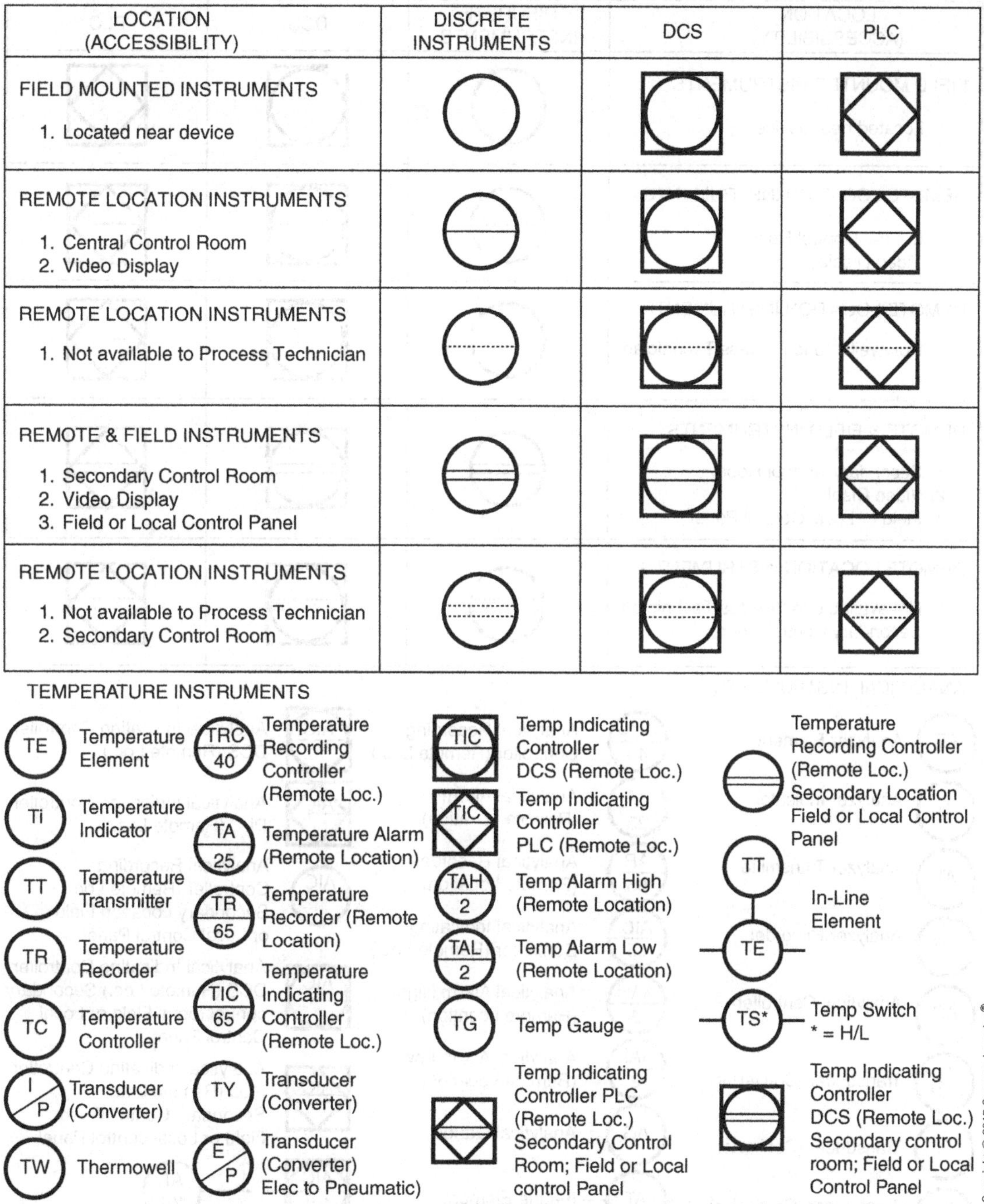

LOCATION (ACCESSIBILITY)	DISCRETE INSTRUMENTS	DCS	PLC
FIELD MOUNTED INSTRUMENTS 1. Located near device			
REMOTE LOCATION INSTRUMENTS 1. Central Control Room 2. Video Display			
REMOTE LOCATION INSTRUMENTS 1. Not available to Process Technician			
REMOTE & FIELD INSTRUMENTS 1. Secondary Control Room 2. Video Display 3. Field or Local Control Panel			
REMOTE LOCATION INSTRUMENTS 1. Not available to Process Technician 2. Secondary Control Room			

TEMPERATURE INSTRUMENTS

TE Temperature Element

Ti Temperature Indicator

TT Temperature Transmitter

TR Temperature Recorder

TC Temperature Controller

I/P Transducer (Converter)

TW Thermowell

TRC 40 Temperature Recording Controller (Remote Loc.)

TA 25 Temperature Alarm (Remote Location)

TR 65 Temperature Recorder (Remote Location)

TIC 65 Temperature Indicating Controller (Remote Loc.)

TY Transducer (Converter)

E/P Transducer (Converter) Elec to Pneumatic)

TIC Temp Indicating Controller DCS (Remote Loc.)

TIC Temp Indicating Controller PLC (Remote Loc.)

TAH 2 Temp Alarm High (Remote Location)

TAL 2 Temp Alarm Low (Remote Location)

TG Temp Gauge

Temp Indicating Controller PLC (Remote Loc.) Secondary Control Room; Field or Local control Panel

Temperature Recording Controller (Remote Loc.) Secondary Location Field or Local Control Panel

TT TE In-Line Element

TS* Temp Switch * = H/L

Temp Indicating Controller DCS (Remote Loc.) Secondary control room; Field or Local Control Panel

Figure 12.11 *Temperature Instrument Symbols*

Analyzers provide an open window on the molecular level to the chemical composition of a substance. Like other control loops, an analytical control loop can be designed to automate a manual process. For example, the blow-down on a cooling water basin can use a conductivity meter as the primary element, a transmitter, a controller, a transducer, and a control valve located on the blow-down line. When the parts per million exceeds operational

LOCATION (ACCESSIBILITY)	DISCRETE INSTRUMENTS	DCS	PLC
FIELD MOUNTED INSTRUMENTS 1. Located near device			
REMOTE LOCATION INSTRUMENTS 1. Central Control Room 2. Video Display			
REMOTE LOCATION INSTRUMENTS 1. Not available to Process Technician			
REMOTE & FIELD INSTRUMENTS 1. Secondary Control Room 2. Video Display 3. Field or Local Control Panel			
REMOTE LOCATION INSTRUMENTS 1. Not available to Process Technician 2. Secondary Control Room			

ANALYTICAL INSTRUMENTS

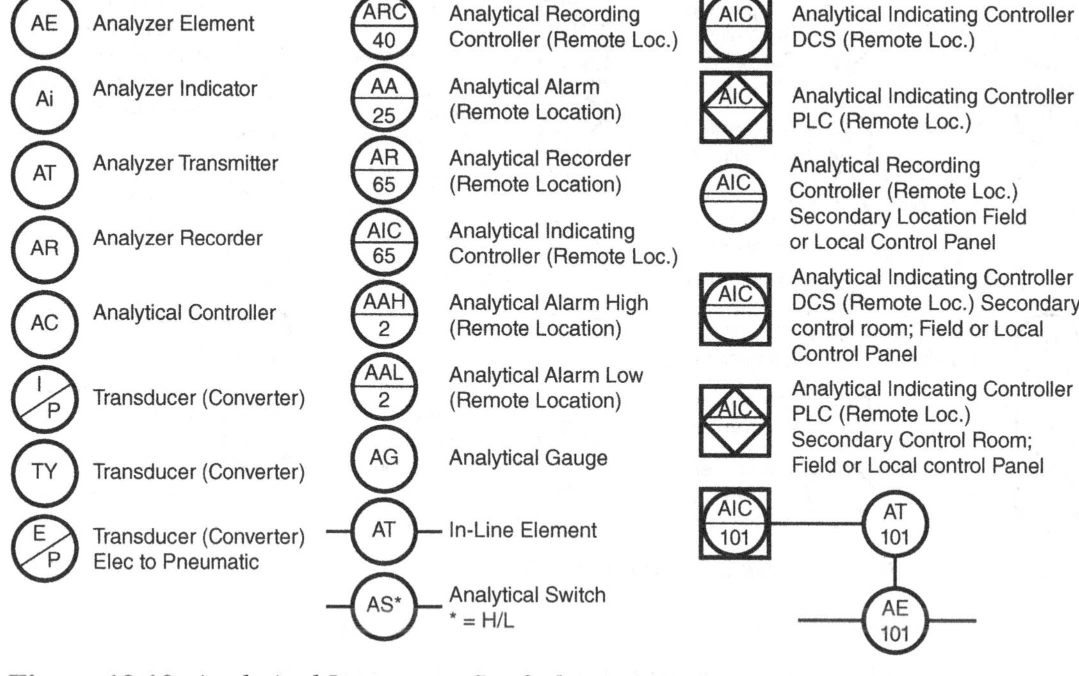

Figure 12.12 *Analytical Instrument Symbols*

specifications, the control valve opens and allows a specific volume of water to go to the holding lagoons. This lost water is replaced by fresh water from the make-up system. The symbol for an analyzer used in this way is "AE," or analyzer element. Figure 12.12 illustrates the basic symbols used to describe analyzers.

Since a variety of analytical variables are controlled with an analytical control loop (AIC), each of the five elements of the control loop will be analysis-related and labeled appropriately. It is also possible to have other analytical symbols like analytical recorders (AR), high-analyzer alarms (HAA), and low-analyzer alarms (LAA).

Equipment Symbols

Pumps and tanks come in a variety of designs and shapes. Process symbols are designed to graphically display the process unit. Common pump and tank symbols can be found in Figure 12.13. Tanks may be designed as bin, drum, dome roof, open top, internal floating roof, double wall, cone roof, external floating roof, and a variety of spherical shaped tanks. The materials used in these tanks can be plastic, carbon steel, stainless steel, specialty alloys, or any variety of materials. Tanks have special designs that are shown on a P&ID and on some symbol charts.

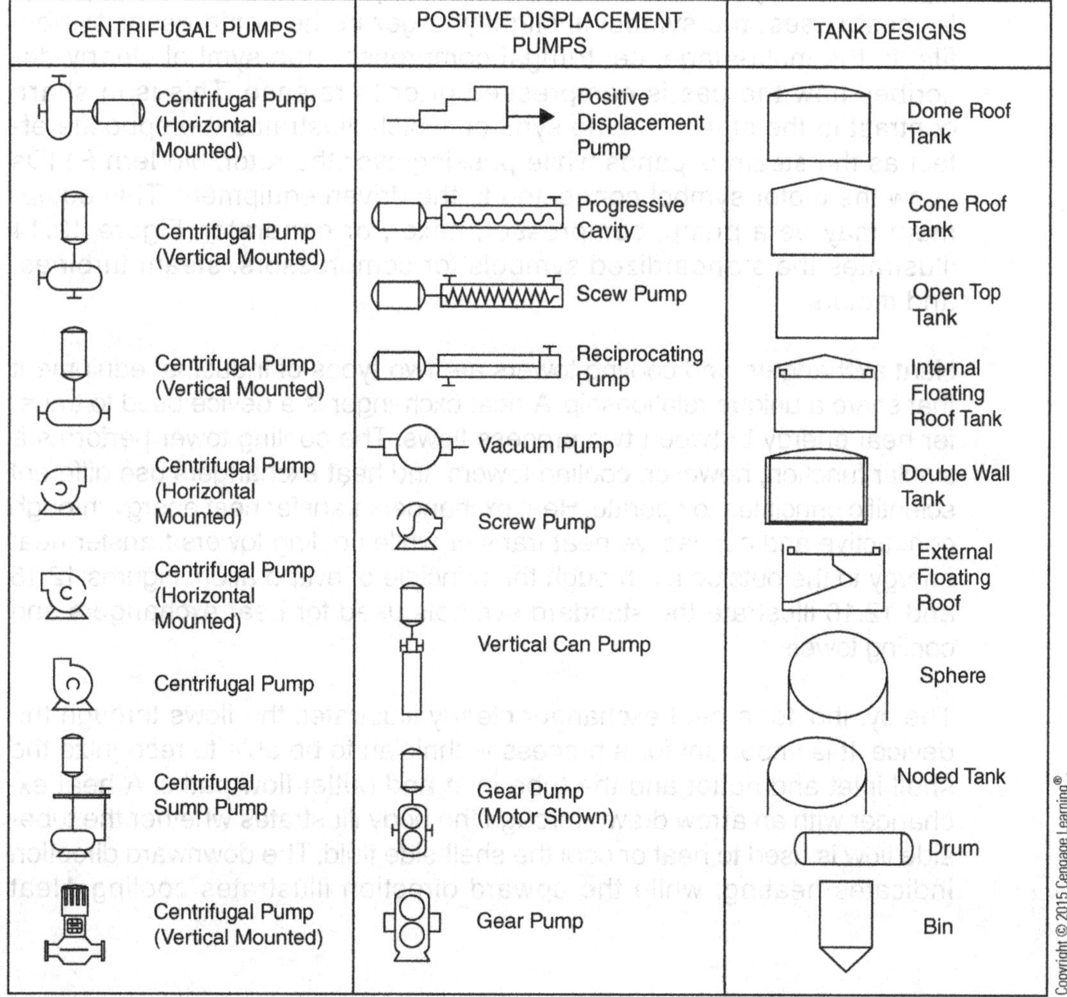

Figure 12.13 *Pump and Tank Symbols*

Each tank is different and has a variety of pressure and temperature requirements.

Pumps come in two basic designs: dynamic and positive displacement. Dynamic pumps include the very popular centrifugal pump and axial pumps. Positive displacement pumps have two categories: rotary and reciprocating. Reciprocating pumps typically include piston, plunger, and diaphragm. These operate with a back-and-forth motion, displacing a predetermined amount of liquid. Rotary pumps use a rotary motion that displaces a specific amount of fluid on each rotation. Rotary pumps include the following designs: lobe, gear, vane, and screw.

A pump and tank system includes piping, valves, instruments, and control loops as well as the pump and tank. These systems are designed to be circulated.

Compressors and pumps share a common set of operating principles. The dynamic and positive displacement families share common categories. The symbols for compressors may closely resemble a pump. In most cases, the symbol is slightly larger in the compressor symbol file. In the multi-stage, centrifugal compressor, the symbol clearly describes how the gas is compressed prior to release. This is in sharp contrast to the steam turbine symbol, which illustrates the opposite effect as the steam expands while passing over the rotor. Modern P&IDs show the motor symbol connected to the driven equipment. This equipment may be a pump, compressor, mixer, or generator. Figure 12.14 illustrates the standardized symbols for compressors, steam turbines, and motors.

Heat exchangers and cooling towers are two types of industrial equipment that share a unique relationship. A heat exchanger is a device used to transfer heat energy between two process flows. The cooling tower performs a similar function; however, cooling towers and heat exchangers use different scientific principles to operate. Heat exchangers transfer heat energy through conductive and convective heat transfer while cooling towers transfer heat energy to the outside air through the principle of evaporation. Figures 12.15 and 12.16 illustrate the standard symbols used for heat exchangers and cooling towers.

The symbol for a heat exchanger clearly illustrates the flows through the device. It is important for a process technician to be able to recognize the shell inlet and outlet and the tube inlet and outlet flow paths. A heat exchanger with an arrow drawn through the body illustrates whether the tube-side flow is used to heat or cool the shell-side fluid. The downward direction indicates heating, while the upward direction illustrates cooling. Heat

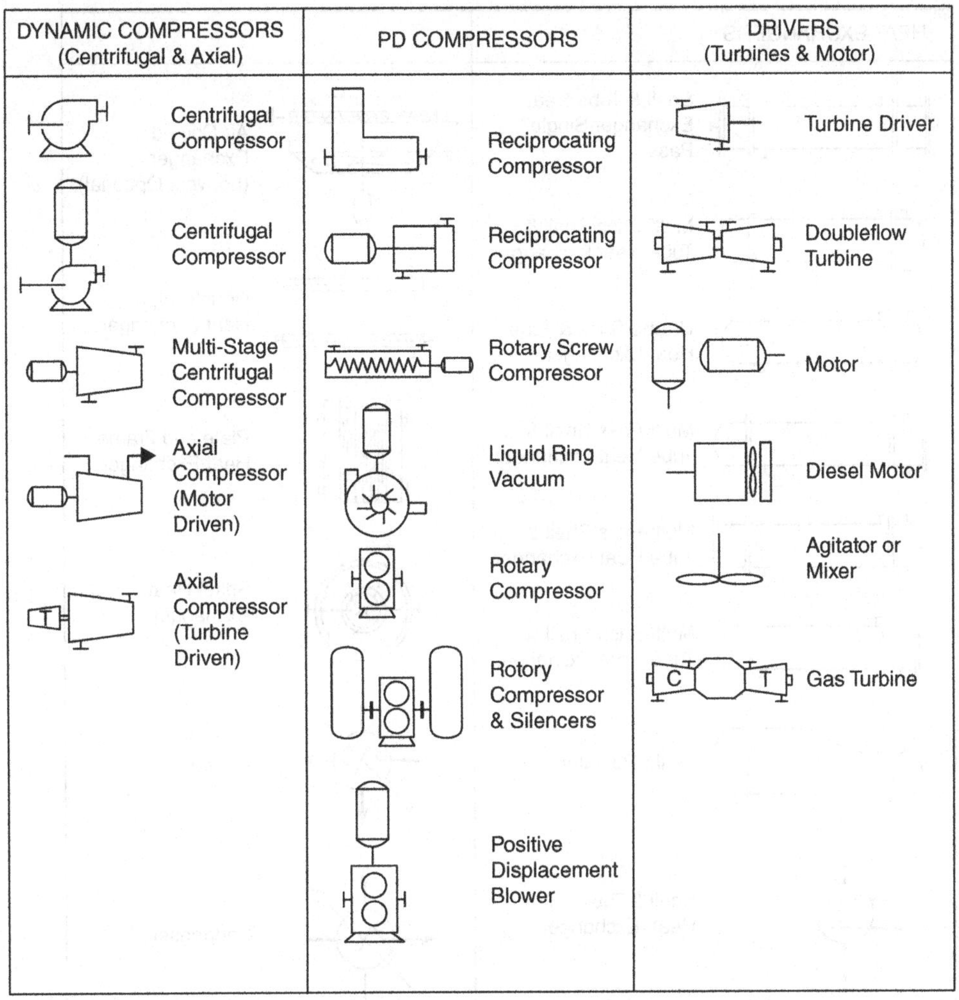

DYNAMIC COMPRESSORS (Centrifugal & Axial)	PD COMPRESSORS	DRIVERS (Turbines & Motor)
Centrifugal Compressor	Reciprocating Compressor	Turbine Driver
Centrifugal Compressor	Reciprocating Compressor	Doubleflow Turbine
Multi-Stage Centrifugal Compressor	Rotary Screw Compressor	Motor
Axial Compressor (Motor Driven)	Liquid Ring Vacuum	Diesel Motor
Axial Compressor (Turbine Driven)	Rotary Compressor	Agitator or Mixer
	Rotory Compressor & Silencers	Gas Turbine
	Positive Displacement Blower	

Figure 12.14 *Compressor, Steam Turbine, and Motor Symbols*

exchangers come in a variety of designs from simple to complex. Standard symbols have been developed for shell and tube heat exchangers, air-cooled heat exchangers, spiral heat exchangers, and plate and frame heat exchangers.

The symbol for a cooling tower is designed to resemble the actual device in the process unit. Cooled product flows out of the bottom of the tower and to the processing units, while hot water returns to a point located above the fill. Cooling towers can be found in the following designs: forced draft, natural draft, induced draft, and atmospheric. Each of these designs has similarities and differences with the others. The symbol will not show all of the various components of the cooling tower system; however, it will provide a technician with a good foundation in cooling tower operation with enough information to clearly see the process.

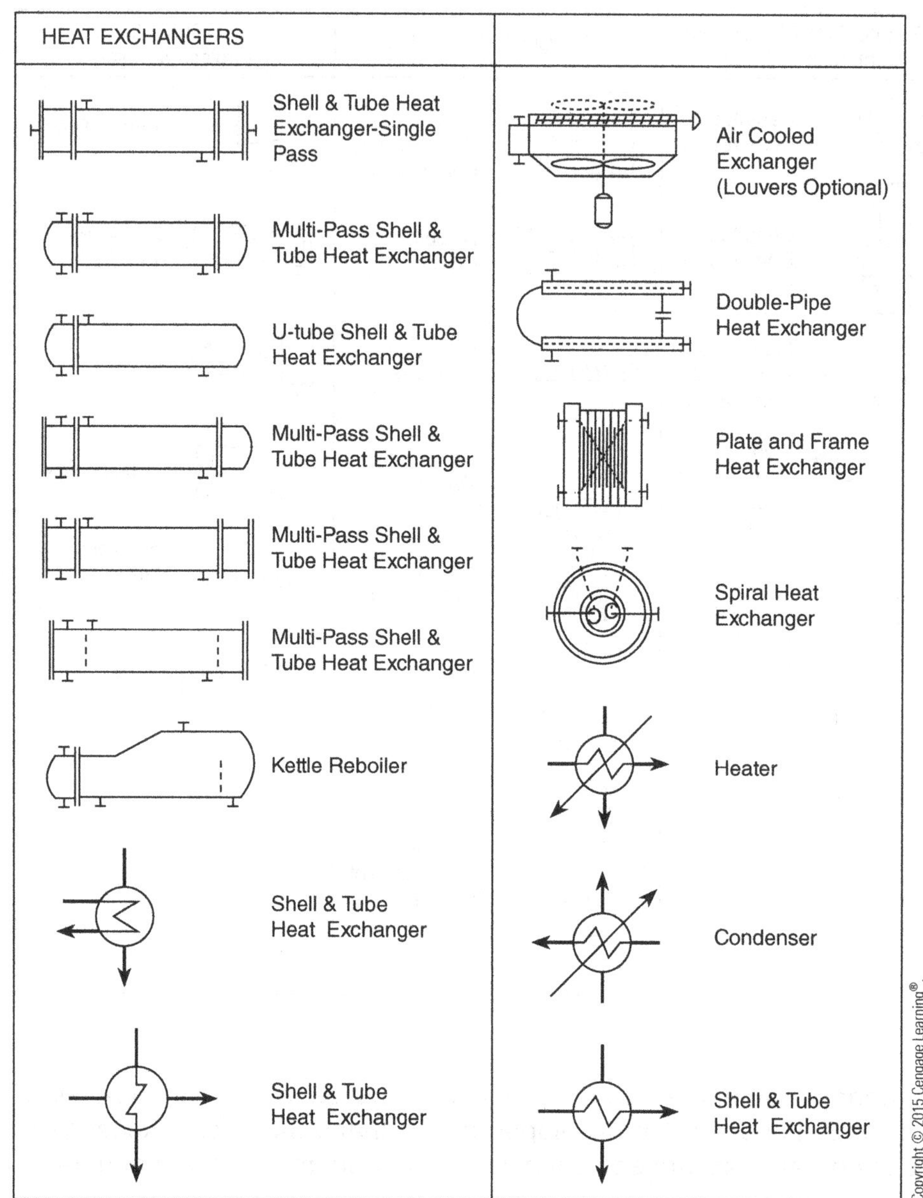

Figure 12.15 *Heat Exchanger Symbols*

On a typical P&ID, distillation columns, reactors, boilers, and furnaces will be drawn as they visually appear in the plant. The standard symbols file for these devices can be found in Figure 12.21. If a proprietary process includes several types of equipment not typically found on a standard symbol file, the designer will draw the device as it visually appears in the unit. Figure 12.17 illustrates standard symbols used for boilers and furnaces. The basic components and operation of boilers and furnaces are covered in equipment textbooks. A furnace is designed to heat up large volumes of

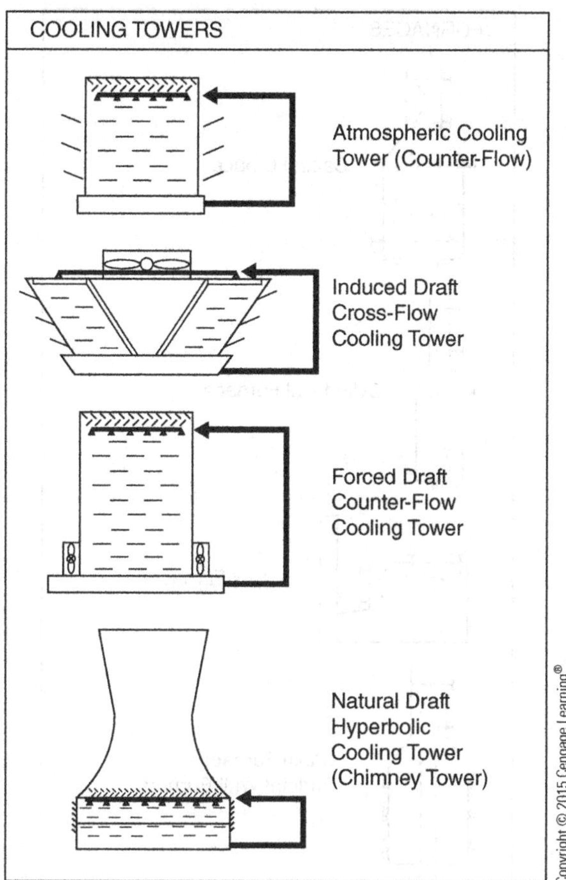

Figure 12.16 *Cooling Tower Symbols*

feedstock for chemical processing in a reactor or distillation column. Feed enters a furnace in the cooler convection section. Heat transfer is primarily through radiant and convective processes. The hottest spot in the furnace is the firebox. Tubes in the firebox are referred to as radiant tubes. A boiler is similar to a furnace; however, it is designed to boil water for steam generation. This process changes the internal arrangement of the boiler. Water can absorb a tremendous amount of heat. When it changes state, it expands to many times its original volume. A typical water tube boiler has a large upper, steam-generating drum; a lower mud drum; and a series of tubes designed to provide natural circulation of the steam back to the generating drum. Figure 12.17 shows the symbols used for steam generators and furnaces.

Distillation columns come in two basic designs: plate and packed. Flow arrangements vary from process to process. The symbols allow the technician to identify primary and secondary flow paths. The two standard symbols for distillation columns can be found in Figure 12.18. Distillation is a process designed to separate the various components in a

Figure 12.17
Boilers, Furnace Symbols

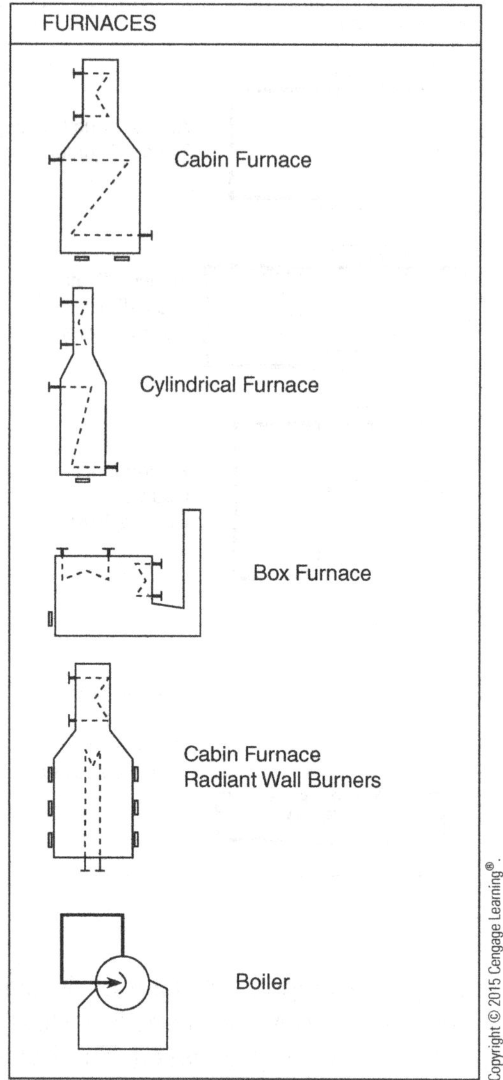

mixture by boiling point. A distillation column is the central component of a much larger system. This system typically includes all of the equipment symbols found in this chapter. Plate distillation columns include sieve trays, valve trays, and bubble-cap trays. Packed columns are filled with packing material, rings, saddles, sulzer, intalox, teller rosette, or panapak.

Reactors are stationary vessels and can be classified as batch, semi-batch, or continuous. Some reactors use mixers to blend the individual components. A reactor's design is dependent upon the type of service it will be used in. Some of these processes include alkylation, catcracking, hydrodesulfurization, hydrocracking, fluid coking, reforming, polyethylene,

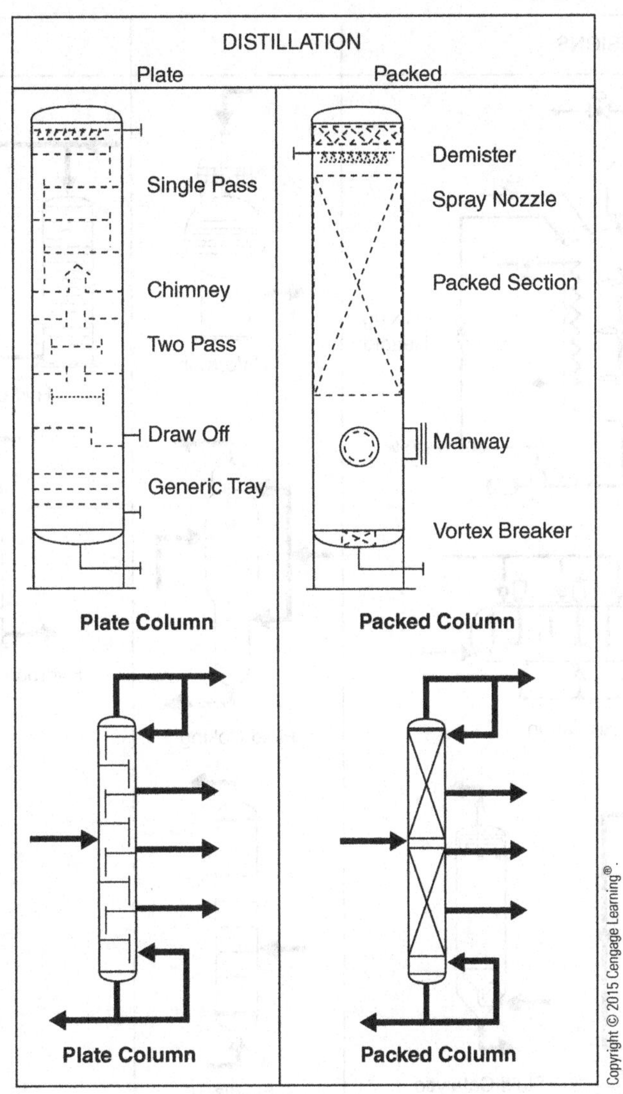

Figure 12.18
Distillation Symbols

mixed xylenes, and many other processes. A reactor is designed to allow chemicals to mix together under specific conditions to make chemical bonds, break chemical bonds, or make and break chemical bonds to form new products. The process it is used for will determine the shape and design of a reactor. Figure 12.19 shows several examples of reactor symbols.

Process and Instrument Drawings

A P&ID is a complex representation of the various units found in a plant (Figure 12.20). It is used by people in a variety of crafts. The primary users of the document after plant startup are process technicians and instrument and electrical, mechanical, safety, and engineering personnel.

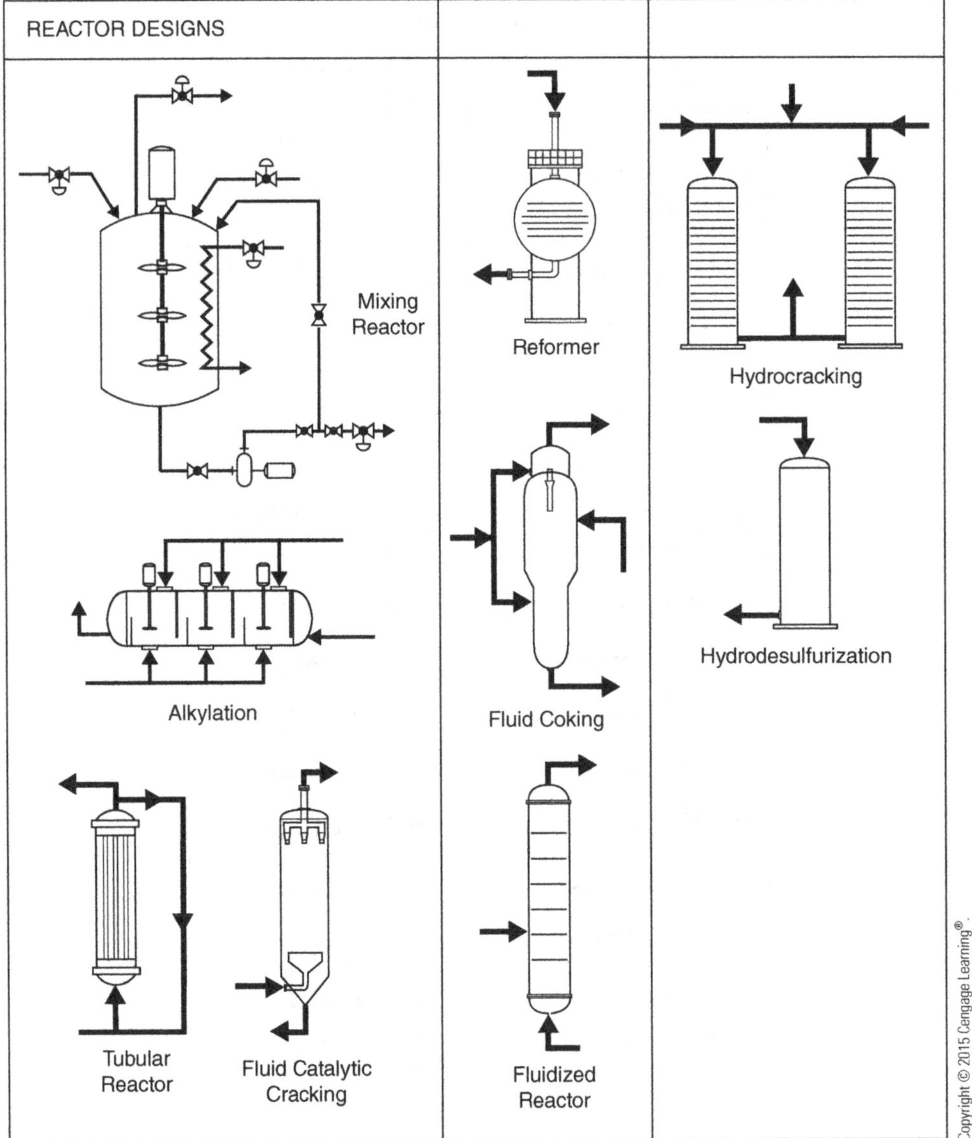

Figure 12.19
Reactor Symbols

In order to read a P&ID, the technician needs an understanding of the equipment, instrumentation, and technology. The next step in using a P&ID is to memorize your plant's process symbol list. This information can be found on the process legend. Process and instrument drawings have a variety of elements, including flow diagrams, equipment locations, elevation plans, electrical layouts, loop diagrams, title blocks and **legends**, and **foundation drawings**. The entire P&ID provides a three-dimensional look at the various operating units in a plant.

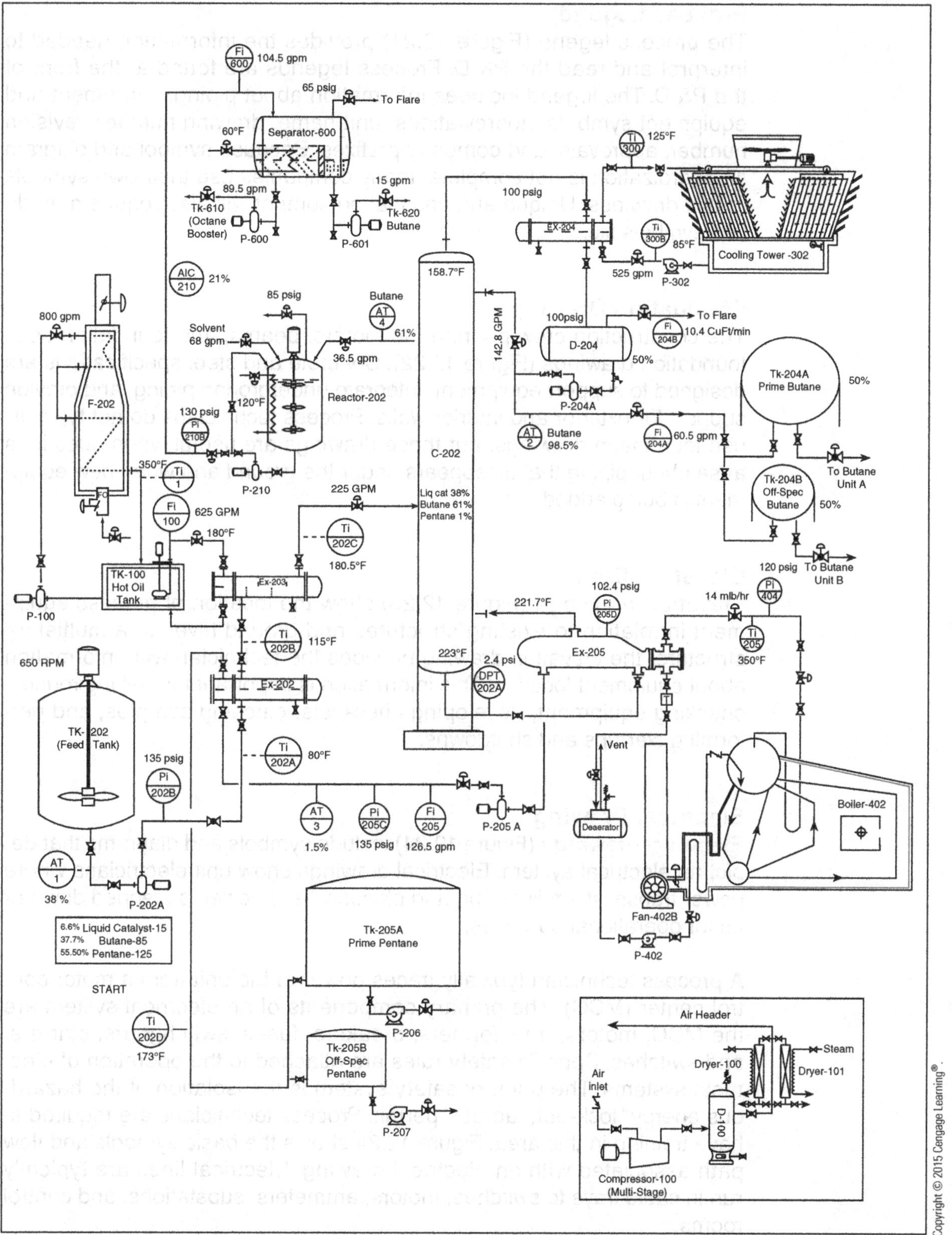

Figure 12.20 *Process and Instrument Diagram (P&ID)*

Process Legend

The process legend (Figure 12.21) provides the information needed to interpret and read the P&ID. Process legends are found at the front of the P&ID. The legend includes information about piping, instrument and equipment symbols, abbreviations, unit name, drawing number, revision number, approvals, and company prefixes. Because symbol and diagram standardization is not complete, many companies use their own symbols in unit drawings. Unique and unusual equipment will also require a modified symbols file.

Foundation Drawing

The construction crew pouring the footers, beams, and foundation uses foundation drawings (Figure 12.22). Concrete and steel specifications are designed to support equipment, integrate underground piping, and provide support for exterior and interior walls. Process technicians do not typically use foundation drawings, but these drawings are useful when questions arise about piping that disappears under the ground and when new equipment is being added.

Elevation Drawing

Elevation drawings (Figure 12.23) show the location of process equipment in relation to existing structures and ground level. In a multistory structure, the elevation drawing provides the technician with information about equipment location. This information is important for making rounds, checking equipment, developing checklists, catching samples, and performing startups and shutdowns.

Electrical Drawing

Electrical drawings (Figure 12.24) include symbols and diagrams that depict an electrical system. Electrical drawings show unit electricians where power transmission lines run and places where power is stepped down or up for operational purposes.

A process technician typically traces power to the unit from a motor control center (MCC). The primary components of an electrical system are the MCC, motors, transformers, breakers, fuses, switchgears, starters, and switches. Specific safety rules are attached to the operation of electrical systems. The primary safety system is the isolation of the hazardous energy "lock-out, tag-out" permit. Process technicians are required to have training in this area. Figure 12.24 shows the basic symbols and flow path associated with an electrical drawing. Electrical lines are typically run in cable trays to switches, motors, ammeters, substations, and control rooms.

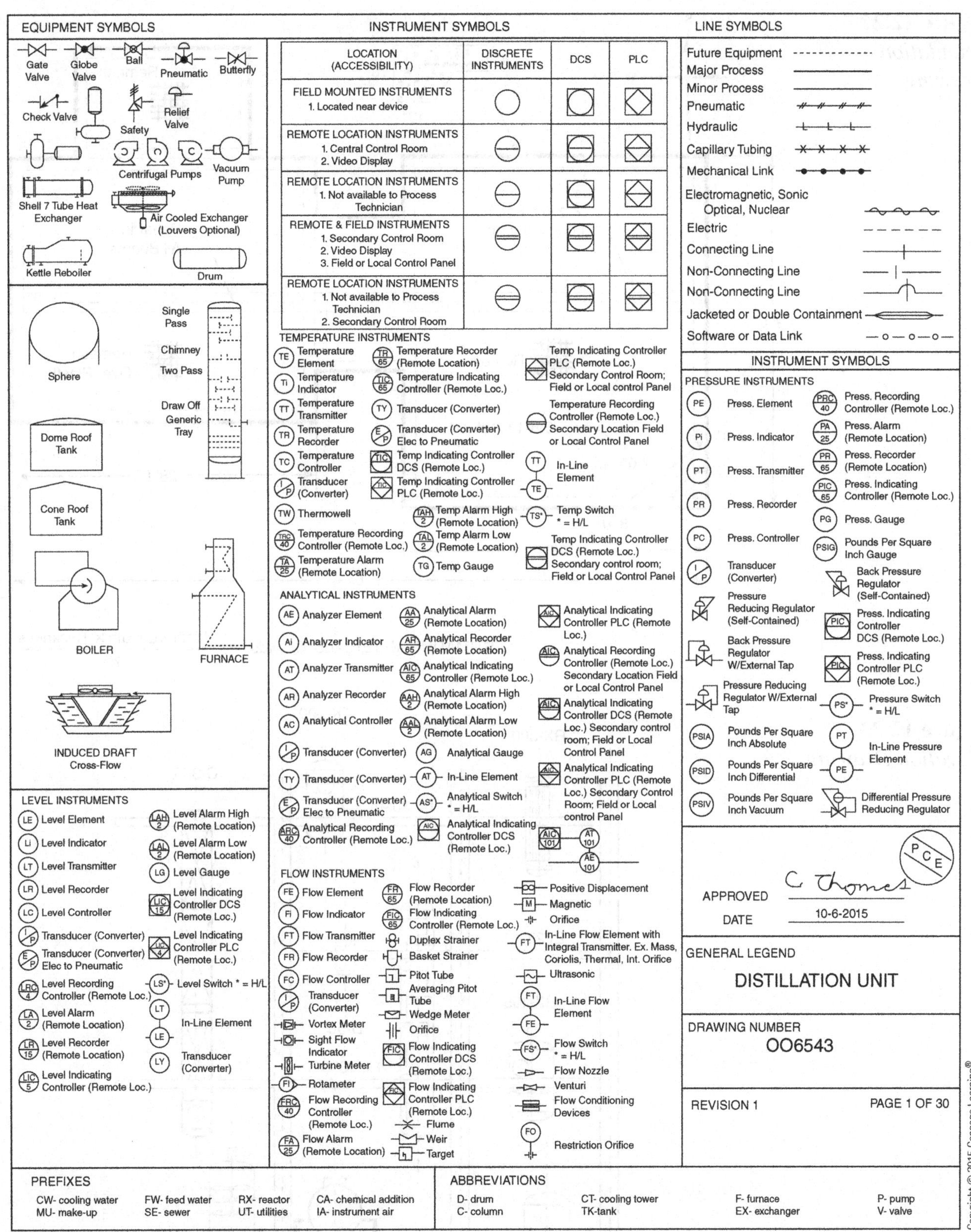

Figure 12.21 *Process Legend*

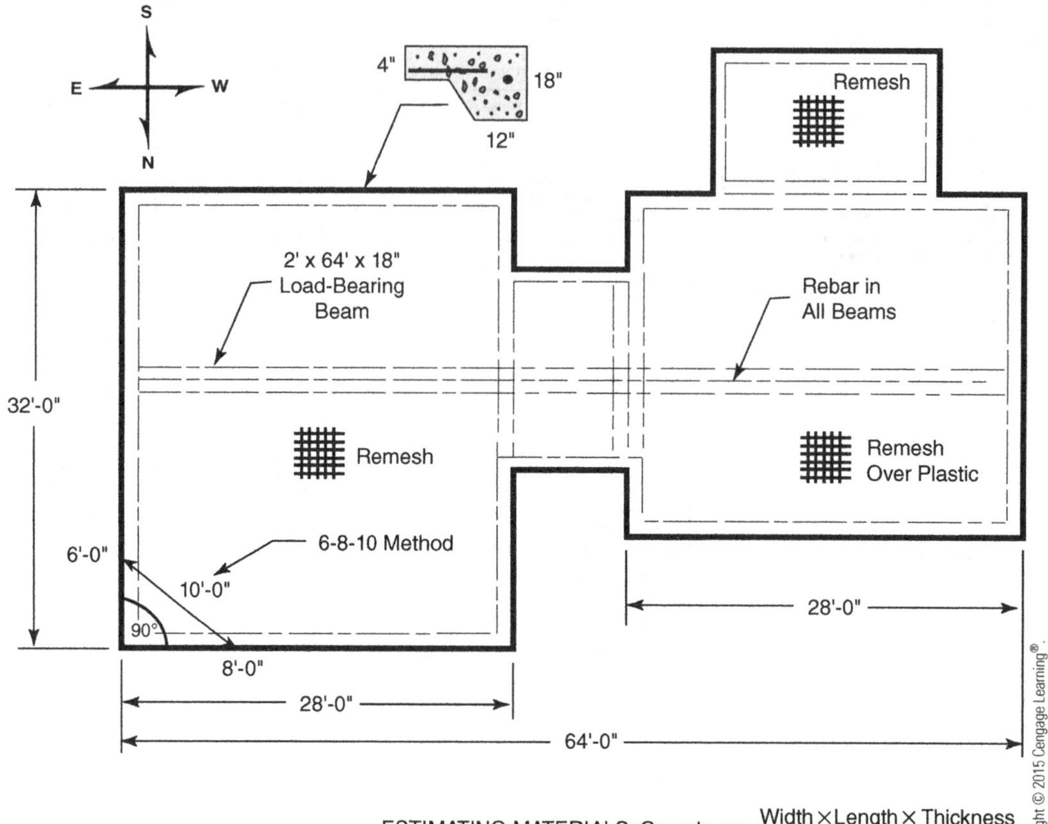

Figure 12.22
Foundation Drawing

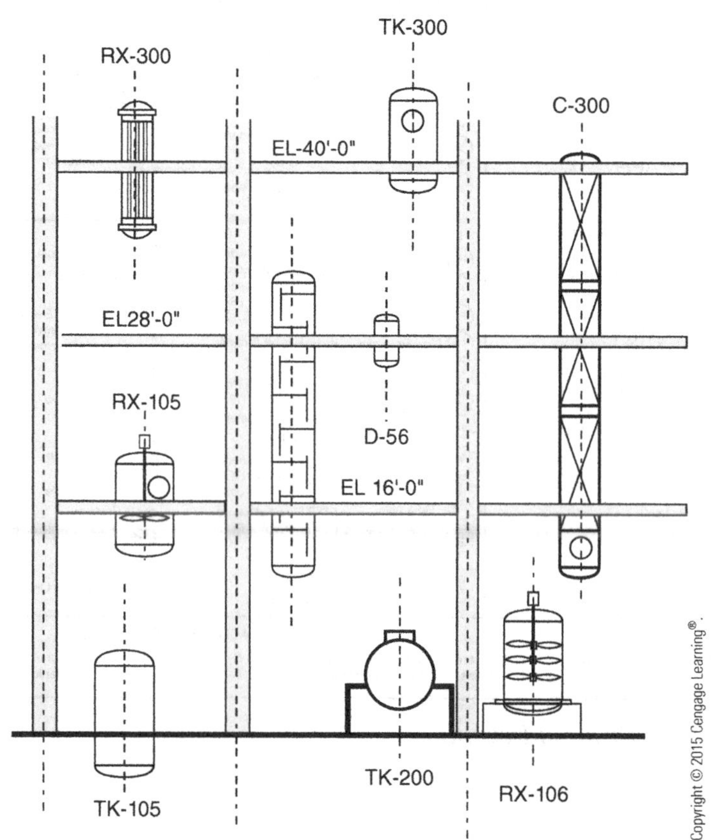

Figure 12.23
Elevation Drawing

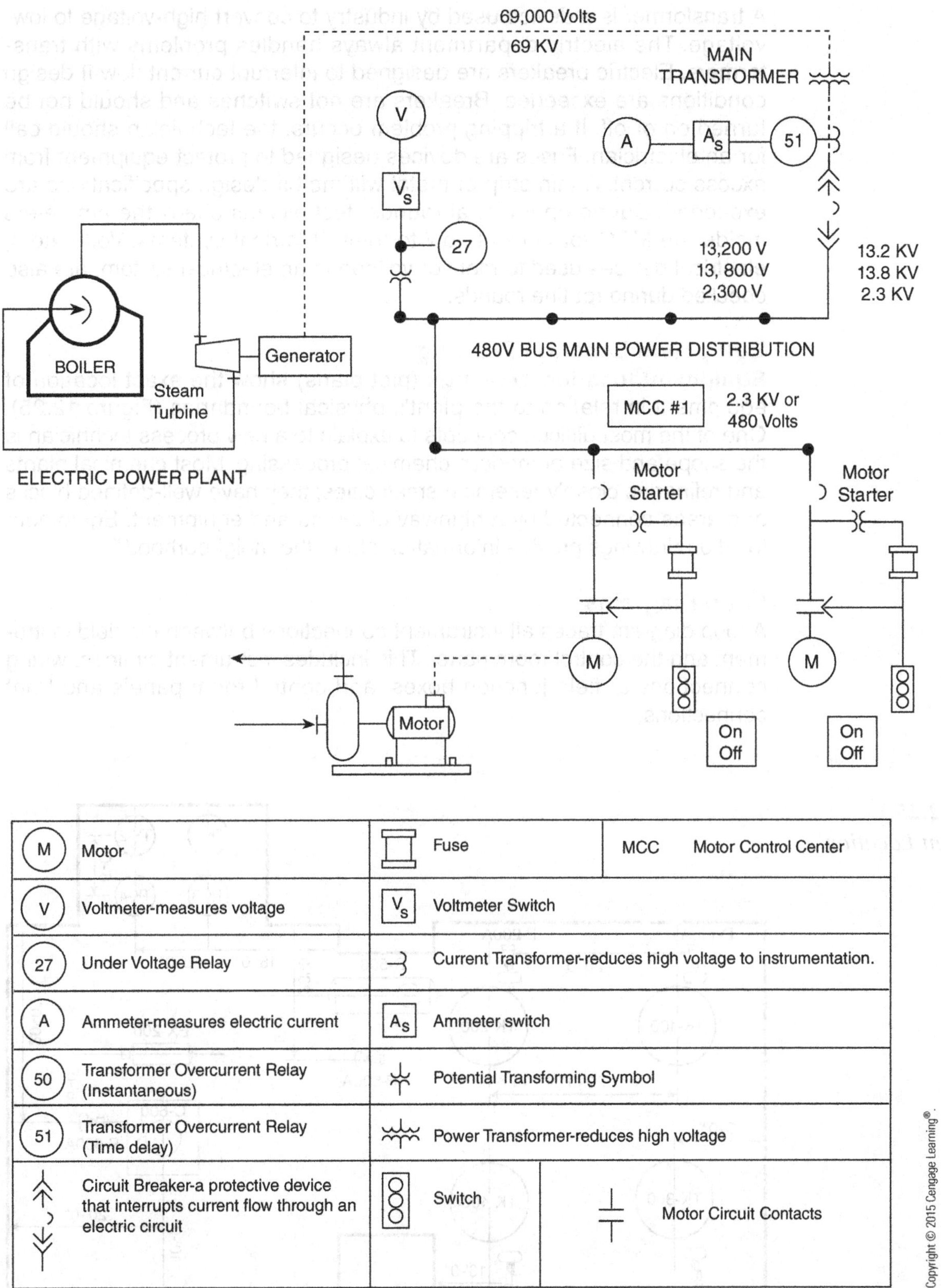

Figure 12.24 *Electrical Drawing*

A transformer is a device used by industry to convert high-voltage to low-voltage. The electric department always handles problems with transformers. Electric breakers are designed to interrupt current flow if design conditions are exceeded. Breakers are not switches and should not be turned on or off. If a tripping problem occurs, the technician should call for an electrician. Fuses are devices designed to protect equipment from excess current. A thin strip of metal will melt if design specifications are exceeded. During operational rounds, technicians check the ammeters inside the MCC for current flow to their electrical systems. Voltmeters, electrical devices used to monitor voltage in an electrical system, are also checked during routine rounds.

Equipment Location Drawing

Equipment location drawings (plot plans) show the exact location of equipment in relation to the plant's physical boundaries (Figure 12.25). One of the most difficult concepts to explain to a new process technician is the scope and size of modern chemical processing. Most chemical plants and refineries closely resemble small cities; they have well-defined blocks and areas connected by a highway of piping and equipment. Equipment location drawings provide information about the "neighborhood."

Loop Diagrams

A loop diagram traces all instrument connections between the field instrument and the control room panel. This includes instrument air lines, wiring connections at field junction boxes, and control room panels and front connections.

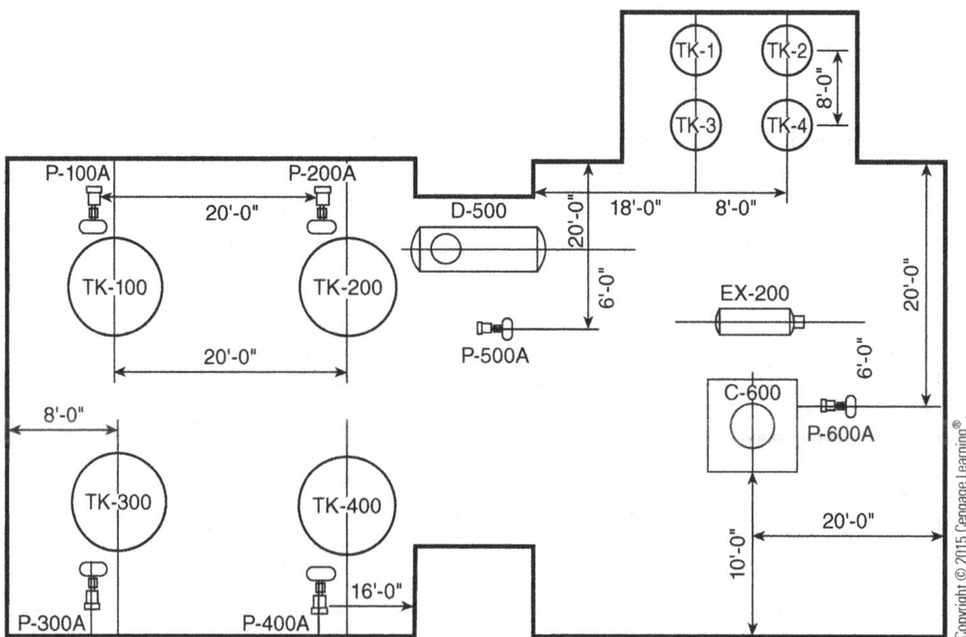

Figure 12.25 *Equipment Location Drawing*

Electrical One-Line Diagrams

Like the piping in process systems, the wiring in a unit follows a path. Electrical diagrams show a flow path for distributing power throughout the unit and to all electrical equipment. These diagrams show the different voltage levels in the unit, electrical equipment such as transformers, circuit breakers, fuses, and motors and horsepower required. It also includes start/stop switches, emergency circuits, and motor control centers. Process technicians can use these diagrams to trace a system from the power source to the load.

Sources of Information for Process Technicians

Information used by process technicians comes from a variety of sources. Some of these sources are:

- Operating training manuals
- Process descriptions
- Process control manuals
- Equipment summaries
- Safety, health, and environment regulations
- Operating procedures
- Startup and shutdown procedures
- Emergency procedures
- Process diagrams
- Technical data books
- Detailed equipment vendor information

Process Variables

The process variables that a technician is responsible for controlling include temperature, flow, level, pressure, and analytical variables. It stands to reason that a study of the scientific principles associated with these variables is required for any new trainee. This study would also include the instrumentation specific to the process and the equipment associated with it inside the system. Technicians can spend a lifetime studying the unique features associated with the many processes found in the chemical processing industry.

Pressure Control Loop

A pressure control loop design uses the five elements of the control loop. The area to be controlled is a special vapor-disengaging cavity that allows vapors or gases that are compressible to be controlled at a set point. The one area that changes consistently is the first primary elements and sensors. Pressure control loops use devices to detect pressure changes. These primary elements are typically expansion-type devices. Primary pressure elements include C-bourdon tubes, helical, spiral, bellows, pressure capsule, or diaphragm. Figure 12.26 includes a pressure element, pressure transmitter, controller, transducer, and control valve. An electric

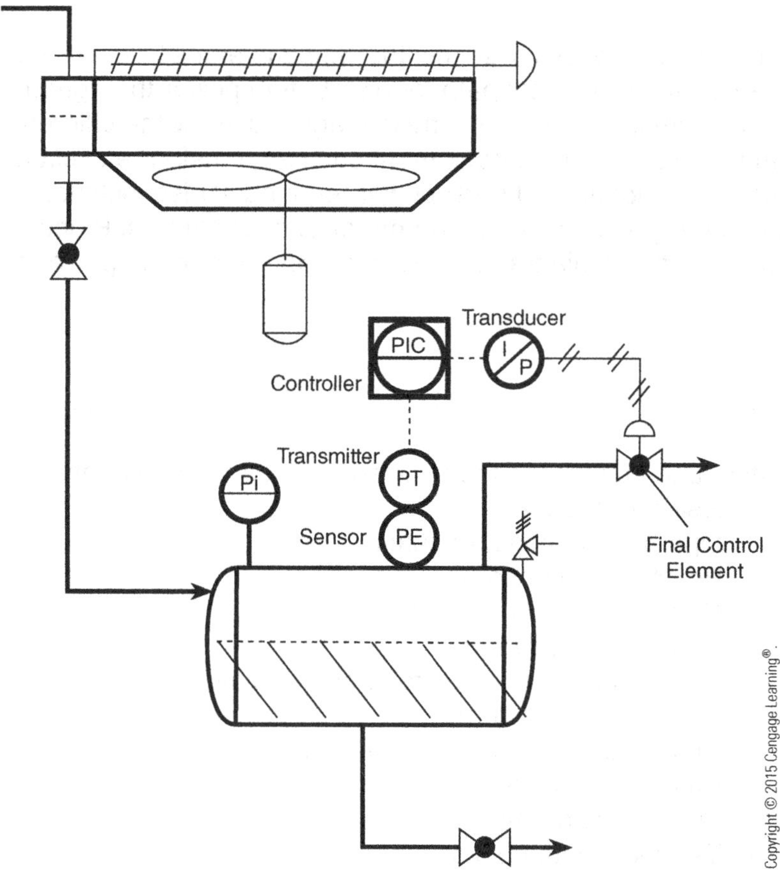

Figure 12.26 *Pressure Control Loop*

signal (4–20 mA) connects the PE, PT, and PIC to the transducer where the signal is changed to a (3–15 psig) air signal that corresponds to changes required by the controller's set point.

Pressure changes can affect temperature, level, and flow. Pressure changes the boiling point of chemicals, reaction rates, and the speed at which fluids flow through piping. These changes can impact product quality so a variety of instruments have been invented and designed to monitor and control pressure. The instruments discussed in this chapter include the following:

- Pressure gauges—psia and psig
- Vacuum gauge expressed in inches of mercury (in Hg)
- Manometers
- Pressure elements—sense changes in pressure and convert to mechanical motion
  - — Bellows bourdon tube
  - — C type bourdon tube
  - — Diaphragm capsule pressure elements
  - — Helical bourdon tube
  - — Pressure transmitter
  - — Spiral bourdon tube

Temperature Control Loop

Temperature control loop can control the amount of heat or cooling a substance is receiving, for example, steam to a kettle reboiler, hot oil to a heat exchanger (preheater), or cooling water to a condenser (heat exchanger). It can also control the amount of natural gas flowing to a burner. The burner provides heat to the boiler or furnace. A hundred other applications could be described for controlling the temperature of a substance; however, the most common is by controlling the flow. Figure 12.27 is a simple layout of a temperature control loop. Notice the location of the primary element and the transmitter and how the electric signal is sent to the controller. The controller compares the signal to an incoming electric or pneumatic set point. If a change is required, it is sent to the final control element. The primary sensors used to detect temperature are RTD's, thermocouples, and thermistors, often called temperature elements. Temperature elements are linked to transmitters. A 4 to 20 mA (milliamp) signal is sent to a controller that compares it to a setpoint. Controllers may be located in the field near the

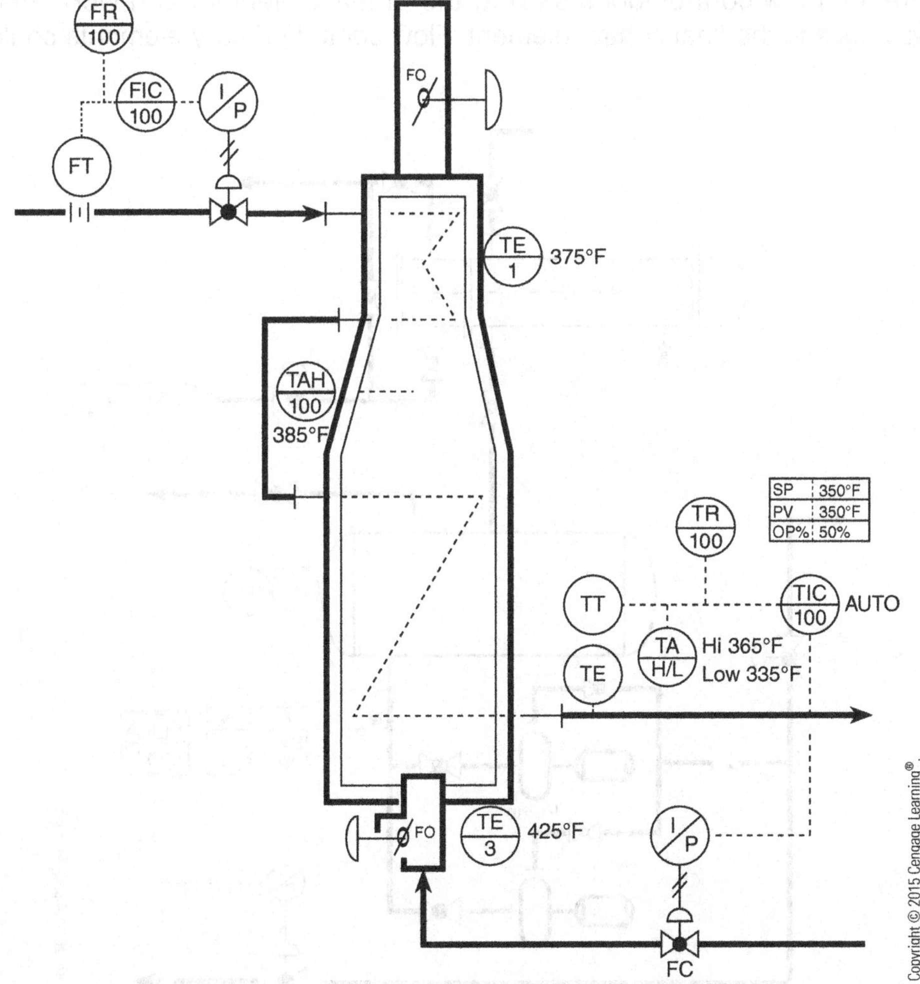

Figure 12.27
Temperature Control Loop

equipment or in a control room. The controller sends an electric signal to a transducer that is located on or near the final control element or valve. This will eliminate a process lag or delay. The transducer converts the signal to a 3–15 psig pneumatic signal. The pneumatically actuated control valve opens and closes depending upon the signal. Temperature is controlled by reducing or increasing the opening on the valve.

Level Control Loop

The following example of a level control loop can easily be applied to level control on any vessel or tank. Level control uses floats and float gauges, displacers, tapes and tape gauges, differential pressure transmitters, bubblers, load cells, capacitance probes, electromagnetic measuring devices, and nuclear measuring instruments. Figure 12.28 illustrates what a standard control loop looks like.

Flow Control Loop

Flow control loops are typically designed so that a measurement of the flow rate is taken first and then the flow is interrupted or controlled down stream. Flow control loops start at the primary element and work their way back to the final control element. Flow control primary elements could

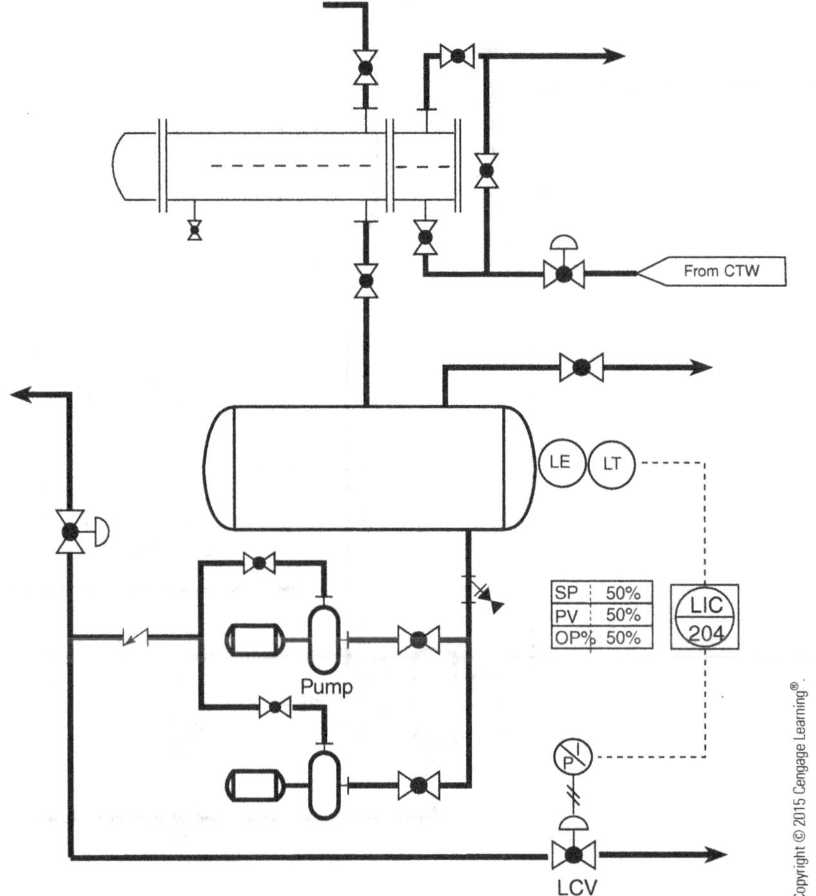

Figure 12.28
Level Control Loop

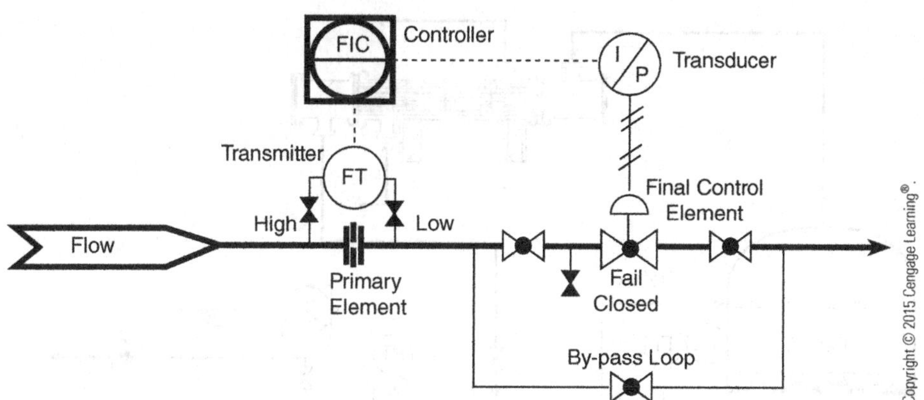

Figure 12.29
Flow Control Loop

include orifice plates, flow nozzles, pitot tubes, annubar pitots, magmeters, turbine meters, mass flow meters, nutating disks, oval gears, venturi tubes, vortex meters, target flow meters, and integral orifice flow meters. The most common primary element is the orifice plate. Orifice plates artificially create a high-pressure/low-pressure situation that can be measured by the transmitter. Primary elements are typically used in conjunction with a transmitter. Figure 12.29 shows an example of a flow control loop. Although it appears that the primary element is interrupting the flow, this is not the case. Increased velocity across the orifice plate compensates for the restriction. The transmitted signal is sent to a controller that compares the incoming signal with the desired set point. If a change is required, the controller will send a signal to a final control element.

Cascaded Control Loop

A cascaded control loop utilizes a series of unique features in modern process control. In a cascaded control loop, two different control variables work together to control a critical variable. For example, in some cases temperature control is more important than flow. In this case, a cascaded control loop can be used. A cascaded control loop has one "primary" or master controller and one secondary or slave controller. The master controller is wired to the secondary controller and has the ability to override the initial or original setpoint. During operation the secondary controller operates with all five elements of the control system. The master controller has a (1) primary element, (2) transmitter, and (3) master controller electrically connected to the other control loop. Figure 12.30 shows how a typical cascaded control loop looks.

Analytical Control Loop

An analytical control loop is typically designed to use flow to control some compositional variable. For example, the pH of a cooling water basin may be controlled by adding acid or caustic. When the PPM in the water basin exceeds operational guidelines, the analytical control loops removes a specified percentage of the basin water and replaces it with fresh water. Analytical control loops can be applied to a wide assortment of operational processes. Figure 12.31 shows an example of a series of compositional control loops.

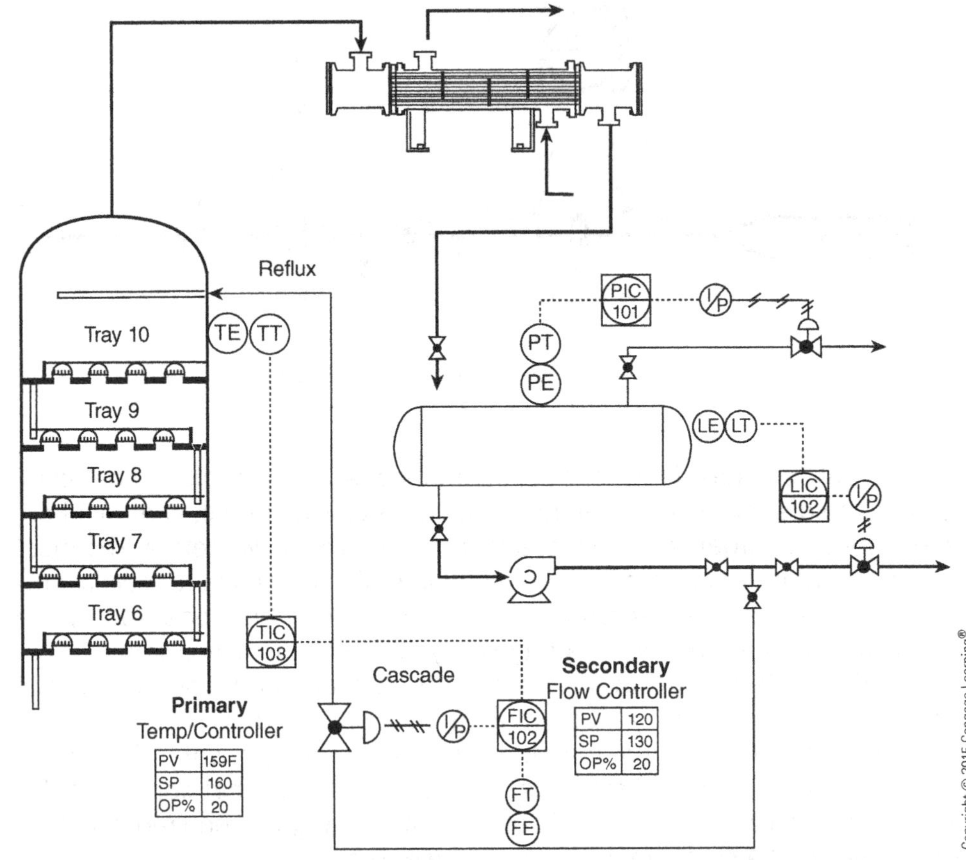

Figure 12.30
Cascaded Control Loop

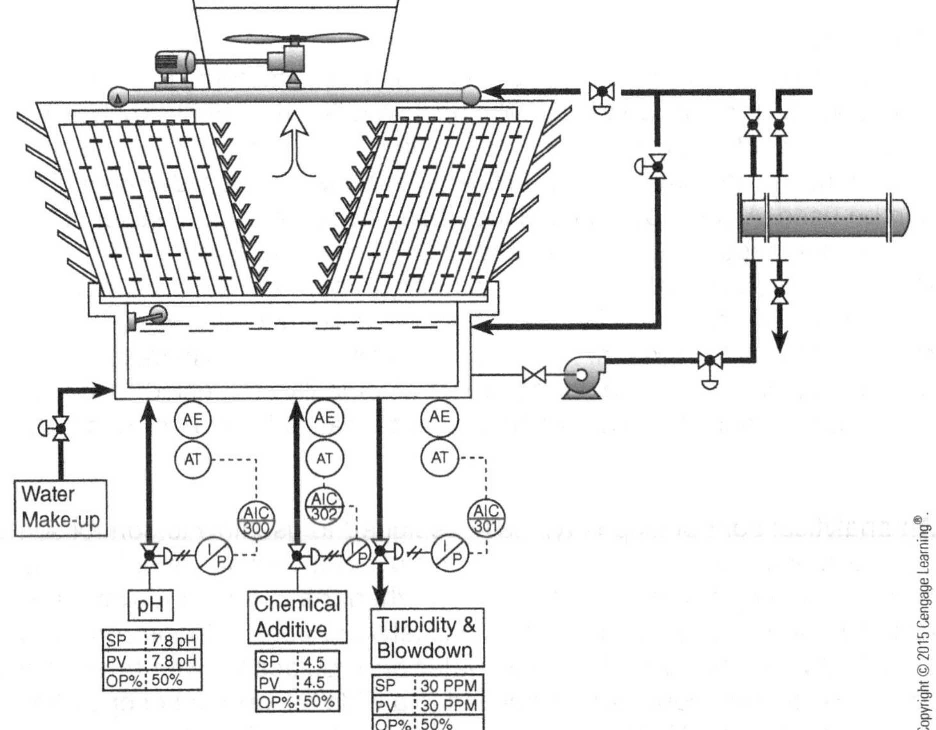

Figure 12.31
Analytical or Compositional Control Loop

Summary

Process flow diagrams (PFDs) and process and instrument drawings (P&IDs) are used to outline or explain the complex flows, equipment, instrumentation, electronics, elevations, and foundations that exist in a process unit. A PFD is a simple flow diagram that describes the primary flow path through a unit. A P&ID is a complex representation of the various units found in a plant. Standardized symbols and diagrams have been developed for most pieces of industrial equipment, process flows, and instrumentation.

Review Questions

1. Describe the basic components of a process flow diagram.
2. Draw the symbols for a gate, globe, and automatic valve.
3. Draw the symbols for a centrifugal pump and positive displacement pump.
4. Draw the symbols for a blower and a reciprocating compressor.
5. Draw the symbols for a steam turbine and centrifugal compressor.
6. Draw the symbols for a heat exchanger and a cooling tower.
7. Draw the symbols for a packed distillation column and plate distillation column.
8. Draw the symbols for a furnace and a boiler.
9. Sketch a simple process flow diagram of a pump-around system.
10. What information is obtained from a loop diagram?
11. What information is available on electrical one-line diagrams?
12. What information is contained on a plot plan drawing?
13. Describe the basic components of a block flow diagram.
14. Examine the basic components of a P&ID and explain how it can be used by a technician to learn how to operate a process unit.
15. Draw an analytical control loop on a stirred reactor.
16. Draw a series of cascaded control loops on a process system.
17. Sketch a simple process flow diagram of a shell and tube heat exchanger and heat transfer system.
18. Sketch a simple process flow diagram of a stirred reactor system; include a pump-around system.
19. Sketch a simple process flow diagram of a distillation system.
20. Draw a P&ID of a reactor/distillation system.

Utility Systems

Objectives

After studying this chapter, the student will be able to:

- Describe the basic components of a steam system.
- Describe raw water and fire water systems.
- Describe boiler feedwater treatment.
- Describe the basic components of a cooling water system.
- Describe the basic components of an air system.
- Describe the basic components of a nitrogen system.
- Describe the basic components of a gas system.
- Describe the basic components of an electrical system.
- Describe the basic components of steam systems and traps.
- Describe the basic components of an industrial sewer system.
- Describe the key elements of a process refrigeration system.
- Describe a relief and flare system.

Key Terms

Anions—negatively charged ions.

Blowdown—the process by which water is removed from the boiler system.

British thermal unit (Btu)—energy measurement unit; one Btu is the energy needed to raise one pound of water one degree Fahrenheit.

Carryover—contamination of steam with boiler water solids.

Cations—positively charged ions.

Flare—a device to safely burn excess hydrocarbon vapor.

Flare header—a pipe that connects the plant to the flare.

Flocculation—the bridging together of coagulated particles.

Isolation valve—see *Block valve*.

Knockout drum—a tank located between a flare header and the flare; used to separate liquid hydrocarbons from vapor.

Latent heat—heat that cannot be measured; the heat required to change a liquid to vapor (latent heat of evaporation) or vapor to a liquid (latent heat of condensation).

SHP—super-high-pressure (SHP) steam that operates between 1,000 and 1,200 psig.

Introduction to Process Systems

Chemical plants and refineries include unique combinations of equipment and systems that are used to separate materials into pure or relatively pure products. Examples of process systems include reactions, distillation, absorbing/stripping, adsorption/drying, crystallization, evaporation, extrusion, extraction, filtering, screening, and utility systems. Plant utility systems typically include steam generation, compressed air and gases, water treatment, and cooling tower operations.

Steam Generation Systems

A steam system is a complex arrangement of equipment and piping that works together to produce and circulate steam at a variety of pressures and temperatures. Because steam is used in so many industrial applications, most chemical plants and refineries have their own steam generation facilities. In a closed steam system, a condensate return header is used to remove slugs of water or condensate from the live steam line. This treated water collects in low points in the system, where devices called steam traps direct flow back toward the boiler. A vessel called a *deaerator* is used to collect fluid from the condensate return header. The deaerator is designed to remove excess air from the system. Demineralized makeup water is often added to the deaerator. The cooled condensate is fed back into the boiler, where it is changed into steam.

A typical steam system will include the following equipment:

- Boiler (fire tube or water tube)
- Steam header system (low, medium, and high pressure)
- Condensate return system (steam traps, valves, piping, pumps)
- Deaerator system

Steam systems can include super-high-pressure (SHP) steam generation and distribution, high-pressure (HP) steam, medium-pressure (MP) steam, and low-pressure (LP) steam. Each system is described in the following paragraphs.

Super-High Pressure (1,000–1,200 psig)

Super-high-pressure steam is generated in fire-tube or water-tube boilers. This allows the boiler plant operator to deliver steam to remote locations of the plant; however, many process systems are not designed to operate at super-high pressures. In this situation, the pressure is reduced to the correct operating condition. A relationship exists between the pressure and temperature of the steam. The higher the pressure, the higher the temperature of the steam. Process technicians use steam tables to identify the temperature of steam at a given pressure.

The SHP steam is generated by a boiler and is supplied to the SHP steam header. The typical pressure of the SHP steam is 1,000 to 1,200 psig. SHP steam is supplied to turbines, which drive compressors. In order to control the pressure in the high-pressure steam header and to transfer excess amount of high-pressure steam to the high-pressure steam header, a letdown (pressure reducing) station is provided between the SHP steam header and HP steam header.

High-Pressure Steam System (400–800 psig)

HP steam is supplied as required by lowering the pressure from the SHP header. High-pressure steam extracted from steam turbine drives can be used to feed the HP steam header. HP steam may be imported from other units into the high-pressure steam header to maintain overall steam balance. Extraction steam and imported steam flows are regulated to maintain the header pressure at 400–800 psig. The pressure of the imported steam supplied through a pressure control valve is slightly higher than HP header pressure. In order to maintain pressure and steam balance in the medium-pressure steam header, a letdown station is provided between the HP steam header and MP steam header.

Medium-Pressure Steam System (180–200 psig)

Nominal steam pressure in the MP steam header is 180 psig. Medium-pressure steam is supplied by lowering the pressure from the high-pressure header or by extracting exhaust steam from steam turbine operations. As high-pressure steam enters a turbine, it is partially converted to useful mechanical energy. Noncondensing operations utilize the exhaust steam in operations where lower pressures are required. MP steam is supplied partially by exhaust from small back-pressure turbines and extraction steam from larger turbines. In case of insufficient supply from the above sources, the letdown station between the HP and MP headers makes up the balance of steam required at the MP header.

Low-Pressure Steam (50–60 psig)

LP steam is supplied by lowering the pressure from the MP header, or by extraction from a steam turbine continuous blowdown drum. Nominal steam pressure in the low-pressure header is 50 psig. Steam to the LP steam header may be supplied from the exhaust of steam turbine driven pumps. The letdown station between the MP header and the LP header maintains the balance of steam in the LP header. An automatic pressure-controlled dump system with silencers is provided to dump excessive steam from the LP header to the atmosphere during plant operation upset conditions.

Condensate Recovery

The condensate recovery system is separated into clean and suspect (dirty) condensate sections. Clean condensate is routed to the deaerator. Suspect condensate is collected from steam users where the process side pressure is higher than the steam pressure. This condensate is considered to be suspect because a mechanical failure, such as tube leakage or failure, could cause contamination; and the reuse of contaminated condensate could seriously damage the steam generators and turbines. Suspect condensate collected in a drum is closely monitored for hydrocarbon contamination. Depending upon the results of analysis, the condensate is pumped through a carbon filter and then routed to the deaerator.

Raw-Water and Fire-Water Systems

A storage reservoir, supplied from municipal sources or local rivers and lakes, normally supplies the plant systems that require raw water (RW), which is used for all purposes throughout the plant, and fire water (FW), which is used only in specific units such as boilers (which usually must be treated for purity). Raw water is supplied to the units in the plant by pumps. All pumps can be remotely started from a utility's unit control service. Normally, one or two pumps are in operation to maintain 125 to 150 psig pressure on the RW system. Figure 13.1 illustrates how the raw water system operates. All raw water lines are located above grade. Plants use raw water in utility station water hoses, process unit fire water (FW) hose reels,

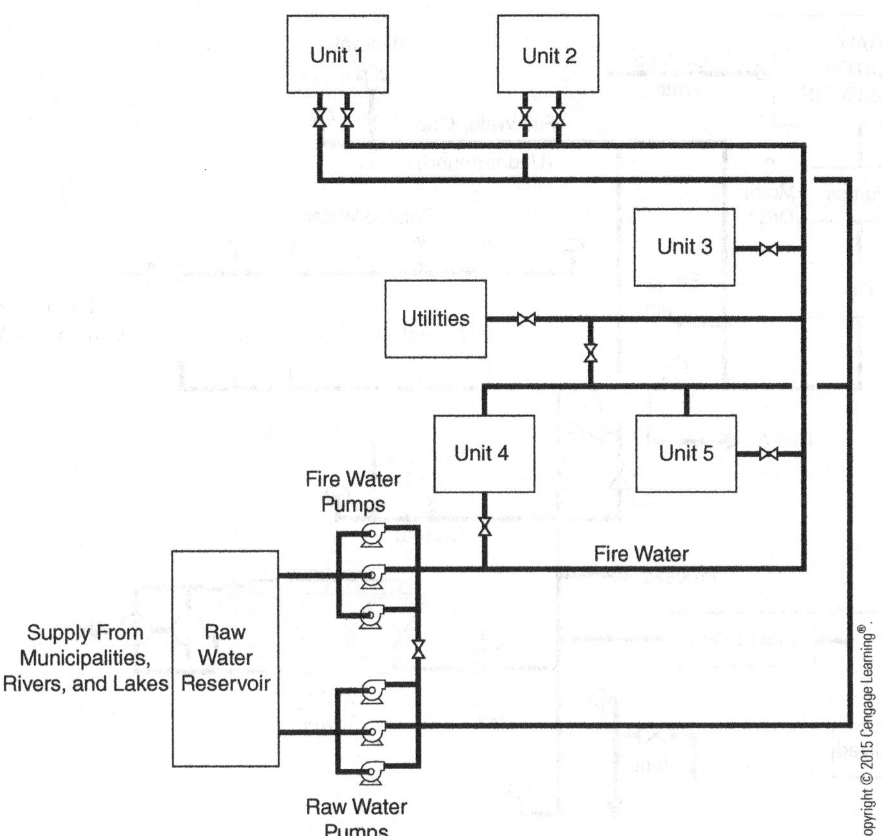

Figure 13.1
Raw Water and Fire Water

the cooling water supply, the supply to the fire water system, and boiler feedwater makeup.

Fire water is supplied to the plant by pumps, which must be capable of continuing to run if power to the plant is lost. These pumps are typically diesel driven to ensure operation if electric power is lost. The FW system pressure can be supported by cross-connecting from the raw water system. (*Note:* Operations at some plants can differ greatly from those described on these pages.)

The fire-water pumps are instrumented to start automatically when the fire-water system pressure declines below 110 psig. The first backup pump starts at 110 psig; the second diesel-driven pump starts at 100 psig; the third diesel-driven pump starts when the system pressure drops to 90 psig. Some plants operate at a fire-water system pressure as high as 300 psig. These pumps can also be remotely started from the utility's unit control center.

All lines in the FW system are underground, with a minimum of 2 feet of cover. Each process unit has a loop system, which supplies FW to the unit. Loop **isolation valves** are installed in order to isolate any leaks or broken lines in the loop. Fire hydrants are located on road shoulders on all sides of

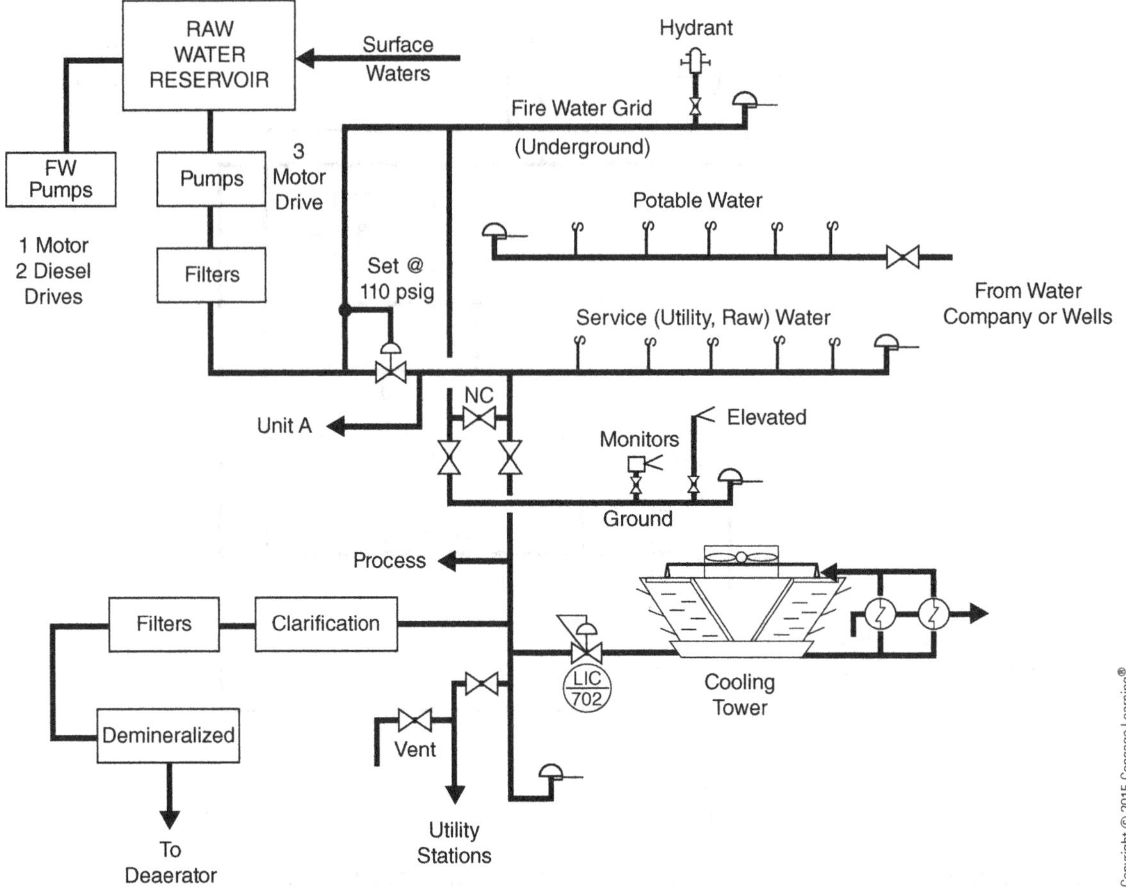

Figure 13.2 *Water Systems*

the units and inside the unit. Deluge systems on storage vessels are also supplied from the FW systems. Figure 13.2 shows how RW and FW water systems are integrated.

Boiler Feedwater Treatment

Boiler feedwater treatment is the process by which suspended and dissolved solids are removed from the waste used in a plant. Clean condensate and demineralized makeup water are deaerated in order to supply boiler feedwater (BFW) to multiple BFW pumps. Two or three pumps are used in the pump-around operation. The primary pump typically has one or two backup pumps that are placed in standby mode.

It is essential that all impurities be removed; otherwise, harmful scales or excessive corrosion will occur in the boilers. High concentrations of dissolved solids in boiler water will cause foaming of the water within the boiler. Foaming produces **carryover** of impurities in the steam, resulting in

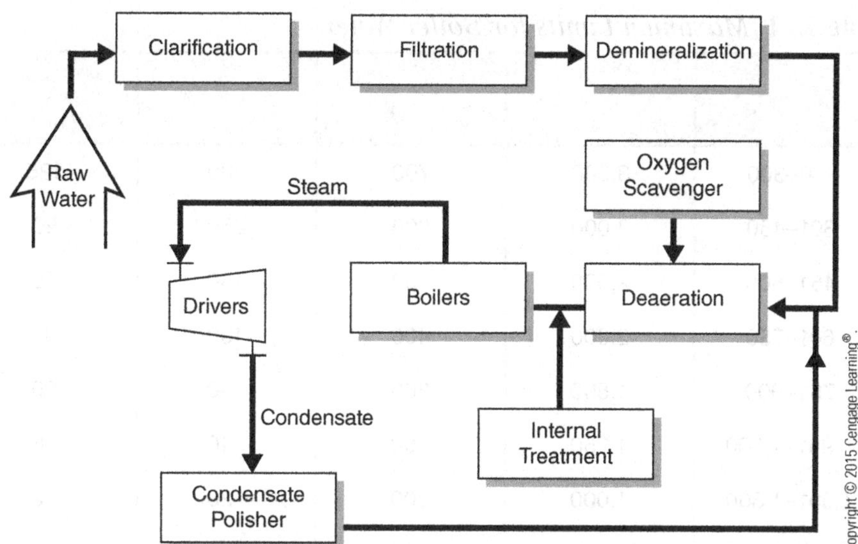

Figure 13.3
Boiler Feedwater

deposits on the turbine wheels. Figure 13.3 illustrates the equipment necessary to produce the quality of water required for a 900 psig steam system.

Impurities in the Water

The impurities that are present in water supplies can be divided into suspended and dissolved solids. Suspended solids are those that do not dissolve in water and that can be removed by filtration. Examples of suspended solids are mud, clay, and silt. Dissolved solids are those that are naturally dissolved in water and, therefore, cannot be removed by filtration. Examples of dissolved solids are sodium chloride (common table salt), calcium carbonate (limestone), silica (sand), calcium sulfate (gypsum), magnesium sulfate (Epsom salts), magnesium carbonate (dolomite), hydrated sodium sulfate (Glauber's salt), aluminum, iron, manganese, fluorides, and other trace substances. Impurities must be removed from water before it can be used in industrial applications.

Table 13.1 shows the standard for boiler operation established by the American Boiler Manufacturers Association (ABMA).

Physical Properties of Water

Water is one of the most abundant resources found on earth. In its purest form, water is colorless, odorless, and tasteless. It is the only inorganic substance that occurs in all three states in the earth's temperature range: ice (solid), steam (gas), and water (liquid). A molecule of water has two hydrogen atoms and one oxygen atom; its chemical formula is H_2O. Water boils at 212°F (100°C) at atmospheric pressure. Pressure increases cause the boiling point to rise, and pressure decreases cause it to fall. For example, at a pressure of 200 psig, water has to be heated to 388°F before it will boil. When atmospheric pressure is removed, water will boil at temperatures as low as 35° to 40°F. At 3,200 psig, water approaches the critical pressure, at which

Table 13.1 *Maximum Limits for Boiler Water*

Boiler Pressure (psig)	Total Solids (ppm)	Alkalinity (mg/L)[a]	Suspended Solids (ppm)	Silica (ppm)
0–300	3,500	700	300	125
301–450	3,000	600	250	90
451–600	2,500	500	150	50
601–750	2,000	400	100	35
751–900	1,500	300	60	20
901–1,000	1,250	250	40	8
1,001–1,500	1,000	200	20	2.5
1,501–2,000	750	150	10	1.0
Over 2,000	500	100	5	0.5

[a]Milligrams per liter.

water is converted to steam without a volume change. Processing units for boiler operation and a wide variety of other applications use steam.

It takes one **British thermal unit (Btu)** to raise the temperature of one pound of water one degree Fahrenheit. It takes 970 Btu to change one pound of water to steam (Figure 13.4). This heat energy is stored in the

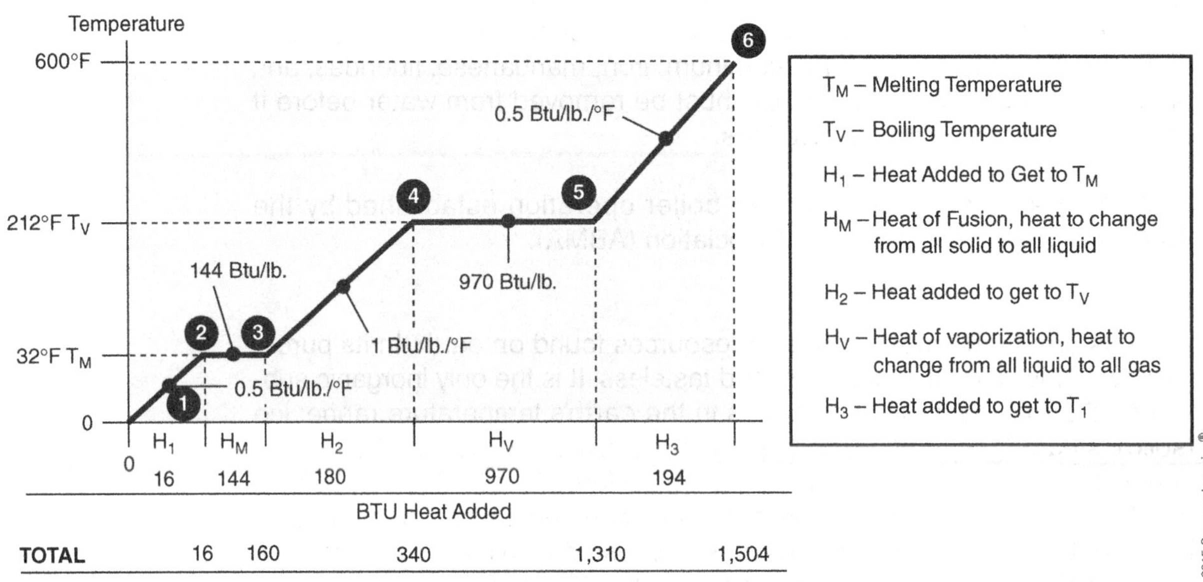

Figure 13.4 *Solid, Liquid, and Vapor*

steam and will not be released until the steam begins to cool and condense. Heat stored in steam is often called the **latent heat** of vaporization. One pound of boiler feedwater at 72° (22.2°C) requires 140 Btu to reach 212°F (100ºC) and an additional 970 Btu to change to steam. A total of 1,110 Btu is stored in the steam. In the reverse process, the energy released when steam changes to water is the latent heat of condensation.

Local water reservoirs and storage systems have a wide range of natural water quality, including various amounts of impurities. At higher temperatures, the dissolving nature of water is enhanced, so the amount of dissolved solids may be increased.

Water hardness is associated with the presence of calcium and magnesium compounds. Salt water varies from fresh water in the salt mineral content of the water. Salt water contains 300 pounds of salt for every 1,000 pounds of water.

As boiler feedwater temperature increases, impurities precipitate (fall out) and adhere to the hot metal surfaces of the boiler, forming scale. Typically, dissolved bicarbonates of calcium and magnesium break down and form insoluble carbonates. Deposits in a boiler are composed of a variety of compounds, including calcium carbonate, calcium phosphate, sulfate or silicate, magnesium hydroxide, iron, copper oxides, and silica. Phosphate deposits are soft brown or gray and can easily be removed. Carbonate deposits are granular and porous. Sulfate deposits have a crystallized structure that forms a dense and very hard deposit. Silica deposits are very hard, with a fine crystallized structure that resembles porcelain. Iron deposits are characterized by a dark brown color. The primary problem associated with deposits is tube failure.

Corrosion

The process of corrosion is a complex electrochemical reaction that can attack a large area or a small pinpoint area. Corrosion in a boiler is typically associated with the reaction of oxygen with the metal; however, pH (a measure of acidity), stress, chemical compounds, and other factors can contribute to corrosion. Corrosion may occur when the boiler or boiler feedwater has excess acidity or alkalinity, high levels of oxygen and carbon dioxide, high temperature, stress, or high levels of ammonia- or sulfur-bearing gases.

Corrosion fatigue occurs when thermal warp, caused by rapid heating or cooling, affects a weakened area, or it may be caused by cracks created internally from the action of applied cyclic stresses. Caustic cracking is characterized by metal stress, embrittlement, caustic water, presence of silica, and a small leak that focuses stress on a small area.

Corrosion prevention is characterized by maintaining pH levels in boiler feedwater, removing dissolved oxygen from water, using chemical treatments, and following established out-of-service and blowdown procedures. Poor water quality will produce poor-quality steam that results in boiler water foaming, which produces carryover (contamination of steam with boiler water solids). Deposits of solids can damage the steam system. The primary causes of boiler water carryover are foaming (froth that is carried out with steam), aquaglobejection (water spray carried out), load changes or priming (like uncapping a bottle of soda), and steam contamination (leakage of water). Avoiding high-water levels and boiler load changes and maintaining proper concentration of solids prevent carryover.

Coagulation and Flocculation

Coagulation neutralizes the negative charges of suspended solids, thereby allowing them to clump together and settle out. **Flocculation** is a term used to describe the bridging together of coagulated particles (that is, *floc*). Common coagulants include aluminum and iron salts. Chemical names for these coagulants include ferric sulfate, ferric chloride, aluminum sulfate or alum, and sodium aluminate. Aluminum and iron have three positive charges that counter the negative charges of the suspended solids. Coagulants form a floc net that collects suspended material.

Chemical Precipitation

Chemical precipitation uses a chemical to react with dissolved minerals in feedwater to form a solid precipitate that drops harmlessly out of solution. Lime soda is frequently used as a water-softening agent. Calcium hydroxide reacts with soluble calcium and magnesium bicarbonates to form a solid precipitate. These reactions are illustrated by the following equations:

$$\underset{\text{Lime}}{Ca(OH)_2} + \underset{\substack{\text{Calcium}\\\text{Bicarbonate}}}{Ca(HCO_3)_2} \rightarrow \underset{\substack{\text{Calcium}\\\text{Carbonate}}}{2CaCO_2} + \underset{\text{Water}}{2H_2O}$$

$$\underset{\text{Lime}}{2Ca(OH)_2} + \underset{\substack{\text{Magnesium}\\\text{Bicarbonate}}}{Mg(HCO_3)_2} \rightarrow \underset{\substack{\text{Magnesium}\\\text{Hydroxide}}}{Mg(OH)_2} + \underset{\substack{\text{Calcium}\\\text{Carbonate}}}{2CaCO_3} + \underset{\text{Water}}{2H_2O}$$

Settling and filtration are used to remove the sludge, or solid precipitate. Soda ash is also used in the water-softening process. It is used to reduce sulfate hardness.

Modern manufacturers use two approaches when adding lime-soda for water softening: batch and continuous. Batch uses only one tank. Continuous softening uses multiple compartmented tanks, continuous additive addition equipment, and retention time for chemical reaction, sludge removal device, and clear water draw-off.

Ion Exchange Process: Demineralization

Ion exchange resins are used to remove electrically charged particles (ions) from water. When minerals are dissolved in water, they form ions. Ionic resins come in two types: **anion** (negatively charged ions) and **cation** (positively charged ions). Anion resins react with negatively charged ions such as sulfate (SO_4^{-2}). Cation resins react with positively charged ions such as calcium ions (Ca^{+2}) and magnesium ions (Mg^{+2}). Ion resins resemble small beads and are placed in tanks that raw water is passed through. Resin beds are usually several feet thick. Demineralization is a process that passes raw water through both anion and cation exchange resins. Effluent from this process is essentially pure water.

Deaeration

A deaerator is a device used to remove air from water. The deaerator is typically located at the end of the condensate return header and is used as the primary boiler feed source. Dissolved oxygen is a serious problem because it leads to boiler system corrosion. The deaerator is designed to remove dissolved oxygen by heating the water with steam. Dissolved oxygen passes out the vent with the steam.

Deaerators come in two basic designs: spray and tray. Spray deaerators inject steam directly into the water. In the tray design, steam enters at the bottom and water enters at the top. Water drops over a series of sieve trays that provides good contact between liquid and heated vapor.

Water Treatment in Boiler

Water treatment is classified as external or internal. External treatment takes place before raw water enters the boiler. Internal water treatment complements the external process and helps protect the internal components of the boiler system. In some systems, external treatment is limited. The primary location for water treatment is at the boiler. Common tests on boiler feedwater include pH and the presence of particulate matter (measured in ppm), iron, oil, and silica.

Blowdown

Blowdown is a term used to describe the process in which water is removed from the boiler system. As suspended solids concentrate in the water, chemical treatment reaches a point at which it is no longer effective. Removal of a portion of the treated water is necessary. Blowdown depends upon the amount of suspended solids the boiler can handle. It is not unusual for blowdown to equal 5% to 10% of the total capacity. Blowdown is controlled through an instrument designed to measure the electric conductivity of the water. Blowdown ports are typically located on the bottom of the boiler so accumulated sludge can be removed (Figure 13.5). Table 13.2 lists common boiler water problems.

Figure 13.5 *Clarifier*

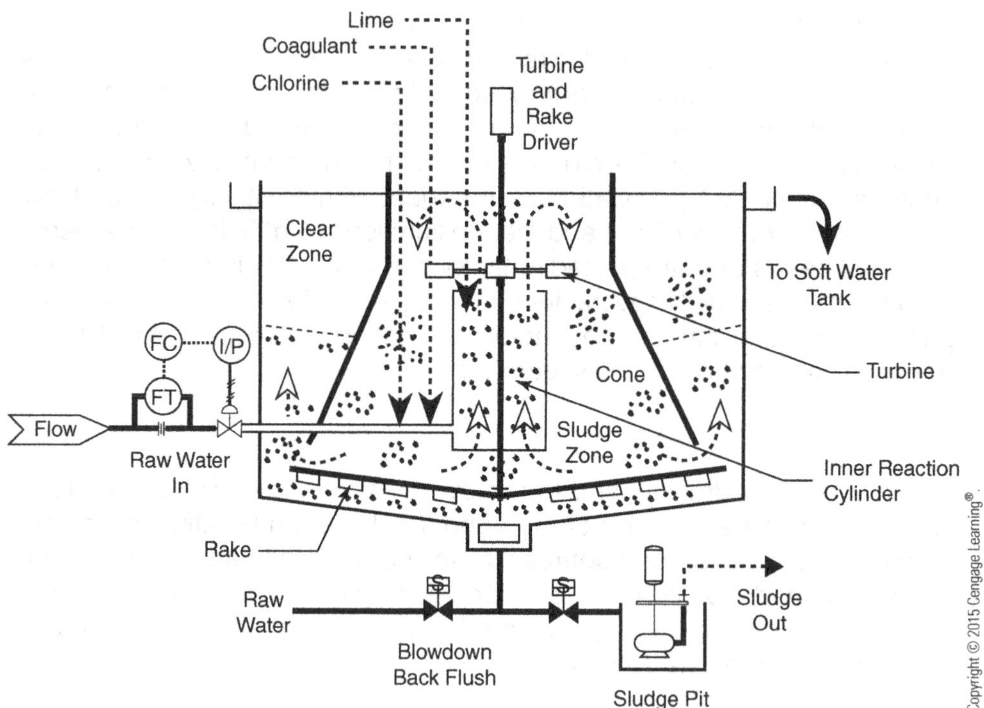

Table 13.2 *Major Impurities, Their Effect, and Treatment*

Characteristic or Constituent	Problem	Solution
Turbidity	Deposits	Coagulation
Color (indicates type of deposit)	Hinders precipitation	Coagulation
Hardness	Scale	Softening
Alkalinity	Foaming, embrittlement	Lime-soda
Acid	Corrosion	Alkalies
Carbon dioxide	Corrosion	Deaeration
pH	pH variations	Acids/alkalies
Sulfate	Scale	Demineralization
Chloride	Corrosion	Demineralization
Silica	Scale	Anion resin
Iron	Deposits	Coagulation
Manganese	Deposits	Coagulation
Oil	Scale, sludge, foam	Coagulation
Oxygen	Corrosion	Deaeration
Hydrogen sulfide	Corrosion	Chlorination
Ammonia	Corrosion	Cation resin
Conductivity	Corrosion	Demineralization
Dissolved solids	Foaming	Lime softening
Suspended solids	Deposits	Coagulation
Nitrate	Solids	Demineralization

Cooling Water System

The cooling water (CW) system is used for cooling the various pieces of equipment in the facility in a closed cooling system. The cooling water system is divided into two complementary circuits. One CW circuit operates on a once-through basis; the other allows the reuse of water to cool a second train after use for cooling the first train.

The cooling water system is designed for a supply temperature of 90°F (32.2°C) and a maximum return temperature of 120°F (48.88°C). A cascade arrangement is used to minimize the amount of circulation water while maintaining the design temperature rise. This cascade arrangement takes advantage of process users that are cool enough to maintain a temperature rise between 5° (−15°C) and 10°F (−12.2°C).

The cooling water is supplied from the cooling tower to the users by cooling water circulation pumps. Cooling water pumps are typically centrifugal and vertically mounted. The driver can be an electric motor or steam turbine.

Makeup water to the cooling tower is supplied from the raw water system. Raw water storage must have sufficient capacity to withstand interruptions in the water supply.

The recirculating cooling water concentration ratio is designed for a minimum of five cycles and is controlled by blowdown. Continuous blowdown is used to prevent buildup of dissolved components. The cycles of concentration are controlled at the blowdown valve on the basis of the recirculating water conductivity.

Sidestream filters may be used to remove airborne solids such as dust and sand and contaminants introduced via the cooling tower makeup water such as silt, organic matter, and water-borne organisms. Each filter is sized to handle the full flow while one filter is backwashing or out of service. The sidestream filters operate automatically under gravity flow and are of the self-contained backwash type.

The cooling water chemical feed system consists of a nonchromate-based corrosion inhibitor, a dispersant or antiscalant, and chlorine for the control of microbiology fouling.

Air and Nitrogen Systems

Air Systems

Utility and instrument air are both supplied from centrally located air compressors. Drivers for the compressors are arranged so that the loss of power or steam will not interrupt the supply of critical instrument air.

Figure 13.6 *Utility Air and Instrument Air System*

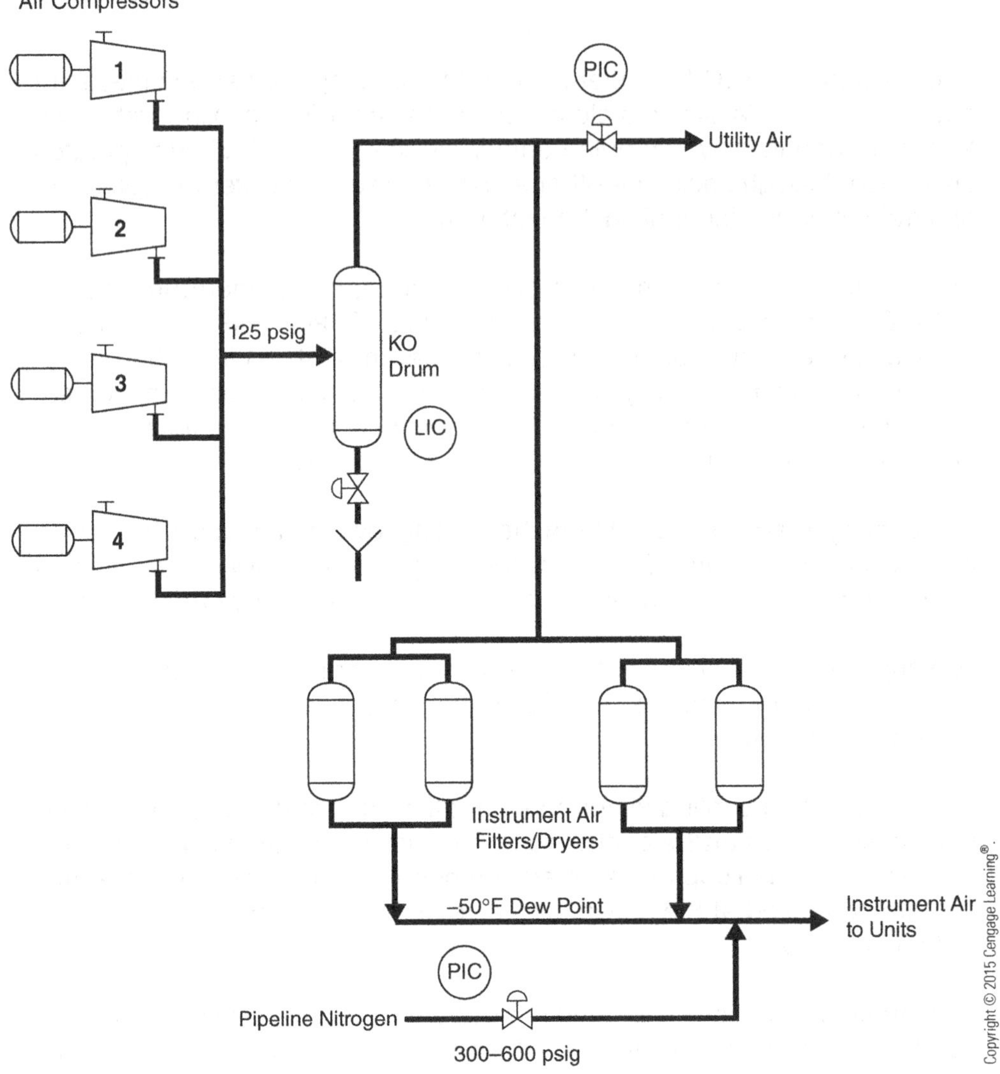

Some plants have additional diesel-driven compressors in case the two utilities are not available at the same time.

Normal plant demand may require only one or two air compressors to be in operation. The air compressors are set up to produce approximately 125 psig air pressure into the system (Figure 13.6). Because the instrument air (IA) system is the more critical system, pressure is controlled to 100 to 105 psig. Utility air (UA) has no pressure control.

Plant Air Systems

A central utilities unit supplies utility air for the entire plant. Free water is removed at a separator drum that is located off the air compressor discharge but is not dried. Utility air is used for unit hose stations (general use), air-driven tools, and decoking unit furnaces.

The largest use of utility air is in the decoking unit furnace, which is operated intermittently. Usually, an extra air compressor is put online during these periods. Utility air should never be used for unplugging lines or sewers or for removing liquid hydrocarbons from trucks or vessels because of the hazard of creating combustible mixtures.

Instrument Air System

Instrument air is dried to approximately −50°F (−45.55°C) dew point with activated alumina desiccant. Instrument air system pressure is protected against low pressure (as a result of excessive utility air usage) by an upstream pressure controller in the utility air system. This control valve will close off if the instrument air header pressure falls below 95 psig. It is important to maintain the instrument air system pressure at its normal controlled pressure of 105 psig, because some control valve actuators begin to malfunction when the pressure drops below 60 psig.

At some locations, the plant instrument air system is protected against pressure failure by a tie-in to the nitrogen (N_2) system. Because of safety concerns, nitrogen is admitted into the instrument air system only during an emergency. An alarm or alert system is in place to warn all unit personnel whenever nitrogen is added to the instrument air header.

Nitrogen Systems

Nitrogen gas is usually generated in the plant or purchased from a supply company. If purchased, it is usually delivered by high-pressure pipeline. Pipeline pressures run from 300 to 600 psig. The supplier may own and operate the meter station, which forms the billing basis for the purchase of nitrogen.

High-pressure nitrogen is delivered to the individual units in a plant (Figure 13.7). The pressure is reduced at each unit to the desired working pressure, which is usually 120 to 150 psig. Nitrogen is used primarily as an inert gas for blanketing vessels and lines or for conveying combustible materials and sometimes as a backup for instrument air.

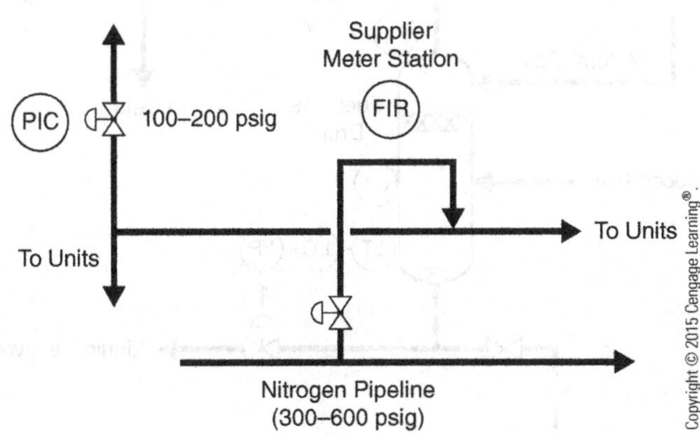

Figure 13.7
Nitrogen System

Nitrogen hose connections are located at utility stations in the process units. These connections are used for nitrogen only and are different from the connections used for air, water, or steam. The nitrogen connections (Natural Standard Coupling) are welded to prevent them from being changed to other types of fittings, namely air or water.

Gas Systems

Most industrial manufacturers purchase natural gas (NG) from suppliers. A high-pressure pipeline generally delivers gas to the plant (Figure 13.8). A pressure letdown station located near one of the plant entrances meters flow to the plant. The letdown station consists of multiple control valves, two of which are normally in service for backup. Additional valves are provided for peak loads such as unit startups. The letdown/meter control valve is in operation at all times. A second valve opens when the flow rate is high and closes at low flow rates, thereby making possible more accurate metering of the gas flow.

The plant natural gas distribution header normally runs around 200 to 250 psig. Each process unit has its own pressure-control station that usually maintains the unit pressure between 60 and 100 psig.

Natural gas is used to supplement the fuel gas produced at a unit in order to meet furnace and boiler demands. Natural gas is also used as a purge medium in unit startup and shutdown. Fuel gas generally consists

Figure 13.8
Natural Gas System

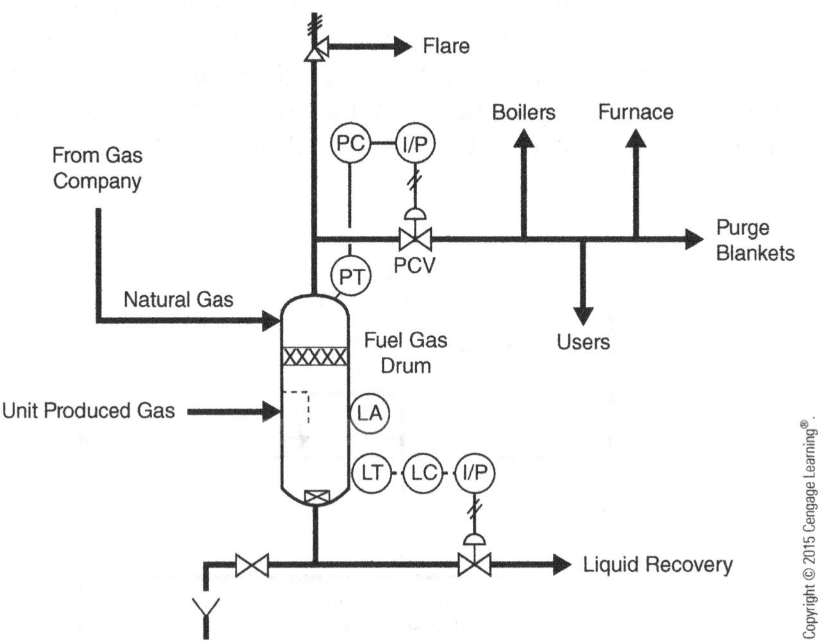

of plant-produced process gas that is augmented with purchased natural gas to satisfy the unit's energy demand. Fuel gas required is typically produced internally during normal unit operations. During a startup mode or emergency mode, where no internal fuel gas is available, gas is imported; the imported gas is measured in a meter station before entering the unit. During normal operation, any excess fuel gas produced is exported by a compressor system.

Electrical Systems

The power received from electric supply companies is generally at voltage levels of 100 to 160 kV. At many plants, this is transformed down to 13.8 kV for distribution within the plant. At the units, the 13.8 kV is stepped down to 4,160 V for use on large motors (200–10,000 hp). The 13.8 kV is also transformed to 480 V, the most common voltage in many plants, and finally to 240 V and 120 V for heating, lighting, and fractional horsepower motors.

Motors

Most of the motors in the plant are the three-phase induction type. Each of the pairs of poles is wound with one of the phases of the AC supply. The effect of this winding is to produce a rotating magnetic field. This field, in effect, pulls the rotor behind it, thus imparting mechanical energy to the machine.

The induction motor always rotates slightly slower than the magnetic field. Some motors are made to rotate at the same speed as the magnetic field. These are called synchronous motors. Further discussion is beyond the scope of this book, except to say that synchronous motors are generally more economical than induction motors for very large applications.

Motor Protection

In addition to mechanical energy, electric energy can also be converted to heat. This is what happens in a toaster or an electric stove. The amount of heat produced is proportional to the product of the current squared (I^2) times the resistance (R): I^2R. The household fuse operates on the principle that it is the weakest link in the circuit. When the current gets large enough to nearly overload the circuit, the fuse burns out and interrupts the circuit.

Large-scale motors have the same type of protection, generally called circuit breakers. These breakers, located in the motor control center (MCC), actually afford two types of protection. One type protects against the instantaneous very large overload, such as might occur if a short circuit develops in the system. The other is a thermal overload that protects against continuous operation at a high motor loading. Either of these devices can shut down a motor. Knowing which of these devices was activated can be a clue as to what is causing the problem.

Voltage Measurement: Voltmeters

Voltmeters are connected in parallel with wires in an electric circuit. Normally, we are not concerned with the internal construction of voltmeters and do not see anything except the dial. Actual volts are indicated on the dial, and no conversions or calculations are necessary.

Current Measurement: Ammeters

Ammeters are similar to voltmeters, except that they are connected directly in one line of the electric circuit. Ammeter readings are read directly from the face of the meter.

Backup Power Systems

Backup power systems ensure continuous power supply to critical equipment and instruments. In addition to the normal power, an emergency diesel-engine generator and an automatic transfer switch provide an alternate source of power. When normal power is lost, the transfer switch delivers backup power from the emergency diesel-driven generator.

The critical instrument systems are supplied with power through an uninterruptible power system (UPS). This system contains batteries, a battery charger, inverter, static transfer switch, and manual bypass switch. This backup system ensures that there will be no interruptions to alarms, computer control systems, and all other instrument systems.

Many plants generate power in addition to that purchased. This is generally at a voltage level of the supplier or at the highest transformed plant voltage level.

Steam Systems

Steam has been used as a source of heat and a conveyor of energy since the Industrial Revolution. It is produced by boiling water and its temperature or heat value can be easily determined because it generally corresponds to the pressure in the system (Table 13.3). Steam carries large amounts of energy in small mass flow rates; the energy of steam is readily transformed into mechanical or heat energy upon condensation.

As we mentioned earlier in this chapter, steam systems can include super-high-pressure steam generation and distribution, high-pressure steam import, medium-pressure steam, and low-pressure steam. SHP steam is generated by boiler or furnaces and is supplied to the SHP steam header. The pressure of the SHP steam is 1,000 to 1,200 psig. Steam is distributed throughout the plant by progressive letdown from preceding headers and

Table 13.3 *Properties of Saturated Steam*

Gauge Pressure	Temperature (°F)	Heat (Btu/lb)		
		Sensible	Latent	Total
25 psiv	134	102	1017	1,119
20 psiv	162	129	1001	1,130
15 psiv	179	147	990	1,137
10 psiv	192	160	982	1,142
5 psiv	203	171	976	1,147
0 psig	212	180	970	1,150
10 psig	239	207	953	1,160
20 psig	259	227	939	1,166
30 psig	274	243	929	1,172
40 psig	286	256	920	1,176
50 psig	298	267	912	1,179
100 psig	338	309	880	1,189
150 psig	366	339	857	1,196
200 psig	388	362	837	1,199
250 psig	406	382	820	1,202
300 psig	421	398	805	1,203
350 psig	435	414	790	1,204
400 psig	448	428	777	1,205
500 psig	470	453	751	1,204
600 psig	489	475	728	1,203

extraction from turbines (Figure 13.9). High-pressure steam flows are regulated to maintain the header pressure at 400 to 800 psig. The steam pressure in the MP steam header is 180 to 200 psig. The steam pressure in the LP header is 50 psig.

Steam Traps

Two types of heat are present in steam: sensible and latent. Sensible heat is heat that can be sensed with a thermometer. Latent heat cannot be sensed, but it is still present and is given off when steam condenses. A steam trap stops or traps the steam until it gives up its latent heat and condenses. The steam trap then allows the condensate to discharge and return to a boiler, where it is reheated into steam. Without a steam trap, the steam would not give off its latent heat, and the efficiency of the heat transfer in the heat exchanger would be greatly reduced. This reduction in efficiency can be translated into increased energy costs. Avoiding these increased costs is the primary reason steam traps are used.

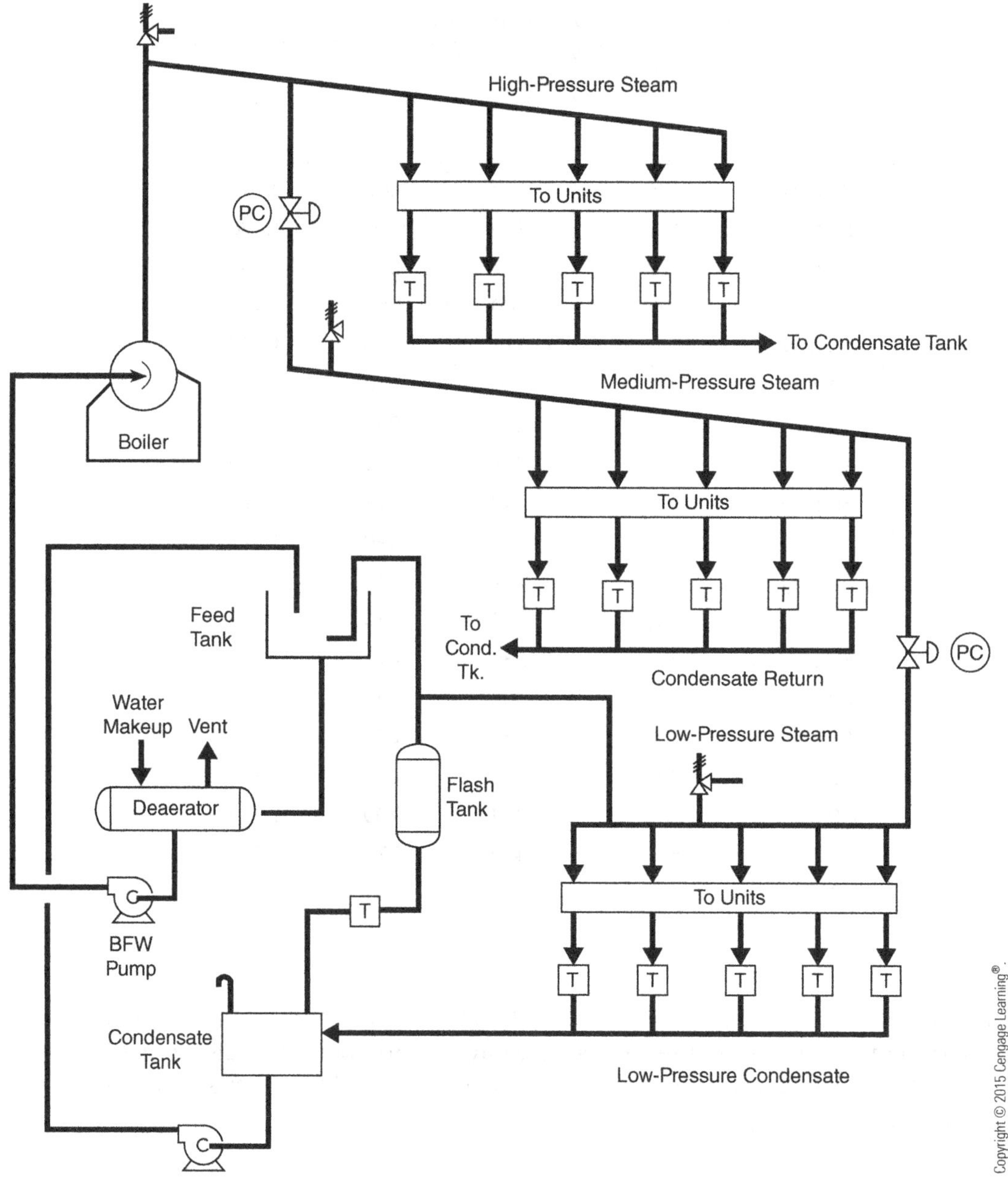

Figure 13.9 *Steam Distribution System*

Steam traps are also used to remove air and dirt from steam systems. Trapped air is especially dangerous in steam systems because oxygen in the air can oxidize steam lines. Over an extended period of time, this oxidation process will eat through the steam pipes and create severe steam leaks.

All steam traps are designed to distinguish between steam and condensate. To be effective, a trap must stop (or trap) steam and release condensate.

The variety of differences between the physical properties of steam and of condensate has led to a variety of approaches in steam trap design. Some traps, for instance, are designed to distinguish between the density of steam and of condensate. Other traps are designed to distinguish between the temperature of steam and of condensate. Still others recognize pressure differences between steam and condensate. Three commonly used types of steam traps are the inverted bucket, impulse (disc or float), and thermostatic (bellows) traps (already discussed in detail in Chapter 9).

Steam traps are sometimes considered a special type of check valve. They permit flow in one direction only. In addition, they permit the flow of only certain fluids. The basic purpose of a steam trap is to allow hot water (steam condensate) to pass while holding back steam.

Steam traps are sometimes used on steam-heated process equipment to remove large amounts of condensate. Another use is on steam tracing and jacketing. It is important for a trap to work properly. If it does not allow collected condensate to pass, steam flow will stop. When steam flow stops, the temperature is no longer maintained on the process equipment or the tracing. If the trap does not close, live steam will continually pass through the trap without being condensed. Depending on the details of the condensate collection system, the passage of uncondensed steam can be very wasteful in terms of heat energy loss and can cause thermal shock in pipes and systems.

Bucket steam traps are used on drip legs and tracing, with moderate heat loads. Operating pressure differential should be specified. Traps must be primed to function and will pass live steam when dry. Bucket steam traps will fail to open. They should not be used in intermittent service or used where steam is superheated. They can be used in continuous winter service, but the trap must be protected from freezing.

Impulse traps are used on steam tracers and drip legs. They do not operate at pressures above 300 psig or below 15 psig. Impulse traps will fail to open. They are self-draining when installed in the vertical or horizontal position with the cap on the side.

Thermostatic (bellows) traps are used in intermittent and 5 psig or lower service. An 18-inch minimum vertical leg is required for horizontal or vertical installation. Bellows traps are self-draining when the outlet is vertical.

Strainers

Most traps are equipped with strainers or have integral strainers built in (Figure 13.10). Many of the trap internals are very small, and pieces of pipe scale or rust could lodge in the trap and cause it to cease to operate. These strainers are designed to keep these small pieces of trash out of the trap. Where strainers are installed, it is important that they be blown down periodically.

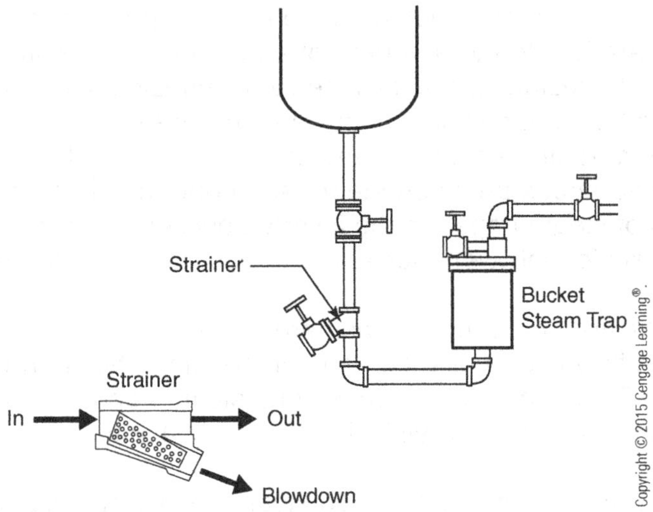

Figure 13.10
Steam Trap

Industrial Sewer System

A plant sewer system collects all drains (water and process fluids) from the process units. All water is collected in a series of ditches and routed to the plant waste treatment area. Units that have hydrocarbons that will remain in the liquid phase at ambient temperatures have an American Petroleum Institute (API) separator, which separates the oil phase from the water phase before routing the water to the plant sewer system. Removed oil is rerun or used for fuel. Some plastic units have only a simple system to remove oil and pellets before sending the sewer water to the plant system.

The plant sewer collection header runs liquid full all the way to the waste treatment API separator. Running it liquid full prevents any hydrocarbon gases from collecting in the sewer line to form a potential explosion hazard. The effluent water from the API separator flows to the surge basin in the waste treatment area. All contamination in the sewer water is removed in the waste treatment system before the water is returned to surface streams or lakes.

Sanitary sewer systems are generally routed to the waste treatment area, where they are treated to remove contaminants. Sewer drains are trapped via a liquid seal in such a way as to prevent gases from backing up into the drain lines.

Refrigeration System

Heating and cooling are two important aspects in modern process control. Refrigeration systems such as air-conditioning are used to provide cooling to industrial applications. Refrigeration units (Figure 13.11) are composed of

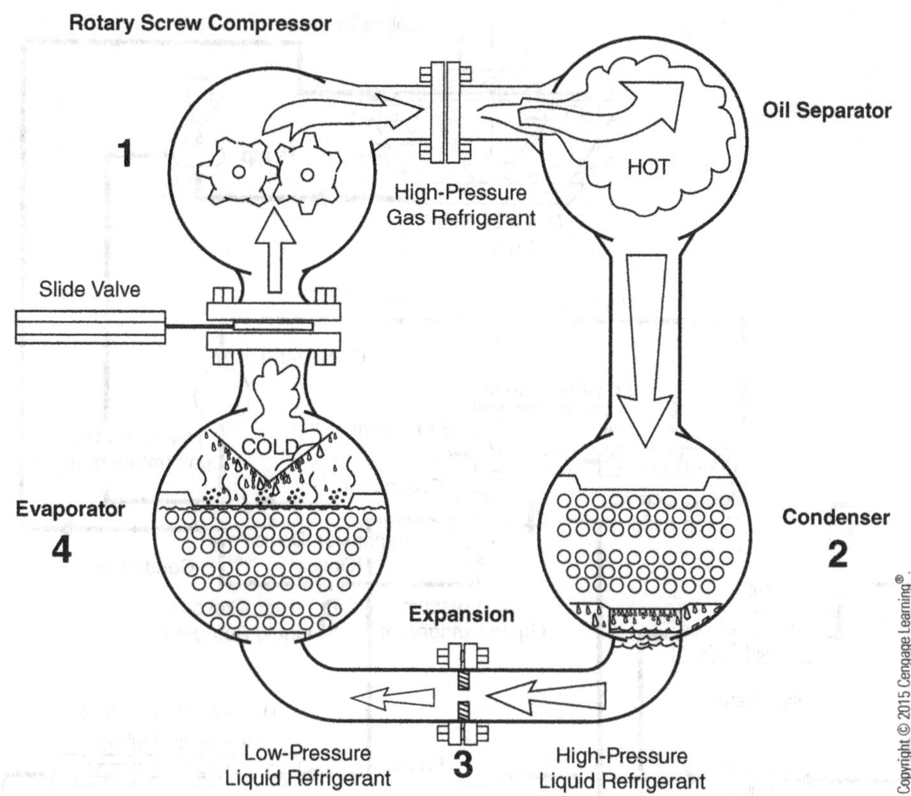

Figure 13.11
Simple Refrigeration System

a compressor (high-pressure refrigeration gas), a heat exchanger (cooling tower), a receiver, an expansion valve (low-pressure refrigeration liquid), and another heat exchanger, called an *evaporator* (low-pressure refrigerant gas).

In the refrigeration process, low-pressure refrigerant gas is drawn into a compressor, converted into high-pressure refrigeration gas, and pushed into a shell and tube heat exchanger. During the compression process, a tremendous amount of heat is generated and must be removed by the exchanger. During the cooling process, the gas condenses into liquid phase and is collected in a receiver. From the receiver, the high-pressure liquid refrigerant is pushed through a small opening in an expansion valve. As the liquid expands, it changes phase. The boiling point of the refrigerant is so low that a cooling effect occurs in the evaporator. As the low-pressure refrigerant leaves the evaporator, it enters into the suction side of the compressor and the process begins again.

One of the most important systems in process units is the refrigeration system. The recovery of ethylene, for example, depends on the efficient operation, reliability, and flexibility of the refrigeration systems employed. Let's review some of the basic principles involved in refrigeration and the equipment utilized. Figure 13.12 is an example of the various types of equipment found in a refrigeration system.

Figure 13.12
Refrigeration System

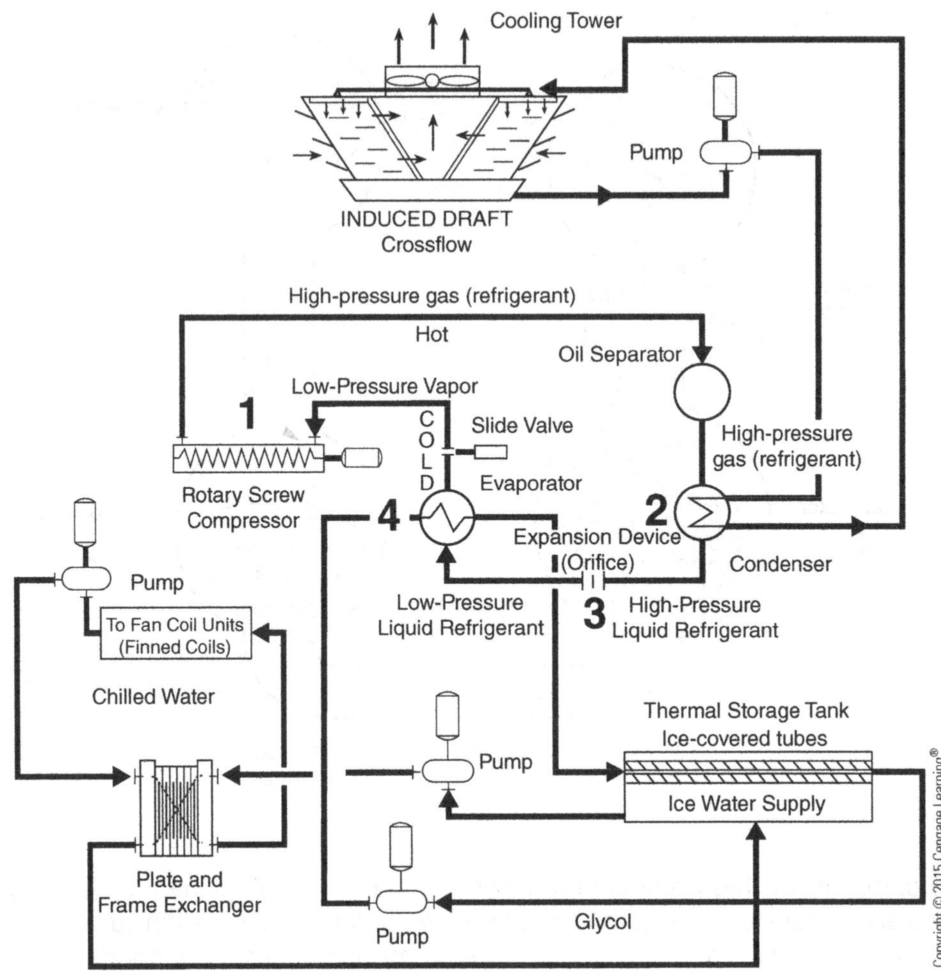

Because heat is transferred from high-temperature areas to low-temperature areas, whenever it is necessary to cool a substance below ambient temperatures, it is necessary to artificially create a lower temperature first. To lower temperatures, process plants use fluids that boil at very low temperatures such as propylene, ethylene, or ammonia—that is, the same principle as the Freon in your refrigerator, car, and house air conditioner. Generally, these refrigerants can be kept in the liquid state at room temperature only if they are kept under high pressure because pressure increases the boiling points of all liquids. An exception to this general rule applies to ethylene and lighter compounds, which are so volatile that it is not practical to store them at room temperature as a liquid; consequently, they are condensed at subambient temperatures. When the pressure on refrigerants is released, however, these liquids boil violently, and their temperatures will drop considerably.

Propylene is a good example of such a liquid. Its boiling point at atmospheric pressure is −54°F. Propylene is normally a gas, but using a compressor can liquefy it. In compressing any gas, work is done, and heat is generated in the gas. In compressing propylene, we add heat to the gas and raise its temperature. Raising the temperature allows us to cool the compressed gas with water, and in this manner the propylene is liquefied.

At this point, we have a supply of liquid at elevated pressure and at about room temperature. To make it boil, all we have to do is release the pressure under which it is stored since room temperature is far above its atmospheric boiling point. One factor to keep in mind is that everything has heat content, unless it is at absolute zero (−273.1°C) (−460°F). Air in a room has plenty of heat; that is why we can cool it with our air conditioners. When a pound of propylene boils away, it absorbs about 188 Btu (called *latent heat*). If we pour a pound of propylene into an open container and watch it boil away, a thermometer inserted into the liquid will read −54°F (−47.77°C). The heat for boiling comes from the container and anything touching the container, including the surrounding air. Moisture in the air will condense and freeze on the container and appear as frost. This is the sequence of events in a simple refrigeration cycle.

The cycle starts with the liquid refrigerant under pressure (Figure 13.13). The liquid flows from the receiver through an expansion (control) valve, where the pressure is decreased, and into an evaporator coil (for example, a chiller). Here the refrigerant picks up heat from the material on the other side of the coil. Beyond the evaporator is the compressor, which puts the vapor under pressure and pumps it to the condenser coils. The heat of vaporization and the heat of compression are both removed by the cooling water in the condenser, leaving the refrigerant once again in the liquid state.

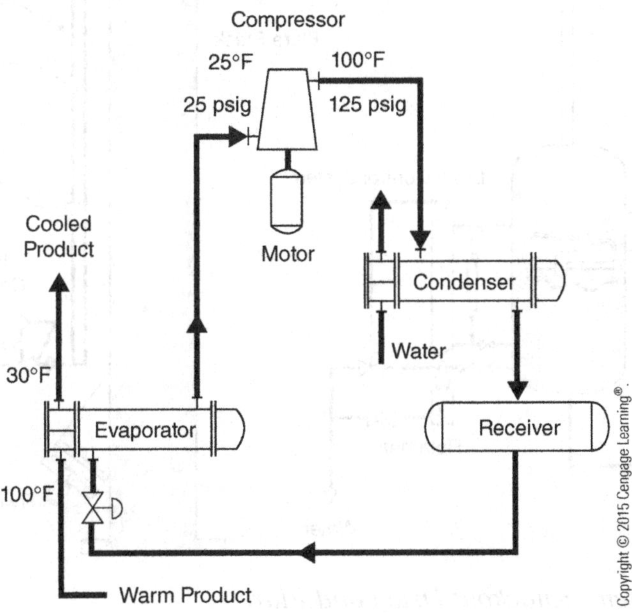

Figure 13.13
Refrigeration Cycle

In the case of an ethylene refrigeration system, owing to ethylene's volatility, propylene refrigerant rather than cooling water is used to condense the compressed ethylene vapors. No matter how complicated the refrigeration system, the ultimate ejection of heat is always directly to air or cooling water.

Most large refrigeration systems operate at several levels of temperature and pressure, called *stages* or *steps* (Figure 13.13). Multiple stages are used because the application of several stages increases the efficiency of a refrigeration cycle and results in lower compressor horsepower. There are practical limits to the optimum number of states; generally three or four steps are employed. It is a good design practice, because of safety considerations, to operate the first or lowest stage compressor suction at a positive pressure to establish the minimum operating temperature. The other temperature levels are set by the design of the particular system.

Relief and Flare Systems

Flare systems are used to safely remove excess hydrocarbons from a variety of plant processes. Flare systems are composed of a complex network of pipes and headers connecting to a **knockout drum** and flare (Figures 13.14 and 13.15). Governmental laws and regulations require the flare to be located a safe distance from the operating units and populated areas.

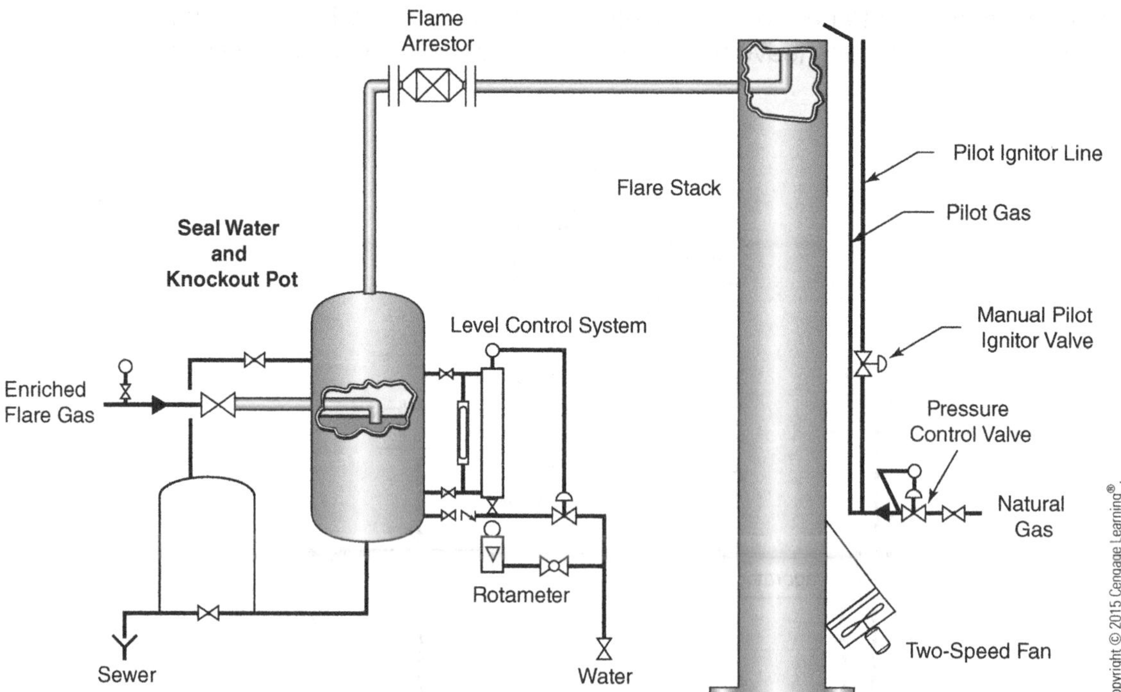

Figure 13.14 *Flare System: Knockout Drum and Flare*

Figure 13.15
Flare System and Stack

Flare systems are part of a plant's safety system. Most process units are lined up to safety relief valves that lift when specified pressures are exceeded. These safety valves discharge into the **flare header**. Unexpected process upsets are dumped to the flare system as a last course of action.

A typical flare system (Figure 13.16) includes a flare (a long, narrow pipe mounted vertically), a steam ring mounted at the top of the flare (used to dispense hydrocarbon vapors), an ignition source at the top of the flare, a fan mounted at the base of the flare (used for forced-draft operation), a knockout drum with water seal, and a flare header.

Relief System

The discharge from all safety valves, pressure relief regulators, and operating vents and blowdown valves in hydrocarbon service are collected in a closed piping system and sent to a flare stack. Steam, air, and nitrogen vapors discharge to the atmosphere. Steam safety valves discharge through tailpipes at a safe distance above grade or above the nearest operating platform. Thermal relief valves are provided to protect against thermal expansion as required in liquid-full service. The flare and relief systems are designed in accordance with applicable regulations and practices to protect equipment and personnel against overpressure situations. Many units that process volatile hydrocarbons have two collecting systems, a warm-wet system and a cold-dry system. This separation is made to avoid freezing, with consequent plugging due to ice formation of the wet vapors.

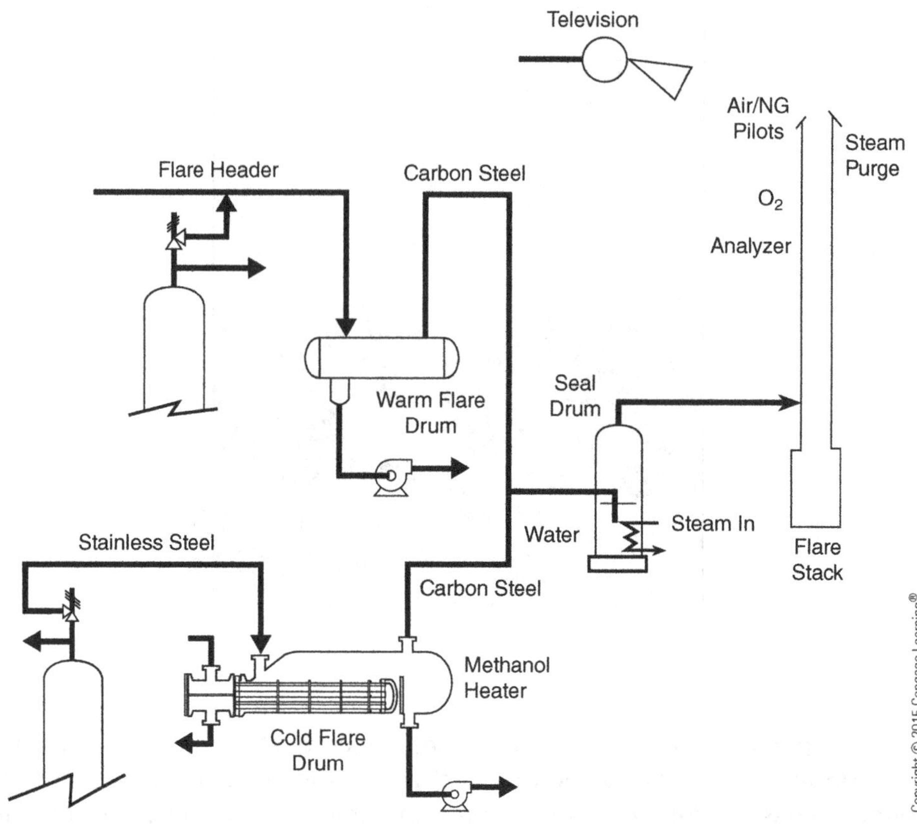

Figure 13.16
Typical Flare System

Both systems go to their respective knockout drums, where the cold vapors are warmed and then combined and sent to the flare stack.

Hot Relief System (Hot Flare)

The hot relief system collects liquids and vapors from units that operate above 32°F (0°C). The flare header branches throughout the plant and terminates in a sloped line leading to the flare knockout drum. The knockout drum is a carbon steel vessel that contains a steam coil to vaporize light materials and to warm up heavy materials so they can be pumped to recovery. The pump can be automated to control level but will automatically shut down on low level in the drum. Low and high alarms in the control room are normally provided.

Cold Relief System (Cold Flare)

The cold relief system collects liquids and vapors below 32°F (0°C) after expansion into the header. It is important for operators to recognize that, although the cold flare system is designed for extremely cold temperatures, there are portions of some connections to the flare that have a higher temperature specification than the cold flare system itself. The various branches throughout the unit terminate in a stainless steel header sloped to drain to the cold flare knockout drum. The cold flare knockout drum is a stainless steel drum containing a heat exchanger that vaporizes the accumulating liquids. The vapor exiting the knockout drum is kept above 32°F (0°C).

Flare Systems

Flare Seal Drum Function

A nitrogen purge is maintained on the flare headers to avoid explosive mixtures in the flare lines. Nitrogen and natural gas connections are provided at the end of the header. Normally, the flare system is operated with a water seal in the flare knockout drum. This water seal ensures a positive pressure on the flare when it is necessary to access equipment connected to the flare for any reason. An analyzer continuously monitors the flare at the stack for oxygen.

Flare Stack and Flare Pilots

The flare stack provides an elevated location for combustion in the flare tip. Flare stack elevation is selected to limit radiation heat at ground level. The tip is designed to allow smokeless flare operation at low discharge flow rates. Some flare systems are designed to handle only partial system or equipment failures without smoking.

The steam flow to the flare is automatically controlled on the basis of the flow rate of gas to the flare. The flare pilots operate on natural gas and air ignited by a flame front generator. Because of the heat radiation from the flare during peak loads, the flame front generator controls are shielded and located away from the base of the flare stack. Entry into the high-flare radiation area must be by permit only and must be coordinated with the operations department. Anyone in the area must be in constant radio contact with the control room.

Flare stacks are provided to burn vent and waste gas from the units. Normally the waste gases are burned as fuel gas, but at times when a unit upset occurs, large quantities of gas must be disposed of at low pressure to the flare. The flare burners and piping are designed to handle these flows. They must burn the waste gas with a minimum amount of smoke and must be high enough so that the surrounding area and its equipment are not endangered by flame, heat radiation, and fumes. Flare stacks range in height from 200 to 350 feet. A self-supporting flare stack looks like a large perpendicular pipe, usually with graduated size, and has a burner at the top. Some of the smaller stacks have guy wires for protection against high winds. Derricks support many smaller flare stacks. Ground flares are low flares, about 40 feet high, and are used only for purging lines, pumps, and vessels in some areas. They are equipped with blowers to supply air for draft at the burner. Because ground flares are used for purging only, the controls are manual and the flares are shut down when purging is completed.

Steam is piped to the burner tip of the unit flares to disperse and provide draft for burning off the gas without smoking and to protect the burner tips from the heat of the flame. Part of the steam flow to the tip is fixed and cannot be shut off; a manual loading station, which operates control valves in the steam line, remotely controls the remainder. These are adjusted as necessary to eliminate smoke at the flare tip.

The flare tips are equipped with pilot burners to ignite the gas as it flows from the tip. These are always burning, but if one goes out they can be relighted from remotely located flame front generators. Air and natural gas are connected to the flame front generator and are admitted in controlled amounts to form a combustible mixture that flows to the pilot burner through a small (1-inch diameter) igniter line. A sparking device in the mixing chamber ignites the gas–air mixture, and the flame travels through the line to the pilot and lights it. Separate natural gas lines supply the fuel to the pilots.

Some flares are provided with seal drums containing water so that flame cannot back into the flare system from the tip. Others depend on some type of flow from the unit to prevent flame from backing into the flare lines. This flow is usually nitrogen-introduced into the flare lines through a meter at some prescribed flow rate. Natural gas can be substituted for nitrogen if necessary. All flares have a knockout drum to collect entrained liquids. A pump is usually provided to pump the liquid off to tankage. The technician assigned to the area controls this pumping manually. Usually, an alarm will give a warning when there is a high level in the drum. A high level in the knockout drum can subject the flare to slugs of liquid that will be carried out the tip in the event of a gas release. This situation has resulted in the eruption of flaming liquid from the flare and onto the surrounding area.

Summary

Utility systems are the combinations of equipment that are used to provide a plant with water, fuel, air, nitrogen, electricity, refrigeration, and other substances or processes it needs to operate. In general, utility systems either separate materials into pure or relatively pure products or distribute supplies to the plant's units in useable amounts, store supplies, and remove waste products.

Water systems include raw water and fire water systems, boiler feedwater treatment systems, cooling water systems, and steam systems.

The condensate recovery system of the steam system is separated into clean and suspect condensate sections. Clean condensate is routed to the deaerator. Suspect condensate is collected in a drum, monitored for impurities, and filtered before being sent to the deaerator.

The cooling water system is designed for a supply temperature of 90°F (32.2°C) and a maximum return temperature of 120°F (48.88°C). The cooling water is supplied from the cooling tower to the users by cooling water circulation pumps. Makeup water to the cooling tower is supplied from the raw water system. Filters may be used to remove airborne solids such as dust and sand and contaminants introduced via the cooling tower makeup water such as silt, organic matter, and water-borne organisms. The cooling water chemical feed system consists of a nonchromate-based corrosion inhibitor, a dispersant or antiscalant, and chlorine for the control of microbiological fouling.

Corrosion in equipment is typically associated with oxygen reacting with the metal, so a deaerator is used to remove air (which contains dissolved oxygen) from water. The deaerator heats the water with steam and dissolved oxygen passes out the vent with the steam.

In demineralization, raw water is passed through an ion and cation exchange resins.

Utility air (UA) and instrument air (IA) are both supplied from centrally located air compressors. Industrial manufacturers purchase natural gas (NG) from suppliers. A high-pressure pipeline generally delivers gas to the plant. Electric power is received from electric supply companies. Relief and flare systems safely remove excess hydrocarbons.

Review Questions

1. Name two types of atmospheric tanks.
2. What is meant by "breathing" action?
3. How does a flame arrestor work?
4. Describe the purpose of a conservation vent.
5. Why are gas-blanketed tanks used?
6. Describe corrosion and the ways by which process technicians prevent it.
7. Why do cooling towers need to be blown down periodically?
8. Describe a fire-water system.
9. Describe boiler feedwater treatment.
10. If it takes 1 Btu to raise the temperature of 1 lb. of water 1°F, how many Btus does it take to convert 1 lb. of water at 35°F (1.66°C) to steam?
11. What are some impurities found in natural water?
12. Why is the purity of boiler feedwater critical to the operation and maintenance of the system?
13. What is the maximum amount of suspended solids permitted in a boiler operating at over 2,000 psig?
14. What are coagulation and flocculation?
15. What is the relationship between water softening and chemical precipitation?
16. How do anion resins and cation resins react with impurities in the water?
17. Describe the operation of a deaerator.
18. What is the justification for hooking nitrogen and instrument air systems together? Do you agree?
19. Describe a flare system.
20. Describe the operation of a three-phase motor.
21. What are natural gas systems used for?
22. What is the primary function of a steam trap?
23. Describe an industrial sewer system.
24. Draw a simple sketch that shows how a refrigeration system operates.
25. What are the basic components of a relief and flare system?

Reactor Systems

Objectives

After studying this chapter, the student will be able to:

- Describe the function of a reactor.
- Describe exothermic, endothermic, replacement, neutralization, and combustion chemical reactions.
- List reaction variables and their effects.
- Balance a chemical equation.
- Describe a continuous and a batch reactor.
- Describe a reactor's function in alkylation.
- Describe a reactor's function in fluid catalytic cracking.
- Describe a reactor's function in hydrodesulfurization.
- Describe a fixed bed reactor's function in hydrocracking.
- Describe a tubular reactor and chemical synthesis.
- Describe a fluidized bed reactor in coal gasification.
- Describe fluid coking.

Key Terms

Alkylation—use of a reactor to make one large molecule out of two small molecules.

Autoclave—see *Stirred reactor*.

Balanced equation—chemical equation in which the sum of the reactants (atoms) equals the sum of the products (atoms).

Catalyst—a chemical that can increase or decrease reaction rate without becoming part of the product.

Chemical equation—numbers and symbols that represent a chemical reaction.

Chemical reaction—a term used to describe the breaking of chemical bonds, forming of chemical bonds, or breaking and forming of chemical bonds.

Combustion—a rapid exothermic reaction that requires fuel, oxygen, and ignition source and gives off heat and light.

Converter—see *Fixed bed reactor*.

Cracking—decomposition of a chemical or mixture of chemicals by the use of heat, a catalyst, or both; thermal cracking uses heat only; catalytic cracking uses a catalyst.

Endothermic—a chemical reaction that must have heat added to make the reactants combine to form the product.

Exothermic—a chemical reaction that gives off heat.

Fixed bed reactor—a vessel that contains a mass of small particles through which the reaction mixture passes; also called a converter.

Fluid catalytic cracking—a process that uses a fluidized bed reactor to split large gas oil molecules into smaller, more useful ones; also called *catcracking*.

Fluid coking—a process that uses a fluidized bed reactor to scrape the bottom of the barrel and squeeze light products out of residue.

Fluidized bed reactor—suspends solids by countercurrent flow of gas; heavier components fall to the bottom, and lighter ones move to the top.

Hydrocracking—use of a multistage fixed bed reactor system to boost yields of gasoline from crude oil by splitting heavy molecules into lighter ones.

Hydrodesulfurization—removes sulfur from crude mixtures.

Material balancing—a method for calculating reactant amounts versus product target rates.

Neutralization—a chemical reaction designed to remove hydrogen ions or hydroxyl ions from a liquid.

Products—the end results of a chemical reaction.

Reactants—the raw materials in a chemical reaction.

Reaction rate—the amount of time it takes a given amount of reactants to form a product or products.

Regenerator—used to recycle or make useable again contaminated or spent catalyst.

Replacement reaction—a reaction designed to break a bond and form a new bond by replacing one or more of the original compound's components.

Stirred reactor—a reactor designed to mix two or more components into a homogeneous mixture; also called an autoclave.

Tubular reactor—a heat exchanger in which a chemical reaction takes place; used for chemical synthesis.

Introduction to Reactions

A reactor is a vessel in which a controlled **chemical reaction** takes place. This chapter discusses five basic types of reactors: stirred tank reactors; **converters**, or **fixed bed reactors**; **fluidized bed reactors**; **tubular reactors**; and furnaces. There are many other types of reaction vessels; however, most of them are simply combinations of those five.

The various things that have an effect on a chemical reaction are called *reaction variables*. The design and operation of a reactor depend largely on how variables affect the chemical reaction that is taking place in the reactor. For example, heat increases molecular activity and enhances the formation of chemical bonds, and pressure increases atomic collisions. Other factors that affect chemical reactions are time, the concentration of **reactants**, and **catalysts**.

Temperature

In a reaction mixture, gas, or liquid, the temperature determines how fast the molecules of the reactants move. At a high temperature, the molecules move rapidly through the mixture, colliding with each other frequently. The more often they collide, the more apt they are to react with one another. A good rule of thumb for chemical reactions is that the speed with which two chemicals will react doubles for each 10°C increase in temperature. This rule assumes, of course, that other variables do not change. For example, if 10% of the chemicals in a reaction form **products** in one hour at 50°C, then 20% will form products at 60°C in one hour. The danger in this type of reaction is that twice as much heat will be given off at the higher temperature. If the extra heat cannot be removed from the reactor as fast as it is formed, the temperature will rise, causing the reaction to proceed at even higher rates. Obviously, you will soon have an uncontrollable reaction and, possibly, an explosion.

Pressure

The effect of pressure on a reaction cannot be generalized as easily as the effect of temperature. In a liquid-phase reaction, pressure can increase or decrease the **reaction rate**, depending upon the readiness of the reactants or products to vaporize. In a gaseous reaction, pressure forces molecules closer together, causing them to collide more frequently. In other words, the higher the pressure there is in a gaseous reaction, the higher the reaction rate.

Reaction Time

Reaction time is simply the length of time that the reactants are in contact at the desired reaction conditions. Contact time refers to the length of time the reactants are in contact with a catalyst. Residence time refers to the length of time reactants remain in a tank before forming a product. Generally, the longer the reaction time, the more products are formed in the reaction. This does not necessarily mean that it is desirable to let a reaction continue for a long time.

Concentration of Reactants

The concentration of reactants in the reactor has a major effect on how fast the reaction will take place, what products will be produced, and how much heat will have to be added to or taken away from the reaction. We generalize here for the sake of simplicity and say that for most reactions, the higher the concentration of reactants the faster the reaction and thus the more heat that is generated or needed.

Agitation provided by mechanical agitators or by the turbulent flow of the reactants may affect concentration. In general, good mixing or good agitation produces an efficient reaction with the desired products. Poor agitation may produce pockets that have a high concentration of one reactant and low concentration of the other. This uneven distribution may produce undesirable as well as unpredictable products.

Catalysts

Many chemicals will not react when placed together at high concentrations and heated or will produce unwanted products. Therefore, an additional substance called a *catalyst* is added to stimulate the reaction and to produce the more desirable product. Most catalysts do not react with the reactants, so usually they can be reused several times until they become dirty or ineffective. Then they must be replaced or regenerated in some way. Other types of catalysts, usually liquids or gases, will react with the reactants, forming an intermediate product. The reaction will continue until the desired product is formed, releasing the catalyst to be used again. It is not unusual for some catalysts to be destroyed in the reaction or discarded because of the difficulty of recovering them.

Catalysts can be classified as adsorption, intermediate, inhibitor, or poisoned. An adsorption-type catalyst is a solid that attracts and holds reactant molecules so a higher number of collisions can occur. It also stretches the bonds of the reactants it is holding, thereby weakening the bonds, which then require less energy to break and rebond. An intermediate-type catalyst forms an intermediate product by attaching to the reactant and slowing it down so collisions can occur. An inhibitor-type catalyst is any substance that slows a reaction. A poisoned catalyst is one that no longer functions or is used up.

Inhibitors

An inhibitor is a substance that prevents or hinders the reaction of two or more chemicals. Sometimes, trace amounts of impurities in the feedstock to a reactor kill a reaction or severely reduce the amount of desirable products formed. The inhibitor may hinder the reaction by reacting with some of the raw materials before the desired reaction can take place, or it may react with the catalyst, making it unstable.

Exothermic and Endothermic Reactions

Exothermic reactions are characterized by a chemical reaction accompanied by the liberation of heat. As the reaction rate increases, the evolution of heat energy increases. Controlling reactant flow rates, removing heat, or providing cooling can control exothermic reactions. This type of reaction is subject to "runaway" if sufficient heat is not withdrawn from the system.

Endothermic reactions absorb energy as they proceed. They must have heat added to form the product. Figure 14.1 is an example of a periodic table and 14.2 illustrates the information found in each cell.

Replacement Reactions

Industrial manufacturers use **replacement reactions** to remove dissolved mineral ions from process water. A number of dissolved minerals can be found in process fluids. A common compound found in process water is calcium chloride. Calcium chloride ($CaCl_2$) forms positive calcium ions (Ca^{+2}) and negative chloride ions (Cl^{-2}) when it is dissolved in water.

A replacement reaction can remove the (Ca^{+2}) ions and the (Cl^{-2}) ions with synthetic resins. Resins come in a variety of shapes and designs. Sometimes resins take the form of plastic strands rolled into balls and charged with ions. A hydrogen (H^+) ion on a resin ball is replaced by the Ca^{+2} ion as the process fluid moves through the resin bed. The replacement reaction will take place until all of the Ca^{+2} is removed from the fluid or the H^+ is used up on the resin balls.

Resin balls can be treated with positively or negatively charged ions and used for replacement reactions. For example, resin balls charged with hydroxyl ions (OH^-) can be used to replace the chloride ion (Cl^{-2}).

Neutralization

Neutralization reactions remove hydrogen ions (acid) or hydroxyl ions (base) from a liquid. Neutralization reactions are designed to neutralize the acidity or alkalinity of a solution. Hydrogen ions and hydroxyl ions neutralize each other.

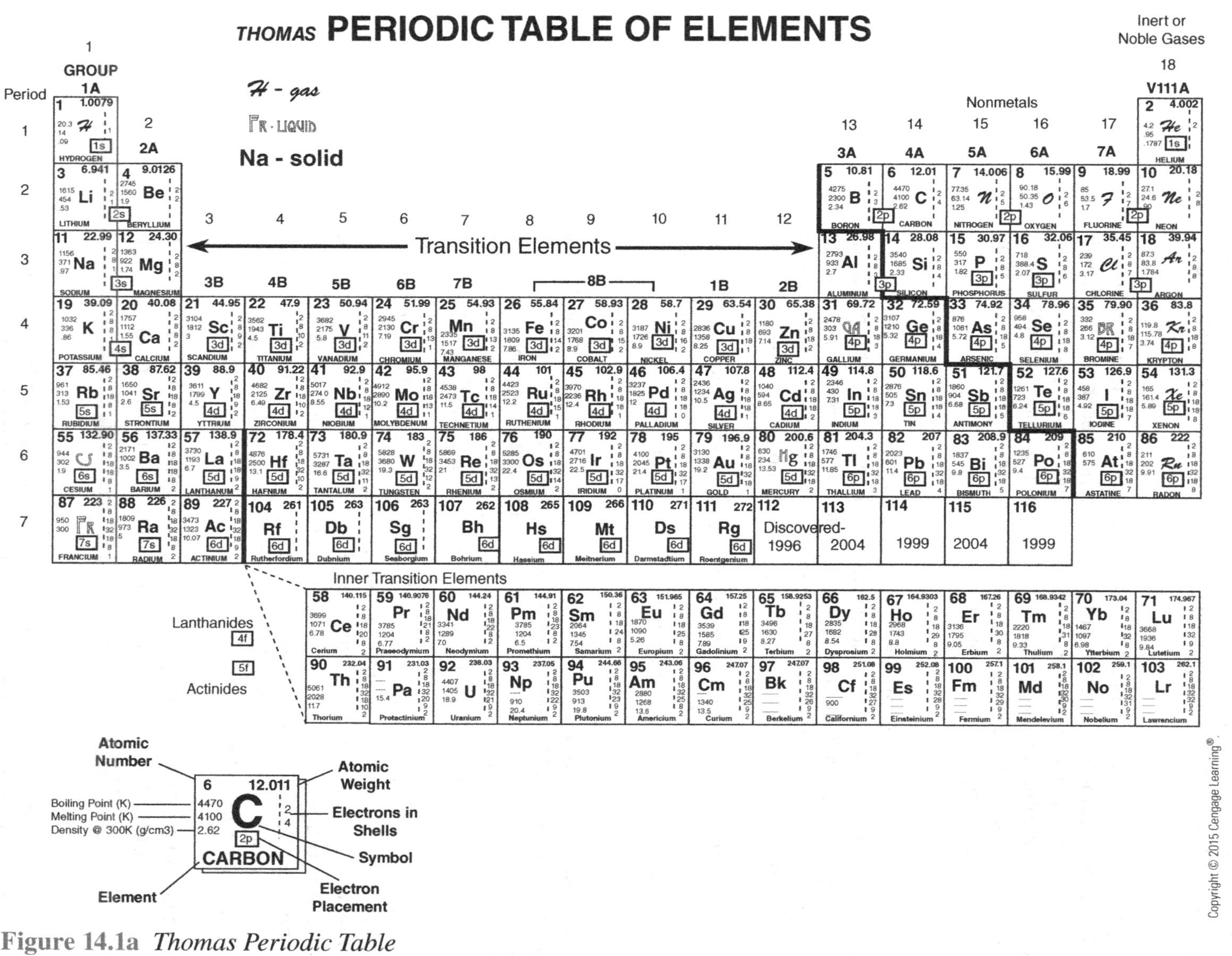

Figure 14.1a Thomas Periodic Table

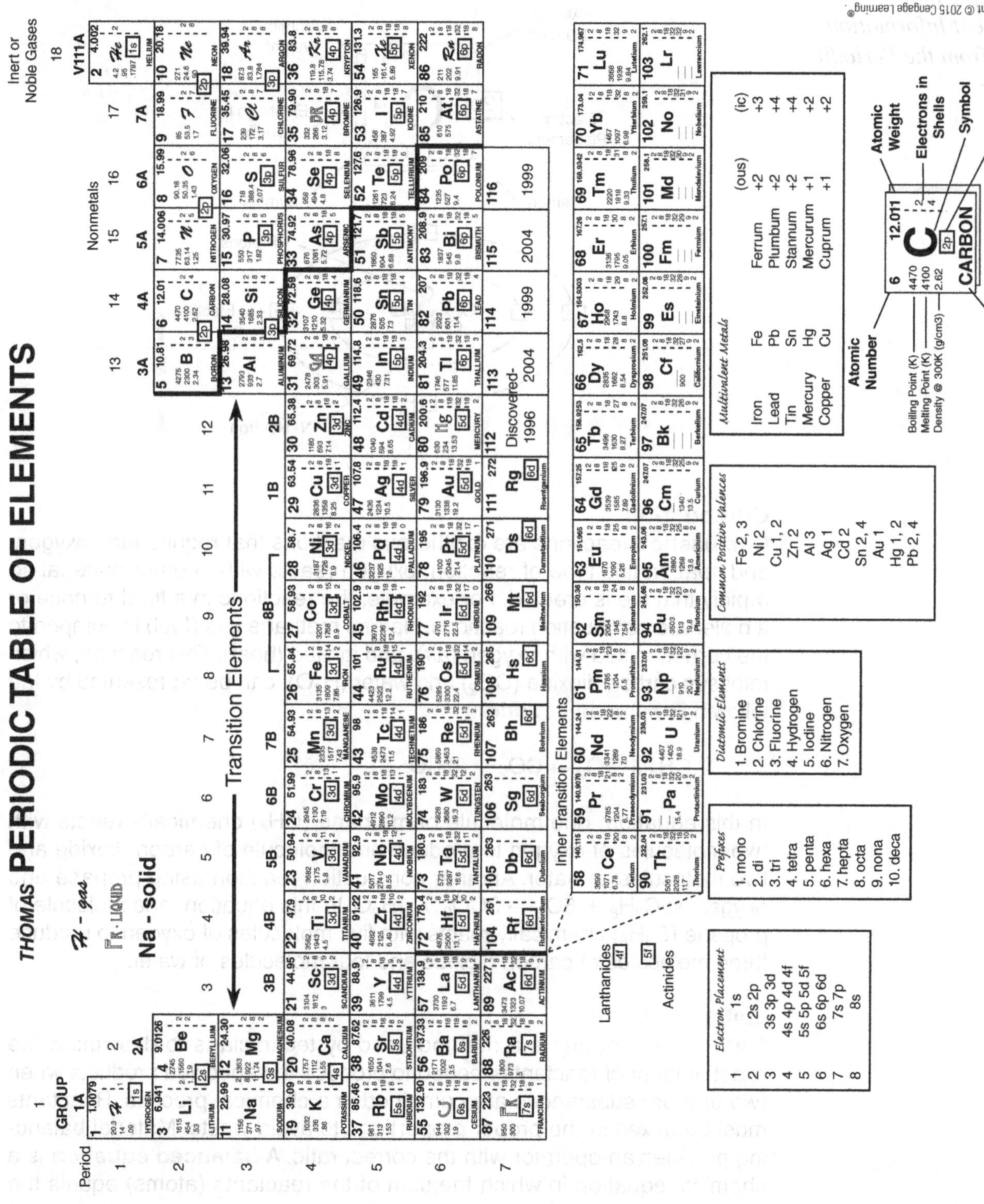

Figure 14.1b Thomas Periodic Table

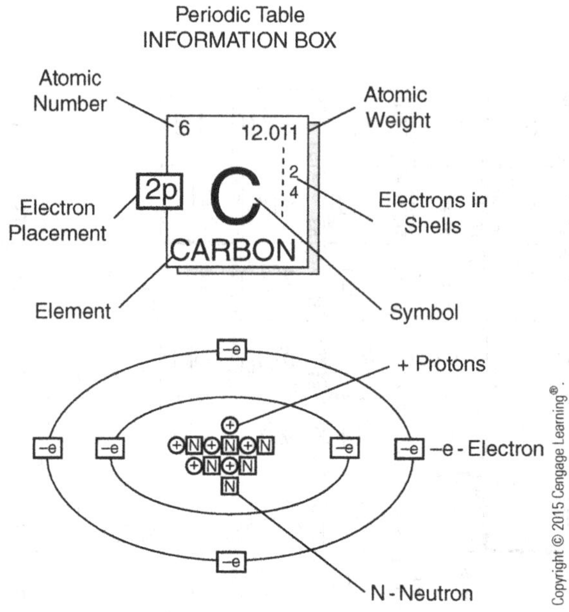

Figure 14.2 *Example of an Element Information Cell from the Periodic Table*

Combustion

Combustion reactions are exothermic reactions that require fuel, oxygen, and heat. In this type of reaction, oxygen reacts with another material so rapidly that fire is created. For example, the reactions in a fired furnace or a boiler are combustion reactions. Natural methane gas (fuel) is pumped to the burner, mixed with oxygen (air), and ignited (heat). This reaction, which releases carbon dioxide (CO_2) and water (H_2O), can be represented by the following **chemical equation**:

$$CH_4 + 2O_2 \rightarrow CO_2 + 2H_2O$$

In this equation, one molecule of methane (CH_4) chemically reacts with two molecules of oxygen to produce one molecule of carbon dioxide and two molecules of water. Another combustion reaction using propane and oxygen is: $C_3H_8 + 5O_2 \rightarrow 3CO_2 + 4H_2O$. In this equation, one molecule of propane (C_3H_8) chemically reacts with five molecules of oxygen to produce three molecules of carbon dioxide and four molecules of water.

Material Balance

Material balancing is a method used by technicians to determine the exact amount of reactants needed to produce the specified products when two or more substances are combined in a chemical process. Reactants must be mixed in the proper proportions to avoid waste. Material balancing provides an operator with the correct ratio. A **balanced equation** is a chemical equation in which the sum of the reactants (atoms) equals the sum of the products (atoms).

The steps in checking a material balance are: (1) determine the weight of each molecule; (2) ensure that reactant total weight is equal to product total weight; and (3) determine the numbers of reactant atoms. The number of atoms is given by numerical prefixes and subscripts. For example, $2H_2$ equals 4 hydrogen atoms. Weight can be expressed in atomic mass units (AMU): 1 AMU = 1 gram, 1 pound, or 1 ton. AMUs are weight ratios, so the units can be anything you want. For example, check the material balance of this chemical equation:

$$H^+ + OH^- \rightarrow H_2O$$

Step 1. Determine the weight of each molecule.

$$H^+ \text{ (1 AMU)} + OH^- \text{ (17 AMU)} \rightarrow H_2O \text{ (18 AMU)}$$

$$H^+ \text{ (1 g)} + OH^- \text{ (17 g)} \rightarrow H_2O \text{ (18 g)}$$

Step 2. Ensure that reactant total weight is equal to product total weight.

Step 3. Determine the numbers of reactant atoms to be sure they match. The reactant side of the equation has two hydrogen atoms and one oxygen atom. The product side of the equation has two hydrogen atoms and one oxygen atom. The equation is balanced.

Determining whether a chemical equation is balanced can also be done by simply listing the reactants and the products and making sure the totals for each element are the same on each side of the equation.

$$Na_2O + 2HOCl \rightarrow 2NaOCl + H_2O$$

Reactants	Products
2Na	2Na
3O	3O
2Cl	2Cl
2H	2H

This chemical equation is balanced. Now, let's look at another equation:

$$2H_3PO_4 \rightarrow H_2O + H_4P_2O_8$$

Reactants	Products
6H	6H
2P	2P
8O	9O

This chemical equation is *not* balanced.

When balancing an equation the following principles and criteria are very helpful:

- Determine if the equation is balanced or not.
- Never change the subscripts. For example in H_2O, changing the subscript 2 will alter the composition and the substance itself.

- Focus on the coefficients in order balance an equation. Work from one side to the other. Typically it is easier to start with one side and then balance the other. Most operations move left to right, trial and error.
- Ensure that you have included each source for a particular element that you are attempting to balance. It is possible that two or more molecules contain the same element.
- Adjust the coefficient of monoatomic elements last.
- Adjust the coefficient of polyatomic ions that are acting as a group in self-contained groups on both sides of the equation. For example,

Not balanced $(NH_4)_2CO_3 \rightarrow NH_3 + CO_2 + H_2O$

Balanced $(NH_4)_2CO_3 \rightarrow 2NH_3 + CO_2 + H_2O$

Reactants		**Products**	
Nitrogen	= 2	Nitrogen	= 2
Hydrogen	= 8	Hydrogen	= 8
Carbon	= 1	Carbon	= 1
Oxygen	= 3	Oxygen	= 3

Each problem will present its own mystery and most can be solved using the "inspection method." You can try this yourself by balancing the following equations:

1. $Fe + O_2 \longrightarrow Fe_2O_3$
   Answer: $4Fe + 3O_2 \longrightarrow 2Fe_2O_3$

2. $Sn + Cl_2 \longrightarrow SnCl_4$
   Answer: $Sn + 2Cl_2 \longrightarrow SnCl_4$

3. $Fe + Cl_2 \longrightarrow FeCl_3$
   Answer: $2Fe + 3Cl_2 \longrightarrow 2FeCl_3$

Continuous and Batch Reactors

Industrial manufacturers use two basic methods of reactor operation: batch and continuous. In a batch operation, raw materials are weighed carefully and added to a reactor. The raw materials are mixed and allowed to react. After a predetermined amount of time, the batch is dumped and a new batch is mixed.

Continuous reactor operation adds raw materials incrementally to the reactor. Finished products flow out while raw materials flow in and are exposed to catalysts, pressure, liquids, or heat.

Stirred Reactors

A common type of reactor is the mixing, or **stirred reactor** (Figure 14.3). The basic components of this device will include a mixer or agitator mounted to a tank. The mixer will have a direct or an indirect drive. The reactor shell is designed to be heated or cooled and withstands operational pressures, temperatures, and flow rates. The design may include tubing coils wrapped around the vessel, internal or external heat transfer plates, or jacketed vessels. Hot oil, steam, or cooling water may be the heating or cooling medium.

The stirred reactor is designed to mix two or more components into a homogeneous mixture. As these components blend together, chemical reactions occur that create a new product. Blending time and exact operating conditions are critical to the efficient operation of a stirred reactor.

Stirred reactors are equipped with a number of safety features: pressure relief systems, quench systems, process variable alarms, and automatic shutdown controls. Quench systems are designed to stop the reaction process. Pressure relief systems are sized to handle and contain any release from the reactor. The relief system may be designed for liquid, vapor, or a liquid-vapor combination. Safety relief systems allow process releases to go to the plant flare system. Sometimes a chemical scrubber system is used before a product is sent to the flare system.

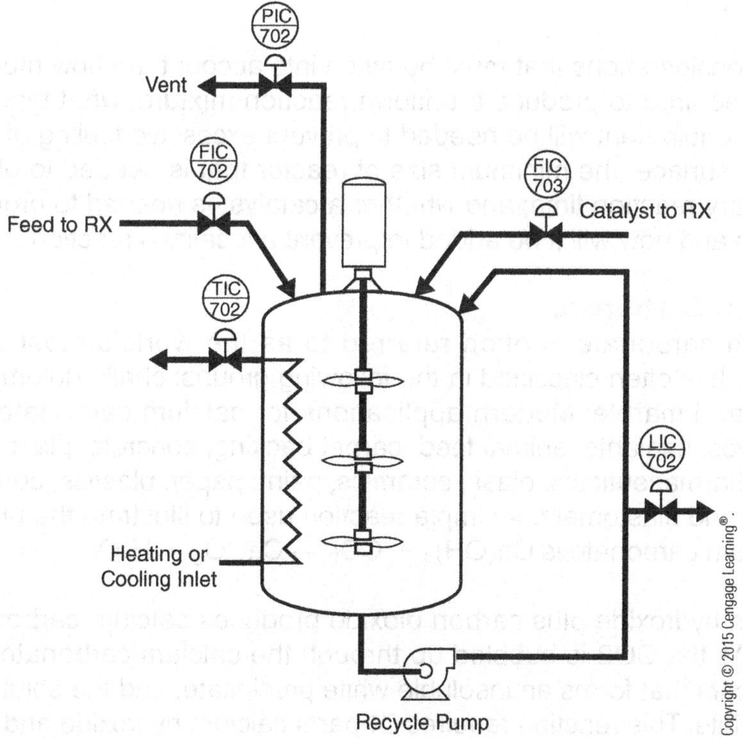

Figure 14.3
Stirred Reactor

Process variable alarms will be activated by analytical (composition), pressure, temperature, flow, level, and time variables. Rotational speed on the agitator may be fixed or variable. A series of interlocks, permissives, and alarms will engage during operation and will provide a support network for the technician. A series of process video trends will be displayed to track each of the critical variables. Samples are taken frequently to ensure product quality. Stirred reactors are connected to off-specification (off-spec) systems that allow flexibility in switching between prime and off-spec situations. The automatic shutdown allows a technician to push one button and shut the unit down in the event of an emergency or runaway reaction.

A stirred tank reactor is often referred to as an **autoclave** or batch reactor. This type of reactor is used in both batch processes, such as the suspension vinyl resins unit and the phenolic resins unit, and continuous feed processes, such as the solvent vinyl resins unit and the polyethylene unit. The basic features of a stirred reactor are designed to provide long reaction times, mechanical agitation, and a method of cooling or heating the reaction. Stirred tank reactors are used for liquid-liquid reactions, gas-liquid reactions, and liquid-solid reactions. Stirred tank reactors are used for reactions that require a relatively long reaction time and for reactions of slurry or thick liquid in which mechanical agitation must be used to provide a uniform reaction mixture. Such reactors are also used when the production rate is to be variable and several different reactions or products are to be made in the same reactor. The early research and development of most processes is conducted in batch stirred reactors because of their versatility.

Other considerations that must be taken into account are how much agitation is required to produce a uniform reaction mixture; what type of heat transfer equipment will be needed to prevent excessive fouling of the heat transfer surface; the minimum size of reactor that is needed to obtain the necessary reaction time; and whether a catalyst is needed to promote the reaction and how will it be added to prevent a localized reaction.

Calcium Carbonate

Calcium carbonate is often referred to as the world's most versatile mineral. It is often classified in the following groups: chalk, dolomite, limestone, and marble. Modern applications for calcium carbonate include adhesives, sealants, animal feed, carpet backing, concrete, plasters, fertilizers, pharmaceuticals, glass, ceramics, paint, paper, plastics, composites, rubber, and elastomers. A simple reaction used to illustrate the production of calcium carbonate is $Ca(OH)_2 + CO_2 \rightarrow CaCO_3 + H_2O$.

Calcium hydroxide plus carbon dioxide produces calcium carbonate and water. As the CO2 is bubbled up through the calcium carbonate, a reaction occurs that forms an insoluble white precipitate, and the solution turns milky white. This reaction requires 74 parts calcium hydroxide and 44 parts carbon dioxide to form 100 parts calcium carbonate and 18 parts water.

It is possible for a second reaction to occur if we continue adding carbon dioxide to the white solution. The chemical equation for this reaction is $CaCO_3 + H_20 + CO_2 \rightarrow Ca(HCO_3)_2$. In this reaction, calcium bicarbonate is produced and the solution changes from milky white to clear. Figure 14.4 illustrates the basic equipment used in this stirred reaction system, and Figure 14.5 shows a variety of stirred reactor designs.

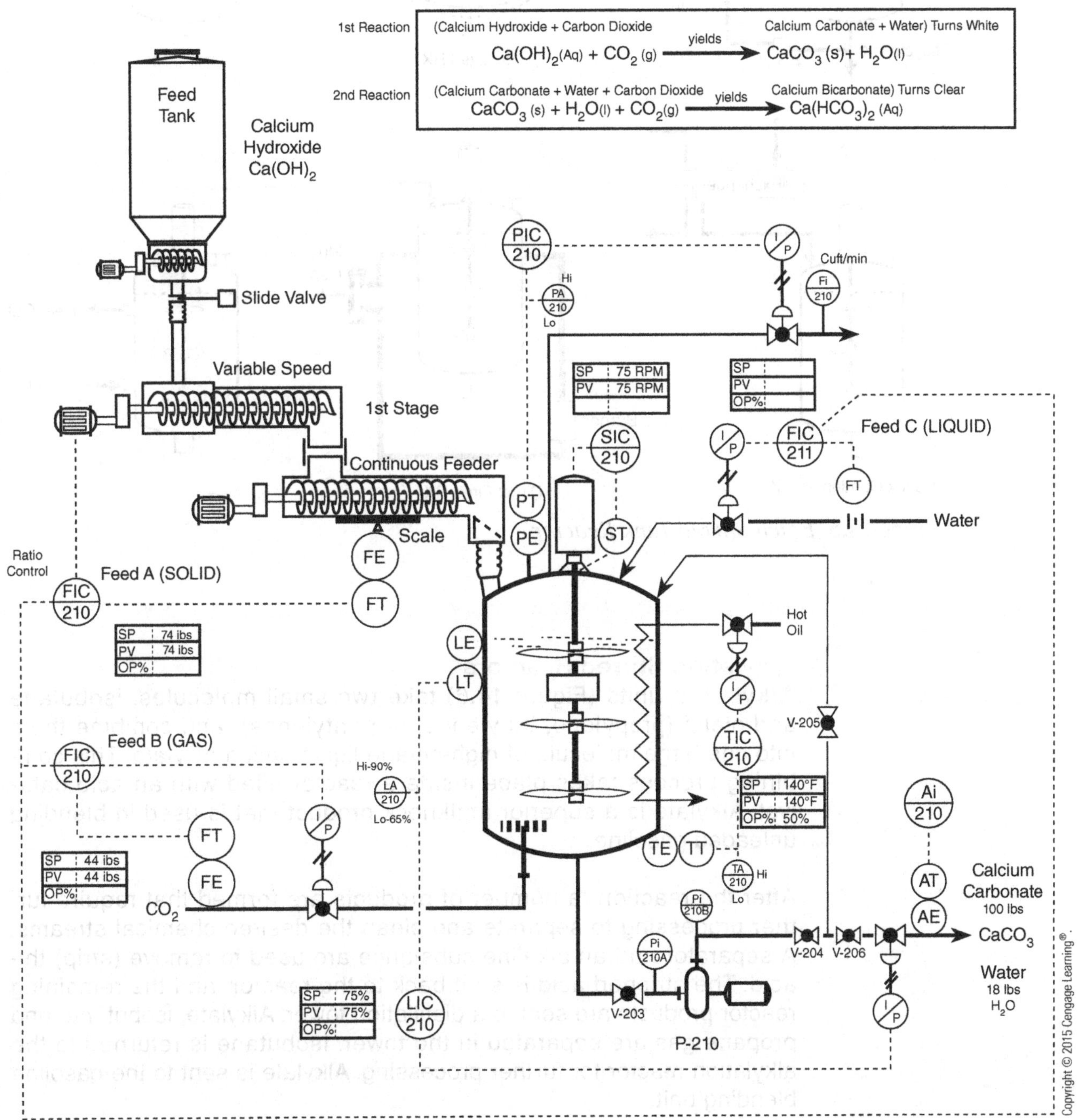

Figure 14.4 *Calcium Carbonate Reaction*

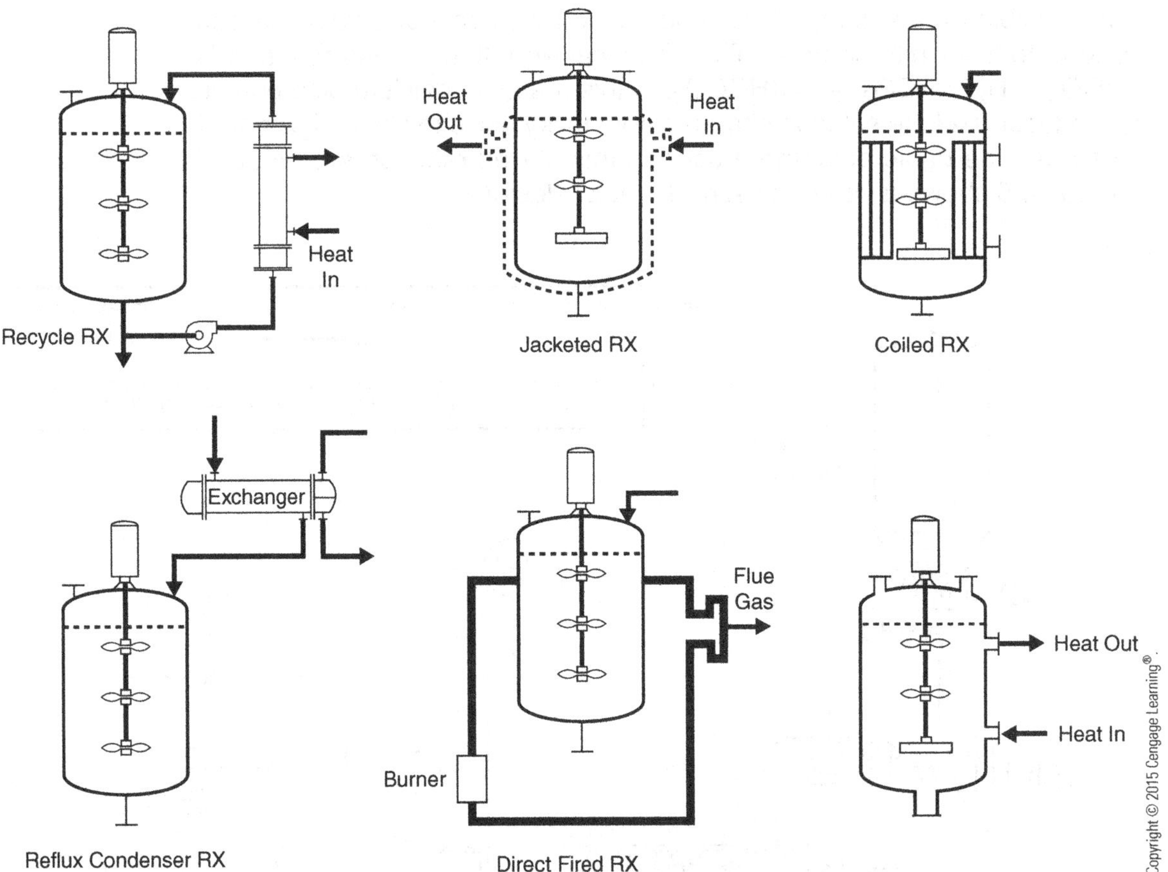

Figure 14.5 *Batch Stirred Tank Reactor*

Alkylation Stirred Reactors

Alkylation units (Figure 14.6) take two small molecules, isobutane and olefin (propylene, butylenes, or pentylenes), and combine them into one large molecule of high-octane liquid called *alkylate*. This combining process takes place inside a reactor filled with an acid catalyst. Alkylate is a superior antiknock product that is used in blending unleaded gasoline.

After the reaction, a number of products are formed that require further processing to separate and clean the desired chemical streams. A separator and an alkaline substance are used to remove (strip) the acid. The stripped acid is sent back to the reactor, and the remaining reactor products are sent to a distillation tower. Alkylate, isobutane, and propane gas are separated in the tower. Isobutane is returned to the alkylation reactor for further processing. Alkylate is sent to the gasoline blending unit.

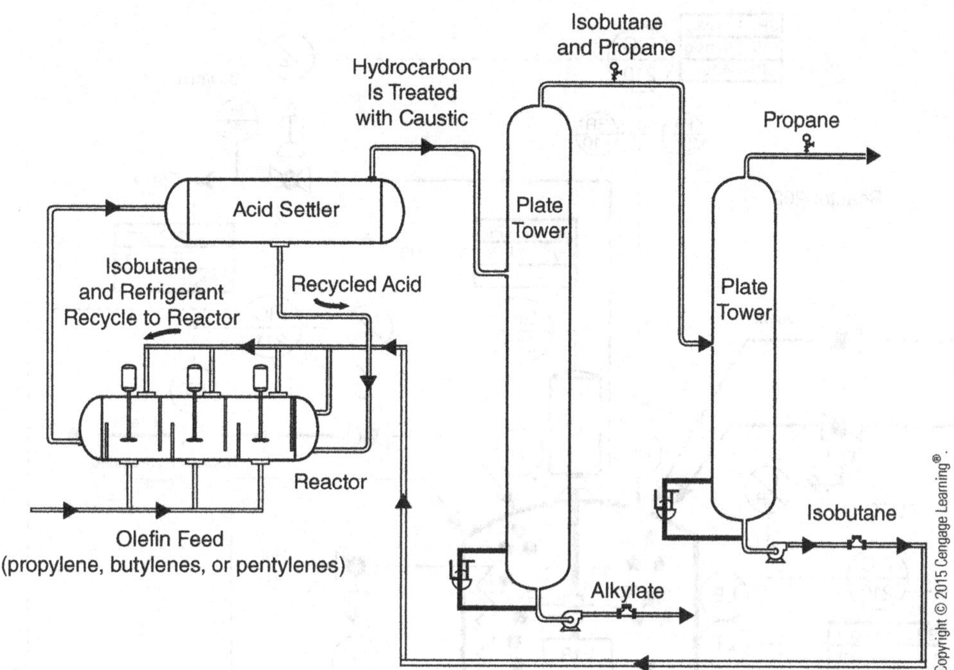

Figure 14.6 *Alkylation Stirred Reactor*

Sodium Phosphate Reaction

A common reaction used as the foundation for the production of sodium phosphate is 98 parts phosphoric acid H_3PO_4 and 120 parts sodium hydroxide 3NaOH produce 164 parts sodium phosphate Na_3PO_4 and 54 parts water H_2O. Figure 14.7 shows the basic components of this stirred reaction and the basic equipment involved.

Fixed Bed Reactors

A fixed bed reactor—often called a *converter*—is a vessel that contains a mass of small particles, usually 0.1 to 0.2 inch in diameter, through which the reaction mixture passes. The mass is usually a packed "bed" of catalyst, which promotes the reaction. The mass could also be column packing such as Raschig rings, Berl saddles, or marbles. In this case, the packing creates more heat transfer surface to give better heat distribution in the converter. The bed of packing, whether catalyst or inert particles, is positioned in the direct path of the process flow and can be arranged as a single large bed, as several horizontal beds supported on trays, or as several beds, each in its own shell. It can also be in several parallel packed tubes in a single shell.

As the process medium passes through the catalyst, the reaction occurs. Since the catalyst occupies a fixed position inside the reactor, it is

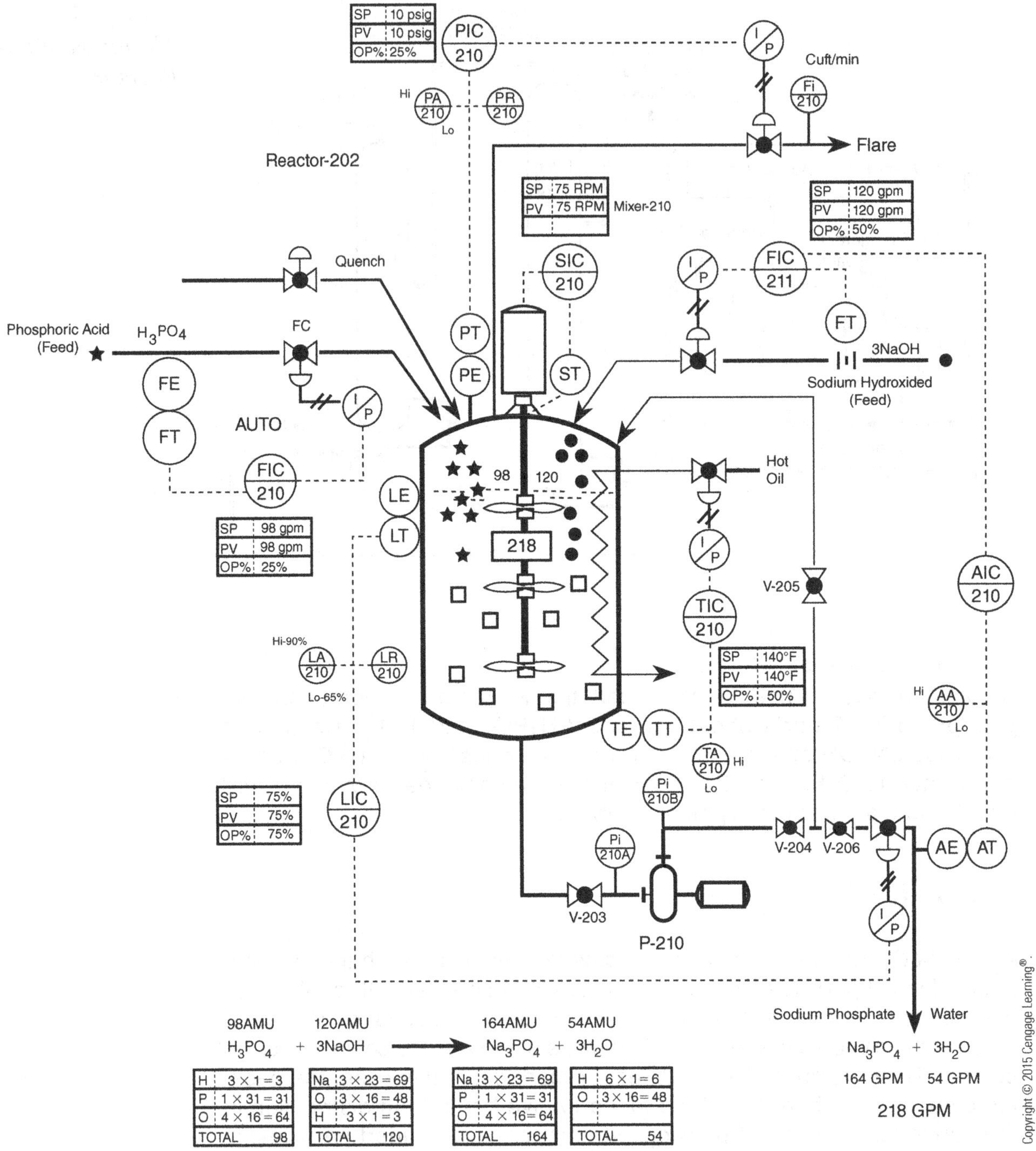

H	3 × 1 = 3
P	1 × 31 = 31
O	4 × 16 = 64
TOTAL	98

Na	3 × 23 = 69
O	3 × 16 = 48
H	3 × 1 = 3
TOTAL	120

Na	3 × 23 = 69
P	1 × 31 = 31
O	4 × 16 = 64
TOTAL	164

H	6 × 1 = 6
O	3 × 16 = 48
TOTAL	54

Figure 14.7 *Sodium Phosphate Reaction*

not designed to leave the reactor with the process fluid. Fixed bed reactors are designed with an inlet line and an outlet line, catalyst bed limiters, distribution and support grids, catalyst removal hatches, process variable instrumentation, and safety systems.

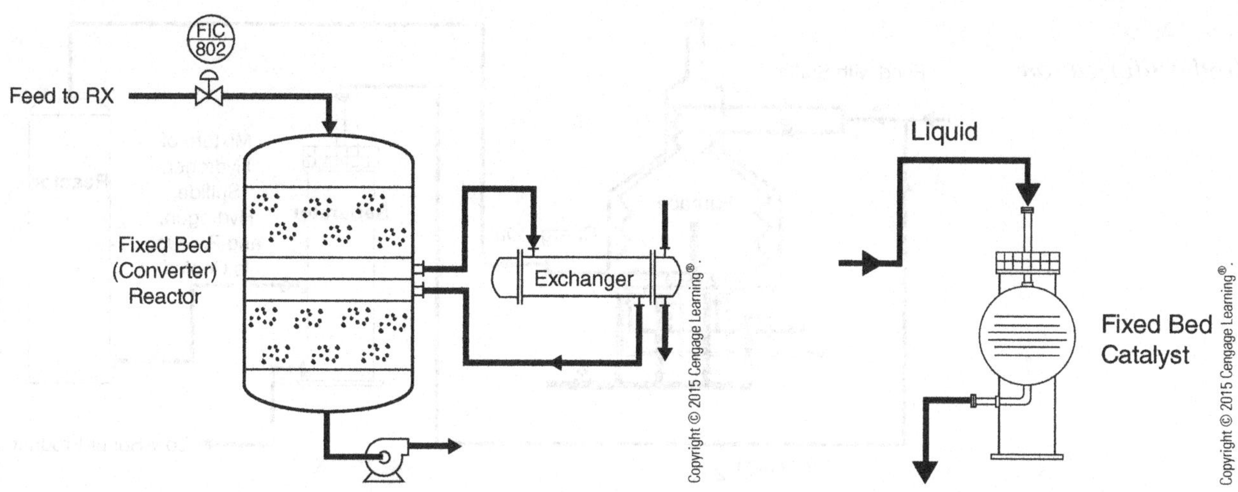

Figure 14.8 *Converter (Fixed Bed) Reactor*

Figure 14.9 *Fixed Bed Reactor*

Many converters have a method of removing or adding heat from the bed. Figures 14.8 and 14.9 illustrate possible converter arrangements. The main reasons for arranging the packed beds in a configuration other than the large single bed are to provide better heat transfer, to provide better distribution of the reaction mixture through the converter so that the gas or liquid feed will not "channel" through the bed, and to reduce the amount of force applied to the packing at the bottom of the bed.

Several design considerations must be taken into account in constructing a converter. The free area of the packing must be great enough to allow an acceptable pressure drop through the reactor at the design feed rates. The packing must be strong enough to resist collapsing at design conditions over a reasonable length of time. If the packing is a catalyst, its active life should be long enough to produce production runs of economic duration. The design of the beds should provide for removal of the heat of reaction so that there are no hot spots in the bed.

Hydrodesulfurization

Crude oil is a mixture of hydrocarbons, clay, water, and sulfur. Some crude mixtures have higher concentrations of sulfur than others. These high-sulfur mixtures are referred to as sour feed. **Hydrodesulfurization** is a process used by industrial manufacturers to "sweeten" the mixture by removing the sulfur.

During operation, hydrodesulfurization units use a fired heater, separator, and a fixed bed reactor (Figure 14.10). Sour feed is mixed with hydrogen and heated in a fired furnace. The heated mixture is sent to a reactor, where the hydrogen combines with the sulfur to form hydrogen sulfide. Lowering the temperature slightly causes the sweet crude to condense, leaving the hydrogen sulfide in vapor state. This vapor and liquid mixture is sent to a separator, where the low-sulfur sweet feed is removed. The hydrogen

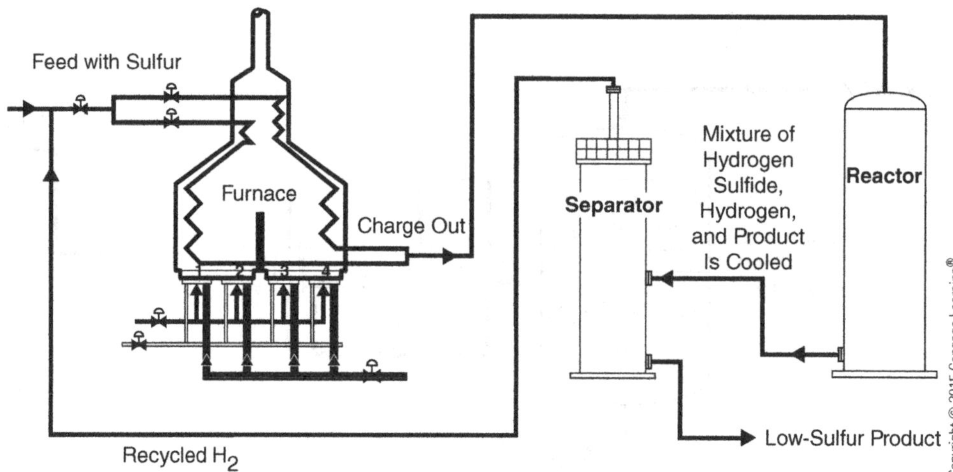

Figure 14.10
Hydrodesulfurization

sulfide and hydrogen are sent on for further processing, and the hydrogen is separated and returned to the original system.

Hydrocracking

Hydrocracking is a process that industrial manufacturers use to boost gasoline yields. The process takes heavy gas oil molecules and splits them into smaller, lighter molecules called *hydrocrackate*. During operation, hydrocracking units use a first- and second-stage fixed bed reactor, a separator drum, and a fractionating tower (Figure 14.11).

The hydrocracking process takes heavy gas oil feed and mixes it with hydrogen before sending it to the first-stage reactor. The reactor is filled with a fixed bed of catalyst. As process flow moves from the top of the reactor to the bottom, the **cracking** reaction takes place. First-stage hdyrocrackate

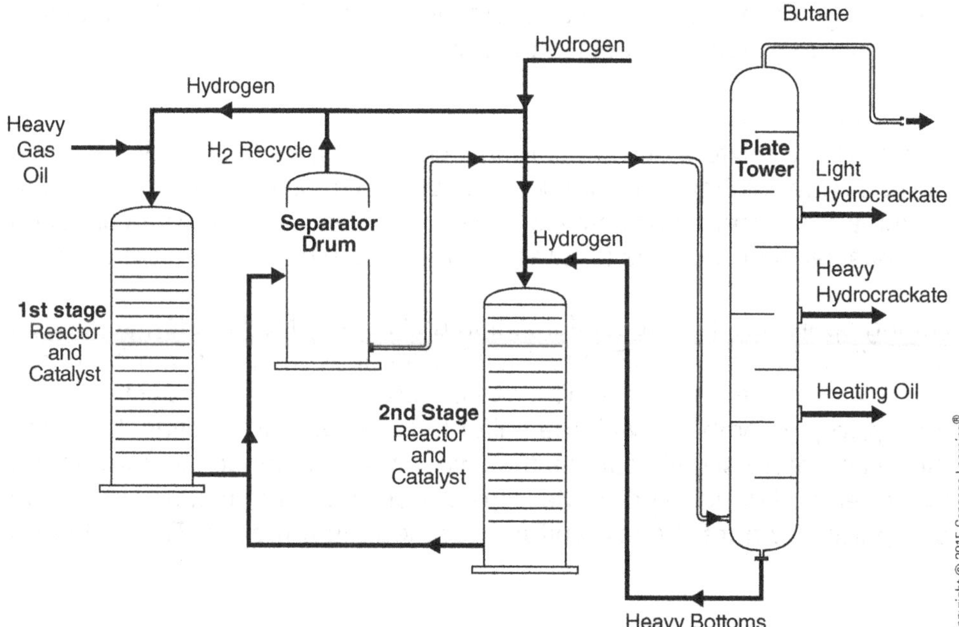

Figure 14.11
Hydrocracking

is sent to a separator drum, where the hydrogen is reclaimed and the hydrocackate is moved on to a fractionation tower.

In the fractionation tower, the hydrocrackate is separated into five cuts: butane, light hydrocrackate, heavy hydrocrackate, heating oil, and heavy bottoms. The heavy bottoms is mixed with hydrogen and sent to the second-stage reactor for further processing. The second-stage reactor reclaims as much of the hydrocrackate as possible before sending it to the separator and tower.

Reformer (Fixed Bed) Reactor

Reformers are designed to change low-octane naphtha into high-octane reformate. The gasoline your car runs off is composed primarily of naphtha. In this process, naphtha is heated up in a large commercial furnace to a specific temperature that brings the liquid to its boiling range. This heated vapor enters the top of a large fixed bed reactor filled with catalyst. When the catalyst and vaporized naphtha come into contact, the larger, low-octane molecules give up some of their hydrogen atoms and change into smaller, high-octane naphtha molecules. Figure 14.12 illustrates how the hot, high-octane naphtha is sent to a heat exchanger to preheat the incoming naphtha before it enters the furnace. A separator is used separate the hydrogen and high-octane naphtha.

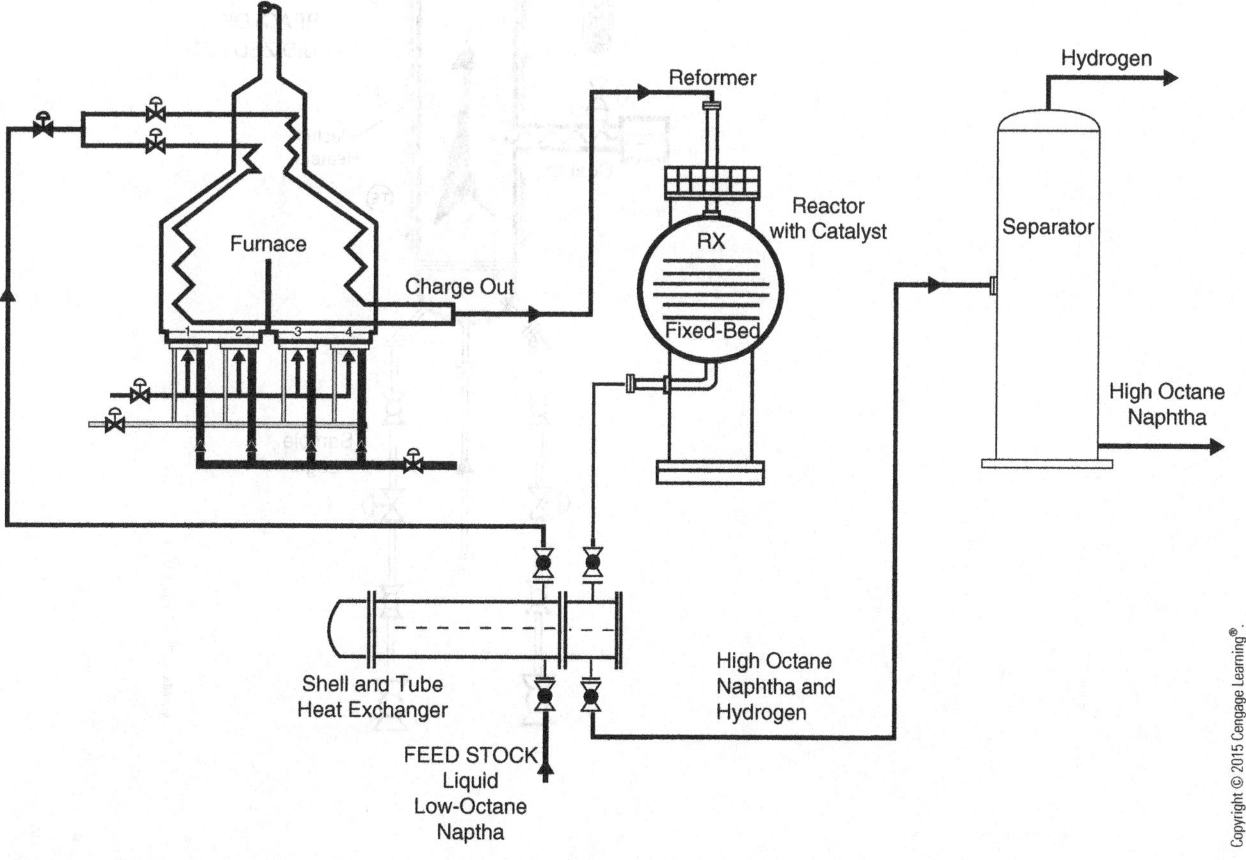

Figure 14.12 *Reformer*

Fluidized Bed Reactors

A fluidized bed reactor (Figure 14.13) suspends solids by the countercurrent flow of gas from the bottom of the reactor. Over time, particles segregate as heavier components fall to the bottom and lighter ones move to the top.

During operation on a coal gasification unit, flows and temperatures are established before the bed is built. The reactor closely resembles a vertical distillation tower. Hydrogen gas inlet flows are mounted on the bottom of the reactor and must be controlled to keep from blowing the bed material

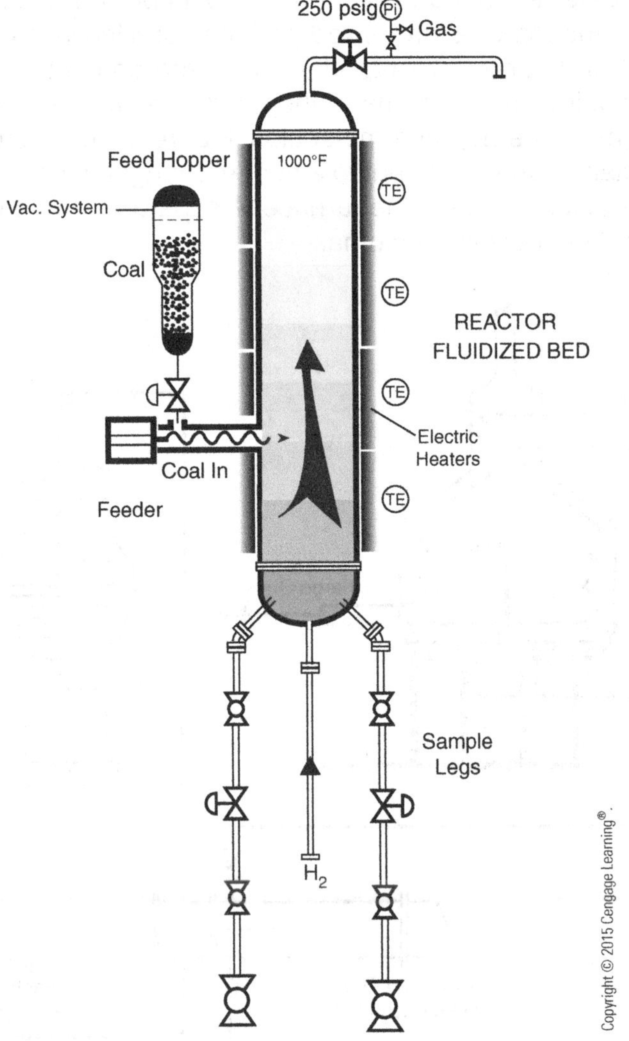

Figure 14.13
Fluidized Bed Reactor

over the top. A solids feeder is mounted a third of the way up the reactor. As the auger turns, coal is fed into the reactor. Methane gas is released during the run.

Fluid Catalytic Cracking

Crude oil comes into a refinery and is processed in an atmospheric pipe still. The sidestream of the pipe still is rich with light gas oil. **Fluid catalytic cracking** (catcracking) units (Figure 14.14) take this gas oil and split it into smaller, more useful molecules.

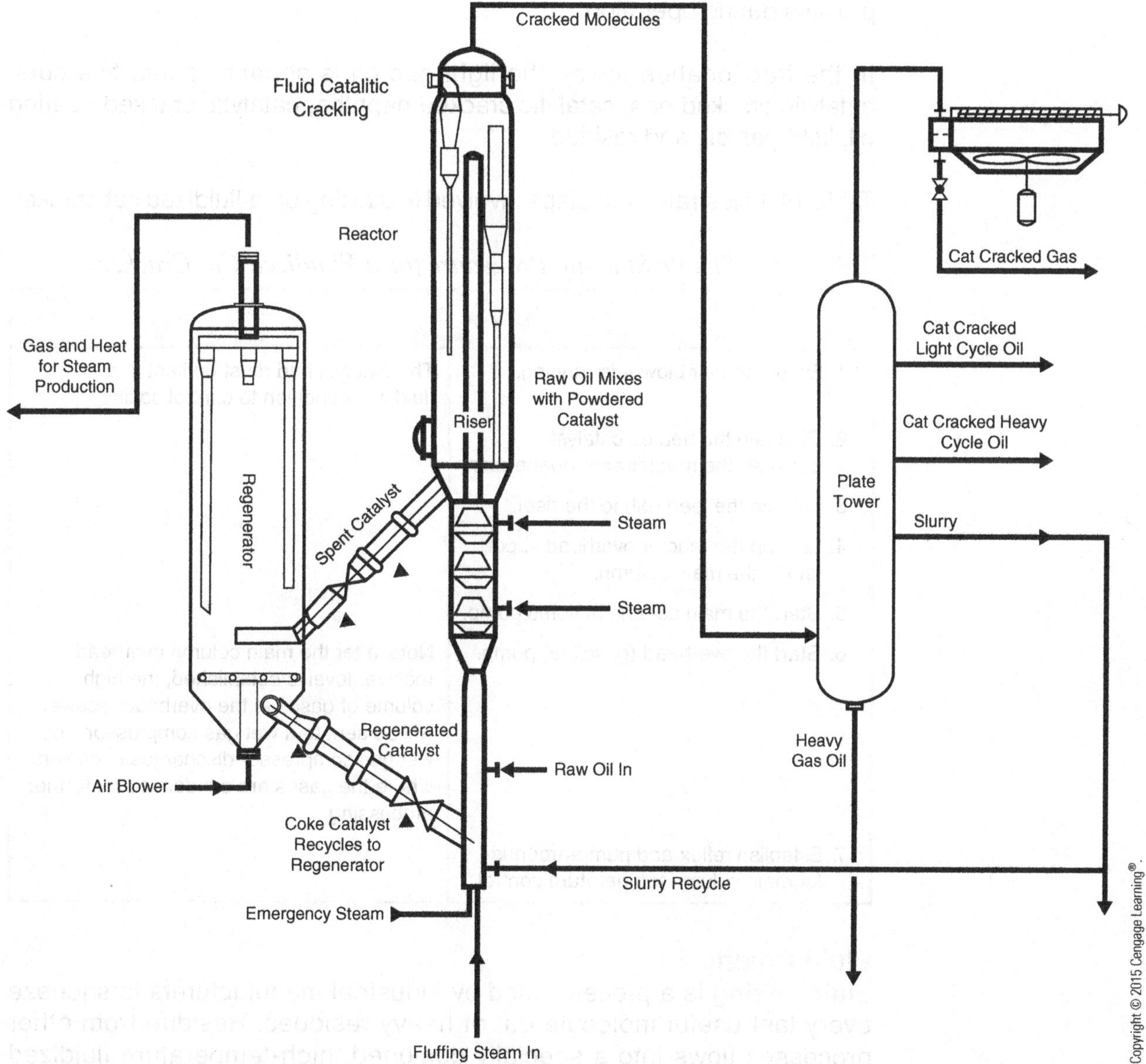

Figure 14.14 *Fluid Catalytic Cracking*

During operation, fluid catalytic cracking units use a catalyst **regenerator**, a fluidized bed reactor, and a fractionation tower. Gas oil enters the reactor and is mixed with a superheated powdered catalyst. The term *cracking* is used to describe this process because, during vaporization, the molecules split and are sent to a fractionation tower for further processing.

The chemical reaction between the catalyst and light gas oil produces a solid carbon deposit. This deposit forms on the powdered catalyst and deactivates it. The spent catalyst is drawn off and sent to the regenerator, where the coke is burned off. Catalyst regeneration is a continuous process during operation.

In the fractionation tower, the light gas oil is separated into five cuts: catalytic cracked gas, catalytic cracked naphtha, catalytic cracked heating oil, light gas oil, and residue.

Table 14.1 illustrates the steps involved in starting up a fluidized cat cracker.

Table 14.1 *Simple Start-up Procedure for a Fluidized Cat Cracker*

Action	Notes
1. Ensure the air blower is running.	The catalyst bed must be kept in a fluidized condition to control coking.
2. Circulate the heated catalyst between the reactor and regenerator.	
3. Line up the feed (oil) to the riser.	
4. Line up the reactor overhead vapor line to the main column.	
5. Start the main column bottoms pump.	
6. Start the overhead (gasoline) pump.	Note: after the main column overhead receiver level is established, the high volume of gases in the overhead receiver will be sent to a wet gas compressor. The wet gas compressor discharges to coolers where the gases are condensed for further processing.
7. Establish reflux and pump-arounds for main column temperature control.	

Fluid Coking

Fluid coking is a process used by industrial manufacturers to squeeze every last useful molecule out of heavy residues. Residue from other processes flows into a specially designed, high-temperature fluidized bed reactor (Figure 14.15). Light products vaporize and flow to a fractionation column. The remaining material is sent to a burner, where further processing takes place. The burner produces three products: coker

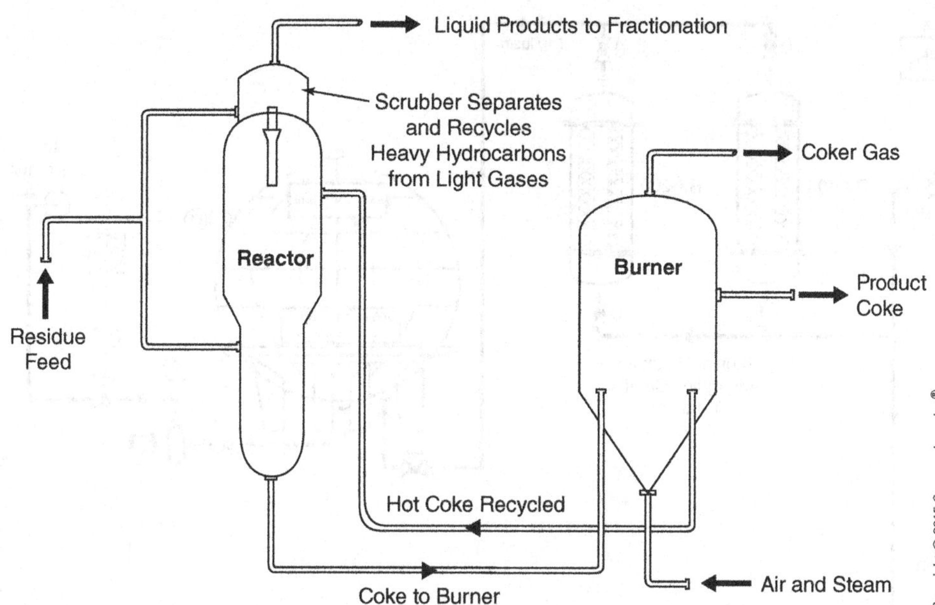

Figure 14.15
Fluid Coking

gas for use in the plant, product coke for sale, and recycled coke for the reactor.

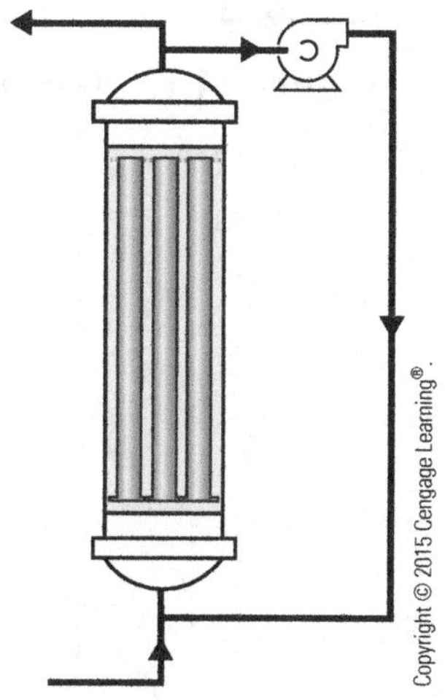

Tubular Reactors

The design of a tubular reactor (Figure 14.16) can vary from the simple jacketed tube to multipass shell and tube reactors. This type of reactor is simply a heat exchanger in which a reaction takes place. Tubular reactors are used for liquid and gaseous reactions. The raw material feed usually must be fairly clean or it may plug the tubes. Agitation in these reactors, if any is needed, is provided by the turbulence of the flow of the feedstock through the tubes. Because the tubular reactor contains only a small volume of reactants at a given time, it is used only for continuous feed processes. The considerations in designing a tubular reactor are the heat transfer surface required to maintain the reaction temperature, the tube volume needed to produce the desired amount of reaction and production rate, and the tube size needed to give an acceptable pressure drop through the reactor. Tubular reactors are needed for chemical synthesis.

Copyright © 2015 Cengage Learning®.

Figure 14.16
Tubular Reactors

Ammonia Process

The production of ammonia is a popular process used by industry on a large commercial scale. The simple equation is 28 parts nitrogen N_2 and 6 parts hydrogen $3H_2$ produce 34 parts ammonia $2NH_3$. Figure 14.17 illustrates how nitrogen and hydrogen are passed through a tubular reactor filled with catalyst.

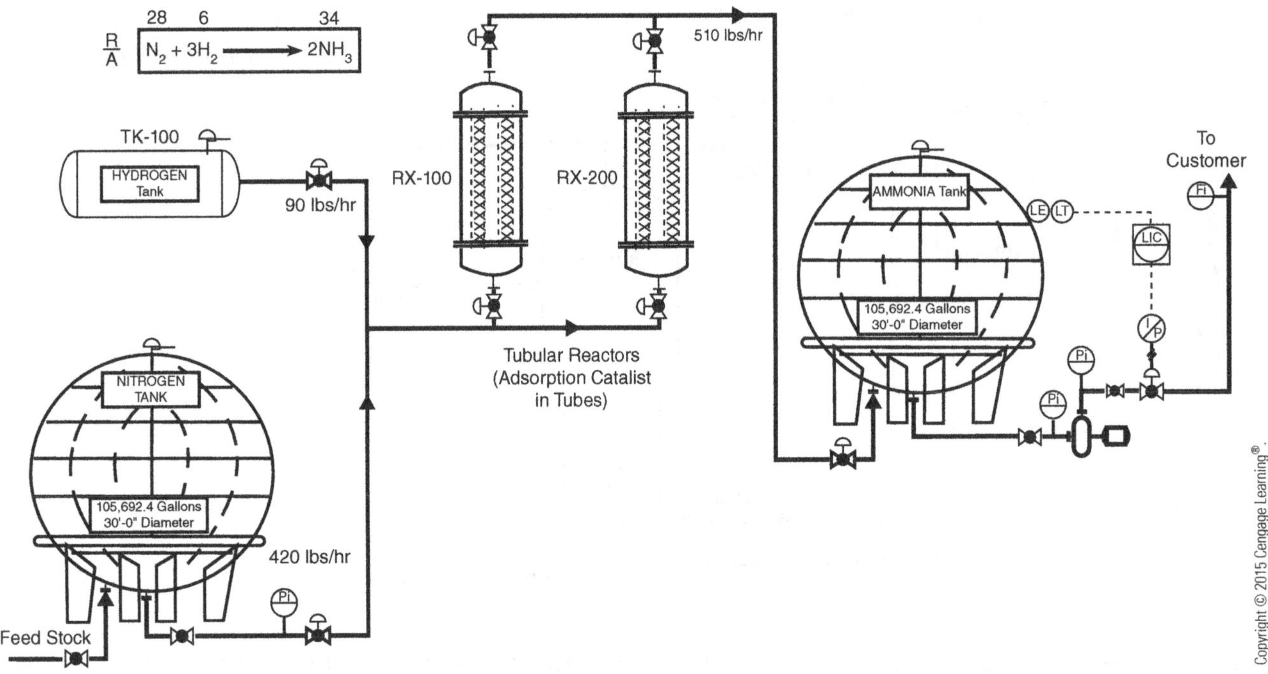

Figure 14.17 *Ammonia Process*

Chemical Synthesis

The purpose of the tubular reactor in chemical synthesis is to combine cyclopentadiene (CPD) and butadiene (BD) to form (synthesize) vinylnorbornene (VNB) (Figure 14.18). VNB is the primary component added to a sodium catalyst and alumina to make ethylnorbornene (ENB), a special additive used to strengthen common rubber products. The equipment used during the operation is feed tanks, three packed towers, a continuous mixing reactor, a tubular reactor, and product tanks.

During operation, a mixture of dicylopentadiene (DCPD) and Aromatic 200 is blended in a feed tank. The mixture is pumped through a preheater and into a fractionation tower. Toluene is added to the tower as a dilutant. A reflux stream also enters the top of the tower and returns CPD, BD, and toluene.

As the feedstock enters the packed tower, it cracks from DCPD to CPD. The heavy, aromatic oil stays in the bottom of the tower and acts as a temperature stabilizer. The overhead product is condensed and sent to a chilled mixing reactor. The product on the bottom—called bottoms—is sent to *slop*, which is best described as a product that is no longer needed for this process, however, it is often recycled and used in another process.

In the continuous mixing reactor, chilled water coils maintain a low temperature. Butadiene enters the bottom of the reactor and mixes with the

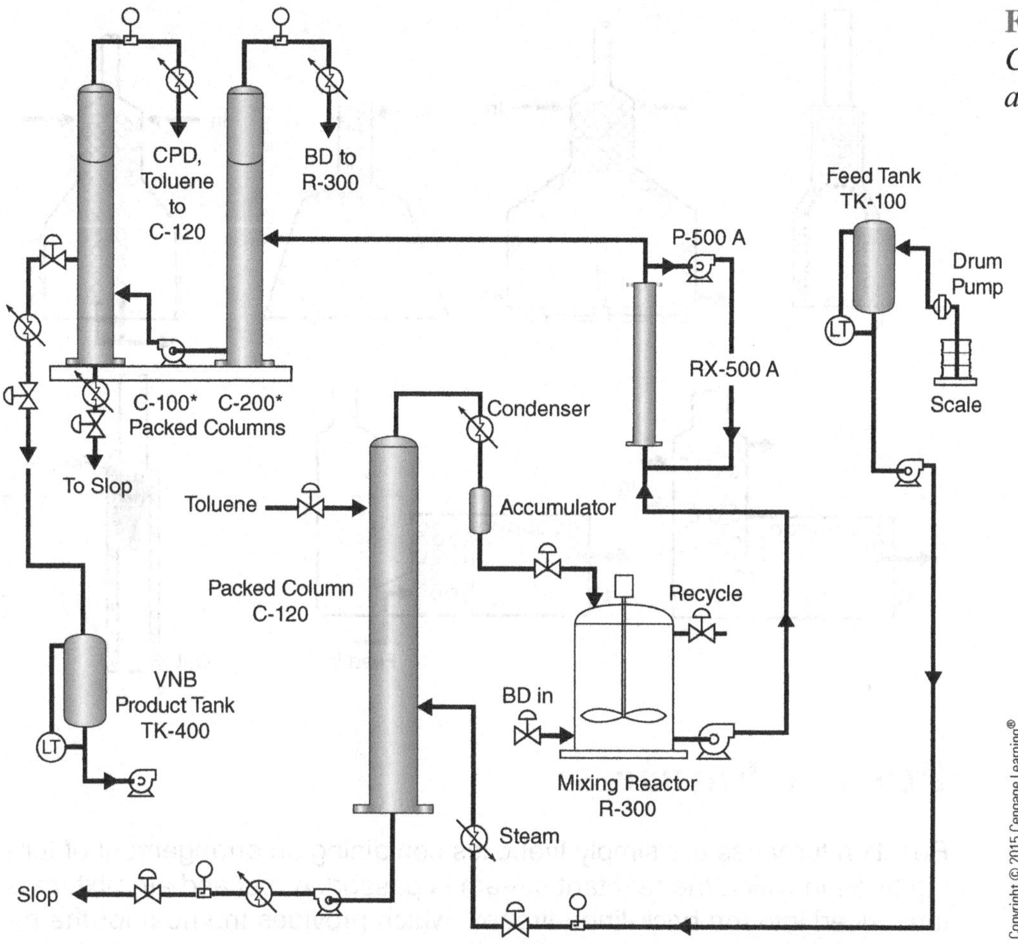

Figure 14.18
Chemical Synthesis and Tubular Reactors

overhead distillate. The mixed feed is sent to a feed preheater before it enters the tubular reactor. The reactor resembles a shell and tube heat exchanger with three large tubes. The outer jacket is filled with hot oil that maintains a specific set point. A large centrifugal pump circulates the feed through the reactor as the synthesis reaction occurs. The reactor is mounted vertically, and flow moves clockwise through the tubes. Special inhibitors are added to the feed to limit the formation of insoluble polymer during operation.

After the reaction occurs, a small percentage of VNB is formed. This mixed stream continuously flows through a reduced overhead line to a second tower. The purpose of the second tower is to separate and recycle the BD. This process is accomplished easily because of butadiene's lower boiling point.

The prime bottoms product is sent to a third (vacuum) distillation column for product separation. In the third fractionation tower, the mixture is separated into three cuts: BD, CPD, and toluene (overhead); VNB (sidestream), and bottoms.

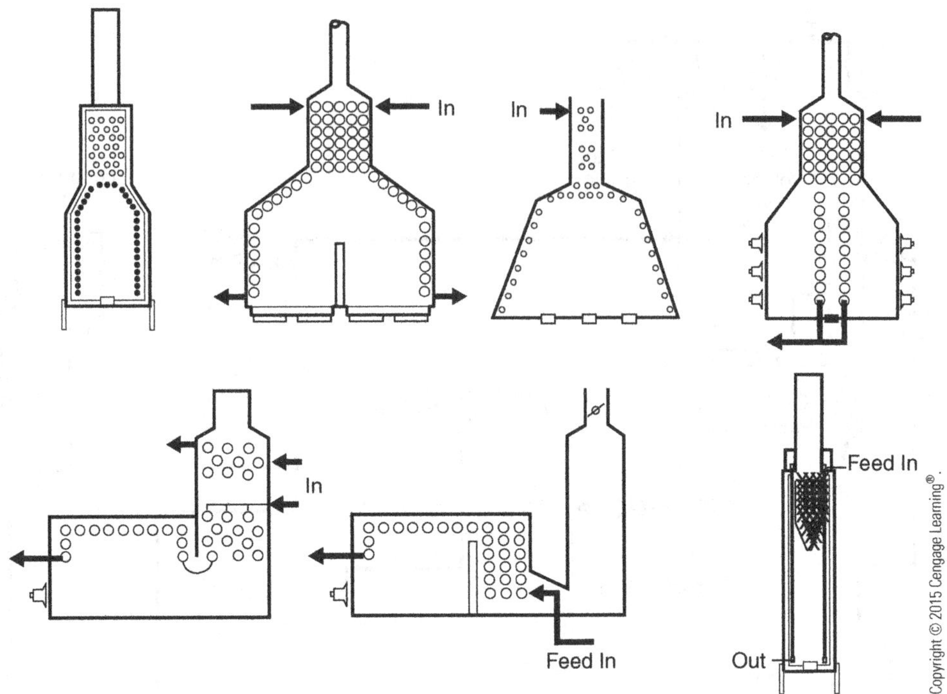

Figure 14.19
Reaction Furnaces

Reaction Furnaces

Reaction furnaces are simply fireboxes containing an arrangement of tubing through which the reactant stream is passed. A fuel and air mixture is introduced into the brick-lined firebox, which provides the heat for the reaction that occurs in the tubes. Most furnaces are used to crack hydrocarbons; however, there are reformer furnaces in which steam and methane are reacted together to get synthesis gas (carbon monoxide and hydrogen). For the catalytic cracking of hydrocarbons, catalytic pellets are packed in the tubes inside the furnace. Some typical flow diagrams of furnaces are shown in Figure 14.19. Notice that the raw material feed to the furnace is preheated by introducing the feed into the furnace through tubes in the vented stack. The hot vent gases heat the feed before it reaches the tubes, which are in direct contact with the flames of the burning fuel gas. Reaction furnaces are used in the olefins and vinyl chloride unit.

General Reactor Design Considerations

Besides choosing a type of reactor and the considerations associated with each type, there are certain design considerations that are common to all reactors. These factors include corrosion, safety devices, heating and cooling media, and instrumentation. At the operating conditions of some reactors, the reactants or products may be corrosive to the common materials of construction. Tests should be made on the reaction system if corrosion data of the system are not known. Corrosion can contaminate the product, inhibit the reaction, or poison the catalyst.

Usually, any reactor or reaction system can have a set of conditions that could cause a hazardous situation if allowed to exist. Safety devices such as high-pressure relief valves and alarms, high-temperature alarms and emergency cooling, fire control systems, and toxic chemical detection systems are available for most hazardous conditions and should be installed as needed.

The choice of the method for heating or cooling the reactor must be made in the design phase. The basis for such a decision is concerned with (1) the availability of steam, brine, cooling water, and the like; (2) the temperature at which the reaction must be controlled; and (3) the cost of each method per pound of production.

Most operating personnel, if allowed to, would have a great array of instrumentation at their disposal to control and monitor a reaction system. On the other hand, many design engineers would like to place a minimum of elaborate instrumentation into a unit to decrease the plant investment. Somewhere in between, the unit obtains a sufficient number of instruments to operate the reactor in an efficient and safe manner. Instruments must be installed to detect and control reaction conditions such as temperature, pressure, feed composition, and flow rates. Other instruments are required to detect hot spots in catalyst beds, excessive pressure drops, and impurities in the feed or product streams. The type and quantity of instrumentation are usually determined by past data and experience and from research and development.

Reactor Systems

A common type of reactor is the mixing or stirred reactor. The basic components of this device will include a mixer or agitator mounted on a tank. The reactor shell is designed to be heated or cooled and to withstand operational pressures, temperatures, and flow rates. The design may include tubing coils wrapped around the vessel, internal or external heat transfer plates, and jacketed vessels. Hot oil, steam, or cooling water may be the heating or cooling medium. The stirred reactor is designed to mix two or more components into a homogeneous mixture. As these components blend together, chemical reactions occur that create a new product. Blending time and exact operating conditions are critical to the efficient operation of a stirred reactor. A reactor is a vessel in which a controlled chemical reaction takes place, depending on the reaction variables. The operation of Reactor-202 (see Figure 14.20), for example, enhances molecular contact between four reactants: pentane, butane, liquid catalyst, and solvent. Feed to the reactor is controlled at 36.5 GPM. The feed composition from

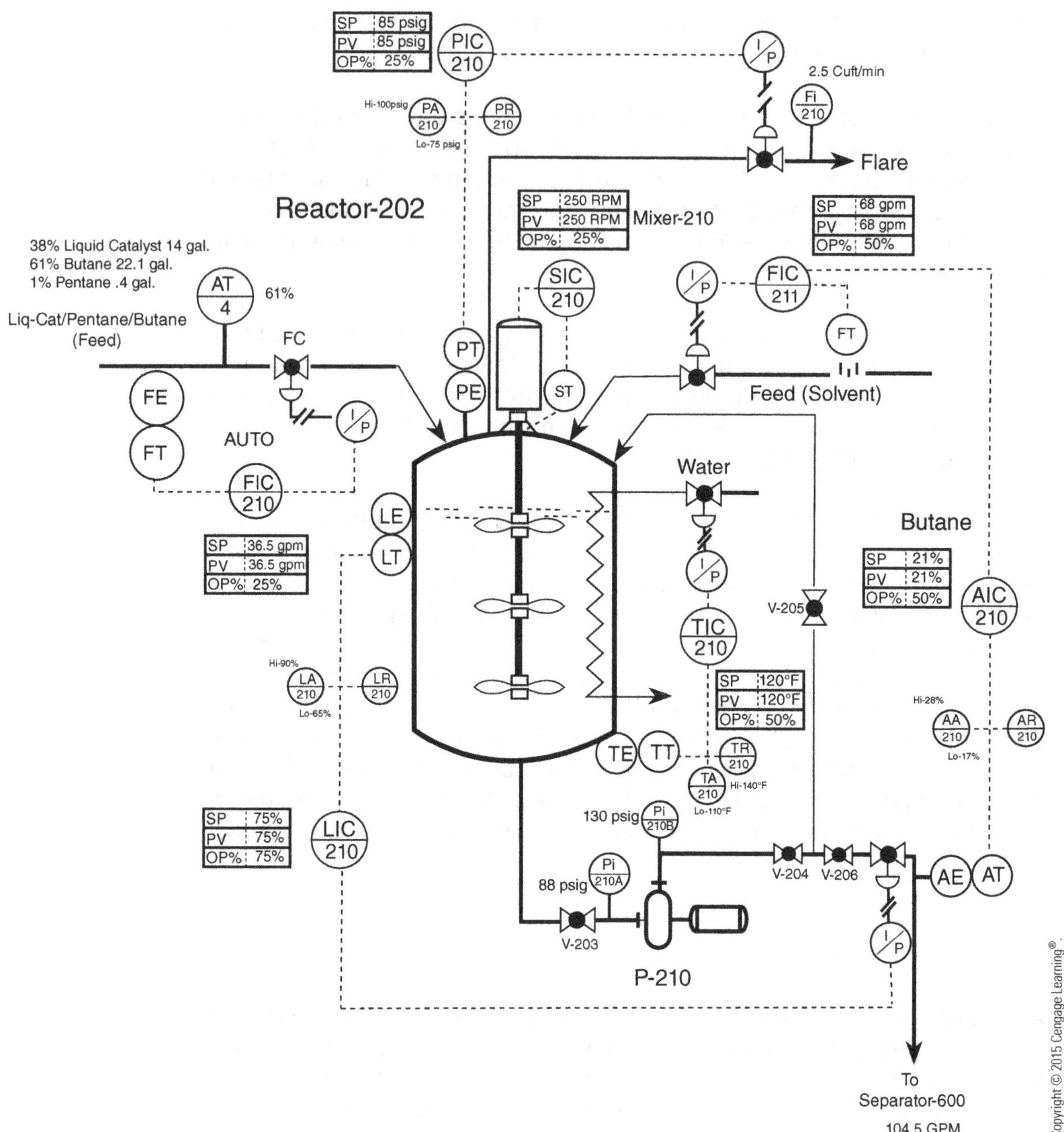

Figure 14.20 *Stirred Reactor System*

the column is 38% liquid catalyst, 61% butane, and 1% pentane. Solvent feed to the reactor is controlled at 68 GPM. The materials in the reactor are chilled to 120°F (48.88°C) at 85 psig. The reactants are designed to form a new product with an excess of pure butane. A separator is used to remove the new product and isolate the butane for storage. The effect of pressure on a reaction cannot be generalized as easily as the effect of temperature. In a liquid-phase reaction, pressure can increase or decrease the reaction rate, depending upon the readiness of the reactants or products to vaporize. In a gaseous reaction, pressure forces molecules closer together, causing them to collide more frequently. Therefore, the higher the pressure in a gaseous reaction, the higher the reaction rate. The concentration of reactants in the reactor has a major effect on how fast the reaction will take place, what products will be produced, and how much heat will have to be added to or taken away from the reaction. Figure 14.20 illustrates the basic components of the stirred reactor system.

Summary

A reactor is a device used to convert raw materials into useful products through chemical reactions. Heat, reaction time, surface area, concentration, pressure, flow rates, and catalysts and inhibitors affect reaction rates. Process technicians are responsible for establishing correct flow or feed rates to the reactor; ensuring correct temperature, pressures, and level; monitoring and controlling reaction rates (time); ensuring that specified mixing or agitation is occurring; and monitoring and maintaining auxiliary equipment.

Reactors combine raw materials with catalyst, gases, pressure, or heat. A catalyst increases or decreases the rate of a chemical reaction without becoming part of the final product. There are five basic types of reactors: stirred tank reactors; converters, or fixed bed reactors; fluidized bed reactors; tubular reactors; and furnaces.

Alkylation units take two small molecules and combine them into one large molecule. Exothermic reactions give off heat. Endothermic reactions absorb heat. Replacement reactions can be used to remove undesired products and replace them with desired ones. Neutralization reactions remove hydrogen ions (acid) or hydroxyl ions (base) from a liquid. Combustion reactions are exothermic reactions that require fuel, oxygen, and an ignition source and give off heat and light. In this type of reaction, oxygen reacts with another material so rapidly that fire is created.

Review Questions

1. Describe the basic design of a converter, or fixed bed reactor.
2. What is the primary purpose of a reactor?
3. What is a catalyst?
4. Describe batch and continuous reactor operations.
5. What is the purpose of an alkylation unit?
6. Describe catcracking.
7. What is hydrodesulfurization?
8. What are the basic components of a tubular reactor?
9. List the five basic types of reactors.
10. What is an exothermic reaction? How do you control it?
11. Describe the different types of chemical reactions.
12. How do temperature, pressure, reaction time, heat, and agitation affect a chemical reaction?
13. What factors affect reaction rates?
14. What is an inhibitor used for in the reaction process?
15. What process instrumentation is included with a reactor system?
16. What is a stirred reactor often referred to as?
17. List the safety features found on most reactors.
18. What effects can corrosion have on a reaction?
19. Balance the following equation:

    $$Fe + Cl_2 \rightarrow FeCl_3$$

20. Draw a simple stirred reactor and control the following variables: temperature, pressure, flow, level, and analytical.

Distillation Systems

Objectives

After studying this chapter, the student will be able to:

- Review the history of the distillation process.
- Describe the principles of distillation.
- Describe the relationship between the boiling point of a hydrocarbon and pressure, temperature, flow, composition, and level.
- Describe the various concepts associated with pressure in a distillation system: vapor pressure, partial pressure, relative volatility, compressibility, liquid pressure, and vacuum.
- Identify the different equipment systems used to make up a distillation system.
- Explain how the methods of heat transfer apply to the distillation process.
- Explain the basic components of a plate column.
- Explain the basic components of a packed column.
- Contrast preheaters and condensers.
- Describe the basic principles associated with column design.

Key Terms

Azeotropic mixture—a mixture of two or more components that boil at similar temperatures and at a certain concentration. The liquid and vapor concentrations of an azeotrope are equal.

Binary mixture—contains two or more components.

Boiling point—the temperature at which a liquid turns to vapor.

Boiling range—the range of temperatures from when the lightest component boils (initial boiling point) to when the heaviest component boils (final boiling point).

Boil-up rate—the balance of the products (vapor and liquid) returned to the column from the kettle reboiler.

Bottoms product—residue; the heavier components of the distillation process fall to the bottom of the tower and are removed.

Bubble-cap trays—a fixed cylindrical tube (riser) and cap mounted vertically to a horizontal tray designed to enhance vapor-liquid contact in a plate column.

Bubble point—the temperature at which a liquid mixture begins to boil and produce vapor.

Bubble point curve—is the lower curve on a Vapor Liquid Equilibrium Curve and identifies the specific temperature where the mixture first produces vapor.

Column loading—the amount of material fed continuously into a distillation column.

Cryogenic distillation—often used to separate oxygen and nitrogen from our atmosphere by using low-temperature drying and distillation methods.

Dew point—the temperature at which a mixture produces the first drop of liquid on a vapor-liquid-equilibrium curve.

Dew point curve—is the upper curve on a Vapor Liquid Equilibrium Curve and is defined as the temperature where the first drop of liquid forms.

Distillate—the condensate taken from a distillation column.

Distillation—the separation of components in a mixture by their boiling points.

Downcomers—downspouts that allow liquid to drop down to lower trays in a column.

Downcomer flooding—occurs when the liquid flow rate in the tower is so great that liquid backs up in the downcomer and overflows to the upper tray. Liquid accumulates in the tower, differential pressure increases, and product separation is reduced.

Entrainment—the upward movement of high-velocity, large-vapor droplets in a distillation column.

Feed distributor—a device used in a packed distillation column to evenly distribute the liquid feed.

Feed tray—point of entry of process fluid in a distillation column, under the feed line.

Final boiling point—the temperature at which the heaviest component boils.

Foaming—often the result of too much heat, impurities in the feed, or active turbulence on the trays above undersized downcomers.

Fractional distillation—separation of two or more components through distillation.

Heat balance—principle that heat in equals heat out.

Initial boiling point—is often described as the temperature where the mixture first starts to boil or evolve vapors.

Jet flooding—occurs when the vapor velocity is so high that liquid down flow in the tower is restricted. Liquid accumulates in the tower, differential pressure increases, and product separation is reduced.

Local flooding—excessive liquid flowing down a column blocks vapor flow up the column in one section.

Material balance—principle that the sum of the products leaving equals the feed entering the distillation column.

McCabe-Thiele—the simplest method used in the analysis of binary distillation. The composition of the mixture on each theoretical tray can be determined by the mole fraction of one of the components.

Overall flooding—local flooding expands to entire column.

Overhead product—the lighter components in a distillation column, which rise through the column and go out the overhead line, where they are condensed.

Overlap—incomplete separation of a mixture.

Overloading—operating a column at maximum conditions.

Packed tower—a tower that is filled with specialized packing material instead of trays.

Partial pressures—the amount of pressure per volume exerted by the various fractions in a mixture of gases.

Puking—occurs when the vapor is so great that it forces liquid up the column or out the overhead line.

Rectification—the separation of different substances from a solution by use of a fractionating tower.

Rectifying section—the upper section of a distillation column (above the feed line), where the higher concentration of lighter molecules is located.

Reflux—condensed distillation column product that is pumped back to increase product purity and control temperature.

Reflux ratio—the amount of reflux returned to the tower divided by the amount of overhead product sent to storage. For example, 120 GPM reflux ÷ 60 GPM overhead to product tank = 2:1 reflux ratio.

Relative volatility—the characteristics associated with a liquid's tendency to change state or vaporize inside a distillation system.

Sieve trays—a flat, thin plate with small holes in it mounted on a horizontal tray designed to enhance vapor-liquid contact in a plate column.

Stripping section—the section of a distillation column below the feed line, where heavier components are located.

Temperature gradient—the progressively rising temperatures from the bottom of a distillation column to the top.

Ternary mixture—three components in a mixture.

Total reflux—a process where the feed to the column is stopped and the total amount leaving the column from the top, sides, or bottom is returned to the column.

Tray columns—devices located on a tray in a column that allow vapors to come into contact with condensed liquids; three basic designs are bubble-cap, sieve, and valve.

Tray drying—occurs due to low feed rates, low reflux rates, or high upward vapor flow rates from tray to tray.

Vacuum distillation—the process of vaporizing liquids at temperatures lower than their boiling point by reducing pressure.

Valve trays—a small round metal disc with three or four legs called risers that allow the valve to lift as hot vapors push through the horizontal tray. Designed to enhance vapor-liquid contact in a plate column.

Vapor-liquid-equilibrium diagram—used to help a technician understand phase behavior of a two-component mixture.

Vapor pressure—the outward force exerted by the molecules suspended in vapor state above a liquid at a given temperature; when the rate of liquefaction is equal to the rate of vaporization (equilibrium).

Weeping—occurs when the vapor velocity is too low to prevent liquid from flowing through the holes in the tray instead of across the tray. Differential pressure is reduced and product separation is reduced.

Overview of Distillation Systems

Petroleum compounds are composed of hydrocarbon molecules of varying sizes and shapes. Molecular weight determines how a chemical reacts during separation. For example, ethane has two carbon atoms and butane has four. During separation, butane remains in the lower section and ethane moves up the tower. The smallest or lightest components in a tower have the lowest **boiling points**. A distillation column is built on the principle that light and heavy molecules have different boiling points.

Industry relies heavily upon this process to produce many of the chemicals we use today. For example, crude oil is a mixture of many of the chemicals used in modern manufacturing, including straight-run gasoline, naphtha, gas oil, various gases, salt, water, and clay. By knowing the temperature at which a chemical vaporizes, an operator can identify these specific components after they are heated, vaporized, and condensed on the different trays in a distillation column.

Distillation involves boiling liquids and condensing the vapors. When water boils, it turns into water vapor. The condensed water vapor is purer than the original mixture because most of the salts, minerals, and impurities do not vaporize at 212°F (100°C), the boiling point of water.

Distillation is a process that separates a substance from a mixture by its boiling point. This simple definition, however, does not explain the complex equipment arrangements that make up a distillation system (see Figure 15.1). The distillation process includes a feed system, a preheat system, the distillation column, the overhead system, and the bottom system.

Figure 15.1
Distillation Systems

The term system is frequently used to describe the various equipment arrangements associated with distillation.

The feed system is designed to safely store, blend, and transport the raw feedstock. This simple system is composed of tanks, pipes, valves, instrumentation, and pumps. Preheating systems are designed to raise the temperature of the raw feedstock before it enters the distillation column. Various heat transfer devices are available for the process. Typically, a heat exchanger or a furnace system is used. Feed rates into and out of this system are carefully controlled.

The distillation column can be a packed or plate type column. Plate columns have a series of trays stacked on top of each other. Packed columns are filled with packing material to enhance vapor-liquid contact. The function of each of these column types is to allow lighter components to rise up the column while heavier components drop down or, in other words, to separate the various components in the mixture by their boiling points. A variety of arrangements is available on the distillation column to condense the hot vapors and collect and separate the **distillate**.

The bottom system is composed of a section in the column designed to allow liquids to boil and roll freely. A reboiler is connected to this section to maintain and add heat energy into the liquid. The term reboil indicates that the liquid was originally boiling. Hot vapors are returned from the reboiler to the bottom of the column, below the bottom tray. A level control loop is used to keep the liquid at a predetermined level. In the bottom system, liquid level, temperature, and composition are carefully controlled. The overhead system is used to cool off the hot overhead vapors. The condensed liquid is used as

reflux or is transported to the tank farm as product. Reflux is returned to the top of the column to control product purity and tower temperature. In packed columns, a liquid **feed distributor** is used to evenly disperse the liquid. In a plate column, the reflux is pumped to the upper tray. Plate columns are not as sensitive to liquid feed distribution as packed columns are.

During the distillation process, a mixture is heated until it vaporizes, then is recondensed on the trays or at various stages of the column where it is drawn off and collected in a variety of overhead, sidestream, and bottom receivers. The condensate is referred to as the distillate. The liquid that does not vaporize in a column is called the residue.

During tower operation, raw materials are pumped to a feed tank and mixed thoroughly. Mixing usually is accomplished with a pump-around loop or a mixer. This mixture is pumped to a feed preheater or furnace, where the temperature of the fluid mixture is brought up to operating conditions. Preheaters are usually shell and tube heat exchangers or fired furnaces. This fluid enters the **feed tray** or section in the tower. Part of the mixture vaporizes as it enters the column, while the rest begins to drop into the lower sections of the tower.

A distillation column is a tower with a series of trays or packing that provides contact points for the vapor and liquid. As vaporization occurs, the lighter components of the mixture move up the tower and are distributed on the various trays. The lightest component goes out the top of the tower in a vapor state and is passed over the cooling coils of a shell and tube condenser. As the hot vapor comes into contact with the coils, it condenses and is collected in the overhead accumulator. Part of this product is sent to storage while the rest is returned to the tower as reflux.

A reboiler maintains the energy balance on the distillation column. Reboilers take suction off the bottom of the tower. The liquid in the tower is circulated through the reboiler. Vaporization occurs in the reboiler and these vapors rise up through the tower. **Boil-up rate** is defined as the balance of the products (vapor and liquid) returned to the column from the kettle reboiler.

History of Distillation

With the advent of the automobile came a technological boom that affected many existing industries and created new ones. The steel and metal fabrication and manufacturing, glass, machining, rubber, petroleum, concrete and stone industries, and many others had to develop new technology to support the growing automobile industry. As the need for petroleum products grew, more efficient ways for producing oil and gas products were developed. The distillation process traces its roots through three distinct operations: batch, continuous batch, and modern **fractional distillation**.

Batch distillation (Figure 15.2) is the oldest distillation process. The raw materials were mixed together and "charged" to the still. During phase 1 of

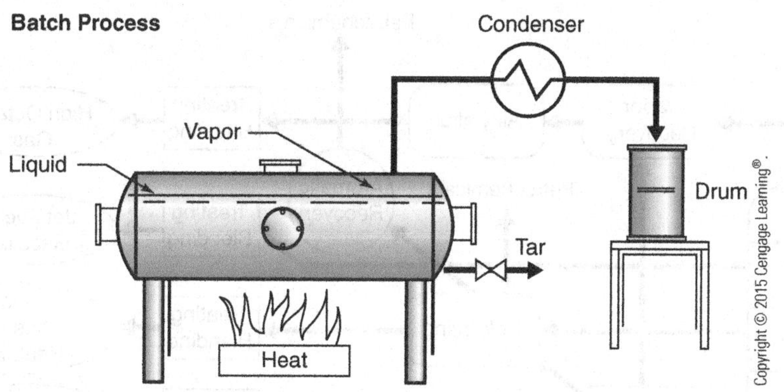

Figure 15.2 *Simple Batch Distillation*

the distillation process, heat was added to bring the mixture to a boil. The overhead vapors were condensed and stored as final product. At the conclusion of phase 1, the overhead line was switched to a new receiver, and the temperature was increased. When the receiver was full, the process was repeated, and the heat was increased incrementally until the batch run was complete. Product quality during batch distillation operations was very poor. Frequent receiver changes wasted time and money.

Continuous batch distillation (Figure 15.3) is a continuation of the batch distillation process. A series of stills are connected to each other to form a battery. The temperature of the first still is lower than the temperatures in the rest of the bank. Feed is fed to the first still at a constant rate. Temperature is increased from one still to the next, producing the same effect as batch operation. The stills are connected to each other through the bottom lines. Product quality from still to still is relatively consistent.

Modern fractional distillation (Figure 15.4) uses a series of trays or packing inside a tower. A continuous feed is heated and fed to the distillation column. Hot vapors rise in the column while liquids drop down the tower. A series of

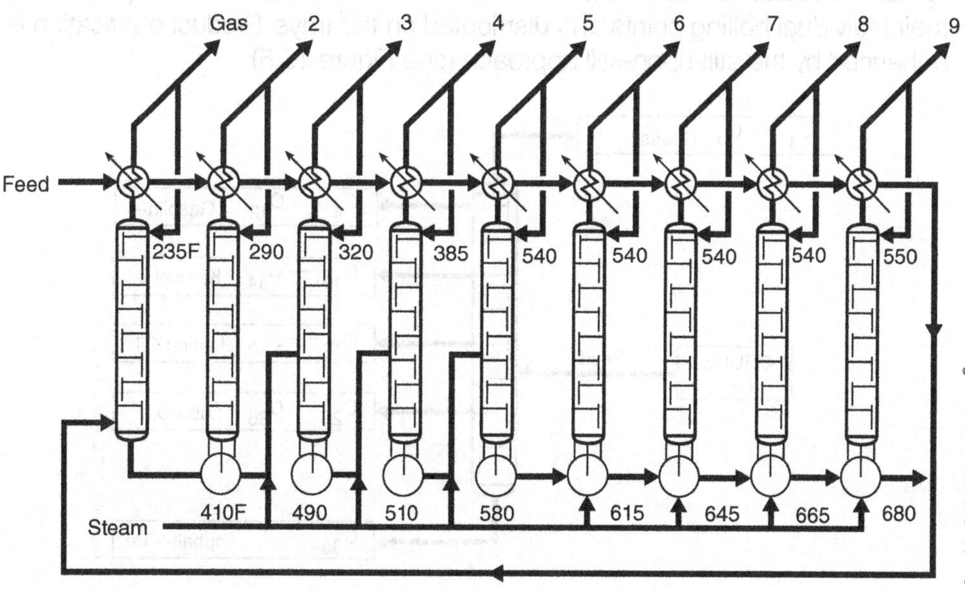

Figure 15.3 *Continuous Batch Distillation*

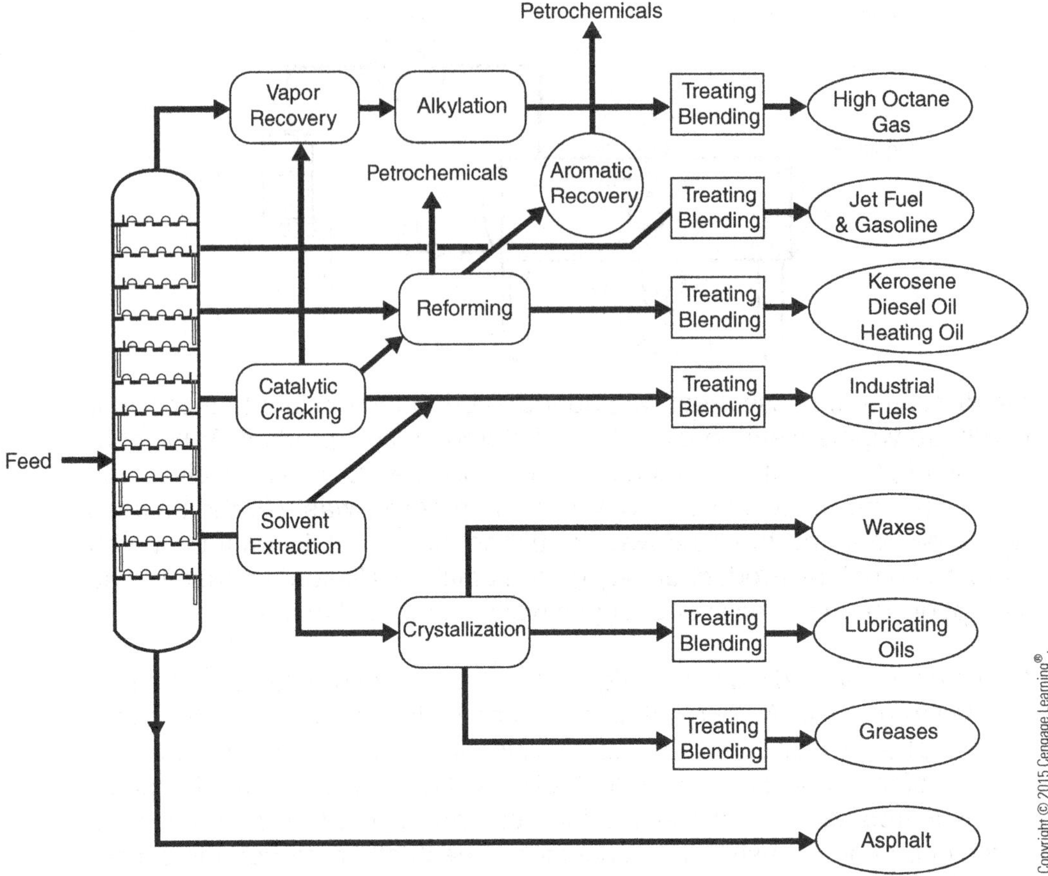

Figure 15.4 *Modern Fractional Distillation*

steps involving vaporization and condensation take place inside the column. Vapor-liquid contact allows the heavier molecules to drop out and the small, lighter molecules to rise up the column. The various fractions are separated by their individual boiling points and distributed on the trays. Product purification is enhanced by the still-upon-still approach (see Figure 15.5).

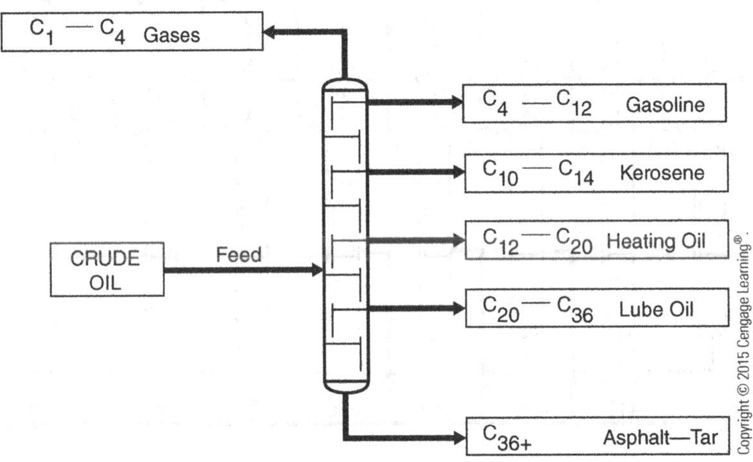

Figure 15.5 *Hydrocarbon Fractions*

Principles of Distillation

The separation of components in a distillation column is based on the differences in volatility or boiling point. As the vapor rises up the tower and contacts the liquid, the concentration of the lower boiling or more volatile components increase. As the liquid descends the tower, the concentration of the higher boiling or less volatile components increases. Increased pressure will result in increased temperatures in the tower. Each section in a trayed or packed column will have its own individual footprint or characteristics: pressure, temperature, composition, velocity, and performance. In the chemical processing industry, distillation and reactor technology are considered to be the two most complex operating areas.

The basic principles of distillation include the study of phase behavior of a binary or two-component mixture, bubble point, dew point, boiling range, initial and final boiling points (IBP and FBP), mole fraction, the McCabe-Thiele method, and the concepts of entrainment, foaming, and weeping. Each of these areas is important for a technician to understand in order to work in and operate a distillation system. Figure 15.6 is often

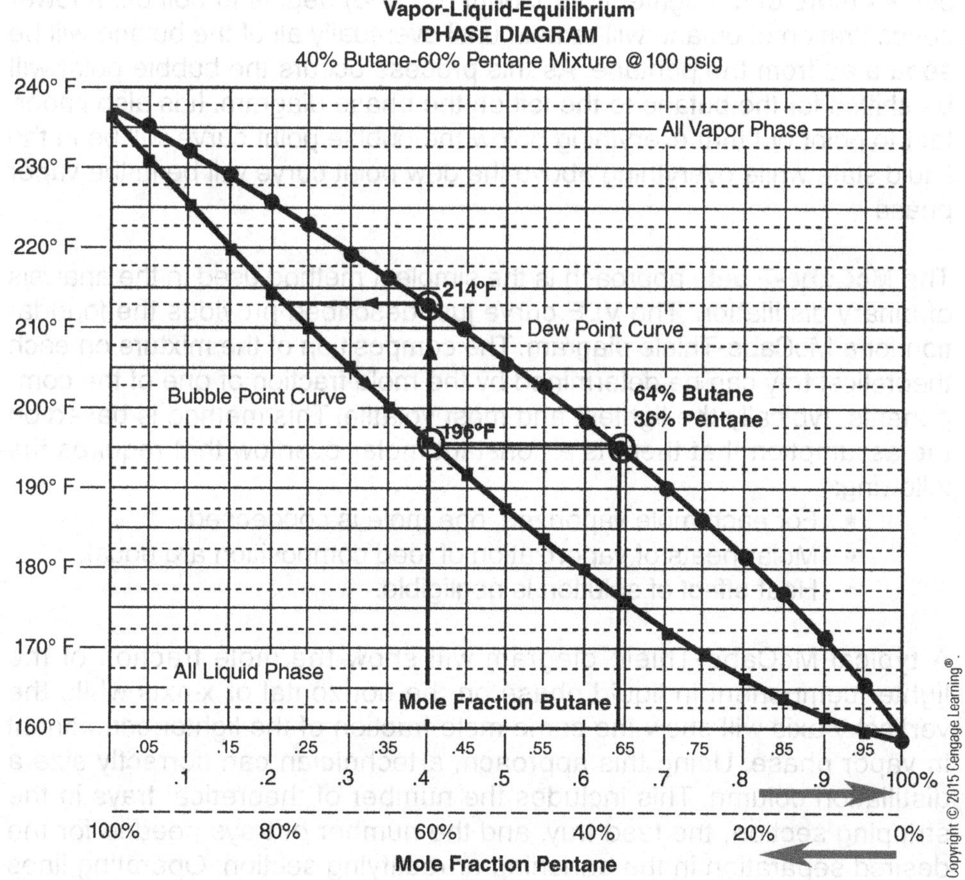

Figure 15.6
Vapor-Liquid-Equilibrium Diagram

referred to as **vapor-liquid-equilibrium (VLE) diagram** or phase diagram. It provides a simple way for a technician to visualize the complex phase behaviors associated with distillation, boiling, and condensing of a two-or-more-component mixture. In this example 40% butane (C_4), and 60% pentane (C_5) will be used. The operating pressure will be controlled at 100 psig. The diagram can be used to show that the mole fraction of butane is 0.4 or 40%.

The starting mole fraction of pentane is 0.6 or 60%. This vapor-liquid-equilibrium diagram is set up with the temperature on the left vertical axis and the mole fraction going across the bottom. The phase diagram also shows the **bubble point curve** and the **dew point curve**. It is important to remember that the **bubble point** is defined as the specific temperature where the mixture first produces vapor. The **dew point** is defined as the temperature where the first drop of liquid forms. Looking at the curve, we see that the bubble point for our mixture is 196°F at 100 psig. If we had a mixture that included a butane mole fraction of 0.6. or 60% and 40% for the pentane, the bubble point would be 180°F. We can also see that a 0.4 butane mole fraction would have a dew point of 214°F.

As this mixture of butane and pentane heats up, vaporization begins to occur. As more of the lighter component (butane) begins to boil off, a lower concentration of butane will remain until eventually all of the butane will be separated from the pentane. As this process occurs the bubble point will be shifted for the butane to the left on the phase diagram. It is also important to point out that everything below the bubble point curve will be in the liquid state while everything above the dew point curve will be in the vapor phase.

The **McCabe-Thiele** approach is the simplest method used in the analysis of binary distillation. The VLE curve just described provides the foundation for a McCabe-Thiele diagram. The composition of the mixture on each theoretical tray can be determined by the mole fraction of one of the components, typically the lightest and most volatile. This method is based on the assumption that there is a constant molar overflow that requires the following:

- For each mole vaporized, one mole is condensed.
- Molar heats of vaporization of feed composition are equal.
- Heat effect of solution is negligible.

A typical McCabe-Thiele diagram will show the mole fraction of the lighter component in liquid phase on the horizontal or x-axis while the vertical y-axis will show the same mole fraction of the lighter component in vapor phase. Using this approach, a technician can correctly size a distillation column. This includes the number of theoretical trays in the stripping section, the feed tray, and the number of trays needed for the desired separation in the enriching or rectifying section. Operating lines can be used to define mass balance relationships between the vapor and liquid phases of the distillation tower. Typically there is an operating line for the rectification or desired overhead composition, bottom stripping section or desired composition, and feed composition. Figure 15.7 is a simplified example of a McCabe-Thiele model. To construct the operating line for the enriching or rectifying section, the engineering target for the overhead purity is located on the VLE diagram and a vertical line that intercepts the plot point is drawn. The slope R/(R+1) is sketched from the intersection point.

- (R) is equal the ratio of reflux flow.
- (L) is distillate flow.
- (D) is the reflux ratio.

The calculation for identifying the operating line for the lower stripping section utilizes the mole fraction of the desired bottoms composition. A vertical line is drawn through the target and a line of slope L_s/V_s is sketched.

- (L_s) is the liquid rate down the lower stripping section.
- (V_s) is the vapor rate up the stripping section.

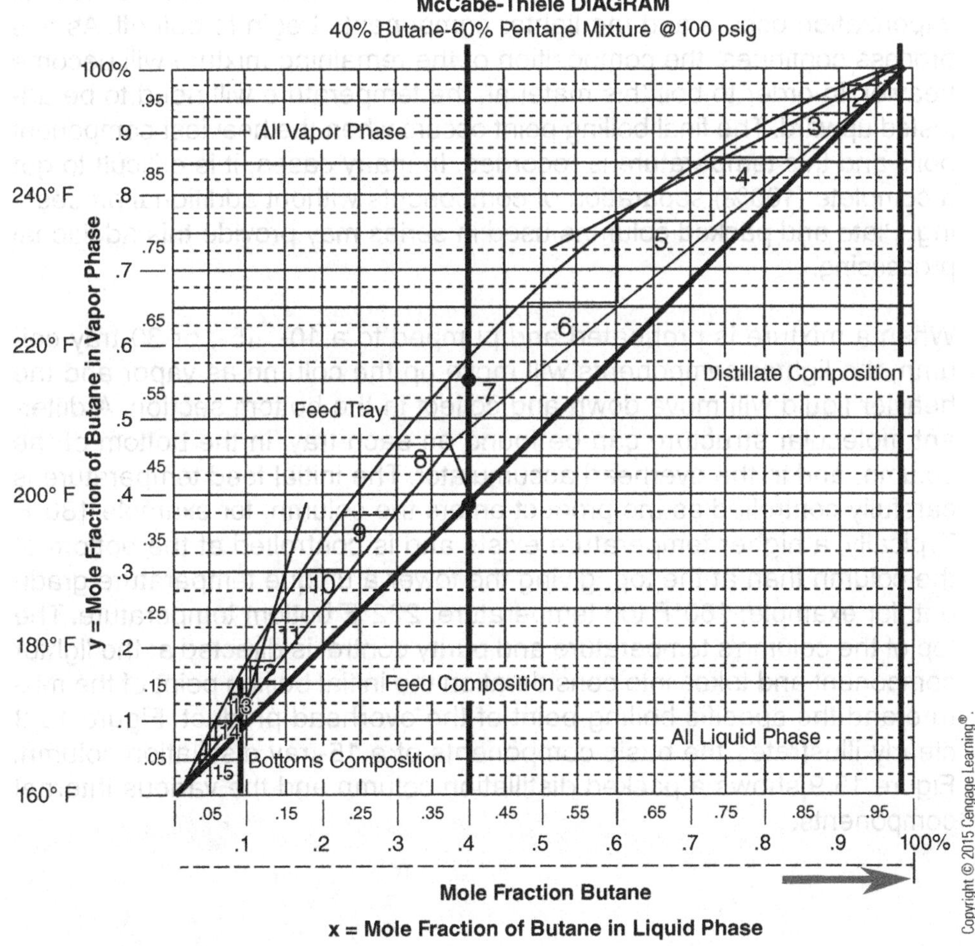

Figure 15.7 *McCabe-Thiele Method*

McCabe-Thiele assumes that the vapor above and the liquid resting on the tray are in a state of equilibrium. Another important feature of McCabe-Thiele is the feed line where the binary mixture should be introduced. The condition of the feed is calculated by the slope of the feed line or q-line. In determining overall column design, the McCabe-Thiele approach can help in identifying the required number of stages, degree of separation, and location of the feed tray. Other important factors include tray spacing, internal configuration, diameter, preheat, reboil, and cooling.

Another important concept that can be applied to this section is the contrast between the **initial boiling point** (IBP) and the **final boiling point** (FBP). Typically, in a multi-component mixture, there is no single boiling point that can be used to vaporize the entire solution. In this example, a **boiling range** would be needed that would take into account each of the components. This would also include the pressure on the system.

The initial boiling point is often described as the temperature where the mixture first starts to boil or evolve vapors. As the mixture continues to boil, vaporization occurs and the lighter components begin to boil off. As this process continues, the composition of the remaining mixture will become heavier. In order to boil this material, the temperature will need to be adjusted upward. The final boiling point occurs when the heaviest component boils and this temperature is recorded. In many cases, it is difficult to get a complete (100%) separation of components without additional processing. Plate and packed columns used in series may provide this additional processing.

When a mixture is preheated and pumped to a 10-, 15-, or 30-tray column, the lighter components will move up the column as vapor and the heavier liquid will move down and collect in the bottom section. A different molecular structure can be found on each tray, in the bottom of the column, and in the overhead accumulator. The initial feed temperature is carefully controlled as the product enters the column, for example 180°F. Typically, a higher temperature exists and is controlled at the bottom of the column than at the top, giving the tower a unique temperature gradient, for example: 160°F top temperature, 222°F bottom temperature. The top of the column's temperature and purity control is directed at the lighter component and takes into consideration the initial boiling point of the mixture and the specific boiling point of the overhead product. Figure 15.8 clearly illustrates the basic components of a 15-tray distillation column. Figure 15.9 shows a packed distillation column and the various internal components.

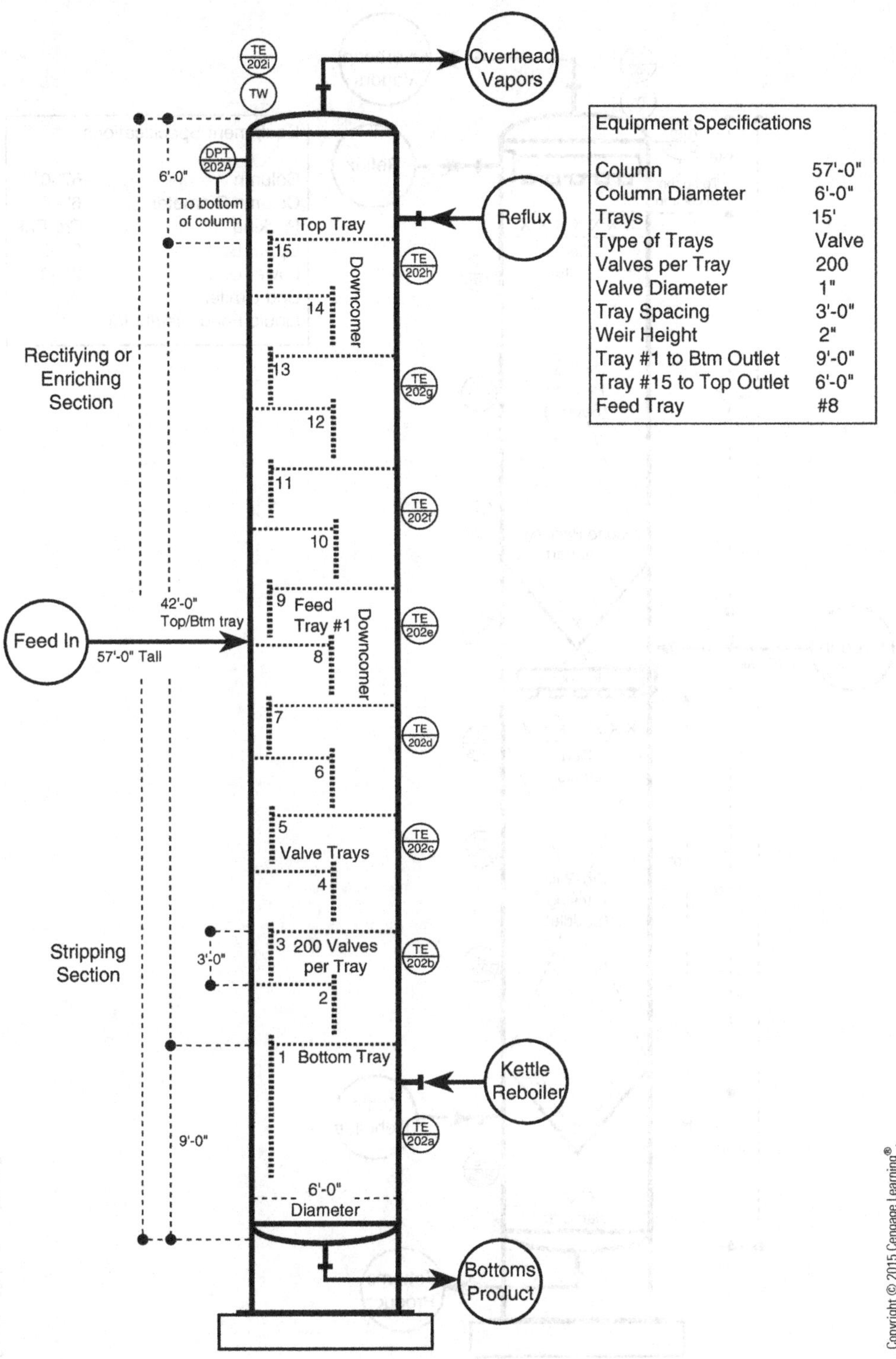

Equipment Specifications	
Column	57'-0"
Column Diameter	6'-0"
Trays	15'
Type of Trays	Valve
Valves per Tray	200
Valve Diameter	1"
Tray Spacing	3'-0"
Weir Height	2"
Tray #1 to Btm Outlet	9'-0"
Tray #15 to Top Outlet	6'-0"
Feed Tray	#8

Figure 15.8 *Components of a Plate Column*

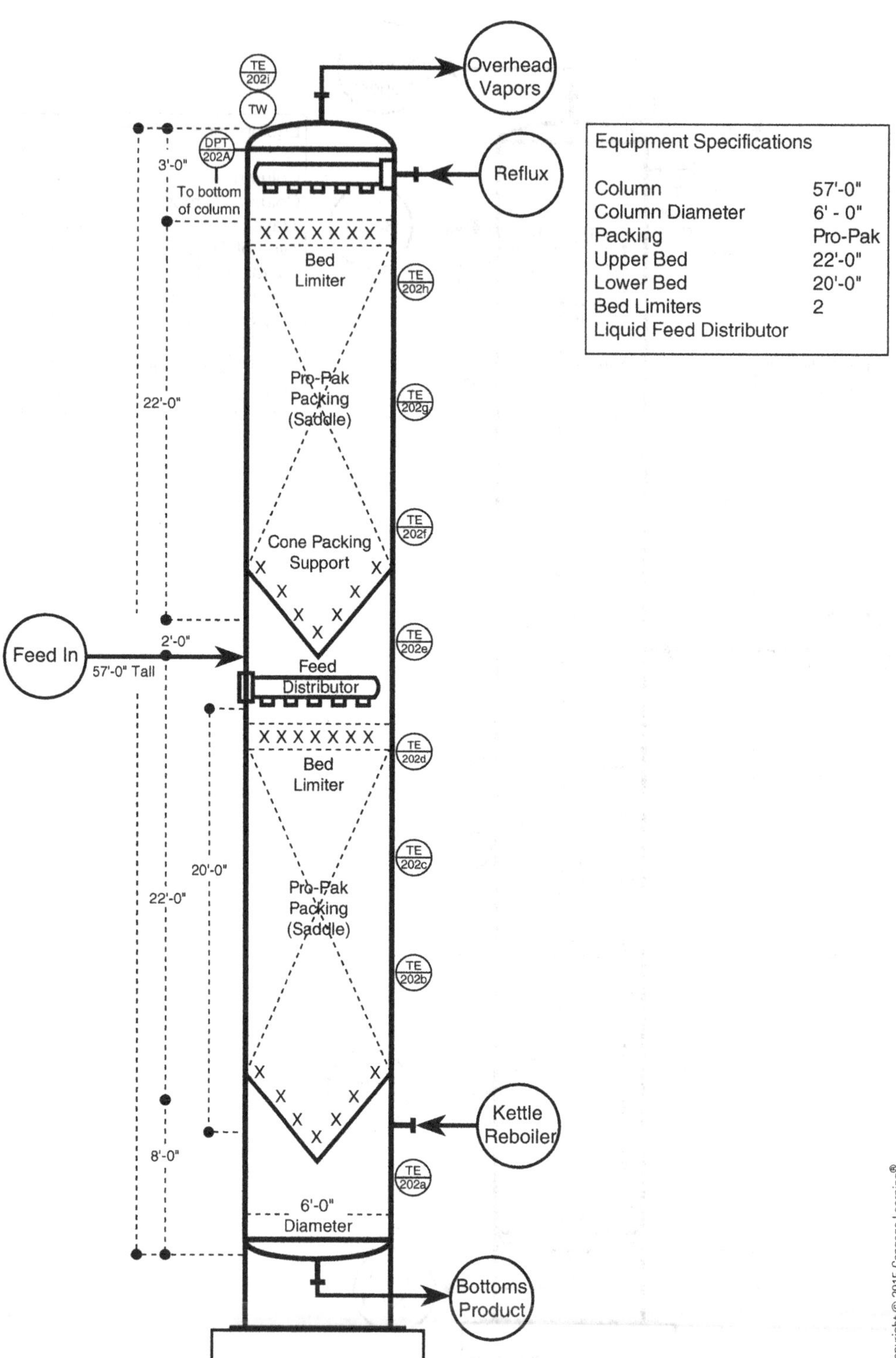

Figure 15.9 *Components of a Packed Column*

Two Distillation Examples

Ethanol and Water Distillation

A mixture of 50% water and 50% ethanol can be separated through distillation. Ethanol boils at 172°F (78°C), and water boils at 212°F (100°C). As the mixture is heated, ethanol vaporizes faster than water. If the overhead vapors were analyzed, ethanol would make up the highest concentration. Some water would be found in the mixture because of product **overlap** and because ethanol and water have similar boiling points. The overhead vapors could be distilled further if needed by sending the overhead to another distillation tower.

Methanol and Water Distillation

Under normal conditions, methanol boils at 148°F (64.4°C), and water boils at 212°F (100°C). A 50/50 mixture heated to 180°F (82°C) would start the vaporization process at atmospheric pressure. The vapor concentration would be 80% methanol and 20% water. This incomplete separation is called overlap. The temperature at which this concentration is produced is referred to as the effective cut point. If the mixture were heated to 212°F (100°C), both methanol and water would vaporize, and a 50/50 vapor mixture would be obtained. Another viewpoint states that the higher concentration of methanol in the vapor is based on Raoult's law, which states that the **partial pressure** of component is the product of the **vapor pressure** and the mole fraction of the component.

Kinetic Energy

Liquids are composed of molecules that are closer together than the molecules of gases. The speed at which these molecules move is proportional to the amount of heat they encounter. As heat is transferred to a liquid, the molecules begin to speed up and move farther apart and the liquid expands. This molecular motion is referred to as kinetic energy. Its temperature determines the kinetic energy of a liquid.

Sensible and Latent Heat

Distillation involves both vaporization and condensation. The heat that produces state changes consists of sensible and latent heat. Sensible heat can be sensed or measured. Latent heat of vaporization, the heat required to change a liquid to vapor, cannot be sensed or measured. For example, suppose you have a feed tank filled with heptane, which boils at 209°F (98°C) under atmospheric conditions. Before heat is added, the temperature is 72°F (22°C). As heptane is pumped to a column, it is heated by a shell and tube heat exchanger. As the temperature of heptane rises, the effect of sensible heat can be measured. As the temperature approaches 209°F, the effect of latent heat can be observed. At 209°F (98°C), the temperature stops rising and a great deal of vapor is produced.

Latent heat is thermal energy that has been added to heptane but does not cause the temperature to change.

Pressure

Pressure is defined as force or weight per unit area. The term pressure typically is applied to gases or liquids. Pressure is measured in pounds per square inch (psi). Atmospheric pressure is produced by the weight of the atmosphere as it presses down on an object resting on the surface of the earth. Pressure is directly proportional to height; the higher the atmosphere, gas, or liquid, the greater the pressure. At sea level, atmospheric pressure equals 14.7 psi.

Three types of pressure are important to the vaporization process: vapor pressure, control pressure, and vacuum. Vapor pressure is the force exerted as molecules escape the surface of the liquid. In a closed container, the effects of vapor pressure can be observed. As the temperature of a liquid rises, more molecules escape the surface of the liquid and the vapor pressure rises. The lower the boiling point, the faster the vaporization process occurs.

Pressure is measured and controlled at the top of the column. When pressure is added to a liquid, molecules are forced back into solution. More heat is required to vaporize the liquid. This concept is important to understand because having to add pressure in a distillation tower could force you to expend more energy. It can also limit tower efficiency and stop vaporization in severe cases.

Boiling Point and Vapor Pressure

The boiling point of a substance is the temperature at which (1) vapor pressure exceeds atmospheric pressure, (2) bubbles become visible in the liquid, and (3) vaporization begins. Molecular motion in vapors produces pressure and increases as temperature is added to the liquid.

The vapor pressure of a liquid is:

- A physical property of the liquid at a given temperature
- Independent of the volume of the liquid
- The same no matter what gas or pressure is above it
- The same even if a vacuum exists above the liquid
- More often associated with closed containers
- Measured when the rate of liquefaction is equal to the rate of vaporization (equilibrium)
- Used to determine the boiling point of a liquid when the temperature and vapor pressure equal atmospheric pressure

The vapor pressure of a substance can be directly linked to the strength of the molecular bonds of a substance. The stronger the bonds, or molecular attraction, the lower the vapor pressure. If a substance has a low vapor pressure, it has a high boiling point. For example, gold changes from a solid to a liquid at 1,947°F (1,064°C) and boils when the temperature reaches 5,084.6°F (2,807°C). Water changes from a solid to a liquid at 32°F (0°C) and boils when the temperature reaches 212°F (100°C).

Liquids do not need to reach their boiling points to begin the process of evaporation. For example, a pan of water placed outside on a hot summer day (98°F, or 36.6°C) will evaporate over time. The sun increases the molecular activity of the water vapor, and some of the molecules escape into the atmosphere. Wind currents enhance the process of evaporation by sweeping away water molecules, which are replaced by other water molecules.

Pressure Impact on Boiling

Pressure directly affects the boiling point of a substance. As the pressure increases, the boiling point increases; the escape of molecules from the surface of the liquid is reduced proportionally; the gas or vapor molecules are forced closer together; and the vapor above the liquid could be forced back into solution. The effects of pressure are important for a process technician to understand. A change in pressure shifts the boiling points of raw materials and products. Pressure problems are common in industrial manufacturing environments and must be controlled.

Vacuum Distillation

Vacuum distillation is used commonly by the chemical-processing industry. Atmospheric pressure is 14.7 psi, and any pressure below atmospheric is referred to as vacuum. Inside a distillation column, vacuum affects the boiling point of a substance in the opposite way that positive pressure does. Instead of moving molecules closer together, it encourages them to move apart. The effect of vacuum pressure is very helpful in situations in which large, heavy hydrocarbons are being broken down in the bottom of a column. Adding heat can damage equipment and product. Vacuum towers offer an excellent option in situations that normally would require high temperatures.

Vacuum systems lower the boiling point of a substance, enhance the molecular escape of molecules from the surface of liquid, reduce energy cost, reduce molecular damage by overheating, and reduce equipment damage.

When heat is added to the distillation system, molecular activity increases and vapor pressure increases. Because a distillation column is a series of stills stacked one on top of the other and pressure changes affect the boiling points, each stage must be considered separately in pressure calculations. The following list identifies only a few of the considerations that need to be reviewed for pressure calculations on a tower.

- The liquid level on each tray puts pressure on the tray.
- Rising vapors encounter resistance at each plate, which slows them down, creating pressure.

- Each tray has a different mixture on it.
- Vapor pressures vary between trays.
- Inlet feed tray pressure varies if feed composition changes.
- Flow rate changes affect tower pressure.
- Downcomer liquid level puts pressure on the plate.
- Reboiler discharge increases pressure where it enters the tower.
- Cool reflux slows molecular activity.
- Each tray has pressure from above, below, and within the stage.

Vacuum distillation allows vapor molecules to escape to the surface of the liquid at lower temperatures. Vacuum towers usually are wider than normal towers because of high vaporization rates. Vacuum distillation allows heavier hydrocarbons to be separated without damaging the molecular structure with excessive heat.

The use of vacuum conditions will result in decreased temperatures in the tower or column. Vacuum distillation is often used to safely separate azeotropic mixtures. An **azeotropic mixture** is defined as a mixture of two or more components that boil at similar temperatures and at a certain concentration. The liquid and vapor concentrations of an azeotrope are equal. Other methods used include the addition of an entrainer that will affect the volatility of one of the components. Examples of this can be found in azeotropic distillation, chemical action distillation, dissolved salt distillation, extractive distillation, and pervaporation and other membrane methods.

Cryogenic Distillation (Oxygen and Nitrogen)

Cryogenic distillation can be used to separate the two major components of the earth's atmosphere, nitrogen and oxygen. Modern manufacturing utilizes a three-phase approach to this operation: a warm end section, a coldbox, and liquid oxygen or nitrogen storage.

Our atmosphere is composed of 78% nitrogen and 21% oxygen. In the cryogenic process, the air in a distillation column is cooled to minus 173°C or 100K to produce high-purity nitrogen. Oxygen starts to liquefy at minus 112°C in the coldbox. While similar equipment is used for each of these processes, the system will either be used for nitrogen or oxygen separation. Two figures are included with this section to help a new technician see the various equipment arrangements used for each operation.

Oxygen Process

During the oxygen cryogenic process, a compressor is used to draw outside air into the warm end section and transfer it to a receiver. The compressed air generates heat and needs to be passed through a chiller before moving into the coalescing filter section. The final step takes the compressed air and sends it through an activated carbon adsorber. This completes the "warm end process." The equipment used in this section includes a compressor, piping, valves, receiver, chiller, instrumentation, filter, and redundant adsorber drums.

The equipment found in the coldbox includes heat exchangers, turbines, and a packed distillation column. As air enters the coldbox, it flows into the main heat exchanger where it is cooled to minus 112°C. At this temperature, some of the oxygen liquefies and enters into the boiler where complete liquification occurs as the cooled liquid oxygen evaporates. Air leaves the boiler and flows into the top of the packed column where it is allowed to flow down through the packing material, giving up the lighter nitrogen gas. The evaporated oxygen steam from the boiler vents into the lower section of the column where it flows upward and contacts the liquefied descending oxygen. As the liquid oxygen flows downward and through the packing, it becomes richer and richer in oxygen content, and is collected at the bottom of the column. From here it can be transferred to the liquid oxygen tank.

In the nitrogen cryogenic process the warm end process is the same; however, upon entering the coldbox the main heat exchanger rapidly cools the air to minus 165°C. Nitrogen collects in the top of the column while oxygen pools at the bottom. Liquid nitrogen is sent to and stored in a tank.

Refrigeration for the oxygen and nitrogen cryogenic process is provided by an air brake expansion turbine physically located at the base of the coldbox. High-pressure gas is expanded to low pressure and cooled in the turbine.

Gas Laws and Distillation Curves

A distillation system is by nature designed to separate the components in a mixture. As the mixture is heated in a furnace or heat exchanger, the piping keeps the liquid confined as it expands, artificially shifting the boiling point. The feed rate to the column is carefully controlled. The heated mixture enters the column on the feed tray and rapidly expands, with the lighter components vaporizing and the heavier liquids cascading down the internals of the column until they gain enough heat energy to flash or vaporize. The different trays in the column have unique molecular components associated with volatility.

If you know the vapor pressure exerted by a specific chemical, you can calculate its partial pressure on the various trays. To do this, we use Dalton's law of partial pressure ($P_{total} = P_1 + P_2 + P_3$), which states that the total pressure of a gas mixture is the sum of the pressures of the individual gases (that is, the partial pressures). Furthermore, Dalton's law states that the total pressure of a gas mixture is the sum of the pressures of the individual gases (the partial pressures). Figure 15.10 illustrates Dalton's law of partial pressures.

To see Dalton's law applied, here are some examples. A mixture of 25% hexane, 50% benzene and 25% heptane will exert a predictable pressure at 175°F. To calculate partial pressures, use the formula:

Partial pressure = Vapor pressure × Percent of fraction

For example, consider the following data:

Substance	**Vapor Pressure @ 175°F (psia)**	**Percent (%)**
Hexane C_6H_{14}	20.6	25
Benzene C_6H_6	14.7	50
Heptane C_7H_{16}	8.8	25

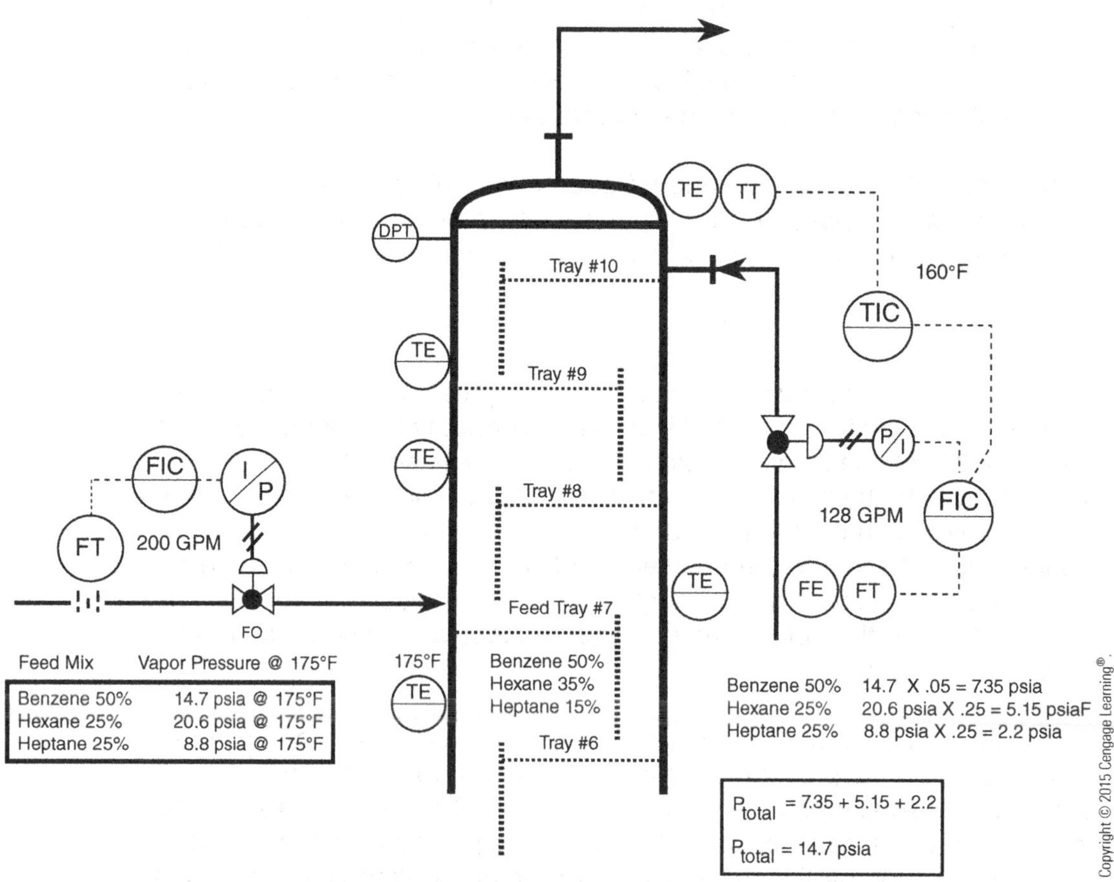

Figure 15.10 *Dalton's Law of partial pressures*

Given the above data, the vapor pressure exerted by a mixture consisting of 25% hexane, 25% heptane, and 50% benzene could be calculated as follows:

Hexane C_6H_{14}	=	20.6 psia	×	0.25	=	5.15 psia
Benzene C_6H_6	=	14.7 psia	×	0.50	=	7.35 psia
Heptane C_7H_{16}	=	8.8 psia	×	0.25	=	2.20 psia
			Total pressure		=	14.70 psia

Using this information, we can calculate the total pressure on tray 6 by adding up the partial pressures. This information also illustrates that the chemical with the highest volatility is hexane (C_6H_{14}), which has a boiling point of 69°C (156.2°F). Heptane has a boiling point of 98°C (208.4°F). Benzene has a boiling point of 80.1°C (176.2°F) and represents the lowest percentage of the three components in vapor state above the tray. The larger the difference between the partial pressures, the easier it is to separate the fractions by boiling point.

		Original Feed%	Boiling Point
Hexane C_6H_{14}	5.15 ÷ 14.7 = .35 × 100 = 35%	(25%)	69°C
Benzene C_6H_6	7.35 ÷ 14.7 = .5 × 100 = 50%	(50%)	80.1°C
Heptane C_7H_{16}	2.2 ÷ 14.7 = .15 × 100 = 15%	(25%)	98°C

The original mixture had a 25% heptane, 25% hexane, and 50% benzene liquid concentration.

The larger the difference between partial pressures, the easier it is to separate the fractions by boiling point. Equilibrium curves (Figure 15.11) are used to determine the composition of a vapor above a boiling liquid and how one mixture differs from another.

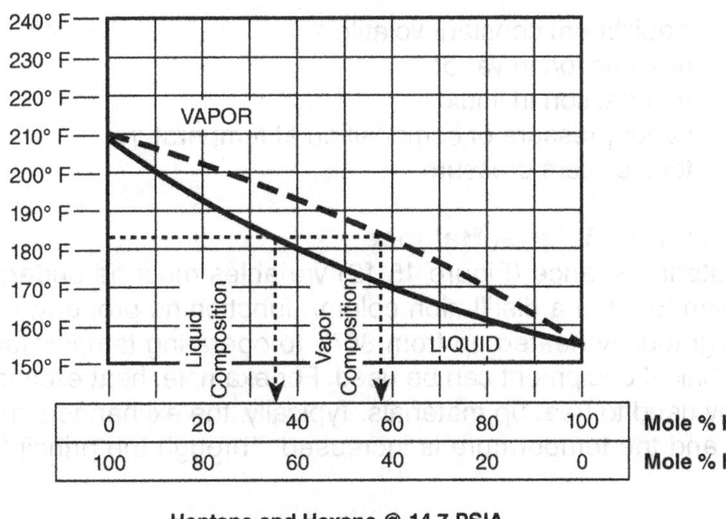

Figure 15.11
Equilibrium Curve

The ideal gas law $PV = nRT$ calculates the pressure, temperature, volume, or moles of any ideal gas, where:

P = pressure of the gas
V = volume
N = moles of gas
T = temperature in Kelvin (K)
R = ideal gas constant: 0.08206 L × atm/mol × K

The combined gas law is used to calculate changes in a gaseous substance from one condition to another:

$$\frac{P_1V_1}{T_1} = \frac{P_2V_2}{T_2}$$

Ideal and Nonideal Systems

Efficient tower operation requires an operator to understand the theory of distillation and the key components of a distillation system. It also requires years of field experience. The distillation theory is related to two types of systems: an ideal (binary or two-component) system and a nonideal (complex, multicomponent) system.

Distillation design theory is based on heat and mass transfers from tray to tray. Distillation calculations are used to determine vapor-liquid equilibrium data. A common equation used to determine performance is Raoult's law:

$$K_1 = y_i = p_i$$
$$x_i = p$$

Where

K_1 = equilibrium constant volatile
y_i = mol fraction in vapor
x_i = mol fraction in liquid
p_i = vapor pressure of composition at temperature
p = total system pressure

Heat Balance and Material Balance

Heat and material balance (Figure 15.12) variables must be understood by a technician to keep a distillation column functioning properly. As the feedstock is gradually heated up from 80°F to operating temperatures a variety of technical equipment can be used. For example, heat exchangers are frequently used to heat up materials. Typically, the exchangers are set up in series and the temperature is increased. Through the principles of

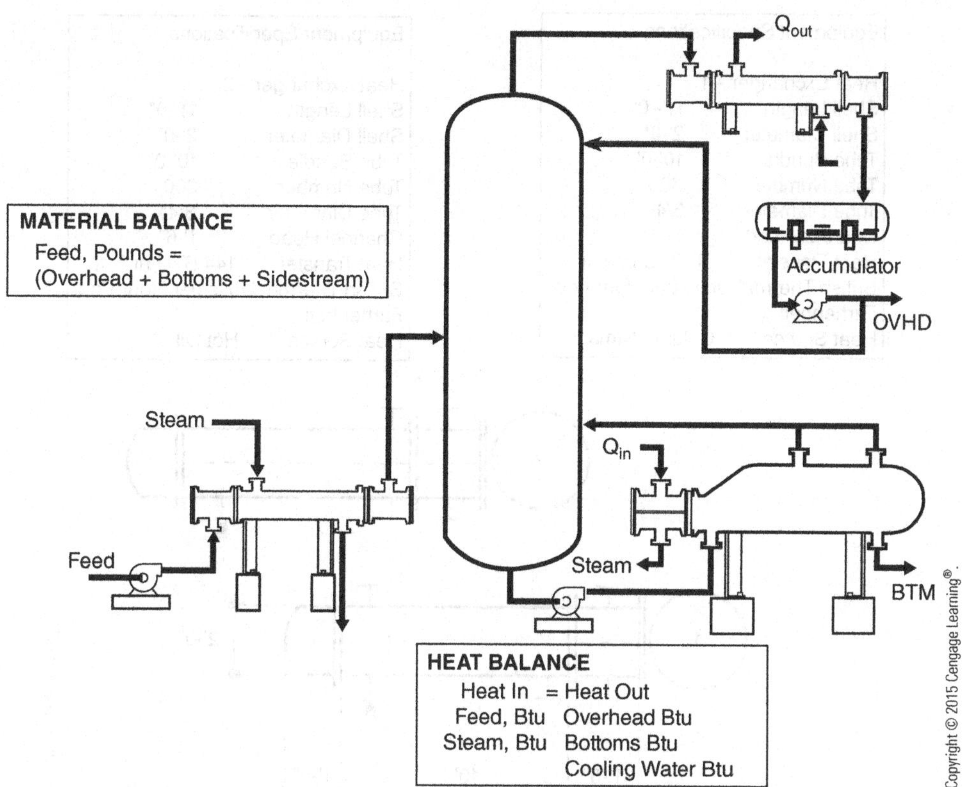

Figure 15.12
Heat Balance

conduction, convection, sensible heat, latent heat, and others, heat transfer occurs. Flow rates measured in gallons per minute are exposed to hot oil or steam that transfer energy through solid walls and into a moving fluid matrix. As these materials enter a distillation column, the pressure of the liquid being released on the feed tray or through the feed distributor immediately gives up heat energy. This heat needs to be sustained and controlled. Figure 15.13 and 15.14 show the detailed specifications of a heat exchanger and a preheat system. All raw materials entering a tower are converted into products. **Material balance** theory states that the sum of the raw materials entering must equal the total of the products leaving the distillation tower. **Column loading** is another term associated with material balance. It is defined as the amount of material fed continuously into a distillation column.

Heat balance states that the heat that goes into a distillation column must be equal to the heat that goes out of it. Maintaining this balance is critical to the proper functioning of the tower.

To better understand how the principle of material balance works, let's consider the flow of raw materials through a column. Flow enters the column at 550 gallons per hour. This is roughly equal to 10 barrels per hour. Flow

Figure 15.13
Heat Exchanger Specifications

Equipment Specifications	
Heat Exchanger - 1	
Shell Length	11'- 0"
Shell Diameter	2'-0"
Tube Bundle	10'-0"
Tube Number	300
Tube Diameter	3/4"
Channel Head	1'-6"
Heat Transfer	30 BTU/hr F. ft
British Thermal Units per Hour Foot Farhenheit	
Heat Source	Column Btms

Equipment Specifications	
Heat Exchanger - 2	
Shell Length	11'-0"
Shell Diameter	2'-0"
Tube Bundle	10'-0"
Tube Number	300
Tube Diameter	3/4"
Channel Head	1'-6"
Heat Transfer	144 BTU/hr F. ft
British Thermal Units per Hour Foot Farhenheit	
Heat Source	Hot Oil

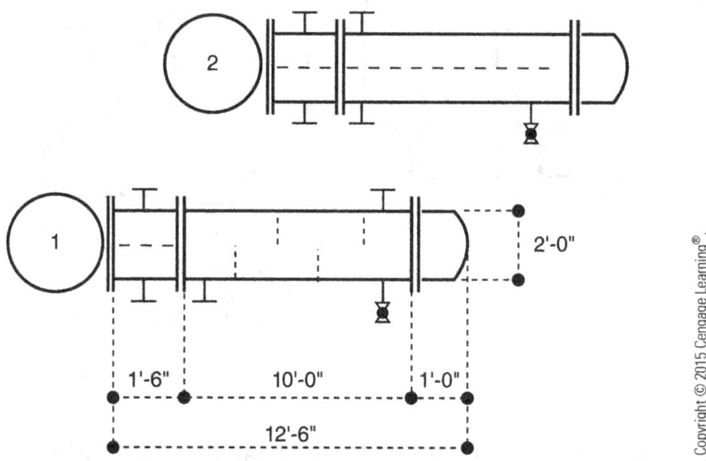

leaves the tower through one of three paths: bottom, sidestream, or top. The **bottoms product** contains heavier molecules than the sidestream or top. To obtain the material balance, add the sums of the three outlet flows. This number should equal the sum of the inlet flow. By knowing the boiling points of the different components in the feed mixture, an operator could determine the approximate material composition and the place on the tower each component will exit.

Heat balance operates under the same principle as material balance. What goes in must come out and must be maintained in order to hold the heat balance. Initially, tower **temperature gradients** are established by engineering. The temperature gradient is the temperature profile that exists on the tower from bottom to top. Tower temperature is maintained so the expected separations occur. Temperature in a tower can be affected by pressure, flow rates, fluid levels, feed composition, reflux rate, and equipment problems.

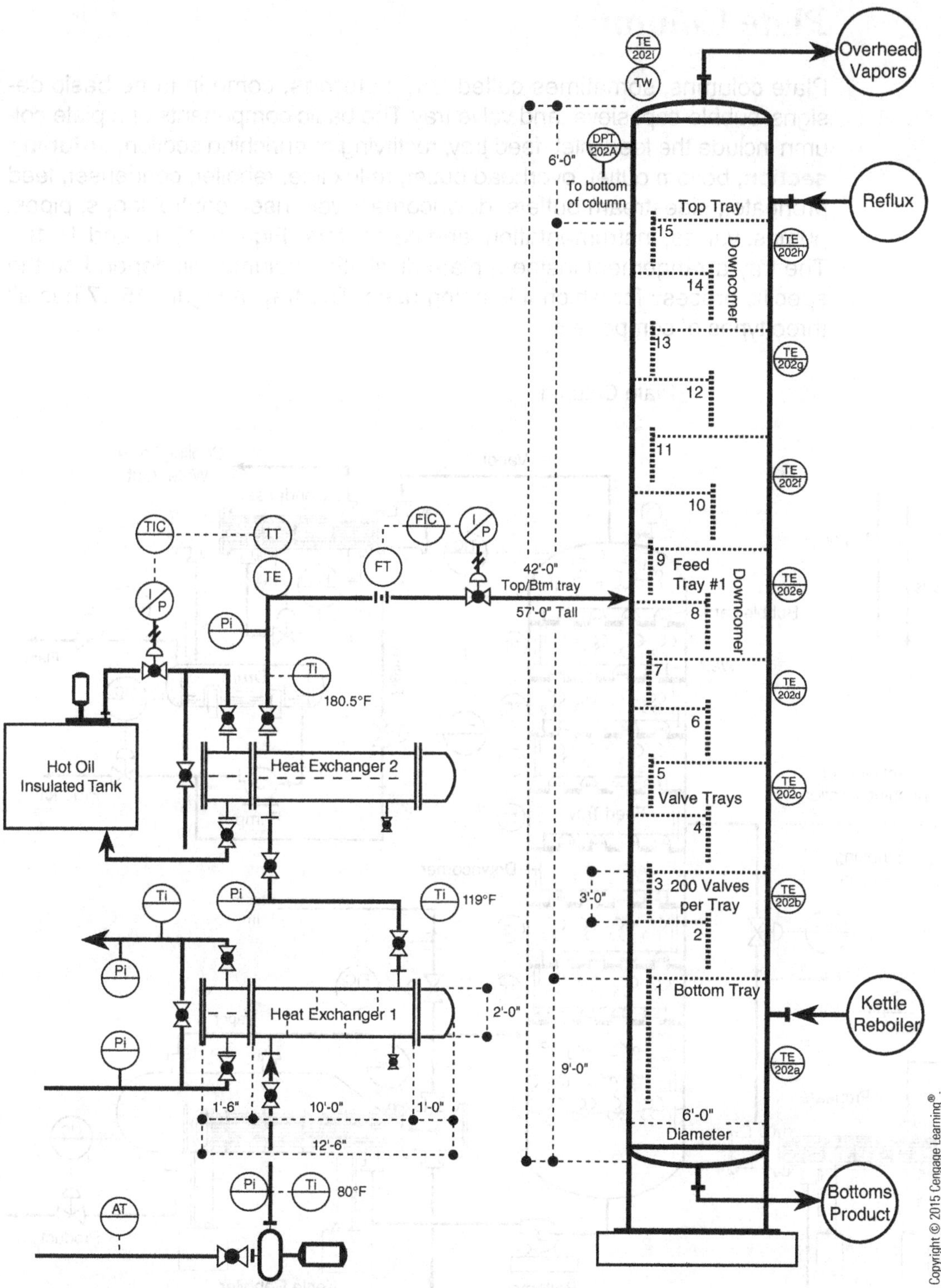

Figure 15.14 *Heat Exchanger in Series Flow*

Plate Columns

Plate columns, sometimes called **tray columns**, come in three basic designs: bubble-cap, sieve, and valve tray. The basic components of a plate column include the feed inlet, feed tray, rectifying or enriching section, **stripping section**, bottom outlet, overhead outlet, reflux line, reboiler, condenser, feed preheater, sidestream outlets, downcomer, weir, riser, control loops, pipes, pumps, valves, instrumentation, and computers (Figures 15.15 and 15.16). The tray arrangement inside a plate distillation column will depend on the specific process for which it is being used. The tray in Figure 15.17 has all three types of components.

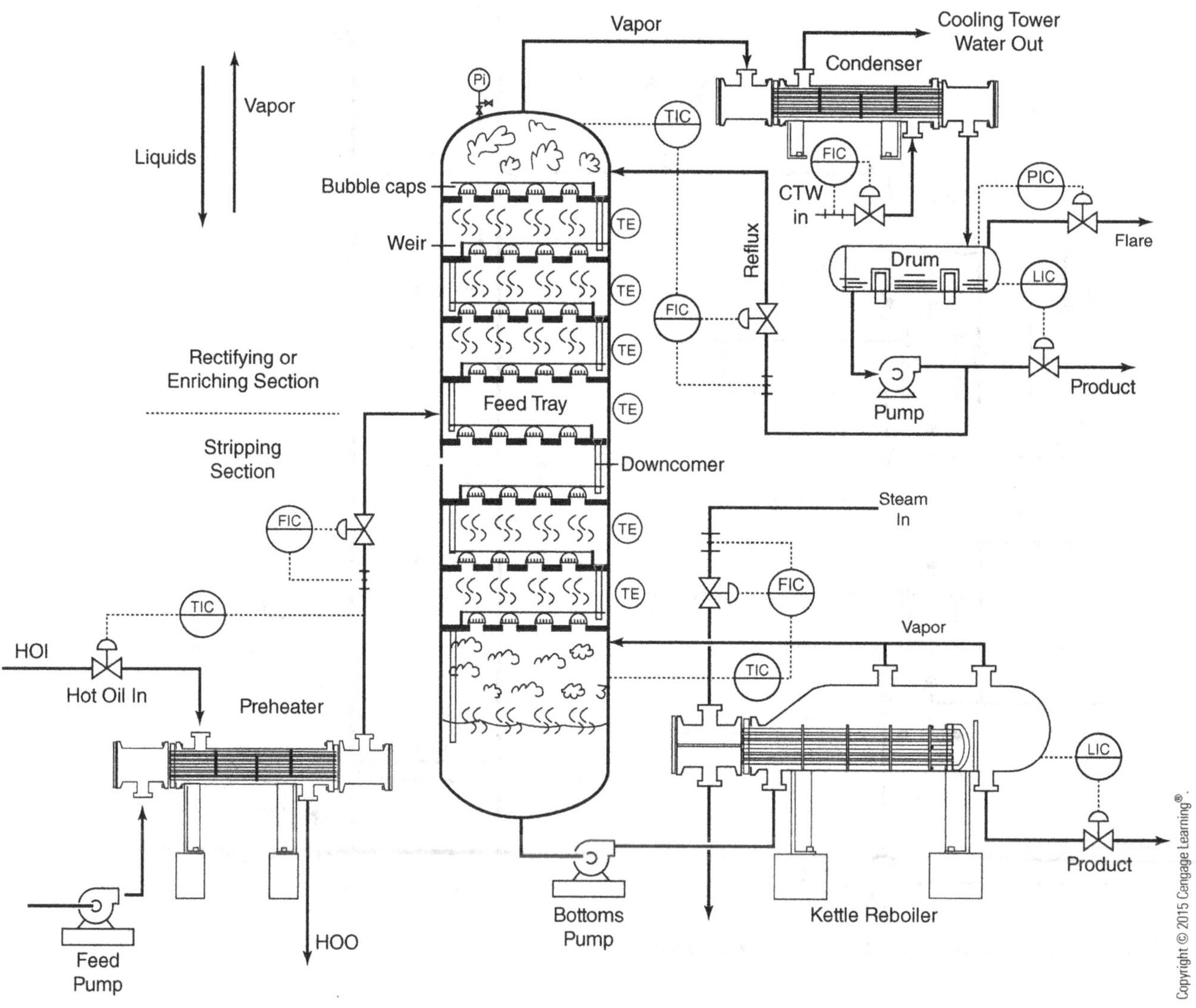

Figure 15.15 *Plate or Tray Column*

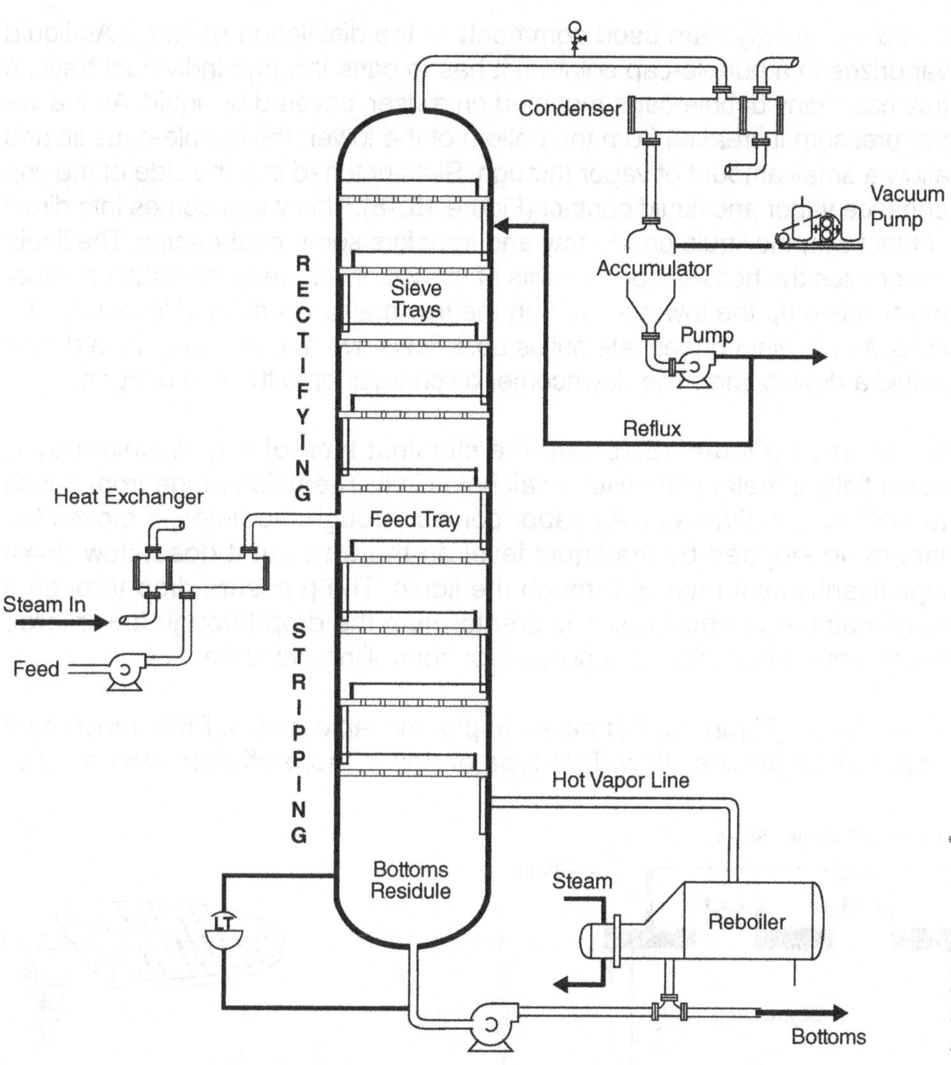

Figure 15.16
Sieve Tray-Plate Tower

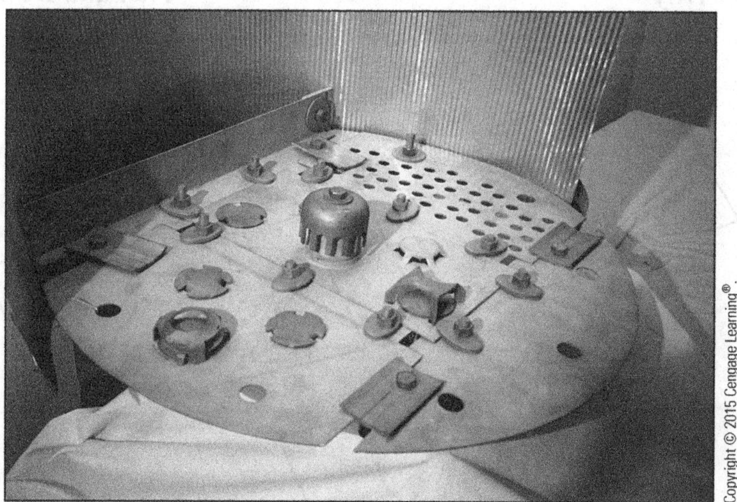

Figure 15.17
Bubble-Cap, Valve, Sieve Tray

Bubble-cap trays are used commonly in the distillation process. As liquid vaporizes in a bubble-cap column, it has to pass through individual trays. A tray has many bubble-caps mounted on a riser, covered by liquid. As the vapor pressure increases from the bottom of the tower, the bubble-caps lift and allow a small amount of vapor through. Slots notched into the side of the cap enhance vapor and liquid contact (Figure 15.18). This vapor comes into direct contact with the liquid on the tray and transfers some heat energy. The liquid condenses the heavier components of the vapor, whereas the lighter components move up the tower. A weir on the tray maintains a liquid level over the caps. As heavier condensate builds up, it flows over the weir and out a device called a downcomer. The downcomer discharges onto the tray under it.

Sieve trays (Figure 15.19) are the simplest type of tray. A sieve tray is essentially a metal plate with small holes in it. The holes range from 1 inch to 1/8th inch in diameter. As vapor comes through the holes, it moves too fast to be stopped by the liquid level on the tray, but it does slow down significantly as it passes through the liquid. The pressure drop through a large number of small holes is greater than the drop through fewer large holes. This information is important for controlling pressure.

Valve trays (Figure 15.20) have simple, movable plates. Plate movement depends on process flow. This type of tray is more efficient over a wider

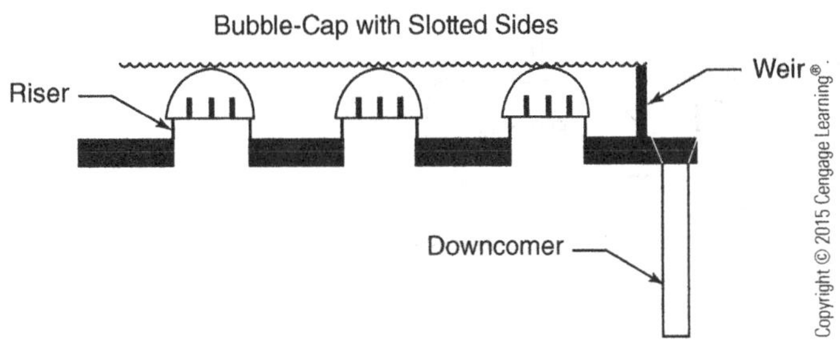

Figure 15.18 *Bubble-Cap Tray*

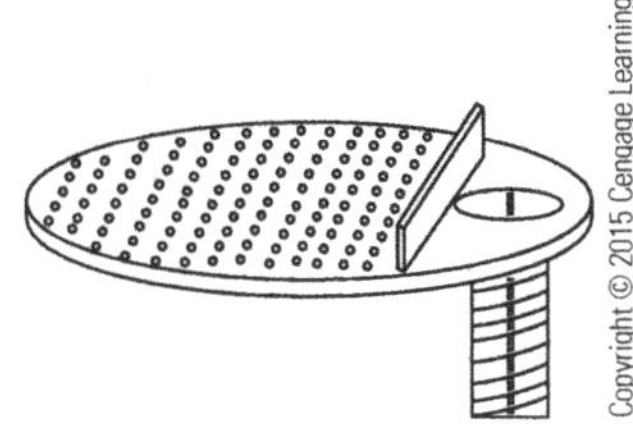

Figure 15.19 *Sieve Tray and Downcomer*

Figure 15.20 *Valve Tray*

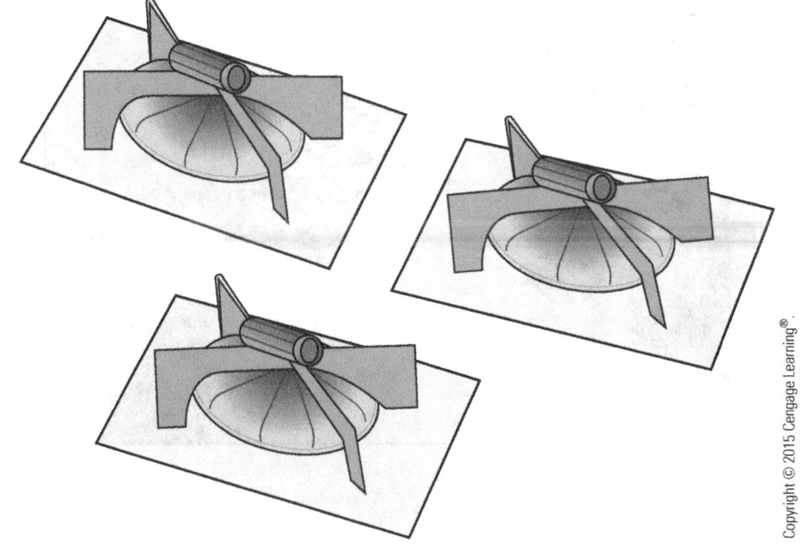

range of flow rates than the sieve tray. Each valve is a metal strip or a small metal disc that lifts under vapor pressure. Flow is directed horizontally on the tray to produce good mixing.

Inlets and Outlets

The feed inlets are located between the rectifying and stripping sections. Distillation columns can have three or more feed lines. The bottoms outlet is used to remove residue from the distillation column and to feed the reboiler. The overhead outlet is the vapor exit point from the distillation column. The overhead outlet allows hot vapors to pass into the condenser.

Feed Tray

The location of the feed tray is determined by the ratio of stages above and below the inlet feed line. The ideal feed tray location can fluctuate so three or more feed lines typically are established on a tower. Experimentation identifies which feed point is the most effective.

Thermal condition is defined as the amount of heat required to vaporize one mol of feedstock, divided by the latent heat of vaporization. If the feed composition changes, the thermal condition of the feedstock will be changed. The composition and condition of the feedstock as it enters the tower affect the number of trays needed, reflux rate, level, pressure, and temperature.

Rectifying or Enriching Section

A tower's rectifying or enriching section is located above the feed tray. **Rectification** is the separation of different substances from a solution by use of a fractionating tower. As a mixture is heated, the vapors rise through the column. Chemicals that boil first are in the top fraction; those that boil last stay in the bottom fraction. The **rectifying section** of a tower has a higher concentration of light components than the stripping section has, so it is said to be enriched.

Stripping Section

The stripping section is located below the feed tray. Feed from the feed tray flows down the tower, stripping the heavier hydrocarbons from the vapor. As the cooler feed drops from the feed tray to the lower trays, it heats up and strips the lighter components. The heavier components flow to the bottom of the tower. Because the trays get progressively hotter as the liquid flows down the tower, lighter hydrocarbons are stripped off the heavier molecules. This process creates a countercurrent movement within the tower as vapor moves up and liquid moves down. The lighter components are vaporized and move up the tower. Temperatures in the stripping section are usually much higher than in the rectifying section.

Reboiler or Steam Injection

Reboilers help control overlap and heat balance. Several types and arrangements of reboilers can be used on a distillation column (Figure 15.21). Kettle reboilers draw feed from the bottom section of the column by gravity flow. An internal weir holds a liquid level over a heated tubular bundle.

Figure 15.21
Reboiler Designs

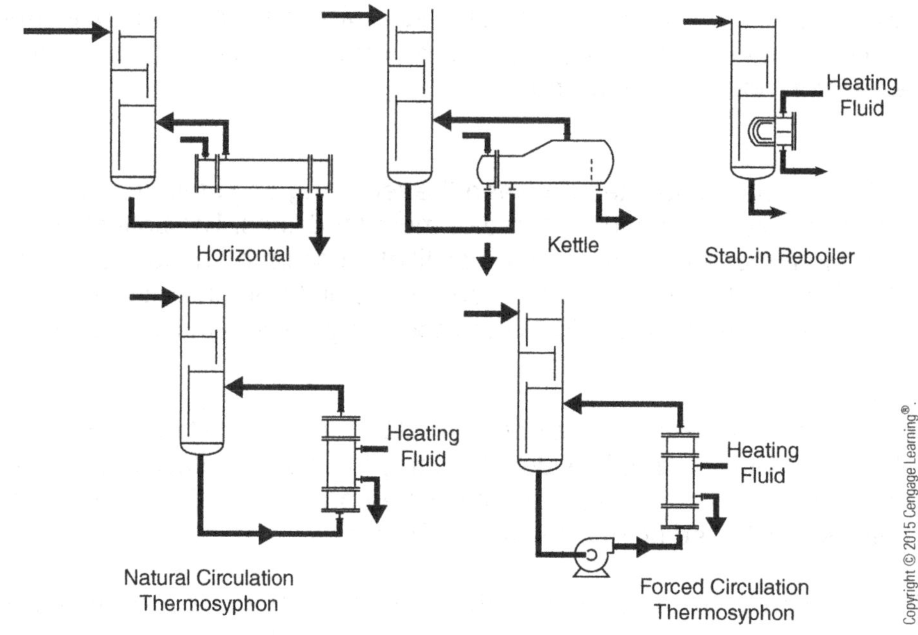

A vapor-disengaging space above the tubular bundle allows the liquid to boil and roll freely and enrich the void with vapors. As the lighter components break free from the larger, heavier molecules, they rise into the vapor disengaging area and exit the top of the reboiler. A vapor inlet port is located below the bottom tray or packing support. An important criterion for a distillation column is that any heat energy lost must be returned. Reboilers are used to restore heat balance and add additional heat for the separation processes. Bottoms products typically contain the heavier components from the column. Reboilers suction off the bottoms product and pump it through their system. Product flow enters the bottom shellside of a reboiler. As flow enters the reboiler, it comes into contact with the tubular bundle. The tube bundle typically has steam flowing through it. As the bottoms product comes into contact with the tubes, a portion of the liquid is flashed off and captured in the dome-shaped vapor cavity at the top of the reboiler shell. This vapor is sent back to the tower for further separation. A weir contains the unflashed portion of the liquid on a reboiler. Excess flow goes over the weir and is recirculated through the system.

A thermosyphon reboiler is a fixed head, single-pass heat exchanger hooked up on the side of a tower to add additional heat to the process. One side of the exchanger is used for heating, usually with steam; the other side takes suction off the tower. Thermosyphon reboilers come in vertical or horizontal designs. The lower leg of the exchanger usually takes suction at a point low enough on the tower to provide a liquid level to the exchanger. A pump is not connected to the tower and exchanger. This system uses buoyancy forces to flash off and pull in liquid. As vapors are flashed off (into the tower), the suction line draws in an equal amount.

Overhead Condenser

Condensers are typically tube and shell heat exchangers used to condense hot vapors into liquid. A condenser can be found at the top of most distillation columns (Figures 15.15 and 15.16).

A heat exchanger that is used to convert the entire overhead vapor to liquid is called a ***total condenser***. This type of condenser is designed to split the distillate into two streams. As vapor flow leaves the column, it enters the overhead condenser, comes into contact with a much cooler environment, and condenses. The condensed liquid collects in an accumulator or drum. A pump takes suction off the accumulator and directs flow to two locations. The primary stream moves the distillate to the product tanks or on for further processing; the secondary stream is pumped back into the column as reflux.

Partial condensers are designed to return all of the condensed liquid back to the top tray of the tower and to function as an external top tray for the tower. Partial condensers also convert 50% of the overhead vapor to distillate and separate vapor from the accumulator, sending it on for further processing. A shell and tube heat exchanger is mounted at the top of a distillation column. Hot vapors pass over chilled tubes and condense. The condensed vapors are collected in an accumulator or receiver. Some columns use this same technique to cool sidestream draws off a tower.

Reflux Line

Distillation columns do not always work perfectly; overhead, sidestream, and bottoms products usually have some composition overlap. Overlap can be controlled or limited by returning part of the overhead product to the tower. This process is referred to as refluxing, and the returned product is called reflux. Cooled reflux keeps a liquid level on the upper trays and condenses heavier molecules. Each component of a mixture pumped to a column has a unique boiling point. The upper rectifying trays, middle feed tray, and lower stripping trays have different temperatures and different products on them.

The principle of **total reflux** exists in a distillation system when the overhead vapors are condensed and returned to the top tray inside the column. When a distillation column is placed on total reflux, feed to the column is stopped, bottom and sidestream flows are stopped, and the **overhead product** is condensed and returned to the column as reflux. This type of operation is commonly used to test the tower or to put the system in standby. It is easier and cheaper to run the tower on total reflux than to shut it down and then restart it.

Downcomers and Risers

Downcomers are directional devices integrated into the tray design to funnel liquid from upper trays to lower trays in a distillation column. Risers are located on the plates. In some systems, they appear to be tubes that allow hot vapors to flow between trays. The top of the riser usually has a bubble-cap or valve attached. Because bubble-caps and valve trays come in over 30 designs, the riser also can be associated with the lower leg of the device that rises and allows vapor to flow up.

Feed Preheater

The feed preheater is usually a shell and tube heat exchanger or a fired furnace. Fired furnaces or reboilers provide the initial heat to a distillation column's feed. Heating the feed avoids thermal shock and matches the temperature set point of the feed tray. Temperature controls are tied to the operation of the feed preheater. In most cases, the flow usually is adjusted to the feed, or in some cases the preheater's steam rate is regulated.

Sidestream Outlets

A sidestream can be located on every tray or stage of a column. Liquid or liquid-vapor mixtures are removed and cooled in a condenser. The cooled liquid collects in an accumulator or receiver and is pumped to a separate product tank.

Pressure, Temperature, Level, and Flow Indicators

Pressure gauges are used to provide valuable information about conditions inside a distillation column. Thermocouples are used to sense temperature changes on a process unit. Level control systems are used to maintain levels in three or more systems. Flow controllers are used to maintain flow rates in and out of the column. Differential pressure between trays or column sections is an important indicator of load and flooding conditions.

Control Loops

The basic elements of a control loop include primary elements and sensors, a transmitter, a controller, a converter, and a final control element. Control loops are used by process technicians to automatically control some of the process variables found in a distillation system. The most common variables controlled by this type of automation are flow, temperature, level, pressure, and composition.

A level control system monitors the level in the bottom of the tower. Part of the stream is sent to a reboiler and part is sent onto a product tank. Level control systems use a sensing device, transmitter, controller, and control valve. This same system exists on the overhead accumulator, the sidestream accumulators, and the feed systems.

A temperature control system monitors tower temperatures and adjusts flow through the reboiler and preheater. As temperature drops in the tower, more feed is allowed into the reboiler. The reboiler is a critical part of the heat balance system. This device returns heat to the tower by flashing some of the liquid into vapor and returning it.

Distributed Control System
The distributed control system is a computer-automated system used to operate process equipment from remote locations.

Packed Columns

Packed distillation columns provide a unique approach to vapor-liquid or liquid-liquid contact during the distillation process. **Packed towers** use specially designed packing materials to provide more surface area for maximum contact between gases and liquids. Liquids wet the surface of the packing as they migrate down the tower. Rising vapors come into contact with the wetted surface and exchange heat. In many packed distillation columns, hot vapors and liquids compete for access through the same points. As they do in any distillation process, lighter components move up the column while heavier ones move down. In a packed column, liquid travels down through feed distributors, hold-down grids, random and structured packing, support plates, redistributors, liquid collectors, structured grids, and finally into the bottom section (Figure 15.22). Hot vapors move countercurrent to the downward flow of liquid.

There are three types of packed distillation columns: random, structured, and stacked. Random packing includes discrete pieces of packing that are randomly dumped or poured into a packed column. Packing provides a surface for good vapor-liquid contact, distillation efficiency characteristics, and predictable pressure drop. Structured packing has specific geometric shapes, like a mesh. It works best in columns requiring high liquid loadings. Stacked packing is uniformly arranged inside a distillation column. Packed columns are designed for pressure drops between 0.20 and 0.60 inch of water per foot of packing medium.

Shell
The cylindrical shell of the distillation column is typically metal, carbon steel, stainless steel, special alloy, or nonferrous. Columns can also be composed of glass, ceramic, plastic, or wood. The type of chemicals that will be used in the distillation column will determine the design material, lining specification, wall thickness, pressure rating, and temperature rating. The inside shell of some packed distillation columns is lined with brick and a special heat- and acid-resistant mortar.

Packing
The packing used in a distillation column provides the heart of the operation. Inside a distillation column, packing is either random or structured. Typically, a combination of both is used in a packed column. Random packing may be dumped into a section of the column or carefully stacked. The random spacing of the packing or areas between the packing gives

Figure 15.22
Packed Distillation Column

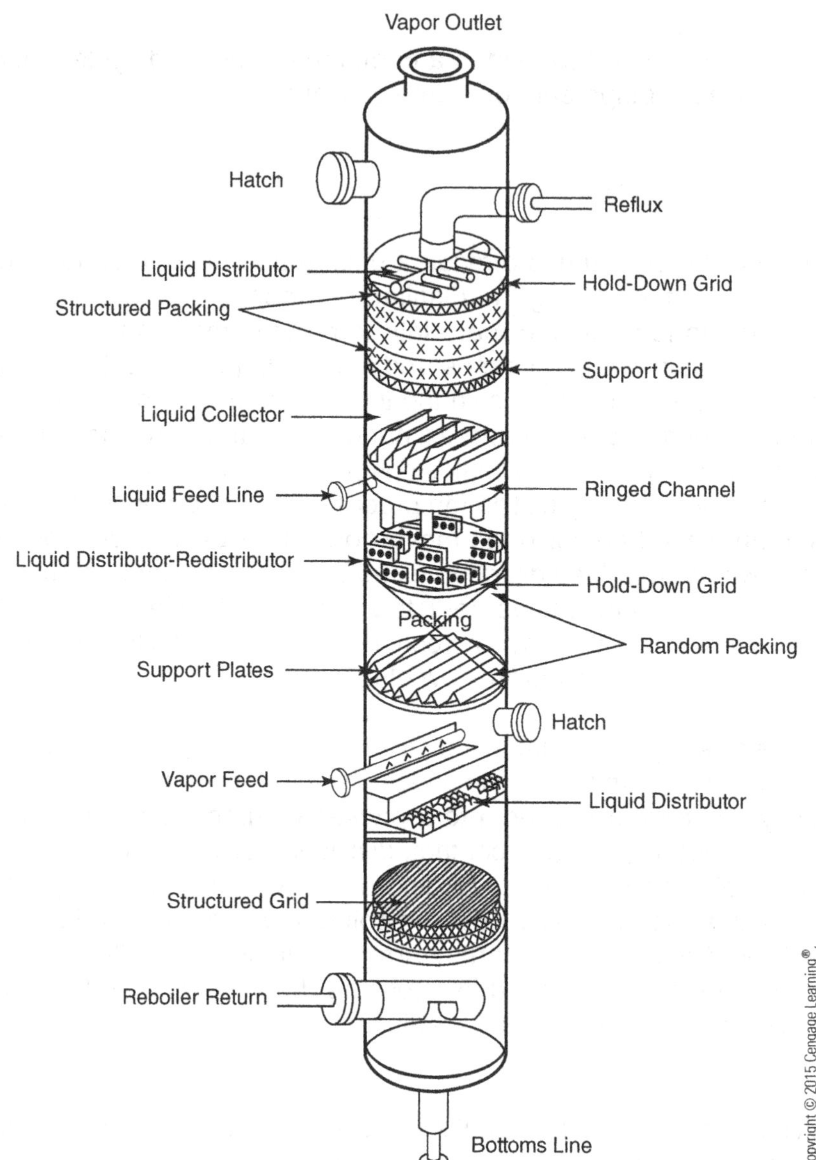

this section its name. Structured packing comes in preformed layers that overlap into a single unit. The uniform spacing between voids characterizes the functionality of this section. Packing comes in a variety of designs and sizes (Figure 15.23). Some of these designs are berl saddle, intalox saddles, rasching ring, cross-partition ring, flexiring, spiral ring, mini-ring, and sulzer packing. A variety of other packing designs can be found in the chemical processing industry, but these are the most popular.

As hot feed is injected into a packed distillation column system through the liquid distributor, it encounters a rapid pressure drop. This pressure drop allows a percentage of the feed to vaporize while the rest of the liquid fraction

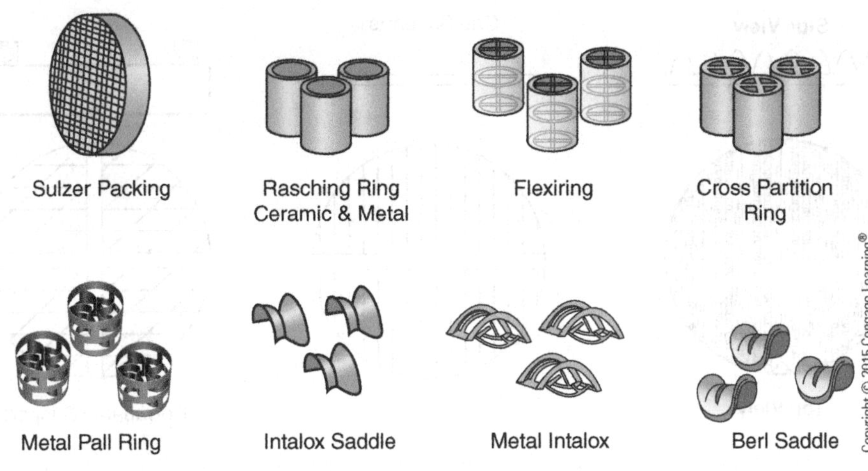

Figure 15.23
Common Packing

is evenly distributed over the packing. Heat balance inside the packed column is maintained with one or more of the following systems:

- kettle or thermosyphon reboiler system
- direct injection of steam
- electric heaters on the column
- steam tracing on the column
- furnace system

A kettle reboiler is a device designed to add heat to the column. Kettle reboilers take suction off the bottom of the column and heat it up until a part of it vaporizes and is sent back to the column. The liquid that does not vaporize is sent to another system for further processing. A thermosyphon reboiler is a vertical, shell and tube heat exchanger mounted near the bottom of the column. Steam is used to reheat the bottom product.

A series of electric plate heaters can be used to heat a packed distillation column. Electric heaters are typically used in small systems that are enclosed in buildings. Another design that does not require environmental protection is steam tracing. Steam tracing uses copper tubing that can easily be wrapped around the outside of a distillation column and insulated. As steam passes through the tubing, heat energy is transferred by conduction into the column. Insulation holds much of the heat in and protects the column from the elements.

A furnace system may be used in large commercial operations where bottom feed is circulated into a small furnace and back to the column. Furnace systems provide large-scale temperature control where flow rates into the column are high. Steam can be injected directly into the bottom of a packed column to heat up the process. Open, or direct, injection allows 100% conversion of Btus into the liquid and helps reduce effective partial pressures of different components. Open injection replaces the need for a reboiler. In a binary system (a system with two components) where the

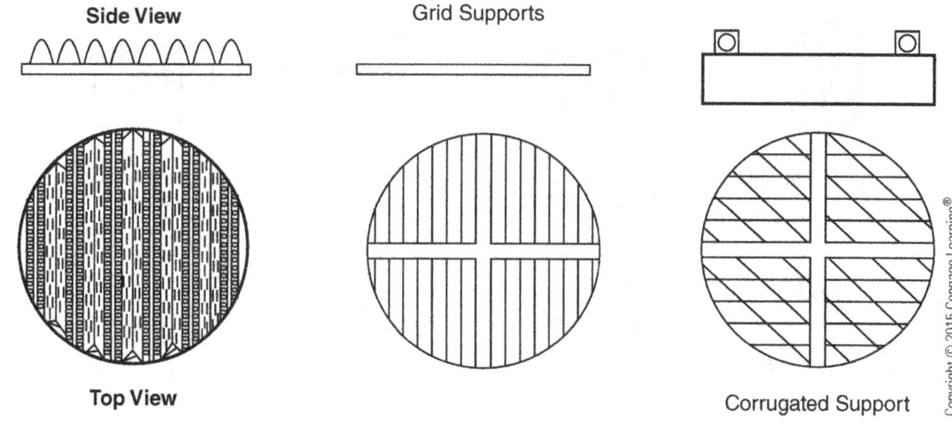

Figure 15.24
Packing Supports

major component is water, open or direct injection of steam is a suitable alternative to using a reboiler.

Packing Supports

The packing inside a distillation column is supported by packing supports. These devices come in a variety of shapes and designs (Figure 15.24). Each support is engineered to hold up the combined weight of the packing and liquid under a variety of abnormal conditions. Some of these conditions include pressure surges, localized flooding, and temperature and flow variations. Packing supports are designed to provide sufficient air-liquid passage and to prevent the packing from migrating into the lower sections of the column. Packing support plates come in the following designs: flat sieve plates, gas-injection support plates, grid supports, cone supports, and corrugated supports.

Bed Limiters and Hold-Down Plates

Bed limiters and hold-down plates (Figure 15.25) are engineered to keep fixed bed packing from migrating or fluidizing out of the section. Hold-down plates rest on top of the packing, whereas bed limiters are attached to the inside wall of the column. Hold-down plates are primarily used on carbon and ceramic packing. Bed limiters are used on metal or plastic packing. During normal operations, a packed column will experience minor fluctuations in the packing. Depending on the **relative volatility** of the various components found in

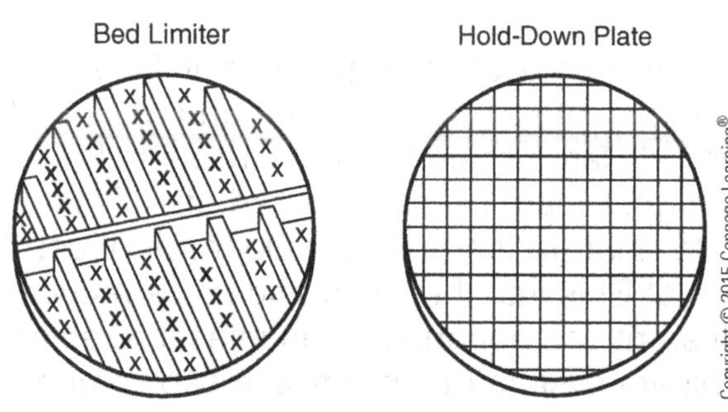

Figure 15.25
Hold-Down Plates and Bed Limiters

a chemical mixture, the packed column can experience significant pressure differential across the upper and lower sections. It is possible for a packed column to experience local and **overall flooding** as the voids between the packing and stages fill up. Flooding causes severe pressure differences between sections in the column. Bed limiters and hold-down plates keep the internal components of the column stable and in operating condition.

Vertical Packed Column

An important part of engineering design is the vertical slant of the column. Most columns are never straight up and down. Shifting of the tower's alignment—which can be caused by soil movement, wind sway, poor foundation construction, or uneven thermal expansion—can severely decrease efficiency in the column. Liquids in a column will flow toward the downward-sloping end of the packed section. This phenomenon, referred to as channeling, will cause uneven distribution of the wetting process and will unbalance the internal vapor-liquid flow. Vapors flow faster through dry sections of packing and slower through the wetted parts. Pilot tests of random packed columns placed on an incline identified efficiency reductions of 5% to 10% for each degree of inclination. In severe cases, 25% to 50% decreases in efficiency have been noted for each degree of inclination. Structured packing provides a more stable efficiency environment with recommended tolerances between 0.2° and 0.5°.

Liquid Distribution and Redistribution

Packed distillation columns use liquid distributors to disperse fluid evenly over the top of the packing. This process plays an important role in the efficient operation of the column. Poor liquid distribution reduces vapor-liquid contact and promotes channeling. Common liquid distributor designs are ladder pipe, spray distributor, perforated ring distributor, notched-trough distributor, tunnel orifice distributor, weir-riser distributor, and orifice pan distributor (Figure 15.26). In most packed columns, the downward flow of liquid is occasionally redistributed. This process tends to pull the liquid off the walls and redirect it into the center.

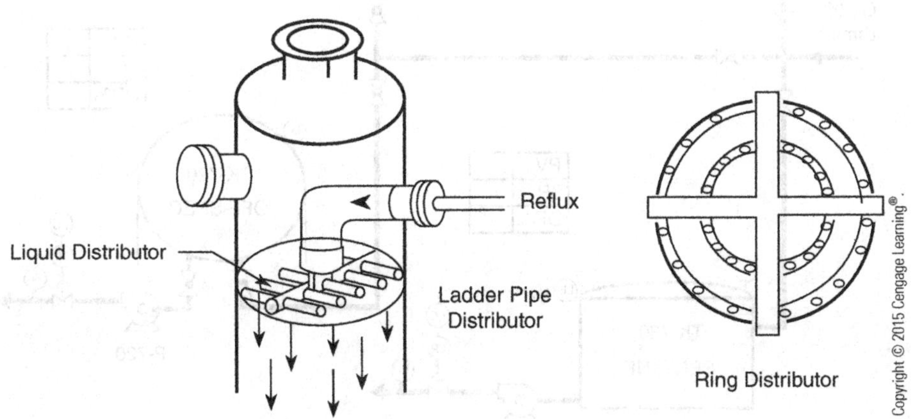

Figure 15.26
Liquid Distributor

Plate Distillation System

This section describes a typical plate distillation system, which involves several subsystems. The primary subsystems are the feed system, the preheat system, the distillation column, the overhead system, and the bottom system. All of these systems work together to produce products that meet or exceed customer specifications.

Feed System

A collection of product storage tanks is often referred to as a tank farm. The tank farm has a complex system of piping, valves, and pumps that allow the technician to line up the various tanks in a number of arrangements. The tank farm in this example has four storage tanks assigned to the debutanizer distillation column (Figure 15.27).

The primary feedstock used for this distillation process is a **binary mixture** of 37% butane and 63% pentane. Tank 702 receives feed from a sister unit on the west side of the chemical plant complex. At the base of Tk-702 are two centrifugal pumps that transfer feed to the preheat section and to the distillation

Figure 15.27
Feed System

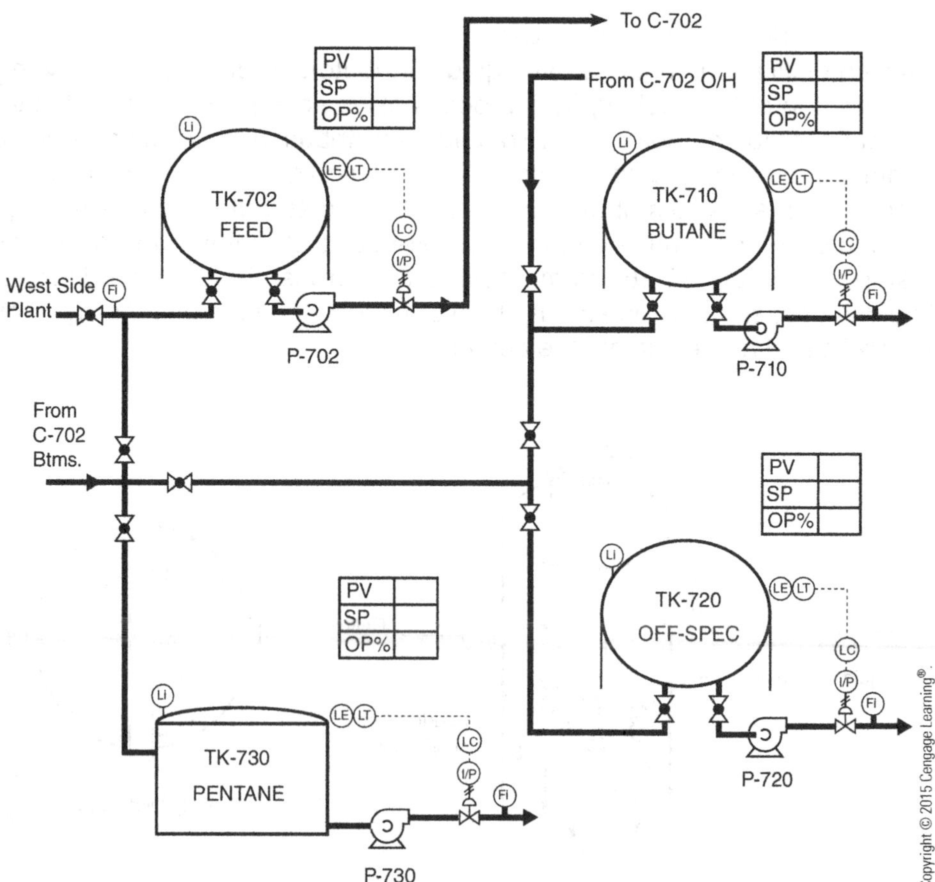

column (C-702). During startup, the off-spec tank-720 is used to store product that does not meet customer specifications. The bottom and overhead lines are lined up into the off-spec system during startup. This mixture can be fed back into the system, or it can be used as a feedstock for a sister unit.

A level indicator provides a readout inside the control room for Tk-702. High- and low-level alarms support the level control system. A flow control loop (FIC-702) is used to regulate the flow of product to the unit. Flow rates are slowly increased in 50-GPM increments, up to 200 GPM. Only one pump at a time is used to transfer feed to C-702. A product analyzer keeps a close eye on the uniformity of the feedstock. Each minute pump P-702 operates, 74 gallons of butane and 126 gallons of pentane are sent to distillation column 702.

Preheat System

The preheat system (Figure 15.28) uses two horizontal shell and tube heat exchangers to heat the butane-pentane mixture, Ex-702 and Ex-703.

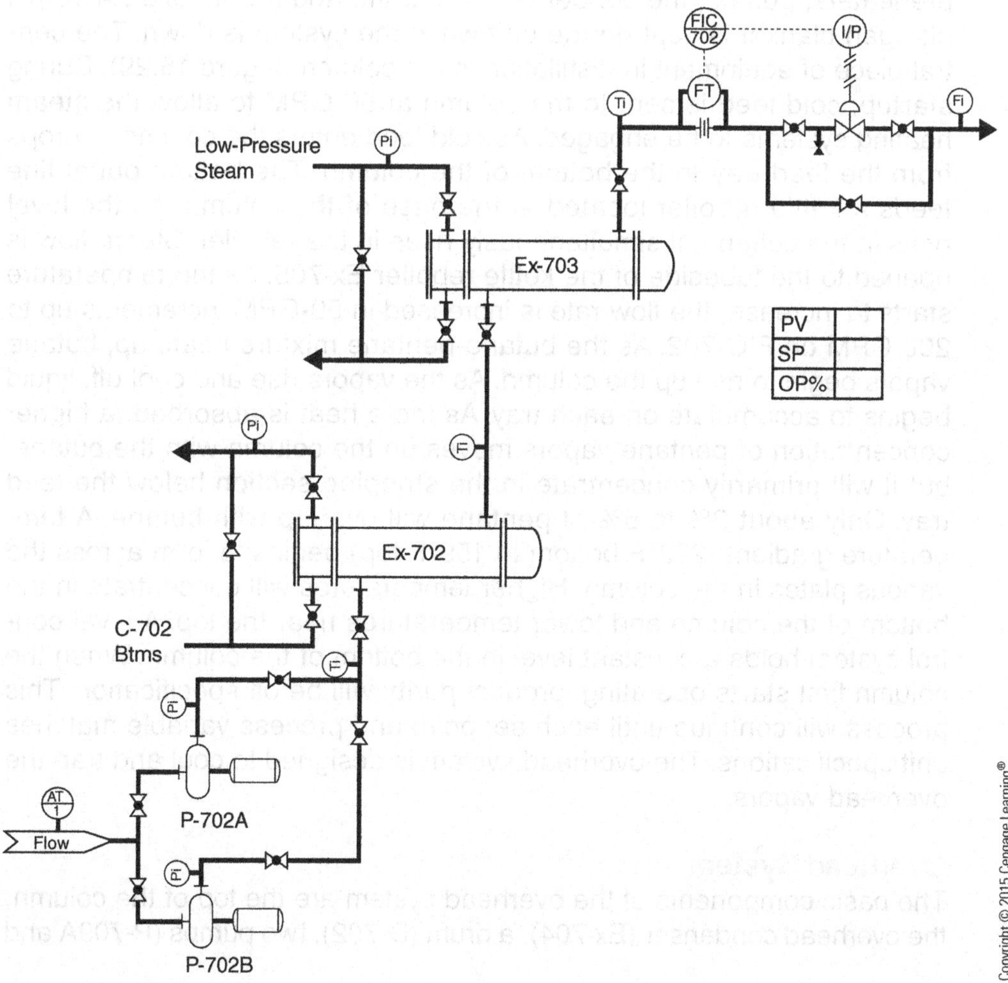

Figure 15.28 *Preheat System*

Each exchanger has a shell inlet and outlet and a tube inlet and outlet. The butane-pentane mixture flows through the shellside of each exchanger. Ex-702 and 703 are set up in series flow. A flow control loop (FIC-702) regulates fluid flow into the column. During operation, feed flows through the system at 200 GPM. The butane-pentane mixture enters Ex-702 at between 70° and 80°F and exits at 120°F. Feed temperature enters Ex-703 at 120°F and exits at 180°F. A low-pressure steam system is used as the heating medium for the tubeside of Ex-703. A thermocouple is located on the inlet feed line to the column. Steam flow to Ex-703 is controlled at 180°F with TIC-701. The hot pentane flowing from the bottom of C-702 is used as the heating medium for Ex-702. This process allows heat to be transferred from the pentane to the butane-pentane feedstock mixture. Correct valve lineup into and out of the exchangers is an important feature of operating the preheat system. A product analyzer continuously monitors feed composition.

Distillation Column

The basic components of the distillation process include the column, the preheaters, pumps, the condenser and drum, and the reboiler, A 10-psi nitrogen blanket is kept on the unit while the system is down. The central piece of equipment in distillation is the column (Figure 15.29). During startup, cold feed is sent to the column at 50 GPM to allow the steam heating systems to be engaged. As cold feed enters the column, it drops from the feed tray to the bottom of the column. The bottom outlet line feeds a kettle reboiler located at the base of the column. As the level rises in the column, it simultaneously rises in the reboiler. Steam flow is opened to the tubeside of the kettle reboiler Ex-705. As the temperature starts to increase, the flow rate is increased in 50-GPM increments up to 200 GPM on FIC-702. As the butane-pentane mixture heats up, butane vapors begin to rise up the column. As the vapors rise and cool off, liquid begins to accumulate on each tray. As more heat is absorbed, a higher concentration of pentane vapors moves up the column with the butane, but it will primarily concentrate in the stripping section below the feed tray. Only about 2% to 5% of pentane will overlap with butane. A temperature gradient (222°F bottom to 159°F top) begins to form across the various plates in the column. Higher temperatures will concentrate in the bottom of the column and lower temperatures near the top. A level control system holds a constant level in the bottom of the column. When the column first starts operating, product purity will be off specification. This process will continue until each set point and process variable matches unit specifications. The overhead system is designed to cool and trap the overhead vapors.

Overhead System

The basic components of the overhead system are the top of the column, the overhead condenser (Ex-704), a drum (D-702), two pumps (P-709A and

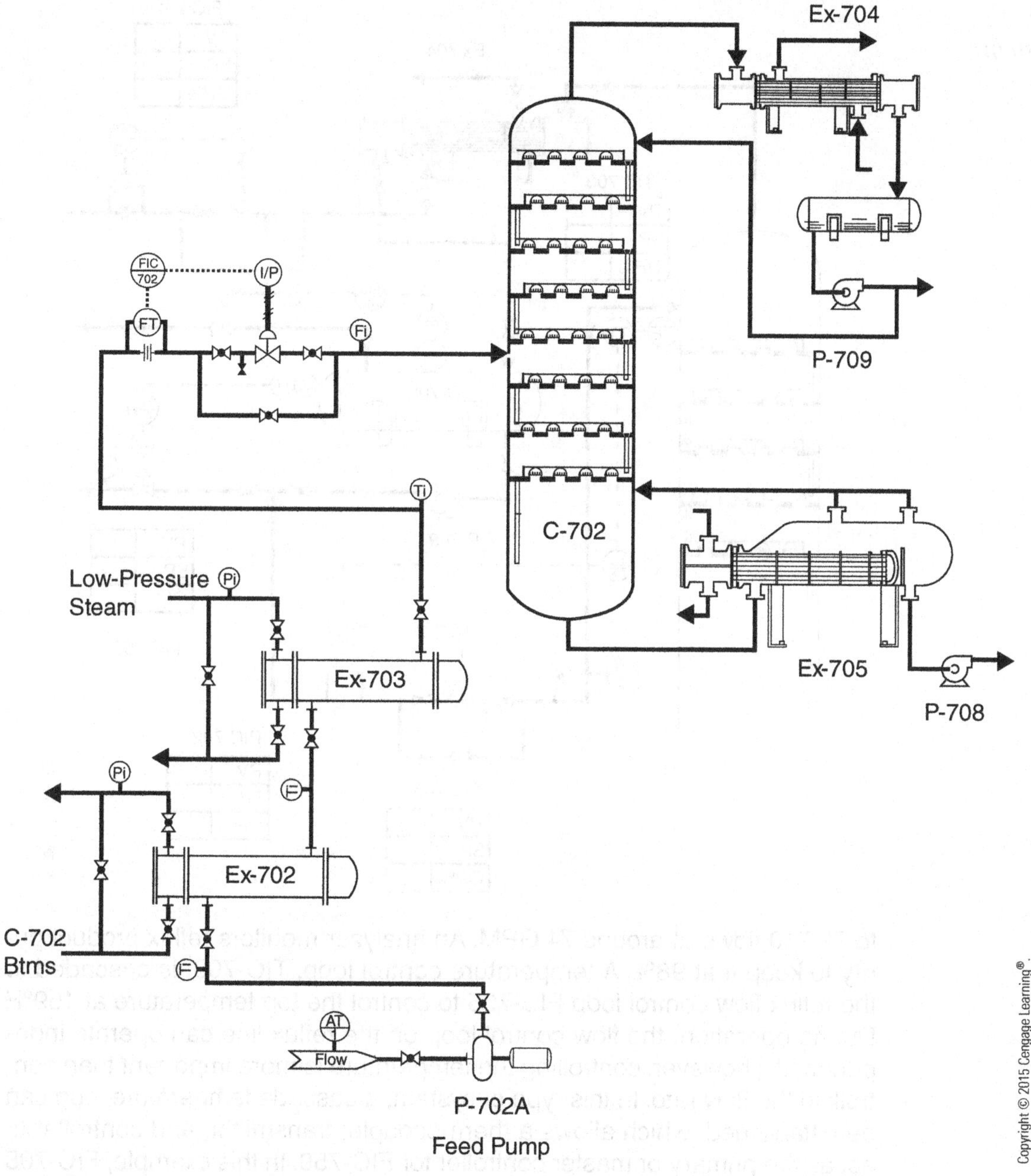

Figure 15.29 *Distillation Process*

P-709B), and a series of control loops (Figure 15.30). As hot butane vapors leave the top of the column, the overhead condenser partially condenses the vapors. Flow control loop FIC-703 automatically maintains 420 GPM cooling water to condenser Ex-704. Liquids are easier to transfer than vapors. A 50% level is typically maintained on D-702 by LIC-703. Pump 709 transfers feed from D-702 to the tank farm (Tk-710) or back to the top of the column as reflux. Reflux rates are maintained at 130 GPM, while product

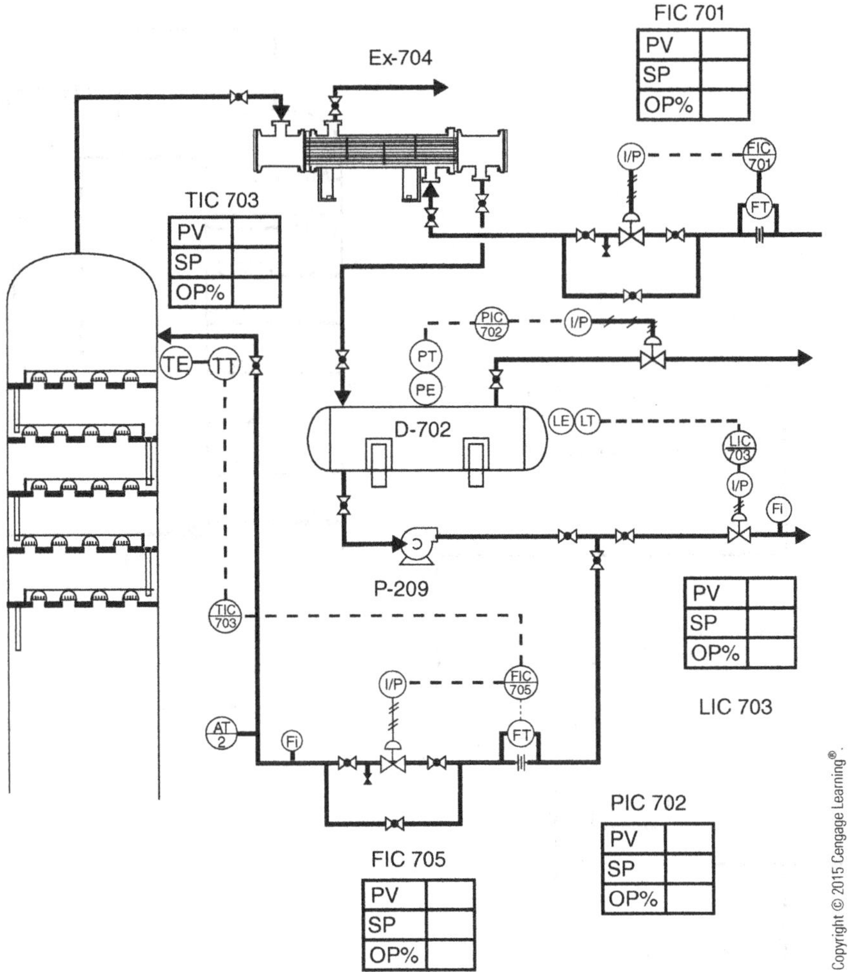

Figure 15.30
Overhead System

to Tk-710 flows at around 74 GPM. An analyzer monitors reflux product purity to keep it at 98%. A temperature control loop, TIC-703, is cascaded to the reflux flow control loop FIC-705 to control the top temperature at 159°F. During operation, the flow control loop on the reflux line can operate independently; however, controlling the temperature is more important than controlling the flow rate. In this type of system, a cascade temperature loop can be established, which allows a thermocouple, transmitter, and controller to act as the primary or master controller for FIC-750. In this example, FIC-705 and all five elements of the control loop act as the final control element for TIC-703. Seventy-four gallons per minute of 98% butane is sent to Tk-710 in the tank farm. A pressure control loop, PIC-702, regulates the pressure on the overhead system at 100 psi. Some minor butane loss occurs through this system. Excess vapors are directed into a flare header system.

Bottom System

Temperatures in the bottom of C-702 are controlled at 222°F. A kettle reboiler, Ex-705, utilizes a natural circulation feed system (Figure 15.31). Gravity flow provides a constant level to the reboiler, where levels are held

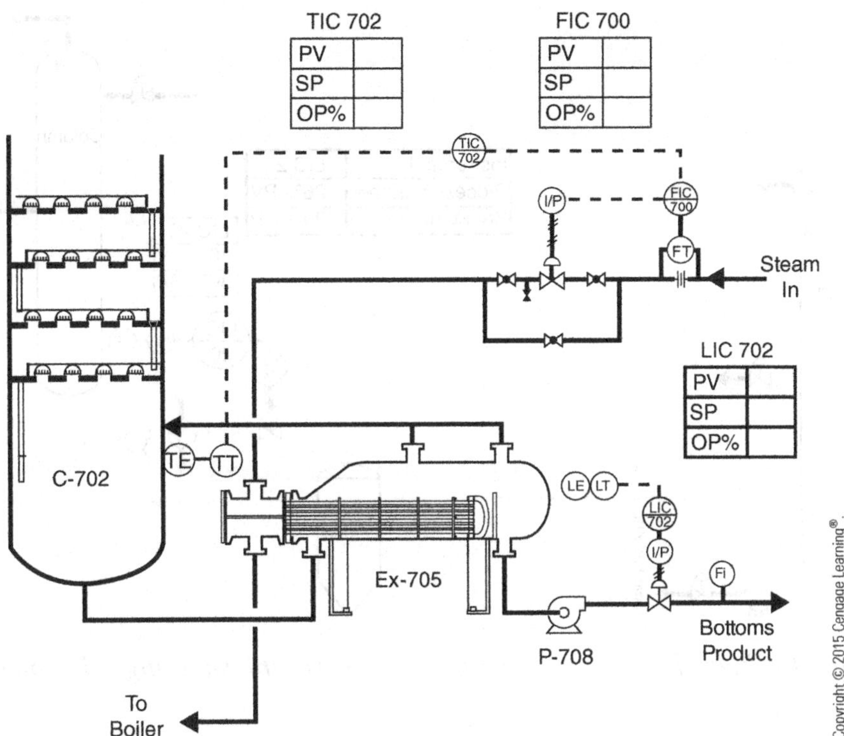

Figure 15.31
Bottom System

at 50% by LIC-702. This level allows the upper two tubes in the reboiler to be exposed above the liquid level to ensure good vapor return to the column. LIC-702 allows the 92% hot pentane to flow to Ex-702 and then to Tk-730 in the tank farm. Bottom flow rates are around 126 GPM. Correct lineups are important from the bottom of the column to the tank farm. FIC-700 controls steam flow rates to the kettle reboiler. TIC-702 has a thermocouple and transmitter mounted above the reboiler vapor inlet line on the bottom of the column. During lined-out conditions, TIC-702 operates as the primary or master controller for FIC-700. An analyzer monitors the composition of the pentane bottoms product.

Troubleshooting a Distillation System

Process technicians troubleshoot process systems in a variety of ways. One of the more common methods of troubleshooting is to check the difference between the set point and the process variable. In Figure 15.32, the system is operating correctly. The process variable and set point match exactly.

A process upset in progress can be identified in Figure 15.33. The set point is 30%, but the process variable is 26%. The difference could be a result of instrument error, pump problems, low bottoms level, or valve problems.

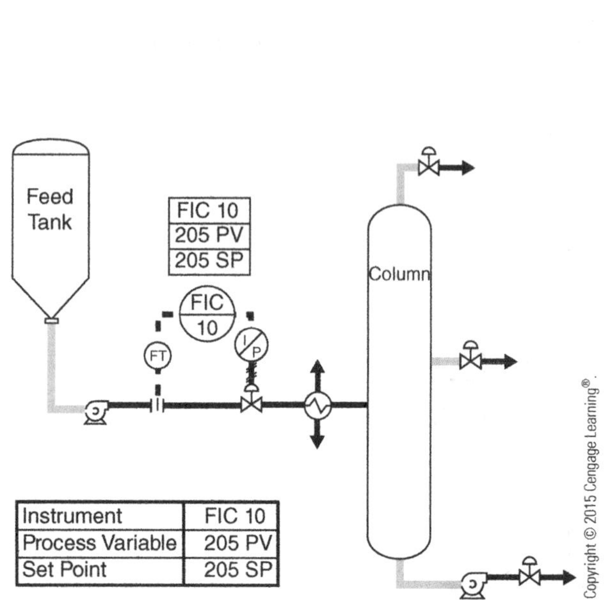

Figure 15.32 *Troubleshooting – Problem 1*

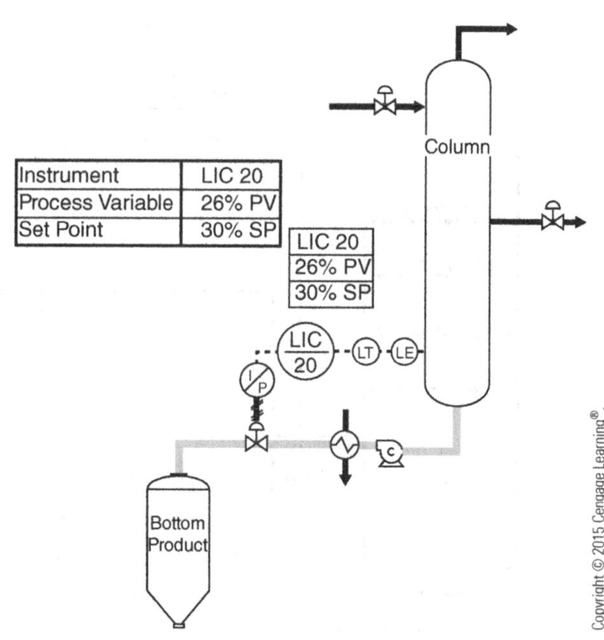

Figure 15.33 *Troubleshooting – Problem 2*

In Figure 15.34, a full-scale process upset is occurring. The initial problem was the partial failure of the steam valve (TIC-2). Look at the other process variables affected by the higher bottom temperature on the column.

Problems in a distillation column typically include feed composition changes, **puking**, flooding, and equipment or instrumentation failure. The feed has a specific composition. It may be a binary or **ternary mixture**. It must be controlled to ensure uniform operation. Puking is a problem that can upset trays, push products up or down the column or out the overhead line. Equipment or instrumentation failure can also damage the operation of a distillation column; temperature, flows, pressure, level, and composition variables are closely tracked. Other problems include flooding, which will result in liquid accumulation in the tower, increased pressure drop, and poor separation of components. Other problems include downcomer flooding, jet flooding, local flooding, weeping, and overloading. **Downcomer flooding** occurs when the liquid flow rate in the tower is so great that liquid backs up in the downcomer and overflows to the upper tray. Liquid accumulates in the tower, differential pressure increases, and product separation is reduced. **Jet flooding** occurs when the vapor velocity is so high that liquid down flow in the tower is restricted. Liquid accumulates in the tower, differential pressure increases, and product separation is reduced. **Local flooding** is when excessive liquid flowing down a column blocks vapor flow up the column in one section. Technicians will typically *break a flood* by decreasing the feed rate, the reflux rate, or the bottom temperature.

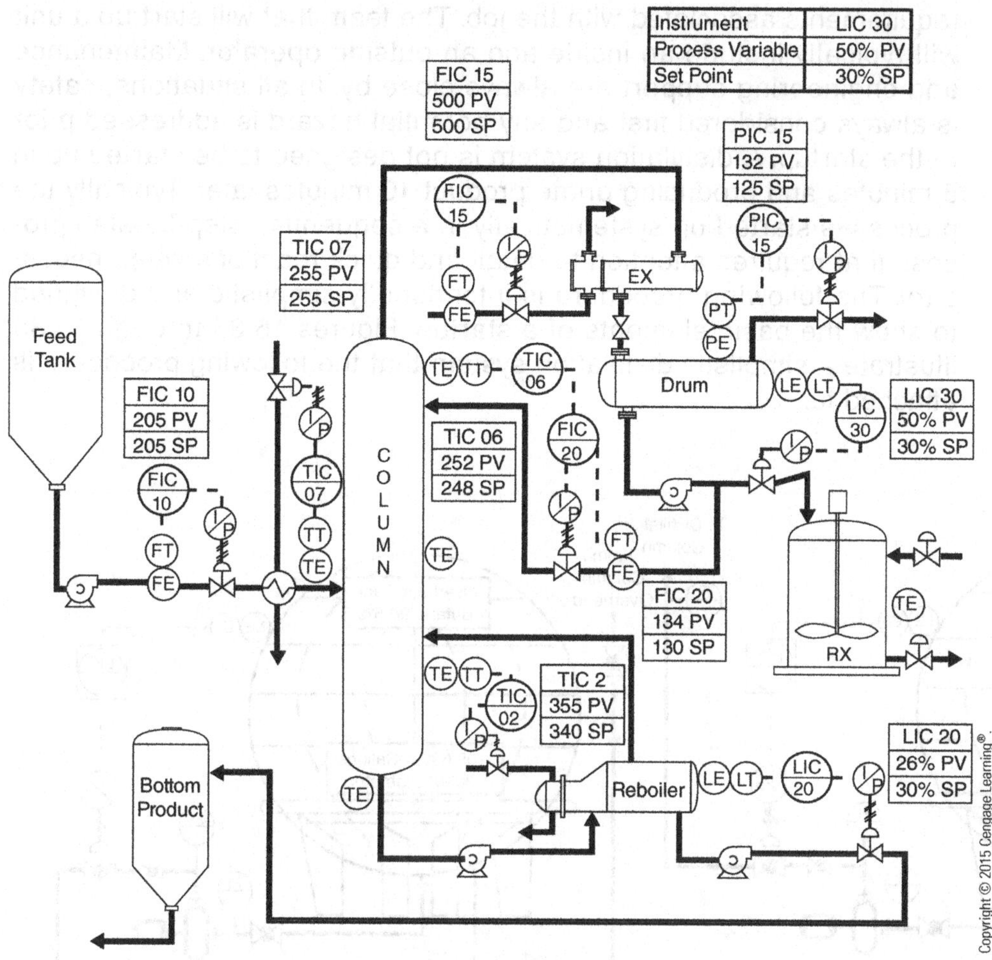

Figure 15.34 *Troubleshooting – Problem 3*

Overloading is best described as operating a column at maximum conditions. Weeping is a phenomenon that occurs when the vapor velocity is too low to prevent liquid from flowing through the holes in the tray instead of across the tray. Differential pressure is reduced and product separation is reduced. Another concept that occurs in distillation is the temperature at which the feed first boils, or initial boiling point, versus the final boiling point of the heaviest components. Final boiling point is the temperature at which the heaviest component boils.

Starting Up a Distillation System

Operating the distillation system requires teamwork and a good understanding of equipment and technology. Each unit has a standard operating procedure (SOP) that provides a systematic approach to unit startup. These procedures are not merely suggestions; rather, they are requirements associated with the job. The team that will start up a unit will typically include an inside and an outside operator. Maintenance and engineering support are always close by. In all situations, safety is always considered first and any potential hazard is addressed prior to the startup. A distillation system is not designed to be started up in 5 minutes and producing prime product 10 minutes later. Typically the process is started up systematically in a sequential, step-by-step process that requires attention to detail and quick reactions when necessary. The following procedure is intentionally simplistic and designed to show the basic elements of a startup. Figures 15.35 through 15.38 illustrate a simplistic distillation system that the following procedure is attached to.

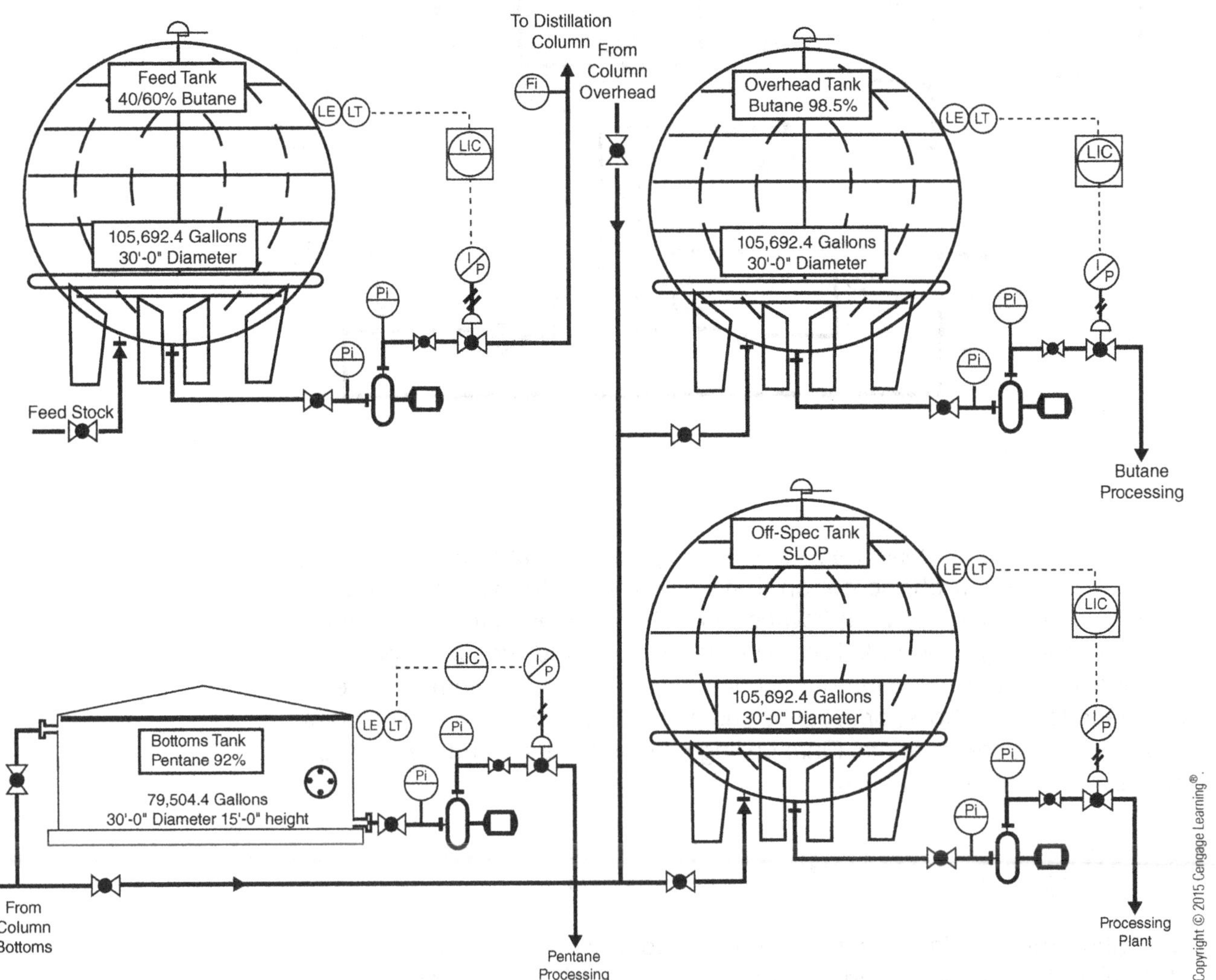

Figure 15.35 *Tank Farm*

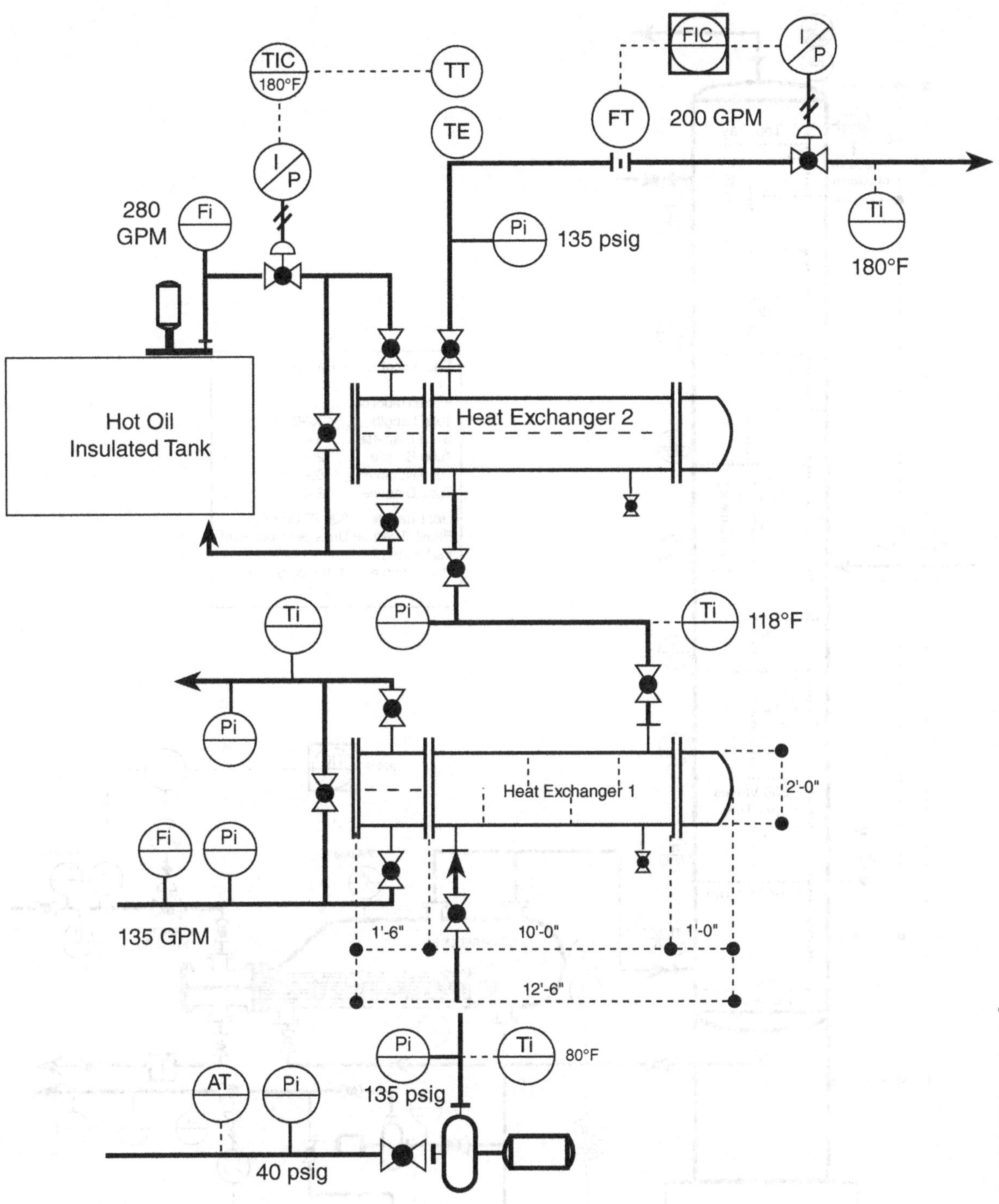

Figure 15.36 *Preheat System*

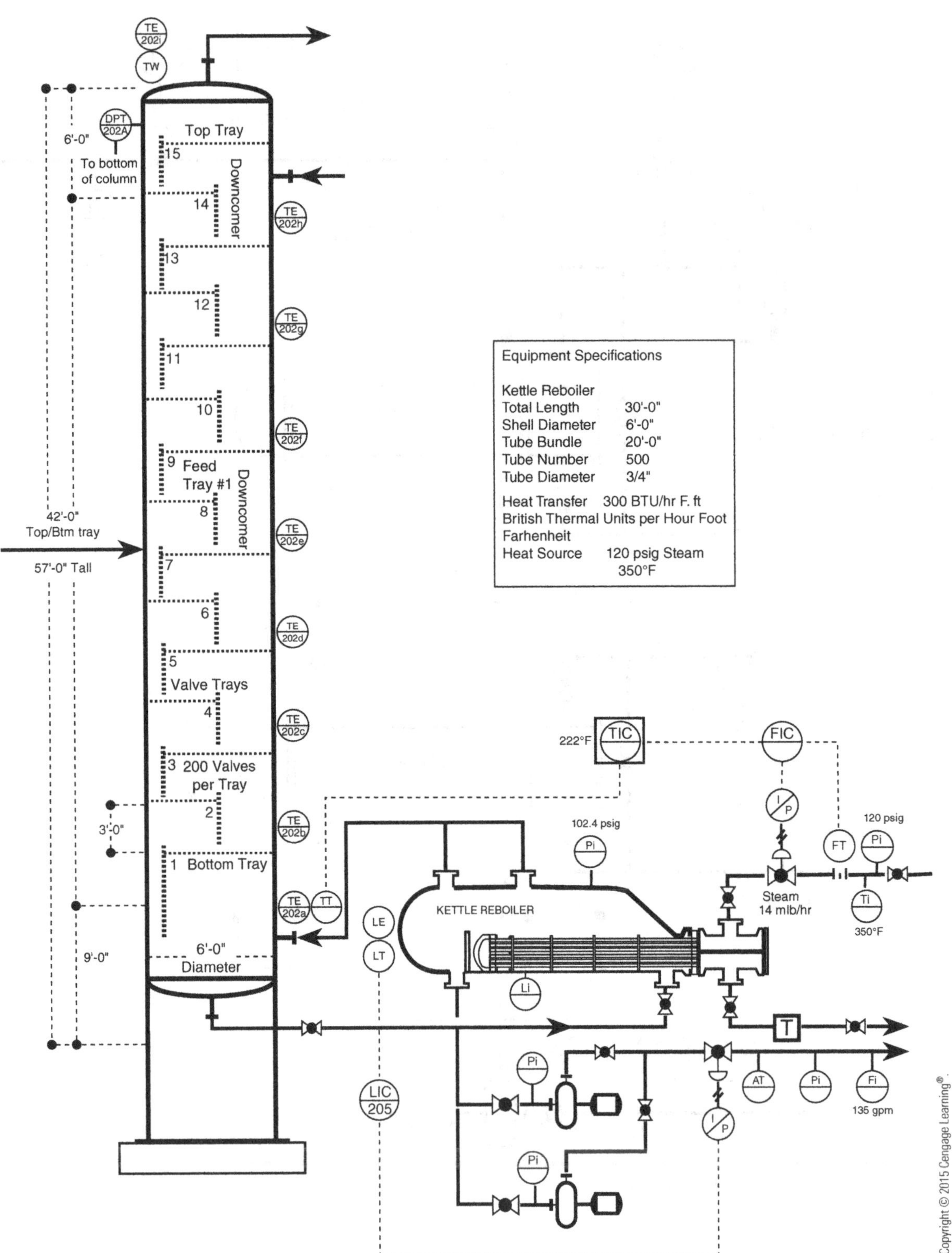

Figure 15.37 *Reboiler and Bottoms*

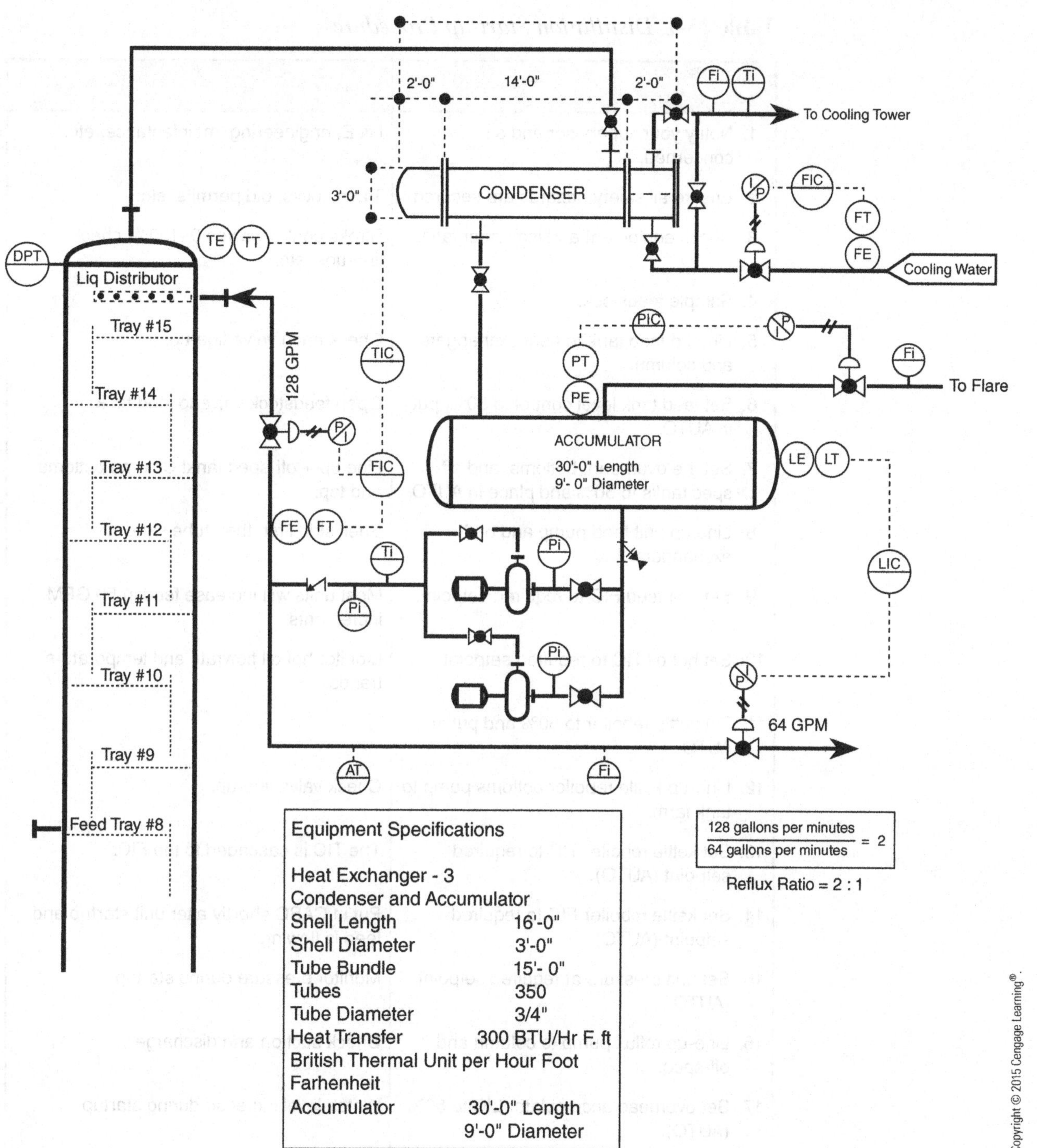

Figure 15.38 *Overhead System*

Table 15-1 is an example of a simple distillation system start-up.

Table 15-1 *Distillation Start-up Procedure.*

Procedure	Notes
1. Notify your supervisor and all concerned.	I & E, engineering, maintentance, etc.
2. Ensure all safety hazards are secured.	Trash, locks, old permits, etc.
3. Check equipment and instrumentation.	Stroke control valves 0–100%, check line-ups, etc.
4. Sample feedstock.	
5. Line up feed tank to heat exchangers and column.	Check each valve line-up.
6. Set feed tank level control to 50%, put in AUTO.	Open feedstock valve to feed tank.
7. Set the overhead, bottoms, and off-spec tanks to 50% and place in AUTO.	Line up-+ off-spec tank! Column bottoms and top.
8. Line up unit feed pump and both exchangers.	Shell side first, then tube side.
9. Set unit feed FIC to required setpoint.	Most units will increase feed in 50 GPM increments.
10. Set hot oil TIC to required setpoint.	Monitor hot oil flowrate and temperature trends.
11. Set kettle reboiler to 50% and put in AUTO.	
12. Line-up kettle reboiler bottoms pump to tank farm.	Check valve line-up.
13. Set kettle reboiler TIC to required setpoint (AUTO).	The TIC is cascaded to the FIC.
14. Set kettle reboiler FIC to required setpoint (AUTO).	Put in CASC shortly after unit startup and feed is flowing.
15. Set unit pressure at required setpoint (AUTO).	Monitor pressure during startup.
16. Line-up reflux pump to column and off-spec.	Check suction and discharge.
17. Set overhead accumulator LIC to 50% (AUTO).	Watch level increase during startup.
18. Set reflux FIC to required set point.	
19. Line up cooling water and set FIC to required setpoint.	Check suction and discharge line-ups
20. Start unit feed pump.	Watch levels in accumulator and reboiler.
21. Cascade kettle reboiler steam FIC to TIC.	

22. Start bottom and overhead pumps.	Do not start without a level.
23. Line up and start off-spec tank pump.	
24. Catch sample on overhead and bottom product.	Take samples to lab.
25. Watch on-line analyzers for top and bottom purity.	Put on-line when good sample comes back from lab.

The following information is a typical example of all of the data, process variables, and so forth recorded, collected, or observed by a process technician.

Sample Specification Sheet And Checklist

Flow			
1. FIC-201	225 GPM	CASC	Feed to Tank 202
2. FIC-202	225 GPM	AUTO	Feed flow
3. FR-202	225 GPM	AUTO	Feed flow to C-202
4. FIC-203	142.8 GPM	CASC	Reflux to C-202
5. Fi-204A	60.5 GPM	gauge	Butane to storage
6. FIC-300	525 GPM	AUTO	Cooling tower water flow
7. Fi-300	525 GPM	gauge	Cooling tower water flow
8. FIC-205	14 mlb/hr	CASC	Steam to reboiler
9. Fi-205	126.5 GPM	gauge	Pentane to storage
10. Fi-204B	10.4 CuFt/min		Butane/Liq.Cat to flare
11. Fi-205A	126.5 GPM	gauge	P-206 discharge
12. Fi-205B	126.5 GPM	gauge	P-207 discharge
13. Fi-204A	60.5 GPM	gauge	P-204C discharge
14. Fi-204B	60.5 GPM	gauge	P-204D discharge
Analytical			
15. AT-1	38%		Butane- feed
16. AT-2	98.5%		Butane- reflux
17. AT-3	1.5%		Butane- bottom
18. AT-4	61%		Butane- reactor
19. SIC-201	650 RPM	AUTO	Agitator motor on Tk-202
Pressure			
20. PIC-204	100 psig	AUTO	D-204 to flare
21. Pi-202A	40 psig	gauge	Suction-P-202A
22. Pi-202B	135 psig	gauge	Discharge-P-202A
23. Pi-202C	40 psig	gauge	Suction-P-202B
24. Pi-202D	130 psig	gauge	Discharge-P-202B
25. PI-100A	35 psig	gauge	Tube inlet Ex-203

(*continued*)

26. Pi-204A	105 psig	gauge	Suction P-204A.
27. Pi-204B	105 psig	gauge	Suction P-204B
28. Pi-204C	160 psig	gauge	Discharge P-204A/B
29. Pi-300A	50 psig	gauge	Lower tube inlet Ex-204
30. Pi-300B	45 psig	gauge	Upper tube outlet Ex-204
31. DPT-202A	2.4 psig	ΔP cell	Bottom and top of column
32. Pi-404	120 psig	gauge	Steam pressure
33. Pi-205A	102.4 psig	gauge	P-205A suction
34. Pi-205B	102.4 psig	gauge	P-205A suction
35. Pi-205C	135 psig	gauge	P-205A/B discharge
36. Pi-205D	102.4 psig	gauge	Ex-205 vapor cavity
Temperature			
37. TIC-100	180°F	AUTO	Tube inlet Ex-203 hot oil system
38. Ti-202A	80°F	gauge	Discharge P-202
39. Ti-202B	115°F	gauge	Discharge shell-side Ex-202
40. Ti-202C	180.5°F	gauge	Discharge shell-side Ex-203
41. Ti-202D	173°F	gauge	Tube outlet Ex-202
42. TR-100	180°F		Temperature recorder hot oil system
43. TAH-100	195°F		High temperature alarm
44. TIC-203	158.7°F	AUTO	Top tray of C-202
45. TE-202i	170.2°F		Tray #5
46. TE-202H	190.2°F		Tray #3
47. TE-202G	210°F		Feed Tray #1
48. TE-202F	222°F		Hat Tray
49. TE-202E	223°F		Bottom of C-202
50. Ti-300	125°F	gauge	Tube outlet Ex-204 Cooling water
51. Ti-205	350°F	gauge	Upper tube inlet on Ex-205 steam
52. TIC-205	221.7°F	AUTO	Below hat tray, bottom of C-202
53. Ti-204	128°F	gauge	P-204A/B discharge
Level			
54. LG-202	50%		Bottom of column
55. LIC-204	50%	AUTO	Drum-204 level control
56. LIC-205	50%	AUTO	Kettle reboiler level
57. LIC-204A	50%	AUTO	Tk-204A level
58. LIC-204B	50%	AUTO	Tk-204B level
59. LIC-205A	50%	AUTO	Tk-205A level
60. LIC-205B	50%	AUTO	Tk-205B level
61. Li-205	50%	gauge	Level over tube bundle in Ex-205
62. LIC-201	75%	AUTO	Level control Tk-202
63. LA-201	Hi-85%/Lo 65%		Located on LIC-201

Reflux ratio

Reflux ratio is best described as the amount of reflux returned to the tower divided by the amount of overhead product sent to storage. Figure 15.38 illustrates how reflux is applied to a distillation system. For example, a reflux flow of 128 GPM ÷ 64 GPM overhead product being sent to storage = 2:1 reflux ratio. Reflux is used to control both temperature in the rectifying section and product purity. As hot vapors exit the top of the column, a condenser is used to partially or totally condense the vapors. The level in the overhead accumulator is automatically controlled at 50%. The accumulator in Figure 15.37 has a 9-foot diameter and is 30 feet long. The condenser is designed to condense most of the vapor; however, some is used to control system pressure in the vapor cavity above the liquid in the accumulator. This pressure control process literally backs up and controls all of the pressure on each tray in the column, top tray to bottom. The system pressure in the accumulator and the liquid feed temperature, combined with the liquid height above the reflux pump, provide adequate pressure to the suction side of the pump. The reflux pump is designed to discharge product to storage and to the top tray of the column. One control valve is mounted on the reflux line and another on the product storage line. Frequently, the reflux flow controller is cascaded to a master temperature controller at the top of the column. The reflux rate and cooling water flow rate to the condenser can significantly impact product purity and temperature. Reflux ratios will vary between different applications and are carefully calculated before the first pump or pipe is installed. It is important for a technician to understand the complex relationship the reflux ratio provides to a successful operation.

Other Problems Encountered in Column Operation

Other problems associated with distillation column operation include; entrainment, foaming, and dry tray. **Entrainment** is best described as the upward movement of high-velocity, large-vapor droplets inside a distillation column. **Foaming** is the result of too much heat, impurities in the feed, or active turbulence on the trays above undersized downcomers. **Tray drying** occurs when feed rates to the column are low, when reflux rates are low, or when high upward vapor flow rapidly moves from tray to tray.

Summary

Distillation separates the basic components or fractions of a mixture by their individual boiling points. During the distillation process, a mixture is heated until it vaporizes, then is recondensed on the trays or at various stages of the column, where it is drawn off and collected in a variety of overhead, sidestream, and bottom receivers. The condensed liquid is referred to as the distillate; the liquid that does not vaporize in a column is called the residue.

A distillation tower can be compared to a series of stills placed one on top of the other. As vaporization occurs, the lighter components of the mixture move up the tower and are distributed on the various trays. The lightest component goes out the top of the tower in a vapor state and is passed over the cooling coils of a shell and tube condenser. As the hot vapor comes in contact with the coils, it condenses and is collected in the overhead accumulator. Part of this product is sent to storage while the rest is returned to the tower as reflux.

A reboiler maintains the heat balance on the tower by reboiling the components that have given up some of their energy during the distillation process. A kettle reboiler or thermosyphon reboiler is typically connected to the bottom of the tower; however, reboilers can be attached to the various sidestreams on the column. A kettle reboiler has a special vapor-disengaging cavity that is designed to remove the lighter components from the mixture and return them in the form of heat back into the column below the lower tray. Heavier components in the liquid flow over the weir in the kettle reboiler and out of the tower system. In contrast, a thermosyphon reboiler will vaporize the liquid and return it to the tower operation.

Equilibrium curves are used to determine the vapor composition above a boiling liquid and how one mixture differs from another. The larger the difference between boiling points, the easier it is to separate the fractions by boiling point.

The temperature in a tower can be affected by pressure, flow, level, feed composition, reflux rate, and upset trays.

Review Questions

1. What is the section of the tower located above the feed tray called?
2. What is the section of the tower located directly below the feed tray called?
3. Describe how the liquid and vapor products on each tray in a tower could run countercurrent to each other.
4. After the level builds up on a tray, it flows over the weir. Where does the liquid go when it flows over the weir?
5. What kind of energy do molecules in motion have?
6. What is the condensed gas taken from a distillation column called?
7. What is the bottoms product on a distillation column?
8. What is the term for the temperature at which a liquid turns to gas?
9. What two components does a distillation column balance?
10. What is the incomplete separation of a chemical in a distillation column called?
11. What is the condensed product that is pumped back to the top tray of a distillation column called?
12. What device is designed to maintain the heat in a column?
13. Name three types of trays found in a distillation plate column.
14. What is heat that cannot be measured called?
15. Name two types of distillation columns.
16. List four conditions that can affect the temperature in a distillation column.
17. Name three types of packing used in a distillation column.
18. Explain how to break a flood on a distillation column.
19. Compare and contrast dry trays and jet flooding.
20. Define "temperature gradient."

Extraction and Other Separation Systems

OBJECTIVES

After studying this chapter, the student will be able to:

- Describe the scientific principles associated with adsorption.
- Describe the extraction process.
- Describe the stripping process.
- Identify the basic equipment associated with extraction.
- Describe the crystallization process.
- Explain how a scrubber works.
- Identify the basic equipment associated with the adsorption process.
- Describe the solvent dewaxing process.
- Explain the basic components and operation of a sidestream-stripping column.

Key Terms

Absorber—a device used to remove selected components from a gas stream by contacting it with a gas or liquid.

Adsorber—a device (such as a reactor or a dryer) filled with a porous solid designed to remove gases and liquids from a mixture.

Concurrent flow—see *Parallel flow*.

Extract—the second solution that is formed when a solvent dissolves a solute.

Extraction—a process for separating two materials in a mixture or solution by introducing a third material that will dissolve one of the first two materials but not the other.

Feed—the original solution to be separated in liquid-liquid extraction.

Raffinate—in liquid-liquid extraction, material that is left after a solvent has removed solute.

Scrubber—a device used to remove chemicals and solids from process gases.

Solute—the material that is dissolved in liquid-liquid extraction.

Solution—a uniform mixture of particles that are not tied together by any chemical bond and can be separated by purely physical change.

Solvent—a chemical that will dissolve another chemical.

Extraction

One of the most frequently encountered problems in chemical process operations is that of separating two materials from a mixture or a **solution**. Distillation is one way of making such a separation, and it is perhaps the most frequently used method. Another useful method is **extraction**. Extraction is a process for separating two materials in a mixture by introducing a third material that will dissolve one of the first two materials but not the other.

There are three basic types of extraction: leaching, washing, and liquid-liquid. In leaching, a material is removed from a solid mass by contacting it with a liquid. Metals are removed from their ores by leaching. In washing, material sticking to the surface of a solid is removed by dissolving it in a liquid and flushing it away. You use this process every time you take a shower. In liquid-liquid extraction, all three materials are liquids, and the mixture is separated by allowing them to layer out by weight or density. This chapter focuses on the liquid-liquid extraction process.

Reasons for Extraction

Each method of separation has its own advantages and disadvantages, and several situations call for the use of liquid-liquid extraction rather than distillation or some other method.

In many cases, it is impractical to separate two chemicals by distillation because the boiling points of materials are too close together. In such a case, it is frequently possible to find a third chemical that will dissolve one of the two chemicals. In this situation, extraction would be a better method of making the separation than distillation.

Many chemicals are highly sensitive to heat and will degrade or decompose if raised to a temperature high enough for distillation. In this case extraction, which can usually be carried out at normal temperatures, would be a practical alternative.

Often, one of the materials to be separated is present in very small amounts. It might be possible to recover such a material by distillation, but it is usually much easier and more economical to do so by extraction.

Finally, the key requirement of any commercial process is that it be economical. In situations in which several alternative means of separating two chemicals could be used, the one that is the most economical is chosen. Because many relatively inexpensive solvents are available, and because the equipment required for an extraction operation is relatively simple, economic considerations often favor liquid-liquid extraction.

Liquid-Liquid Extraction Process

There are basically three steps in the liquid-liquid extraction process: (1) contact the solvent with the feed solution; (2) separate the raffinate from the extract; (3) separate the solvent and the solute. Step 3, recovery of the solvent and solute, is left to be done by some other process such as distillation. In liquid-liquid extraction, the feed is the original solution. The **feed** solution, containing the **solute** (the material that will be dissolved), is fed to the lower portion of the extraction column (Figure 16.1). The **solvent** (the material that dissolves the solute) is added near the top. Because of density differences, the lighter feed solution tends to rise to the top while the heavier solvent sinks to the bottom. As the two streams mix, the solvent

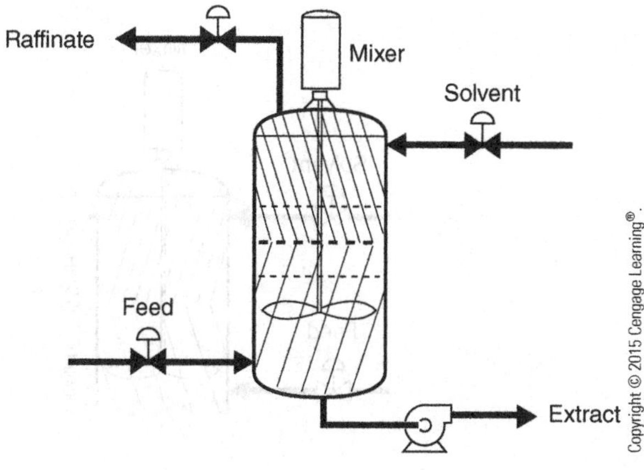

Figure 16.1 *Liquid-Liquid Extraction*

dissolves the solute. Thus, the solute, which was originally rising with the feed solution, actually reverses its direction of flow and goes out with the solvent through the bottom of the column. This new solution, consisting of solvent and solute, is called the **extract**. The other chemical in the feed stream, now free of the solute, goes out the top as the **raffinate**. The raffinate and extract streams are not soluble in each other and will layer out.

Properties of a Good Solvent

The solvent must be able to dissolve the solute, but it should not be a substance that will dissolve the raffinate or contaminate it. It also must be insoluble so that it will layer out. The density of the solvent should vary sufficiently from the density of the raffinate so that they can layer out by the effects of gravity. The solvent must be a substance that can be separated from the solute. It should be inexpensive and readily available, and it should not be hazardous or corrosive.

Equipment

Basically, the equipment for commercial operations is designed to ensure contact between the solvent and the feed and to separate the extract from the raffinate. The simplest extraction apparatus is the single-stage batch unit. In such an operation, the feed and solvent are added to a tank or some other suitable container (Figure 16.2). They are then thoroughly mixed by a mixer in the tank or by circulation in the tank. After the materials have been thoroughly mixed, the mixing is stopped, and the materials are allowed to layer out. The extract and raffinate layers can be removed.

Such a process could be converted to continuous operations by continuously adding the feed and solvent and continuously withdrawing the raffinate and extract. For such a process to be successful, some means of mixing or contacting the materials must be provided while still allowing ample space for the raffinate and extract to layer out. Many simple yet ingenious means have been employed for this purpose.

Most frequently, a single-stage device as described will not provide a perfect separation, and the raffinate must be contacted again with more solvent to

Figure 16.2
Single-Stage Extraction

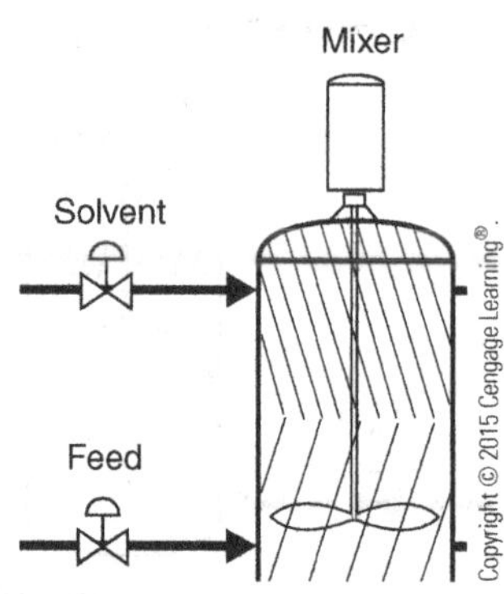

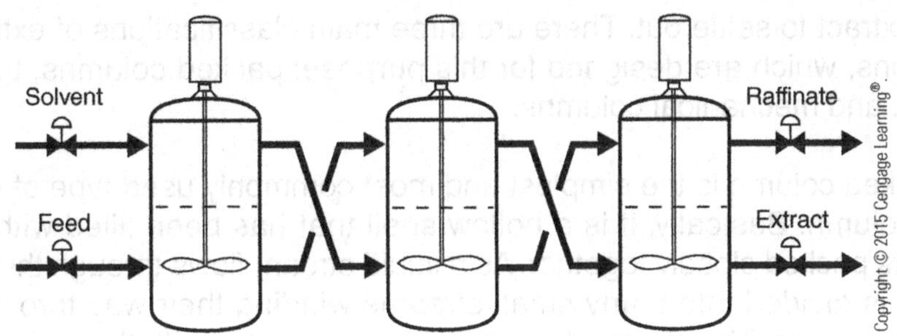

Figure 16.3
Concurrent Extraction

complete the removal of all solute. This problem leads to the concept of the multistage operation. In its simplest form, this could consist merely of a series of single-stage units, coupled close together, as shown in Figure 16.3. In this case, three stages are used, but obviously any number could be used. Notice that the solvent and feed both enter the system on the left and the raffinate and extract are removed at the right. The raffinate from each stage is the feed for the next stage, and the solvent for each stage is the extract from the preceding stage. Such a flow pattern is called **concurrent**; that is, the flows are in the same direction.

In effect, such an arrangement is an attempt to use several stages to accomplish the separation that could be accomplished in a single stage with perfect mixing by remixing the materials again and again. A more efficient approach would be to use a countercurrent flow arrangement, introducing the solvent at the opposite end of the chain of stages from the feed (Figure 16.4). In such a system, the feed to each subsequent stage is contacted with fresher solvent than was in the preceding stage, thus providing a more efficient operation. Such countercurrent flows are almost always used in commercial equipment to provide greater efficiency.

Extraction Columns

From the previous section, we have seen that an efficient extraction column should provide for continuous countercurrent flows. It must provide some means of mixing the solvent with the feed and yet allow the raffinate

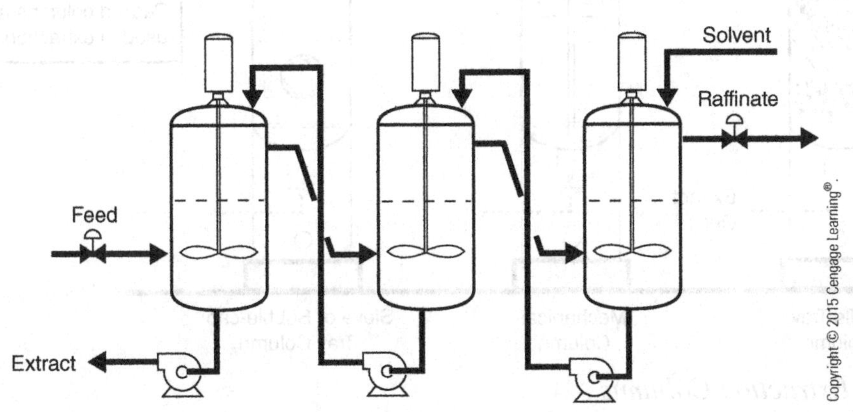

Figure 16.4
Countercurrent Extraction

and extract to settle out. There are three main classifications of extraction columns, which are designed for this purpose: packed columns, tray columns, and mechanical columns.

A packed column is the simplest and most commonly used type of extraction column. Basically, it is a hollow shell that has been filled with small objects packed closely together. As a liquid stream flows through this packing, it is divided into many small streams winding their way through the dense packing. The stream flowing upward competes with the stream flowing downward for the same passageways through the packing, resulting in enhanced surface contact. The types of packing commonly used in this process are rings and saddles.

Tray columns can also be used for liquid-liquid extraction. The tray designs are similar to those employed in distillation operations. Some common types are sieve, bubble-cap, and baffle trays (Figure 16.5). The principle of

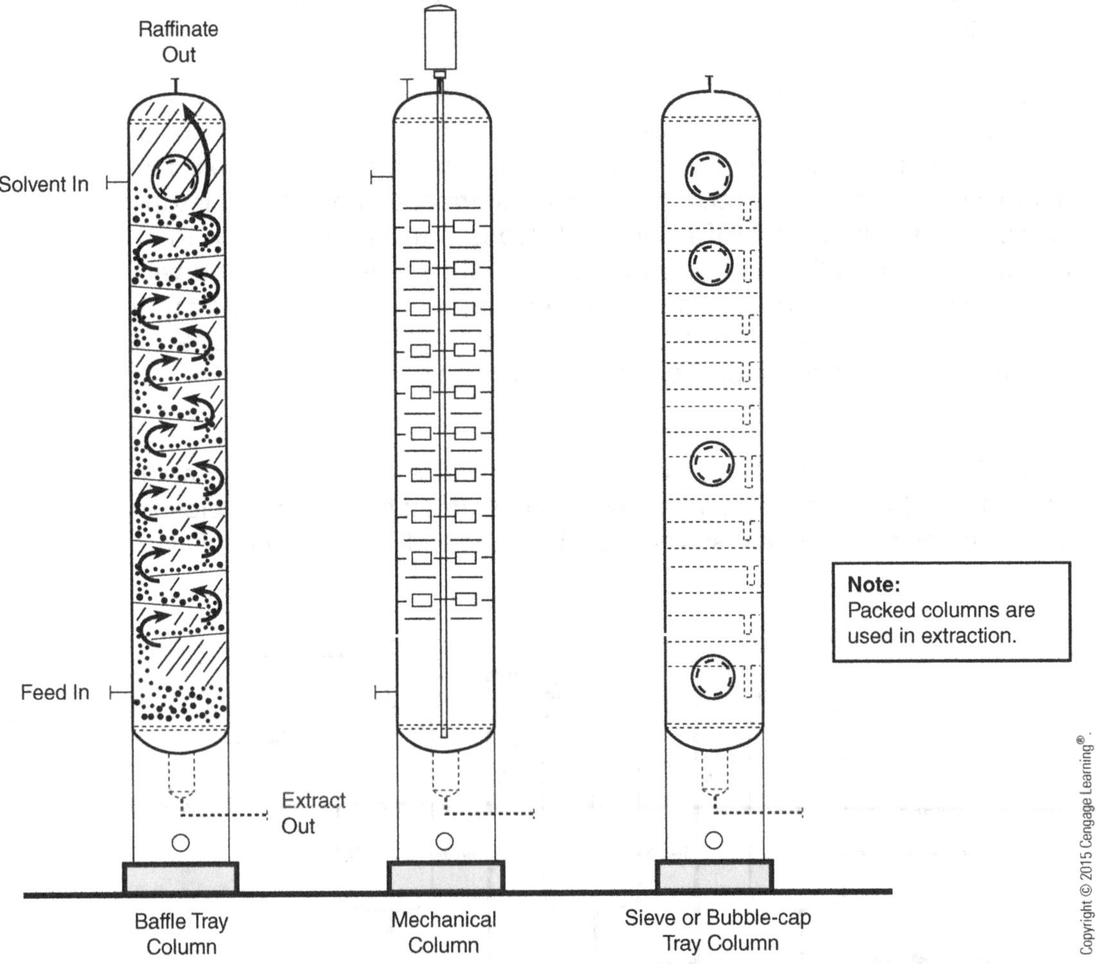

Figure 16.5 *Extraction Columns*

operation is similar to that for distillation. In the case of the sieve tray, one stream is made to flow across the trays while the other flows through the sieve holes. Contact is achieved as the tiny droplets of the rising liquid pass through the flow of the falling liquid across the top of the tray. A bubble-cap tray should perform similarly. A baffle tray is a device for breaking up the countercurrent flows to provide mixing. Baffle trays could take the form of disc and donut trays or crossflow trays. Obviously, contacting is less efficient with such an arrangement, but it is simpler and less susceptible to plugging than are sieve and bubble-cap trays. Finally, the newest and most complicated extraction columns are those with mechanical mixing. These employ some sort of rotating shaft with paddlewheels or other types of mixers affixed to the shaft. Such columns are used primarily when a difficult separation requiring a great deal of mixing must be made. Other mechanical equipment involves the use of ultrasonic vibrations for pneumatic pulsation. The vibrations thus established promote mixing of materials.

Extraction Column Terms and Principles

The contact area between the extract (heavy phase) and the raffinate (light phase) is called the interface. The term dispersed phase is used to describe the "one that bubbles through." The higher the feed rate, the more solvent required. The higher the concentration of solute in the feed, the more solvent required. The product customer specifies partial or total extraction operations. Temperature is not as important in the extraction process as in distillation unless it affects density or solubility or approaches the boiling points. Intimate contact with feed and solvent is required, so good distribution inside the column is needed.

Absorption Columns

An absorption column is a device used to remove selected components from a gas stream by contacting it with a gas or liquid. Absorption can roughly be compared to fractionation. A typical gas **absorber** is a plate or packed distillation column that provides intimate contact between raw natural gas and an absorption medium. Absorption columns work differently than typical fractionators because during the process the vapor and liquid do not vaporize to any degree. Figure 16.6 illustrates the scientific principles involved in absorption. Product exchange takes place in one direction, vapor phase to liquid phase. The absorption oil gently tugs the pentanes, butanes, and so on out of the vapor. In an absorber, the gas is brought into the bottom of the column while lean oil is pumped into the top of the column. As the lean oil moves down the column it absorbs elements from the rich gas. As the raw, rich gas moves up the column, it is robbed of specific hydrocarbons and exits as lean gas.

Figure 16.6 *Absorption and Stripping*

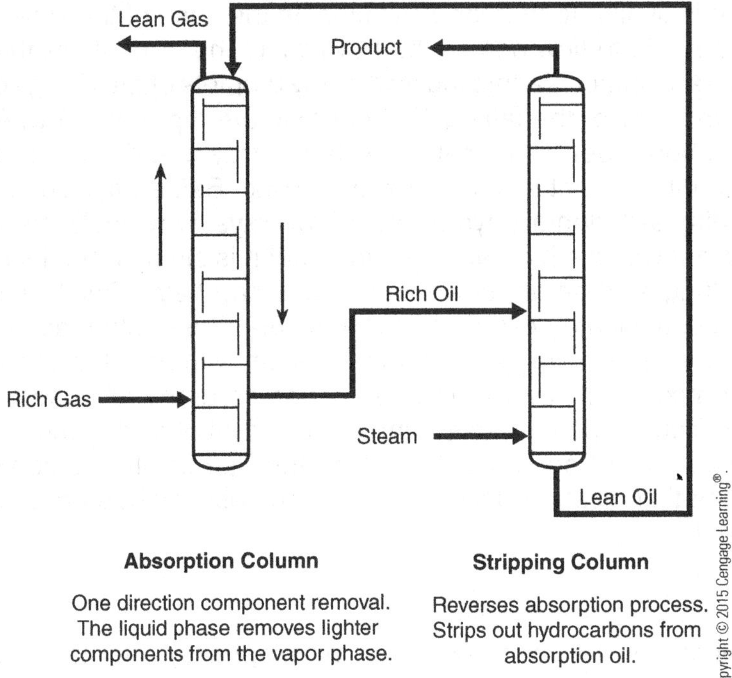

Stripping Columns

Stripping columns are used with absorption columns (see Figure 16.6) to remove liquid hydrocarbons from the absorption oil. To the untrained eye, a stripping column and an absorption column are identical. As rich oil leaves the bottom of the absorber, it is pumped into the midsection of a stripping column. Figure 16.6 illustrates how steam is injected directly into the bottom of the stripper, allowing for 100% conversion of Btus. As the hydrocarbons break free from the absorption oil, they move up the column while the lean oil is recycled back to the absorber.

Adsorption

Adsorption is the process in which an impurity is removed from a process stream by making it adhere to the surfaces of a solid. It should not be confused with absorption.

During the adsorption process (Figure 16.7), a column is filled with a porous solid designed to remove gases or liquids from a mixture. Typically, the process is run in parallel with a primary and secondary vessel. The adsorption material can be activated alumina or charcoal or a variety of other adsorption materials. The adsorption material has selective properties that will remove specific components of the mixture as it passes over it.

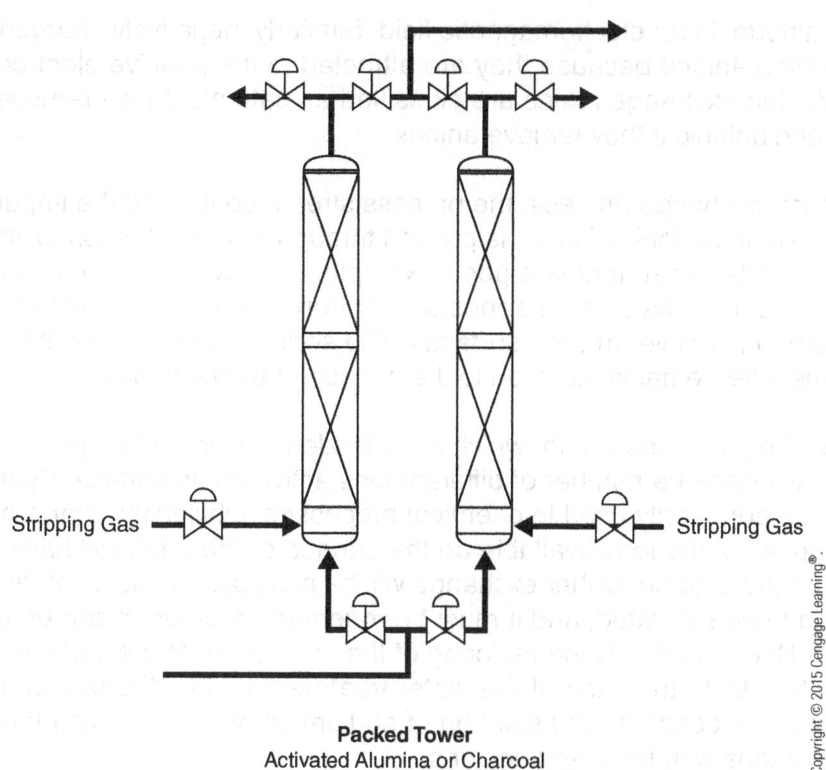

Figure 16.7
Adsorption Process

A stripping gas is used to remove the stripped components from the adsorption material.

During the adsorption process, the mixture to be separated is passed over the fixed bed medium (adsorbent) in the primary device. At the conclusion of the cycle, the process flow is transferred to the secondary device. A stripping gas is admitted into the primary device. The stripping gas is designed to remove or separate the selected chemical from the adsorption material. At the conclusion of this cycle, the stripping gas stops as the process switches back and repeats the process.

Adsorption processes exist in a variety of forms. Ion exchange, molecular sieves, silica gel, and activated carbon are all examples of adsorption. These processes are used in such widely varying applications as water softening and as cigarette filters. We will look at two of these processes, ion exchange and molecular sieves, in more detail.

Ion Exchange

Ion exchange resins are very small, beadlike particles that contain charged ions on their surfaces. You will recall from your chemistry course that ions are atoms that have gained or lost electrons in their outer orbits, thereby obtaining either a positive or a negative charge. A positively charged ion is called a cation because it would be attracted to the negative electrode, the cathode, in an electromagnetic field. Similarly, negatively charged ions are called anions because they are attracted to the positive electrode, or anode. Ion exchange resins are classified as cationic if they remove cations and anionic if they remove anions.

In an ion exchange process, the process stream containing the impurities, which are in an ionized form, is passed through a bed of the ion exchange resins. Water treatment is a good example. Hard water contains salts of metals such as calcium, magnesium, or iron. Water-treating resins have sodium ions active on their surfaces. The sodium ion replaces the "hard" ion; the latter remains attached to the surface of the resin bead.

Depending upon the use for which it is intended, an ion exchange resin may have any one of a number of different ions active on its surface. Hydrogen ions are commonly used in chemical processes. Obviously, over a period of time, all of the ions available on the surface of the resin will have been exchanged, and no further exchange will be possible. At this point, the bed is said to be saturated, and it must be regenerated before it can be useful again. Regeneration is the restoring of the original ion to the surface of the resin beads. In the case of the water treatment resins, the bed could be soaked in a concentrated solution of sodium chloride, replacing the hard metallic ions with fresh sodium ions.

Molecular Sieves

Molecular sieves are an example of a different type of adsorption process. The sieves are small, porous solids containing submicroscopic holes. These holes are actually about the size of an individual molecule. Some molecules will fit inside these holes; other molecules, because of their size or shape, will not. One application is the removal of traces of water from an organic chemical stream; the relatively small water molecule will fit inside the pores while the bulkier organic molecule will not. In the isosieve process, kerosene, containing both normal—or straight-chained—paraffins and isoparaffins with branched chains, is fed to a molecular sieve bed. The normal paraffins fit inside the pores, but the branches on the isoparaffin molecule prevent it from doing so. In this manner, a mixture of normal paraffins and isoparaffins can be separated. This separation cannot be made by more conventional means because the physical properties of the paraffins are too similar.

As in the case of the ion exchange resins, a point is reached when all the pores are filled and the bed is saturated. At this point it must be regenerated. Regeneration can be done by another, smaller molecule.

Equipment

Equipment for adsorption operation is relatively simple. It consists primarily of a tank or vessel containing a bed of adsorbent, be it ion exchange resins, molecular sieve, or whatever (Figure 16.8). In most cases the bed is fixed;

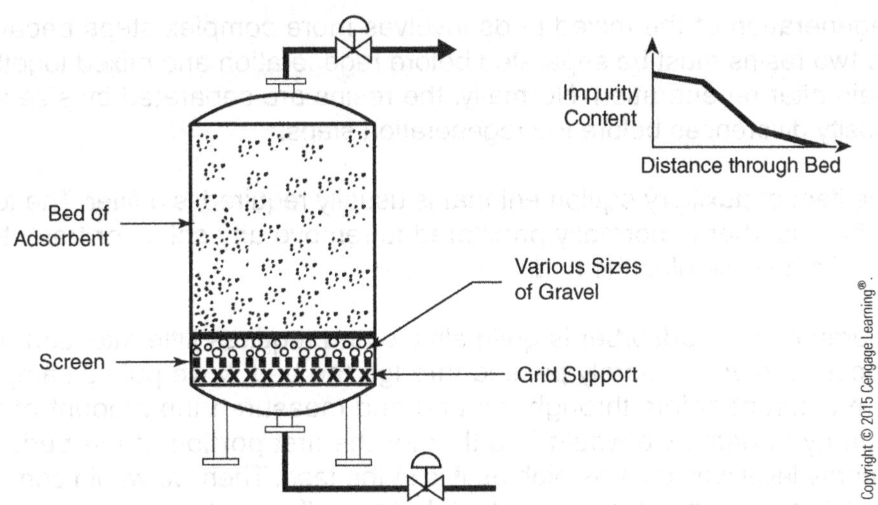

Figure 16.8
Fixed Bed Adsorber

that is, the bed is held in place while the process stream containing the impurities is passed through the bed. In a few cases, the beds are fluidized and are passed through the vessel countercurrently to the flow of the process stream. In such instances, the adsorbent particles are separated as they leave the vessel and are passed through a regeneration step. The regenerated particles are then recycled to the system.

The fixed bed type is by far the more common **adsorber**. It contains some type of bed support mechanism to hold the bed in the vessel. This would typically consist of some sort of grid support covered by a wire-mesh screen. In some cases, the screen is then covered with graduated sizes of gravel or other solid particles to support the adsorbent. Such support facilities are needed to prevent the very small adsorbent particles from being washed out of the bed. The process stream is usually fed through some sort of sparger or distributor to ensure even distribution across the cross-sectional area of the bed. The process stream is removed through a similar type of device, usually covered with a wire-mesh screen to avoid loss of the particles. The process flow through the bed can be from top to bottom or from bottom to top. There are also connections for treating the bed during the regeneration step.

In ion exchange processes, there are generally two types of fixed beds. Single-bed systems remove only one type of ions, either cations or anions. Two-bed systems remove both cations and anions. The type of bed used depends on the process involved and the desired quality of products.

In the case of the two-bed system, two different arrangements are possible. The cation and anion resin can be in separate layers or in separate vessels. This would be equivalent to two single beds in series. In the other arrangement, the cation and anion resins are together in a mixed bed.

Regeneration of the mixed beds involves more complex steps because the two resins must be separated before regeneration and mixed together again after regeneration. Normally, the resins are separated by size and density differences before the regeneration steps.

One item of auxiliary equipment that is usually required is a filter. The feed to the adsorber is normally prefiltered to remove any solids or trash that would otherwise plug the bed.

Operation of an adsorber is quite simple and requires little attention. The process stream is merely passed through the bed. If we pulled samples from different points through the bed and measured the amount of the impurity in each, we would find that for the first portion of the bed, the impurity level would be as high as it is in the feed. Then we would come to a point at which the impurity content dropped off very sharply in a relatively short distance through the bed. Beyond this point, the impurity level would drop off slowly again to essentially zero. If we came back later, we would get the same type of profile, but it would have progressed farther through the bed. Ultimately, we would reach a point when the leading edge of this wave would reach the end of the bed. The make stream begins to show a slight breakthrough of the impurity. From the foregoing discussion, we can expect at this point to have a sudden, major breakthrough of impurity very soon as the main portion of the wave reaches the exit. It is time to regenerate the bed.

Regeneration operations take various forms, depending upon the system. If the period of time required to saturate the bed is relatively short as compared with the time required for regeneration, there will usually be an alternate bed to which the feed is switched. At other times, the bed may just be bypassed or the unit shut down while regeneration takes place.

Regeneration usually takes place in four steps:

1. First, after the bed is bypassed, the process stream is displaced from the bed by draining it or displacing it with another chemical.
2. The bed is sometimes washed or flushed to remove any remaining process chemicals as well as solids, dirt, or trash.
3. The regenerating step itself involves soaking the ion exchange resins or purging the molecular sieves.
4. Finally, after regeneration, the bed must be returned to service. This step involves clearing the bed of the regenerative chemicals and preparing it for normal operations. In some cases, a step called *classification of the bed* is used. This is an operation in which flows are reversed through the bed, partially fluidizing the bed and loosening it up somewhat. At the same time, broken adsorbent particles and trash are washed from the bed.

Abnormal Operations

Because the operation of an adsorber is so simple, there is relatively little that can go wrong with it. Two problems that can arise are the disintegration of the adsorbent particles and the subsequent plugging of the bed. Factors that can cause destruction of the adsorbent particles are sudden or wide changes in temperature or the presence of unanticipated chemicals. Mechanical shock can also cause damage. As the particles break up, the bed gets packed more tightly and the pressure drop rises. Plugging from any other source, such as trash or dirt or scum, would have the same effect. For this reason, the pressure drop measured across the bed is an important indication of the status of the bed. Also, contamination, such as organics in a system designed for water-treating purposes, "blind" the ion exchange resins because the organic molecule is relatively large and will shield the ion exchange resins. For this reason, resins are specified for a specific purpose and any foreign material could completely negate the process.

Drying

A precise definition of drying that differentiates it from other unit operations, such as evaporation, is difficult to formulate. In general, drying means the removal of relatively small amounts of water or other liquids from solids or gases. Drying a solid is usually the final step in a series of operations, and the product from a dryer is often ready for storage or shipment to the customer.

The moisture content of a dried substance varies from product to product. Rarely, a product contains absolutely no moisture and is "bone dry." More often, the product has some moisture. Dried table salt, for example, contains about 0.5% moisture, dried coal about 4%. The moisture in air (i.e., humidity) varies with the temperature. *Drying* is a relative term in a process and simply means a reduction in moisture content from an initial value to a final one.

In many applications, air must be dried before it can be used effectively. If it is not dried, the small amount of water vapor in it can form water droplets and damage, contaminate, or cause a malfunction of the equipment. In other cases, a small amount of a vapor in the air or gas could contaminate the equipment or material with which it comes into contact.

A gas dryer uses the principle of adsorption to remove vapor. A literal definition of adsorption is to "take up and hold on the surface of a solid." The surface of a solid can trap gas and liquid molecules in such a manner that energy must be added to remove them. Different solids have different abilities to trap molecules, which depend upon the type of surface—smooth, rough, porous, and so on—and the type or shape of the molecule. Examples of this phenomenon are fogging up of the inside of an auto windshield and dew forming on grass.

A solid has a limit to how much vapor will adsorb on the surface. It is usually limited to the area where droplets of liquid form. The reason that this is the effective limit is that deposits can be easily dislodged, where a film of liquid normally has to be vaporized to be removed. In a flowing gas stream, dislodged droplets would simply retrain and no effective drying would occur. When this limit is reached, an adsorbent is called *saturated*. That is, the vapor pressure on the surface of the adsorbent is equal to the vapor pressure in the gas stream.

In some cases, when the solid becomes saturated, it is removed, thrown away, and replaced with a fresh, dry solid. In others, the solid can have the vapor removed, usually by heating; this method is called regenerating. The choice of method depends on the economics and time involved.

A regenerating gas dryer usually consists of two vessels (beds) containing a granular solid (Figure 16.9). The solid is specifically chosen for its high adsorbency of the target vapor and ease of regeneration. Piping and valves connect these beds such that the gas being dried enters only one bed; the solids in the other bed are separated from the process so that they can be regenerated. When the first bed becomes saturated, it is switched off the line, the second, regenerated bed is put on the line, and the first bed is regenerated.

Two methods are used to determine when a bed needs to be regenerated. One is to use a strict timed cycle and the other is to measure the moisture content of the outlet gas. When the moisture content reaches a preset value, it sends an alarm or automatically switches the beds.

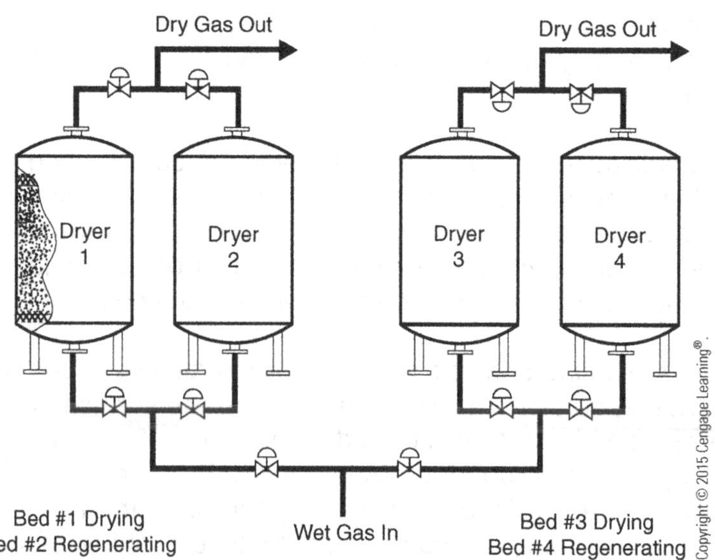

Figure 16.9 *Regeneration and Drying*

Regeneration can be done in several ways. One method is to heat the bed externally and pass a gas through it. Another method is to pass a gas through the bed while it is at a lower pressure. Basically, the liquid on the surface is vaporized into the gas stream and carried away because of the difference in vapor pressure. The regeneration of the solids in a gas dryer can be considered a small-scale operation of drying the granular product from a chemical plant. In the case of a final product, the quantities are usually so large that special equipment is used to handle it, but the principle is the same. The moisture adhering to the solid is vaporized and carried away by a gas stream because the vapor pressure of the moisture on the solid is lower than that in the gas stream.

Scrubber

A **scrubber** is a device used to remove chemicals and solids from process gases. Scrubbers are cylindrical and can be filled with packing material or left empty (Figure 16.10). As dirty gases enter the lower section of a scrubber they begin to rise. As these dirty vapors rise, they encounter a liquid chemical wash that is being sprayed downward. As the vapors and liquid come into contact, the undesirable products entrained in the stream are removed. As the dirty materials are absorbed into the liquid medium, they fall to the bottom of the scrubber, where they are mechanically removed. Clean gases flow out of the top of the scrubber and on for further processing.

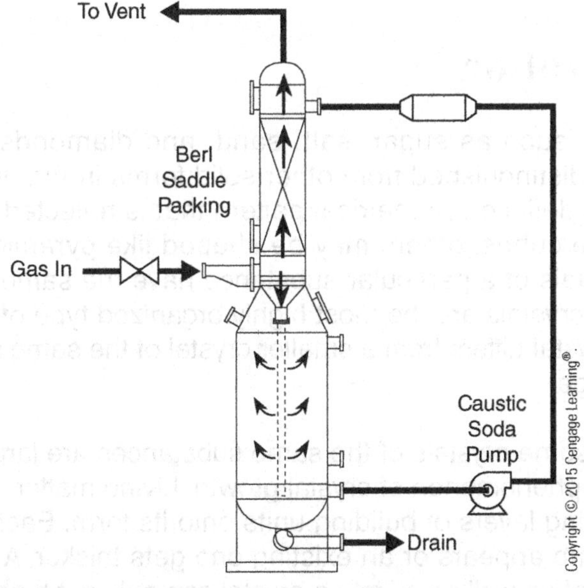

Figure 16.10
Scrubber

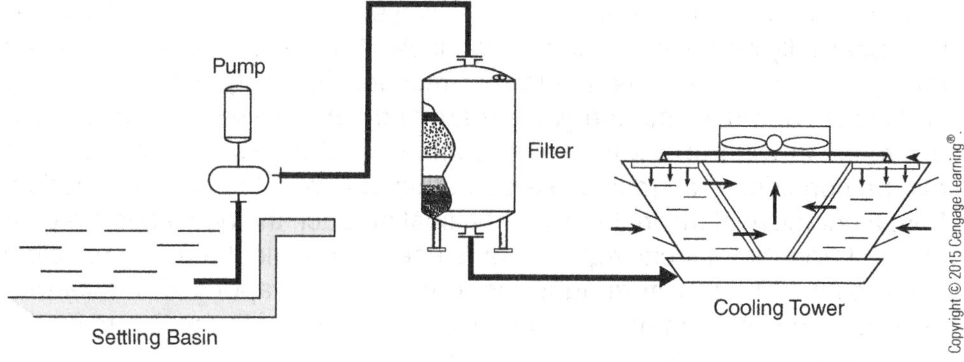

Figure 16.11 *Water Treatment System*

Water Treatment System

In the past, the chemical processing industry pumped a tremendous amount of water out of the ground for industrial applications. This process was stopped after it was discovered that as the water table dropped so did the surrounding countryside, a phenomenon called *subsidence*. The chemical processing industry now uses surface water for most industrial applications. Surface water, water that is drawn in from lakes, rivers, and oceans, must be treated before it can be used. Water treatment systems (Figure 16.11) are designed for that purpose.

A series of large pumps take suction off the basin and send water to a series of filters for additional purification. Some filtered water is sent to demineralizers for additional treatment to remove dissolved impurities.

Crystallization

Some solids, such as sugar, salt, sand, and diamonds, are crystals. They can be distinguished from other solid forms in that their atoms are arranged in a definite symmetrical pattern that is reflected in their shape. Some may be cubes; others may be shaped like pyramids. In the pure state, all crystals of a particular substance have the same shape and no other. In fact, crystals are the most highly organized type of nonliving matter. A large crystal differs from a smaller crystal of the same substance only in that it is larger.

The fact that some crystals of the same substances are larger than others illustrates the phenomenon of crystal growth. Living matter, such as a tree, grows by adding layers of building units onto its form. Each growing season, a new limb appears or an existing one gets thicker. A tree, however, can grow a leaf as well as a limb; a crystal can only get bigger.

Crystallization can be thought of as the exact opposite of dissolving. When salt is added to water, it dissolves into the water; the salt crystals grow smaller and smaller until they disappear. In crystallization, from seemingly nothing, solids form and they get bigger and bigger.

A crystalline substance, however, will not dissolve indefinitely. For example, salt dissolves completely in water only up to a point. If enough salt is added, eventually the crystals stop dissolving. The water has become saturated with the salt. No matter how much more salt is added, no more crystals will dissolve. Something must be changed in order for more crystals to dissolve. The water must become unsaturated by adding fresh water or heating the saturated water.

The exact opposite is done to cause a solution to "undissolve," or crystallize. The liquid is removed by vaporization or the solution is cooled. Either method causes the solution to become supersaturated (i.e., oversaturated) and crystals begin to grow.

Crystal growth is a very complex phenomenon, and many outside factors affect it, including the starting temperature, the starting concentration, the amount of mixing or lack of mixing, the rate of cooling or evaporation, contamination, and the presence of smaller crystals. The crystallization process can be either batch or continuous. The size of the crystals is controlled by adding liquid and removing crystals.

The equipment used in the crystallization process can range from a simple tank to a combination of pumps, agitators, condensers, and tanks (Figure 16.12). The primary purpose of this equipment is to control the operating conditions during the crystallization process.

In crystallization, the separation is possible because of differences in the size and type of molecules and differences in freezing points, melting points, and solubility. In many cases, crystallizers or centrifuges or both are used.

An example of the use of crystallizers is in the production of paraxylene. A feedstock of metaxylene, orthoxylene, and paraxylene is fed into a series of empty vessels. Inside the vessels are rotors that turn constantly. The rotor is attached to a series of mechanical arms (Figure 16.12). These arms reach to the sides of the crystallizer. At the end of each arm is a fibrous scraper blade. As the temperature is lowered in stages, paraxylene freezes. Paraxylene forms crystals that join together, increasing in size as it freezes. Metaxylene and orthoxylene have lower freezing temperatures and remain in a liquid state. The "cake" of paraxylene on the walls of the crystallizer is scraped off and dropped to the bottom, where it is pumped to a tank.

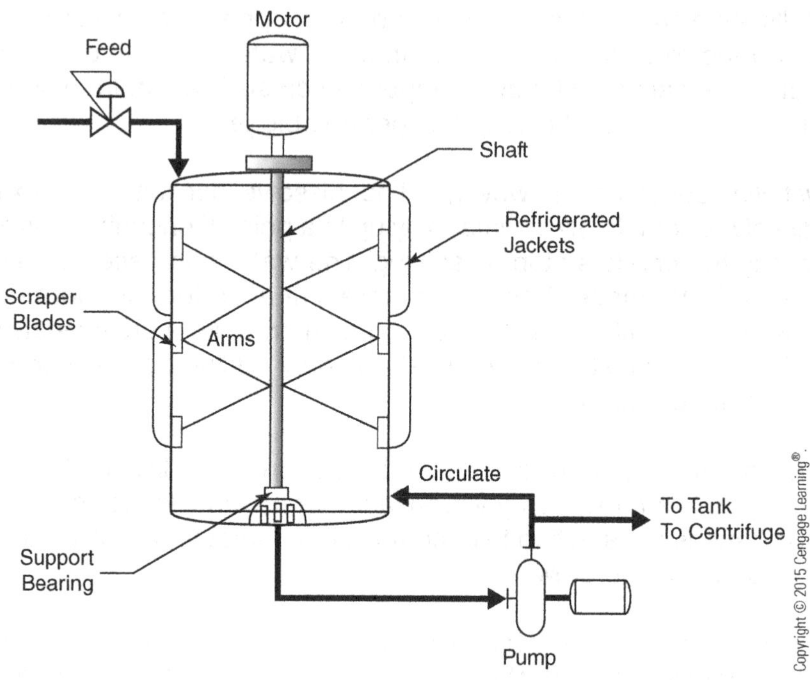

Figure 16.12
Paraxylene Process Crystalizer

From the tank, it is sent to a series of centrifuges. The centrifuges separate the cake paraxylene from the liquid and send it to a refinery tank. The crystallization process is the opposite of the distillation process. It uses refrigeration instead of heat to separate the various fractions. Economic considerations determine whether refrigeration or heating systems are used.

A centrifuge operates on the basis of centrifugal force. A centrifuge consists of a compartment spun around a central axis. Materials in the compartment separate on the basis of density. When a cooling process is used, the products separate easily because they become denser as they cool.

Solvent Dewaxing

Dewaxing is the removal of wax from lubricating oils. The wax content is critical in determining the viscosity of the lubricating oil. In cold weather, it is necessary for lubricating oil to retain its ability to flow and lubricate. Lubricating oils are classified into different grades on the basis of wax content.

Paraffin wax is made up of paraffin hydrocarbons. These hydrocarbons will crystallize when the oil is cooled. The wax then forms and can be separated by the filter in the crystallizer. Besides the filter method, there is also a press method of filtration. The presses, made with a series of cloth filters, eventually become filled with caked wax. The filters are then cleaned. In modern processes, solvents are added to oil to lower the viscosity of

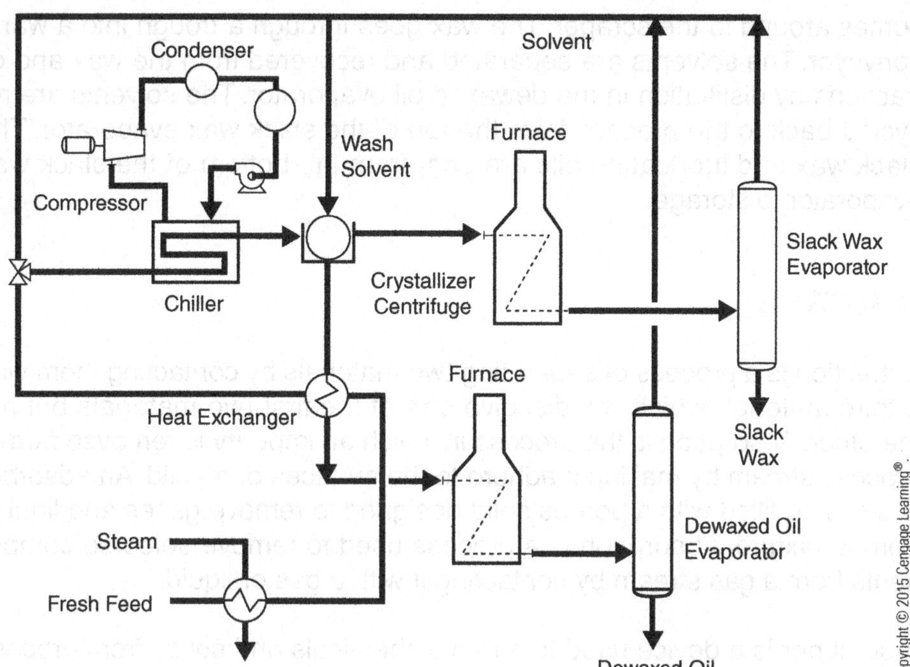

Figure 16.13
Solvent Dewaxing

the wax and make it easy to filter. Because of the low viscosity, rotary vacuum filters are used instead of discontinuous filters. The solvents can be adjusted to the feedstock to maximize the efficiency of the filtering. The two solvents used are toluene and methyl ethyl keytone (MEK). Toluene has the ability to dissolve the oil and keep it fluid when it is cold. MEK helps build wax.

The solvent dewaxing process has three steps:

1. The oil is mixed with the solvent and cooled.
2. The oil is filtered and the wax removed.
3. The solvents are recovered and recycled. Figure 16.13 shows a typical solvent developing process.

In step 1, the solvents are mixed with the waxy oil. The mixture is heated first to ensure thorough mixing and to ensure that the oil stays in solution. Next, the mixture is cooled in separate heat exchangers. The first brings the mixture to about Ð20°C. This cooling is done by exchange with the cold centrifuge filtrate. Next, the mixture is sent to a chiller. The chiller uses refrigerants, such as ammonia or propane.

The filter used in step 2 is a barrel made of a coarse metal mesh. The mesh is covered with cloth. The lower portion of the filter is in the oil and wax mixture. The filter turns slowly. The wax builds up on the cloth. To separate the oil, a vacuum draws the oil and solvent through the barrel. The barrel collects the wax on mesh surfaces. The caked wax is sprayed with cold solvent to wash off the remaining oil. The wax is then scraped off as it comes around to the scraper. The wax goes through a trough into a warm conveyor. The solvents are separated and recovered from the wax and oil fractions by distillation in the dewaxed oil evaporator. The solvents are recycled back to the process from the top of the slack wax evaporator. The slack wax and lubricating oils are sent from the bottom of the slack wax evaporator to storage.

Summary

Extraction is a process of separating two materials by contacting them with a third material, which will dissolve one of the first two materials but not the other. Adsorption is the process in which an impurity is removed from a process stream by making it adhere to the surfaces of a solid. An adsorber is a device filled with a porous solid designed to remove gases and liquids from a mixture. Absorption is a process used to remove selected components from a gas stream by contacting it with a gas or liquid.

A scrubber is a device used to remove chemicals and solids from process gases.

In crystallization, separation is possible because of differences in the size and type of molecules. This process relies on differences in freezing points, melting points, and solubility. In many cases, crystallizers and/or centrifuges are used. The crystallization process is the opposite of the distillation process. It uses refrigeration instead of heat to separate the various fractions.

Dewaxing is the removal of wax from lubricating oils. The wax content is critical in determining the viscosity of the lubricating oil. In cold weather, it is necessary for lubricating oil to retain its ability to flow and lubricate. Lubricating oils are classified into different grades on the basis of wax content.

Review Questions

1. Describe the adsorbtion process.
2. Describe the regeneration process.
3. Describe the scientific principles associated with absorption.
4. Describe the extraction process.
5. Describe the stripping process.
6. What are the three basic types of extractions?
7. Explain how a scrubber works.
8. Define solvent dewaxing.
9. Describe the principles associated with water treatment.
10. What is the difference between adsorption and absorption?
11. Which is heavier, the extract or the raffinate?
12. What are the properties of a good solvent?
13. Describe a mechanical mixing extraction column.
14. What are the three main types of extraction columns?
15. Describe how extraction and distillation work together.
16. What does *layer out* mean?
17. How does a plate extraction column operate?
18. Describe the crystallization process.
19. Crystallization is often referred to as the opposite of what other common process?
20. How does a paraxylene crystallizer operate?

Plastics Systems

Objectives

After studying this chapter, the student will be able to:

- Describe a typical plastics plant's equipment and operation.
- Review the granule storage and feed systems.
- Describe the granule blending systems.
- Describe the extruder system.
- Describe the product drying and storage system.
- Draw a simple PFD of the finishing section and label all equipment.
- Draw a simple PFD of the feed system; include the feed tank granule feeders, homogenizer, and extruder.
- List the various components of an extruder; describe each section.

Key Terms

Classifier—a device that separates good pellets from oversized or misshapen pellets in plastics manufacturing.

Diverter valve—an automatic valve used to divert the flow of solids in plastics manufacturing.

Dry additive—a dry chemical additive mixed with polymer to match customer requirements.

Dryer—a device in plastics manufacturing that dries pellets.

Extruder—a complex device composed of a heated jacket, a set of screws or one screw, a heated die, large motor, gearbox, and pelletizer in plastics manufacturing.

Homogenizer—a device in plastics manufacturing that mixes the streams of granules, peroxide, and additive by stirring with a large spiral auger as the material is moved along by the auger in a trough.

Pelletizer—a device with rotating knives that cut the strands of polymer as they leave the die holes in the extruder, producing pellets.

Screen pack—a group of 27 horizontal mesh filtering screens that traps any particles or foreign objects in a polymer stream.

Star feeder—a solids feeder connected to an air system that is used to transfer plastic.

Plastics

Plastic plants use chemicals such as coal, limestone, petroleum, salt, and water to make hard and soft plastic. Plastic is a synthetic material extracted from chemicals that can be found in any color and molded to any shape or size. The word plastic comes from the Greek word *plastikos*, which means "able to be molded." Plastic products are easy to make, long lasting, attractive, and relatively lightweight. Plastic products can fall into the following categories:

- *Plastic fibers and fabrics*—specially adapted for clothing and textile products
- *Transparent plastics*—such as optical lenses, watch crystals, contact lenses
- Decorative plastics—jewelry, novelties, toys, marbles, wall covering
- *Resistant plastics*—objects that are required to withstand heat and chemicals, such as automobile tires
- *Hard plastics*—football helmets, dinnerware, cameras, washing machines
- *Soft plastics*—flexible toys, squeeze bottles, cushions

Medical Applications

Medical-grade plastics are used for two main reasons: they do not hurt the human body, and they are not affected by acids produced in the digestive system. The medical field has embraced and woven plastics technology into their science. Screws, rivets, and plates are used to join bones. Surgical thread is made of synthetic resins. Plastic and transparent plastic syringes are used commonly. Synthetic intestines are used to replace damaged sections in the intestinal tract. Medical specialists use lightweight plastics to create artificial limbs.

Working with Plastic

Industrial manufacturers use synthetic resins to make paints; lubricants; adhesives; hard and soft plastic; plastic fibers and fabric; and resistant, decorative, and transparent plastic. Various methods are used in this process. In extrusion, solid resins are melted and pushed by a screw or screws through a heated chamber and specially shaped die (like using a hot glue gun). In molding, melted resins are squeezed into a mold (like making waffles). In laminating, sheets of paper or metal foil pass between rollers and are coated with melted resins. The sheets are pressed together to form materials such as electronic circuits (like making a sandwich). In casting, melted resin is poured into a mold (like making gelatin). In calendaring, melted resin is spread over sheets of paper or cloth to form a protective coating over materials such as playing cards (like syrup spreading on a pancake).

Plastics Plant

In today's fast-paced society, plastic products play a valuable role. Polypropylene plants produce plastic that is used by many customers. These products include packaging film for food and clothing, upholstery and carpet backing, disposable baby diaper liners, twine, molded products for appliances, and food containers. Polypropylene plants are composed of four major sections: laboratory, mechanical, polymerization, and finishing.

Laboratory. Laboratory technicians perform a variety of simple and complex tasks related to the verification of product quality. The size and scope of the organization and the variety of product lines will determine the areas of specialization within the lab. Laboratory technicians can be referred to as chemists, chemical technicians, research technicians, quality control technicians, or pilot plant operators.

Mechanical. The mechanical section provides electrical, mechanical, instrument, and machinist support for both the finishing and polymerization sections.

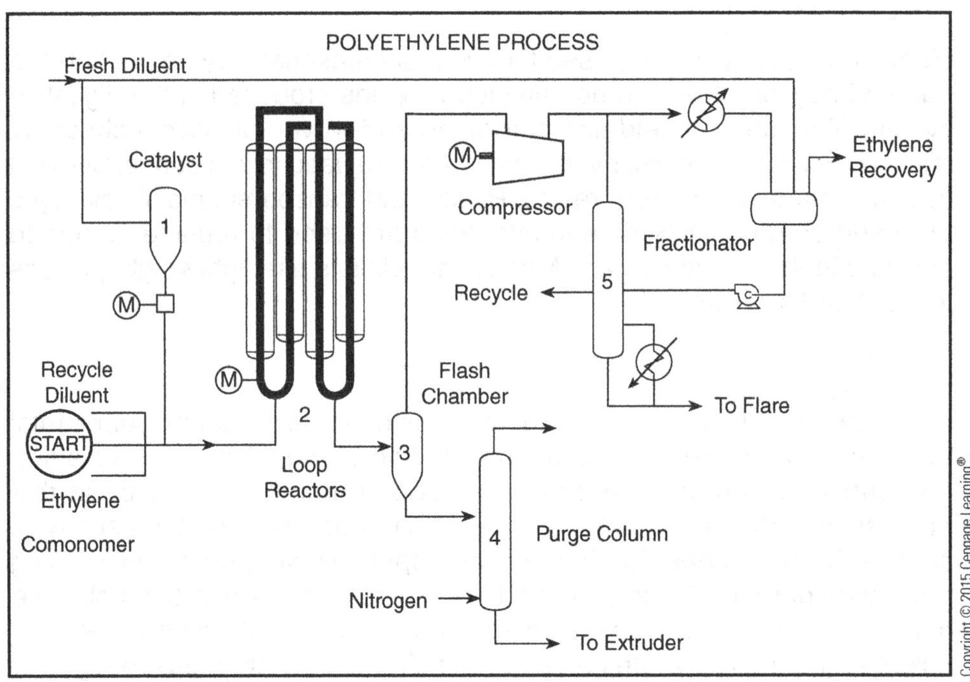

Figure 17.1
A Polyethylene Section

Polymerization. Polymerization facilities use advanced technology to combine chemicals into plastic resins. Molecules of synthetic resins form long chains. Each link of the chain is called a monomer. The entire chain is called a *polymer*.

Creating synthetic resins can be thought of as polymer building. The polymerization unit builds polymers by combining chemical compounds. As these compounds are combined, various reactions occur. These reactions cause certain atoms to cluster together and form the monomer links. These monomers combine to form chains of molecules. This process is referred to as polymerization. Figure 17.1 shows the major equipment found in a polyethylene unit.

Finishing. Controlling customer product specifications is the primary concern of the finishing section (Figure 17.2). Granules are received from the polymerization section and modified through special extrusion techniques to meet the customer's needs. Using a variety of statistical methods and state-of-the-art computer control, the finishing section produces the high-quality products. Plastic plants produce polymer in two forms: granules (small particles that are similar to granulated detergent) and pellets (thin, round, sometimes flat or disc like; similar to a BB).

In the finishing section, granules are prepared for shipment or compounded with additives and peroxide (to control melt flow rate [MFR]) to meet customer specifications. The extruded pellets provide customers with a product that has improved solid flow characteristics.

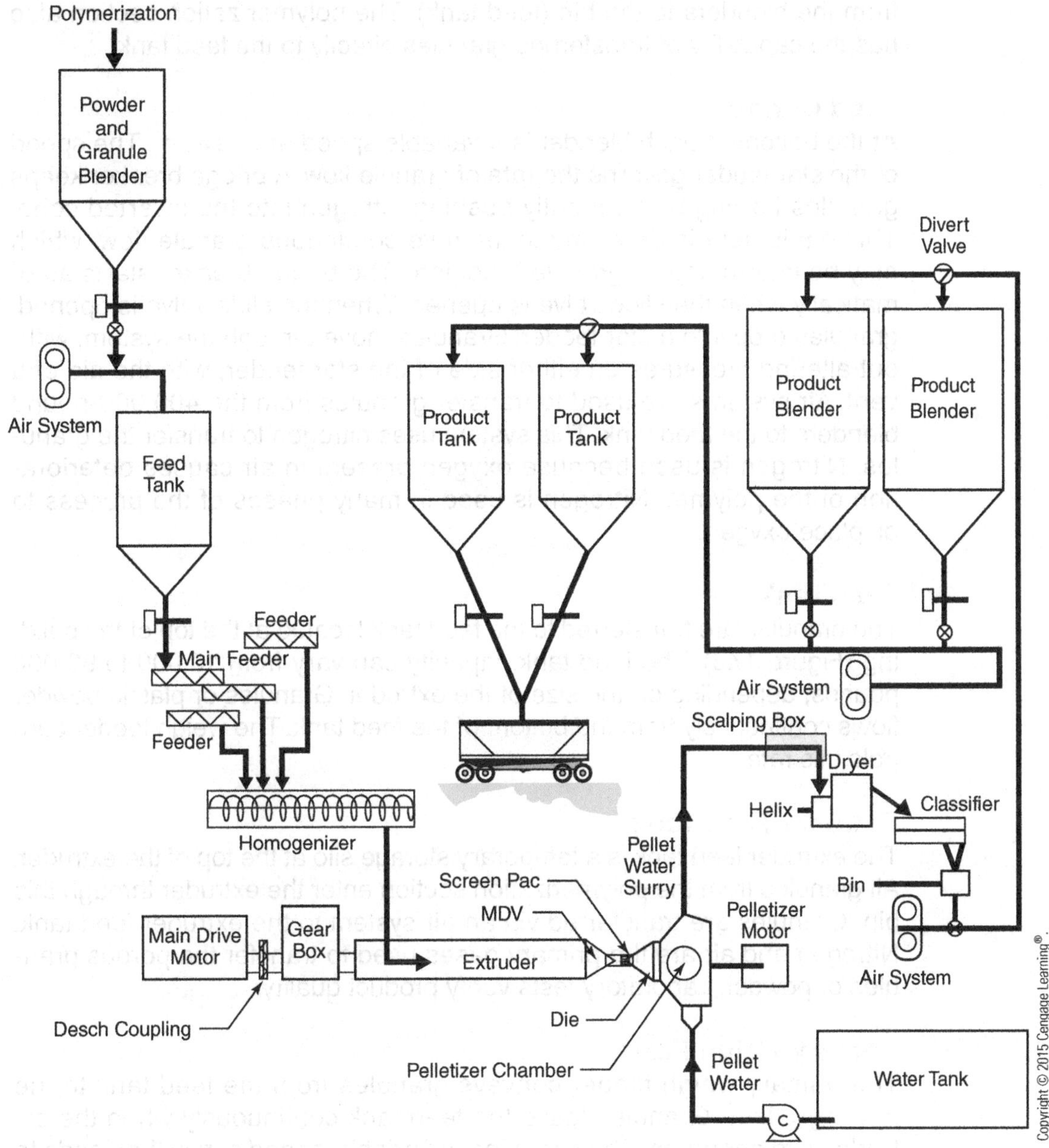

Figure 17.2 *Finishing Section*

Granule Storage and Feed Systems

Granule Blenders

Solid polypropylene granules are received from the polymerization section and stored in the granule blenders. Laboratory tests verify product quality. There are usually two 400,000-pound bins dedicated to supplying the **extruder** with granules. During extruder operation, granules are transferred from the blenders to the bin (feed tank). The polymerization section also has the capability of transferring granules directly to the feed tank.

Feed Control

At the bottom of each blender is a variable speed **star feeder**. The speed of the star feeder governs the rate of granule flow. A bridge breaker keeps granules flowing by frequently pushing nitrogen into the inverted cone. This equipment is designed to improve continuous granule flow, which may be interrupted by granule "bridging." The bridge breaker starts automatically when the slide valve is opened. When the slide valve is opened, granules drop into a star feeder. Granules move through the system, without altering pressures on either side of the star feeder, with the aid of a vent. Air systems are used to transfer granules from the 400,000-pound blenders to the feed tank. This system uses nitrogen to transfer the granules. Nitrogen is used because oxygen present in air causes deterioration of the polymer. Nitrogen is used in many phases of the process to displace oxygen.

Feed Tank

The granules are transferred to the feed tank located at the top of the building (Figure 17.3). The feed tank capacity can vary from 10,000 to 90,000 pounds, depending on the size of the extruder. Granules or plastic powder flows continuously from the bottom of the feed tank. The weigh feeder controls this rate.

Extruder Feed Tank

The extruder feed tank is a temporary storage silo at the top of the extruder. All granules from the polymerization section enter the extruder through this bin. Granules are transferred via an air system to the extruder feed tank. Nitrogen and air are the primary gases used to transfer the porous granules or powder. Laboratory tests verify product quality.

Primary Weigh Feeder

The primary weigh feeder conveys granules from the feed tank to the **homogenizer**. Granules leave the feed tank continuously when the extruder is in operation. They drop into a variable-speed screw that leads to a scale. From the scale, a constant-speed screw delivers granules to a discharge line. The speed of the first screw is adjusted to maintain a constant scale weight. The result is a feed that is constant by weight. The main feeder scale is very sensitive. Standing on it or leaning against it knocks the instrument out of calibration. Even the presence of external dust upsets the delicate balance of the machine. An increase in the pressure downstream of the homogenizer will slow the flow of granules. Layers of powder sometimes develop near the top of the feed tank. When this material enters the weigh feeder, the weigh feeder responds to the change, but the impact is sudden enough to cause a feeder upset.

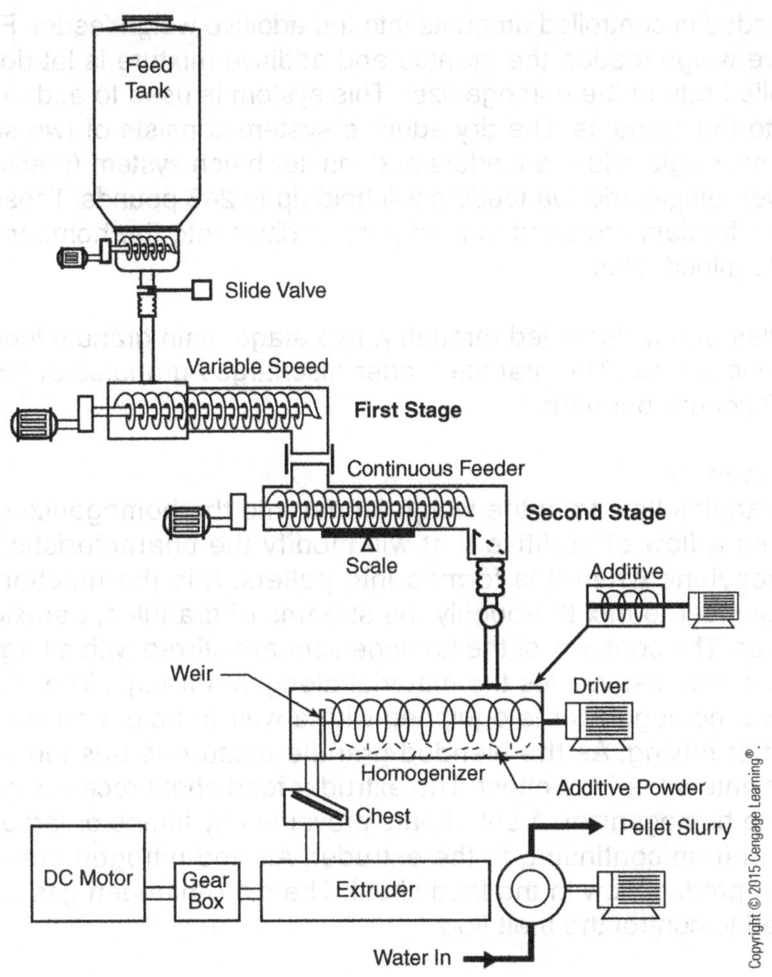

Figure 17.3
Main Feeder System

Blending Systems

Ribbon Blender and Additive Weigh Feeders

The ribbon blender is used to mix additives with granules to form a concentrated master batch. Most mixes total 5,000 pounds, but the overall capacity of the ribbon blender is 10,000 pounds. An automatic valve opens or closes to allow the contents of the ribbon blender to drop periodically into the additive weigh feeder. Pure additive weigh feeders are used to add the master batch to the homogenizer. A set of peroxide pumps meter a peroxide mixture into the homogenizer to control MFR. Pure additive feeders are composed of a hopper and a feeder that discharges into the homogenizer. These steps are optional and depend on customer specifications.

In the master batch system, a controlled amount of granules is collected in a small granules blender. Specific additives are dumped in and then mixed with the granules for about 20 minutes. The blended stream is then added in controlled amounts into the additive weigh feeder. From the additive weigh feeder, the granule and additive mixture is let down at a controlled rate to the homogenizer. This system is used to add **dry additives** to the granules. The dry additive system consists of two separate divisions: single additive feeders and master batch system (granules and additive). Single additive feeders will hold up to 245 pounds. These single additive feeders are used to feed pure additive into the homogenizer at predetermined rates.

Granules or powder is fed through a two-stage main granule feeder into the homogenizer. The granule feeder discharges granules at 25,000 to 35,000 pounds per hour.

Homogenizer

The granules flow from the weigh feeder into the homogenizer. It also receives a flow of additive that will modify the characteristics of the polypropylene when it is formed into pellets. It is the function of the homogenizer to mix thoroughly the streams of granules, peroxide, and additives. The contents of the homogenizer are stirred with a large spiral auger, which also moves the material along in a trough. The discharge side of a homogenizer is equipped with a weir to hold a constant level for better mixing. As the blended granule mixture is pushed over the weir, it enters the feed chest. The extruder feed chest receives granules from the homogenizer. A set of bars prevents any lumps or large pieces of metal from continuing to the extruder. Air and nitrogen are injected into the product flow in the feed chest. The air-to-nitrogen gas ratio can be used to control the melt flow.

The nitrogen-air injection influences the melt flow index of the pelletized polymer. Increasing the air in the ratio increases the melt flow rate. Increasing the nitrogen in the ratio decreases the melt flow rate. Air and nitrogen flow into the feed chest, where they mix with the flow of mixed granules and additive dropping into the extruder feed chest. Gas flow is a constant 3,000 scfh (standard cubic feet per hour). A distributive control system (DCS) set point controls how much air and how much nitrogen are used. The vent system removes dust and peroxide fumes from the homogenizer.

Extruder

The purpose of the extruder (Figure 17.4) is to melt the granule mixture, quickly quench it, and cut it into small pellets that are easier to handle. Melting the granules encapsulates the additives in the polypropylene. In granular form, the molecular weight distribution and swell of the granules are too broad. Various additives, such as peroxide, help narrow this range down. Industrial customers will not accept raw granules because of the danger of dust explosions.

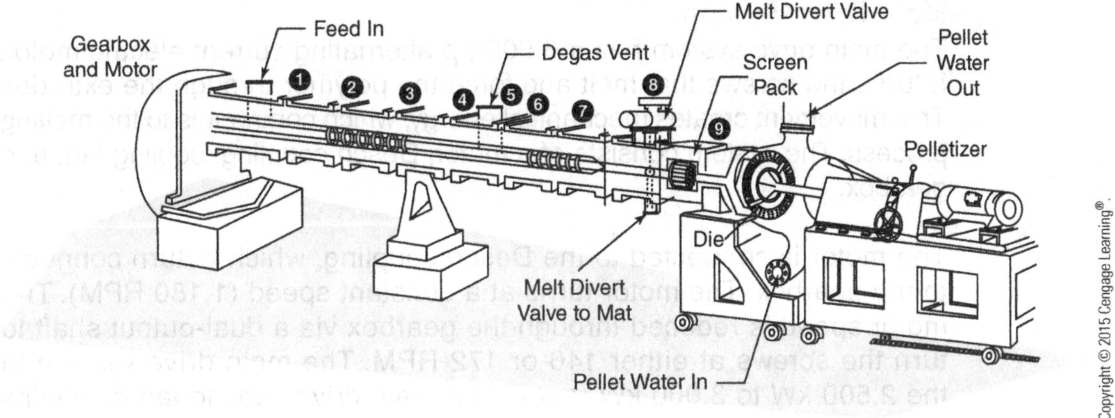

Figure 17.4 *Extruder*

The extruder referred to in this chapter is a German-made Werner & Pfleiderer ZSK-300 machine. (Other types of extruders include Egan, Ferrell, and JSK.) The WP extruder has a twin-screw design that consists of several sections of screw flights and kneading discs. The extruder's twin screws are capable of rotating at 146 RPM or 172 RPM, depending on the chosen gear ratio.

The extruder is driven by a 6,000-hp fan-cooled motor, and power is transmitted from the motor to the two screws by a multistage reduction gear. An auxiliary drive motor and reduction gear are provided to start the extruder. The auxiliary drive motor slowly turns the gears in the gearbox, and the twin screws safely allow the main drive to kick in. Once the main drive motor is started, the auxiliary drive unit is disengaged and shut down. The extruder is operated from a local control panel for startup and is then controlled by the DCS computer inside the extruder control room.

There are two powder seals located between the main drive gearbox and the extruder barrels. These seals are sometimes referred to as the shaft seals. The seals are used to prevent granules from leaking out of the extruder where the shafts enter the back of the extruder. The seals maintain a minimum operating temperature of 340°F (171°C). This high temperature melts granules, forming a seal around the shaft. The temperature set point is maintained with regulated 600-pound steam. Minor leakage of polymer is normal, although the steam pressure set point is set as low as possible to minimize seal leakage.

Two parallel screws turning in the same direction with intermeshing flights carry granules forcefully forward through the extruder barrel. As they do so, they completely mix and melt the polypropylene resin and additive. Barrel 1, the first section in Figure 17.3, has full cooling (no plate heaters) due to polymer bridging. Barrels 2 through 8 are heated electrically to help melt the polymer. Barrels 2 through 8 can be cooled after startup, if necessary.

Main Drive System

The main drive system has a 6,000-hp alternating current electric motor. It turns the screws that melt and force the polymer through the extruder. This movement creates mechanical energy, which contributes to the melting process. The system consists of a motor, Desch coupling, cooling fan, and gearbox.

The motor is connected to the Desch coupling, which in turn connects to the gearbox. The motor turns at a constant speed (1,180 RPM). The motor speed is reduced through the gearbox via a dual-output shaft to turn the screws at either 146 or 172 RPM. The main drive will run in the 2,500 kW to 3,000 kW range. The main drive cooling fan maintains airflow over the motor to keep it cool. The fan must maintain a specified flow to satisfy the interlocks. The Desch coupling is a pneumatic clutch that protects the gearbox from damage. It has a slip monitoring system. Excessive slippage usually indicates there is too much torque or excessive resistance in the gearbox or screws. Lubrication pumps provide lubrication for the main drive inboard and outboard bearings. The complete lubrication system includes an oil level sight glass, oil reservoir, pumps, and filters.

Auxiliary Motor System

The auxiliary drive motor connects to the main drive through the auxiliary drive coupling. The auxiliary drive is designed to roll the screws slowly so that the barrel will be purged of any polymer that remains in it. This purge is necessary because the barrel must be empty, or almost empty, before the main drive will be allowed to start. The auxiliary drive consists of a motor (explosion-proof, 100 hp, 460 V General Electric motor), auxiliary drive coupling, and gear reducer.

Gearbox

The gearbox lube oil system is a force-fed, closed-loop arrangement supplying oil to the pinion shaft bearing assembly, thrust bearing assembly, all bearings, and gear tooth intermesh. The gearbox lube oil system keeps the extruder gearbox cooled and lubricated, thereby preventing damage to the gears. The gearbox pump is a 25-hp, 1,750-RPM oil pump mounted near the auxiliary drive motor, with a suction line connected to the base of the extruder gearbox. The normal operating pressure of the gearbox pump is 90 psig, with 150 psig maximum. The pump circulates oil (120 GPM normal, 131 GPM maximum) through the oil filters. After being filtered, the oil flows to the oil cooler. From the exchanger, the oil flows to the gearbox for lubrication. The lube oil reservoir holds 550 gallons of oil. Oil circulates from the gearbox back into the lube oil reservoir by gravity flow. The temperature of the oil must be maintained at no less than 90°F (32°C) and no more than 150°F (65.5°C). The extruder will shut down if the oil temperature reaches 158°F (70°C). A set point on the DCS prevents the oil from getting too cool. If the oil temperature goes below the set point, the DCS will turn on the lube oil heater. When the temperature of the oil goes above the set point, the heater will turn off.

Screw Elements

Close examination of the two parallel screws shows a complex design. Individual screw elements are aligned along the solid metal shaft to perform four basic functions. Barrels 1 and 2 are the feeding or conveying sections. Barrels 3 and 4 are the kneading and melting barrels. Most of the breaking and mixing goes on in this section. Barrel 5 is the venting or degassing section. The vent plug is in place 99% of the time. Barrels 6, 7, and 8 are the pumping sections that push the polymer through the steam-heated zones toward the die. Each of these sections is designed specially to carry out its individual function.

Screw Barrel

The screw barrel is composed of several sections that are flanged together and held at intervals by support plates that rest on a foundation. These sections are heated by plate heaters attached to the surface of the barrel or are cooled by water that is pumped through passages in the barrel section. The following sections of the barrel are referred to as 600-pound steam-heated zone sections.

Throttling Valve (Zone 8)

The throttling (butterfly) valve is used to create back pressure to get better mixing in the barrel. It helps prevent die hole freeze-off and tends to break the polymer down more.

Melt Divert Valve Section (Zone 8)

The next section is steam heated and is equipped with a **diverter valve**, which sends the stream into a small pan during startup. Then the valve position is changed, and the polymer is pumped through the remaining operating sections of the extruder head and die.

Screen Pack Section (Zone 9)

The melt next passes through the **screen pack** (zone 9), a group of 27 horizontal mesh filtering screens held in a large steel cylinder. The screen pack traps any particles or foreign objects in the polymer stream.

Die Housing (Zone 10)

In the die housing, the polymer stream is split and channeled to the die.

Die (Zone 11)

The die (Figure 17.5), a solid steel plate with 986 holes in it, constitutes the next and last section of the extruder. The pressure of the screws and gear pump actually pumps the melt through the die.

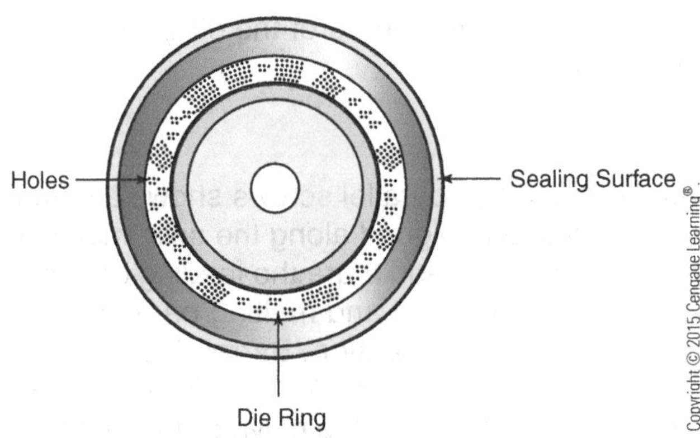

Figure 17.5
Die

Melt Divert Ball Valve System

The purpose of the melt divert valve (MDV) is to direct polymer melt either through the die during operations or to the pan at startup and shutdown. The diverter valve has two positions: dump and process. During the normal on-line operation, the diverter is in the process position. The MDV is in the dump position during the purge or shutdown cycles.

The actuator is a hydraulic motor and pump that turns a shaft. The shaft, in turn, operates a ball valve in zone 8 of the extruder. The flow of polymer at startup is to the pan. When flow is established, the MDV position diverts polymers to the die.

Screen Pack

The purpose of the screen pack (Figure 17.6) is to filter out any contamination, particles, or clumps from the polymer before it flows into the extruder die. The screen pack changer is a precision-made slide plate with two holes in it. The holes are bored out in the exact diameter of the screen packs. This allows for a perfect fit as the primary and secondary screen packs are positioned in the slots.

The slide plate has two positions. One hole is centered in the bore of the extruder; the other is completely outside it. Once reloaded, it is in the

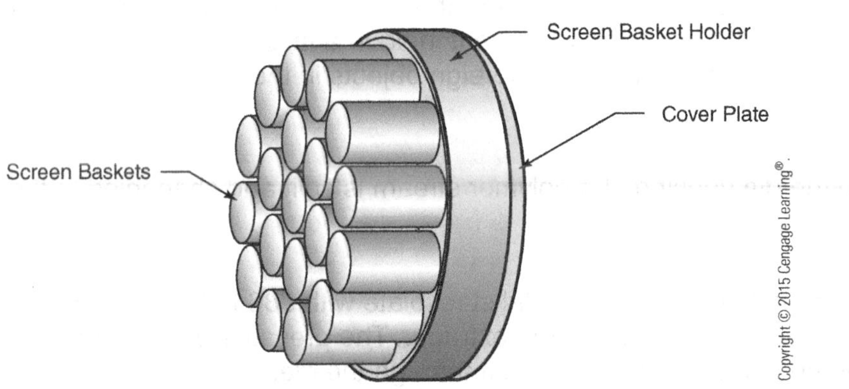

Figure 17.6
Screen Pack

off-line position but ready to replace the on-line primary screen pack. The hydraulic system quickly moves the slide from one position to the other.

Barrel Cooling and Heating System

The purpose of the electric barrel heating system is to melt the polymer granules during startup. Depending upon the grade, the heat generated from the friction of the extruder screws and kneading blocks on the granules is more than sufficient to melt the polymer.

Barrels 2 through 8 are each equipped with electric plate heaters. A set point for each barrel is specified for each product grade. When the extruder is down, the heaters are controlled by a set point.

The barrel cooling system absorbs heat so the temperature of the melt is controlled in extruder barrels 1, 4, 5, 6, 7, and 8. Barrel 1 requires cooling so that granules do not melt prematurely. The seals on the screen pack slide in zone 9 must be cooled to prevent the screen pack changer slide from leaking. A drum holds condensate used as a cooling medium. A pump delivers condensate to the coils inside the barrel jackets. A heat exchanger cools the water on the way to the barrels. Flow from the cooling tower provides the cooling medium. Thermocouples act as temperature sensors for all regions so that cooling and heating can be regulated.

The condensate pump draws condensate from the condensate drum and pumps it to a cooling water header. The pump runs constantly when the extruder is in operation. Steam from the return header condenses in the condensate drum. Diverting part of the flow to a heat exchanger controls the temperature of the water to the header. A local controller controls the temperature of the cooling water. If the thermocouple senses that more cooling is required at the header, the controller throttles the direct flow to the header, forcing a greater proportion of the flow to go through the exchanger.

Pelletizer System

The **pelletizer's** rotating knives cut the strands of polymer as they leave the die holes in the extruder, producing pellets, the product of the extruder post. Either 16, 20, or 24 knife blades are set in a hub (Figure 17.7). The electric pelletizer drive rotates this hub rapidly. The pelletizer assembly is mounted on a wheeled cart that is moved forward on rails to join with the extruder. The pelletizer rotates at 600 to 850 RPM. The newly formed pellets are carried away and kept from sticking together by a flow of water, which is a part of the pellet water system.

The pelletizer hub (Figure 17.8) holds the blades that cut the pellets from the polymer stream extruded through the die. The hub is fastened securely to the pelletizer shaft and is designed to fit against the die face. As the blades sweep across the die face, they cut the liquid polymer, which

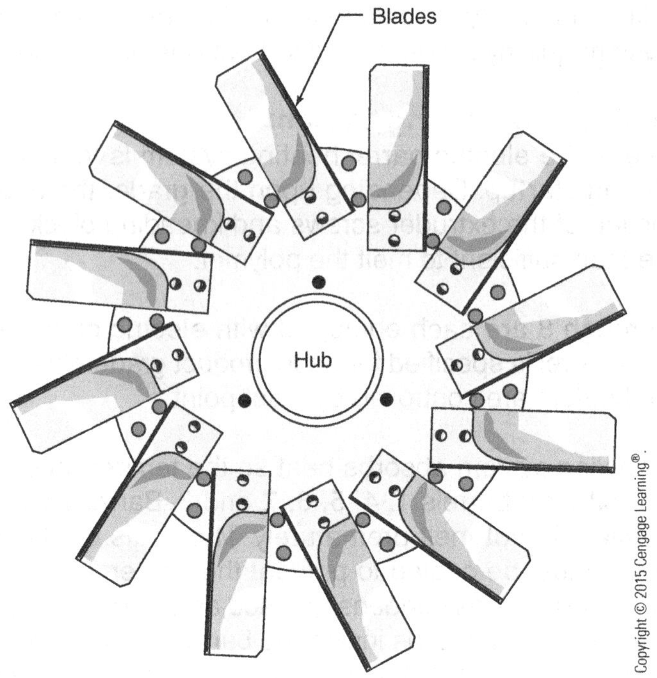

Figure 17.7
Peletizer Hub and Cutters

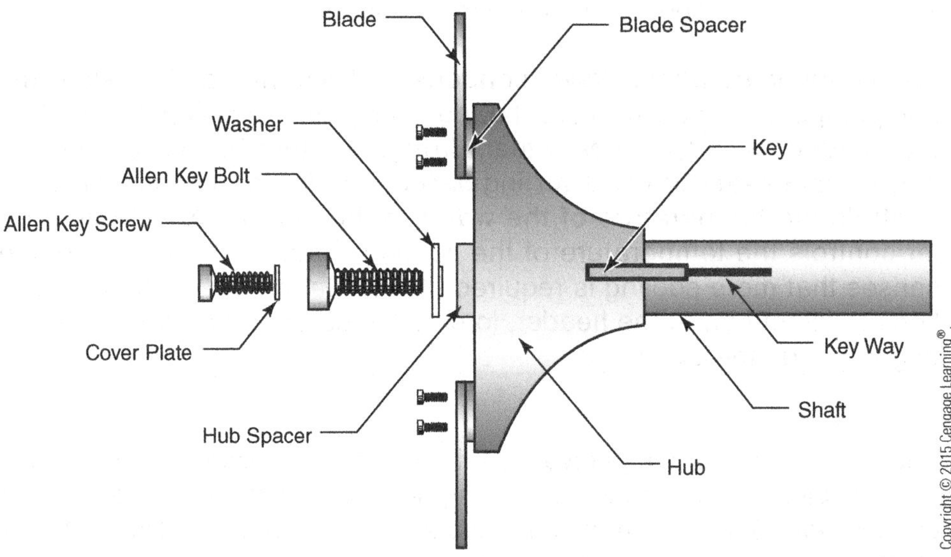

Figure 17.8
Hub Side View

is quenched by the water flowing through the pellet chamber. Tension adjustments on the hub-blade-shaft assembly are critical. If the blades are compressed against the die too tightly, they will damage the die or wear down the blades. Because the hub is attached to the pelletizer shaft, it can be positioned in the fully extended or retracted position. Minor tension adjustment can be made when the shaft is extended against the die. The key components of the pelletizer hub are shown in Figure 17.8.

Pelletizer Chamber and Pellet Water System

The area in which a set of knives cuts the strands of polymer into pellets as they come out of the die is called the pelletizer chamber. The pellets are cooled in the pelletizer chamber to keep them from sticking together or becoming misshaped.

The pellet water system is used to quench the polymer, allowing it to be cut and cooled at the same time. The water is kept between 130°F (54.4°C) and 165°F (73.8°C). The temperature will vary depending upon the grade being run.

The pellet water system consists of the pellet water tank, water pump, heat exchangers, temperature valve (controls the temperature of cooling tower entering or bypassing the exchangers), and two control valves (direct the flow of water either to the pelletizer chamber or directly back to the pellet water tank).

Steam at 150 psig is injected directly into the pellet water tank through the temperature control valve, providing the initial heat-up of the pellet water system when the extruder has been down for an extended time.

Product Drying and Storage System

The **dryer** and **classifier** system (Figure 17.9) removes pellet water from the pellets and classifies them according to size. There are five pieces of equipment in the dryer classifier system: the scalping box, dryer and helix, classifier, dryer vent fan, and pellet divert valve.

The first stage of water separation is the scalping box. Clumps of polymer and trash are diverted to a trash pan by the scalping bars. The slurry of pellets and water that passes through the bars is then on its way to the helix, the first part of the dryer.

The helix is where most of the water and pellets are separated. The water is sent to the pellet water tank, and the damp pellets are sent to the dryer.

Spinning paddles hurl the pellets outward and upward to the top of the dryer. The dryer dries pellets using centrifugal force. The mechanical action of this movement removes the remaining moisture from the pellet surface.

The classifier is a vibrating sieve tub with screens in two stages that permit the desired size of pellets to pass through. Larger pellets or clumps that have managed to pass through the scalping box are eliminated from the product flow here. Between the two screens is a cleaning kit that is used to prevent the lodging of pellets in the individual holes of the classifier.

Figure 17.9 *Dryer and Classifier System*

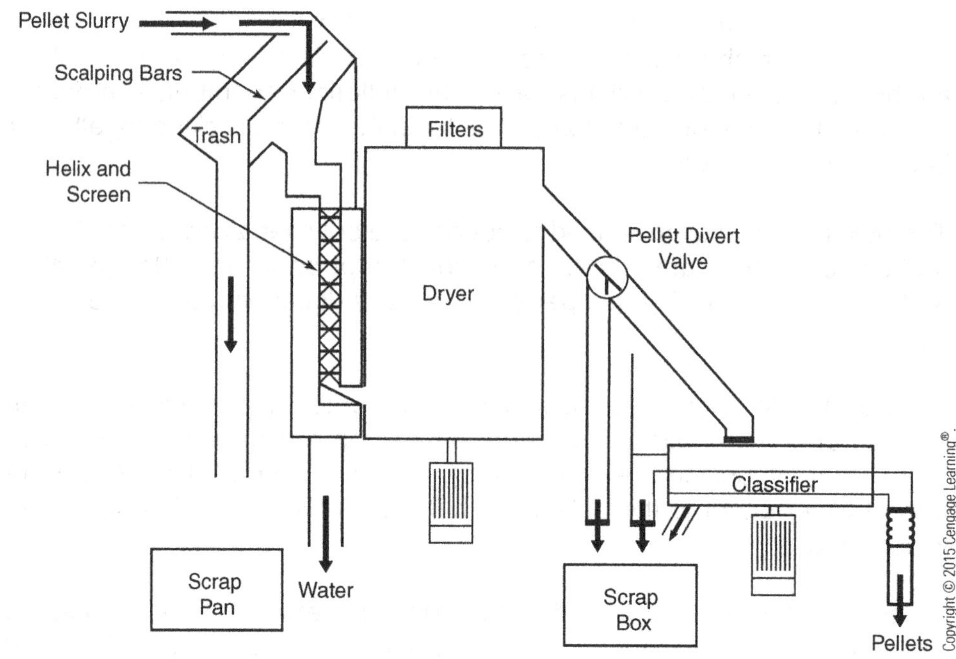

The dryer vent fan removes moisture and heat from the pellets. It also removes fines (fine particles).

The pellet divert valve diverts the pellets from the overs box to the classifier.

The pellet transfer system moves pellets from the classifier to one of two test bins or to one of four blenders. It uses an air system to push the pellets. The pellets come out of the classifier into the bin, where they drop through a star feeder and into the air system. There are typically one off-test bin, one transition bin, and four blenders.

The hold-up bin receives the pellets from the classifier. It serves as a surge bin for the star feeder at its bottom. An air system transfers the pellets from the hold-up bin to one of four receiving blenders and one off-test bin, using computer-controlled diverter valves. The off-test air system moves the pellets from the transition bin to the pellet blenders or off-test system. This movement is operator initiated and computer controlled.

When the pellets reach their bulk-loading destination, they leave the area of responsibility of the extruder technician and become the responsibility of the bulk-loading technician. Depending on the grade, pellets are blended for up to 4 hours. Laboratory testing verifies product quality. On-test bins hold about 40,000 pounds. If the laboratory results are good, the product is transferred to the large 400,000-pound pellet blenders for railroad or hopper truck loadout. The pellets are then loaded into a hopper car or hopper truck or are moved to warehouse bins for packaging in a box, bag, or super sac (i.e., bulk bag) for distribution to the customer. Final samples are verified by the laboratory and retained for future reference. The run data and sample results typically are sent to the customer.

The distributive control system is the main control computer for the extruder and does much of the actual work, such as switching valves and controlling temperatures. The DCS uses a pair of central processing units (CPUs). Information is stored in it for recording flows, production rates, and temperatures. It automatically records DCS changes. It also keeps a historical record, which can be recalled to establish the sequence of events. Computer monitors are located in the control room.

Summary

An extruder is a complex device composed of a heated jacket, a set of screws or screw, a heated die, large motor, gearbox, and a pelletizer. The purpose of the extruder is to melt the granule mixture and quickly quench it and cut it into small pellets. Melting the granules encapsulates the additives in the polypropylene.

Industrial manufacturers use synthetic resins to make paints; lubricants; adhesives; hard and soft plastic; plastic fibers and fabric; and resistant, decorative, and transparent plastic. Various methods are used in this process. In extrusion, solid resins are melted and pushed by a screw or screws through a heated chamber and specially shaped die. In molding, melted resins are squeezed into a mold. In laminating, sheets of paper or metal foil pass between rollers and are coated with melted resins. The sheets are pressed together to form things like electronic circuits. In casting, melted resin is poured into a mold. In calendaring, melted resins are spread over sheets of paper or cloth to form a protective coating over things like playing cards.

Review Questions

1. What is an extruder, and how does it work?
2. Describe the sequence of events through a typical plastics plant.
3. Describe how a granules feeder operates.
4. What is a pelletizer?
5. What is a pellet dryer, and how does it work?
6. Describe a screen pack.
7. Describe a dry additive system.
8. Contrast the auxiliary, gearbox, and main drive systems.
9. What is a melt divert valve?
10. Why do plastic granules need to be melted and extruded?
11. Explain how a classifier operates.
12. What is a star feeder?
13. What is a pelletizer hub used for?
14. What keeps the molten pellets from sticking together?
15. What are the four major sections of a polypropylene plant?

Absolute pressure (psia)—the pressure above a perfect vacuum (zero pressure).

Absorber—a device used to remove selected components from a gas stream by contacting it with a gas or liquid.

Acceleration head—the fluctuations of suction pressure created by the intake stroke of a reciprocating pump.

Accumulation—the pressure difference (ΔP) between initial lift pressure and full lift pressure on a relief valve.

Actuator—a device that controls the position of the flow-control element on a control valve by automatically adjusting the position of the valve stem.

Adsorber—a device (such as a reactor or dryer) filled with a porous solid designed to remove gases and liquids from a mixture.

A-frame furnace—furnace that has an A-frame-type exterior structure.

Aftercooler—a heat-exchange device designed to remove excess heat from the discharge side of a multistage compressor.

Air intake louvers—slats located at the bottom or sides of a cooling tower to direct airflow.

Air preheater—heats air before it enters a furnace at the burners.

Air registers—located at the burner of a furnace, these devices adjust secondary airflow.

Alkylation—use of a reactor to make one large molecule out of two small molecules.

Alloy—a material composed of two or more metals or a metal and a nonmetal.

Angle valve—a valve that operates by admitting fluid flow to the gate or plug and redirecting it 90° out of the discharge port.

Anions—negatively charged ions.

Antiseize compound—lubricant used on exposed valve stem threads.

Approach to tower—the temperature difference between the water leaving a cooling tower and the wet-bulb temperature of the air entering the tower.

Arch—neck-like structure that narrows as it extends between the convection section and stack of a furnace.

Autoclave—see *Stirred reactor*.

Automatic control—allows a control loop to utilize all five elements and work to match the set point.

Axial bearings—devices designed to prevent back-and-forth movement of a shaft; also called *thrust bearings*.

Axial pump—a dynamic pump that accelerates fluid in a straight line.

Azeotropic mixture—a mixture of two or more components that boil at similar temperatures and at a certain concentration. The liquid and vapor concentrations of an azeotrope are equal.

Baffles—evenly spaced partitions in a shell-and-tube heat exchanger that support the tubes, prevent vibration, control fluid velocity and direction, increase turbulent flow, and reduce hot spots.

Balanced equation—chemical equation in which the sum of the reactants (atoms) equals the sum of the products (atoms).

Ball valves—named for the ball-shaped, movable element in the center of the valve.

Basic hand tools—the typical tools process technicians use to perform their job activities.

Basin heaters—designed to keep cooling system water from freezing during the winter months.

Basin—concrete storage compartment or catch basin located at the bottom of the cooling tower.

Bellows pressure element—a corrugated metal tube that contracts and expands in response to pressure changes.

Bellows trap—a thermostatic steam trap that operates by opening or closing a bellows as the temperature changes; this movement opens and closes a valve.

Belt—used to connect two parallel shafts—the drive shaft and the driven shaft—each of which has a pulley mounted on the end; belts fit in the grooves of the pulleys.

Binary mixture—contains two components.

Biocides and algaecides—prevent biological growths from interfering with water circulation.

Blind—a device used in piping to gain complete shutoff.

Block flow diagram (BFD)—a set of blocks that move from left to right that show the primary flow path of a process.

Block valve—any valve that is intended to block flow; also called an *isolation valve*. The term generally refers to gate valves.

Blowdown—the process by which water is removed from the boiler system.

Board-mounted equipment—instruments, gauges, or controllers that are mounted in a control room.

Boiler—a type of fired furnace used to boil water and produce steam; also known as a *steam generator*.

Boiler load—plant demand for steam.

Boiling point—the temperature at which a liquid turns to vapor.

Boiling range—the range of temperatures that the lightest component boils (initial boiling point) to when the heaviest component boils (final boiling point).

Boil-up rate—the balance of the products (vapor and liquid) returned to the column from the kettle reboiler.

Bonding—physically connecting two objects with a copper wire.

Bonnet—a bell-shaped dome mounted on the body of a valve.

Bottoms product—residue; the heavier components of the distillation process fall to the bottom of the tower and are removed.

Bourdon tube—a hook-shaped, thin-walled tube that expands and contracts in response to pressure changes and is attached to a mechanical linkage that moves a pointer.

Box furnace—a square or rectangular furnace with both a radiant and convection section.

Bridgewall—sloping section inside a furnace that transitions between the radiant section and convection section; or the section of a refractory that separates fireboxes and burners.

Bridgewall markings—manufacturer information on the body of a valve.

British thermal unit (Btu)—energy measurement unit; 1 Btu is the energy needed to raise one pound of water one degree Fahrenheit.

Broken burner tiles—located directly around the burner and designed to protect the burner from damage. The furnace rarely needs to be shut down to replace a broken tile unless it is affecting the flame pattern.

Broken supports and guides—tend to fall to the furnace floor. Missing supports or guides will result in tubes sagging or bowing.

Bubble-point—the temperature at which a liquid mixture begins to boil and produce vapor.

Bullet—cylindrical shaped tank with rounded ends that are classified as high pressure.

Burner—used to evenly distribute air and fuel vapors over an ignition source and into a boiler firebox.

Burner alarms—immediately notify technicians when a burner goes out.

Butterfly valves—characterized by their disc-shaped flow-control element, which pivots from its center.

Butt-welded piping—pipe on which the parts to be joined are the same diameter and are simply welded together.

Cabin furnace—a cabin-shaped, aboveground furnace that transfers heat primarily through radiant and convective processes.

Capacity—the amount of water a cooling tower can cool.

Carryover—contamination of steam with boiler water solids.

Cascade control—a term used to describe how one control loop controls or overrides the instructions of another control loop in order to achieve a desired set point.

Catalyst—a chemical that can increase or decrease reaction rate without becoming part of the product.

Cations—positively charged ions.

Cavitation—the formation and collapse of gas pockets around the impellers during pump operation; results from insufficient suction head (or height) at the inlet to the pump.

Cell—the smallest subdivision of a cooling tower that can function as an independent unit. Some cooling tower systems have multiple cells.

Centrifugal compressors—use centrifugal force to accelerate gas and convert energy to pressure.

Centrifugal force—the force exerted by a rotating object away from its center of rotation. Often referred to as a *center-seeking force*, centrifugal force is usually stated as the force perpendicular to the velocity of fluid moving in a circular path.

Centrifugal pump—a dynamic pump that accelerates fluid in a circular motion.

Chain drive—a device that provides rotational energy to driven equipment by means of a series of sprocket wheels that interlink with a chain; designed for low speeds and high-torque conversions.

Channel head—a device mounted on the inlet side of a shell-and-tube heat exchanger that is used to channel tube-side flow in a multipass heat exchanger.

Charge—the process flow in a furnace.

Check valves—mechanical valves that prevent reverse flow in piping.

Chemical equation—numbers and symbols that represent a chemical reaction.

Chemical reaction—a term used to describe the breaking of chemical bonds, forming of chemical bonds, or breaking and forming of chemical bonds.

Classifier—a device that separates good pellets from oversized or misshapen pellets in plastics manufacturing.

Coking—formation of carbon deposits in the tubes of a furnace.

Color chart of steel tubes—shows 10 tube color variations associated with temperature.

Column loading—the amount of material fed continuously into a distillation column.

Combustion—a rapid exothermic reaction that requires fuel, oxygen, and an ignition source and gives off heat and light.

Compression ratio—the ratio of discharge pressure (psia) to suction pressure (psia). Multistage compressors use a compression ratio in the 3 to 4 range, with the same approximate compression ratio in each stage. For example, if the desired discharge pressure is 1,500 psia, a 4-stage compressor with a 3.2 compression in each stage might be used. The pressure at the discharge of each stage would be: 1st = 47 psia, 2nd = 150 psia, 3rd = 480 psia, 4th = 1,536 psia.

Compressors—mechanical devices designed to accelerate or compress gases; classified as positive displacement or dynamic.

Concurrent flow—see *Parallel flow*.

Condensate—moisture produced when steam, hot gases, or vapors condense (change to the liquid state).

Condenser—a shell-and-tube heat exchanger used to cool and condense hot vapors.

Conduction—the means of heat transfer through a solid, nonporous material resulting from molecular vibration. Conduction can also occur between closely packed molecules.

Cone-roof tank—an enclosed tank with a conical shaped roof with vertical walls mounted on a circular concrete pad or directly on the ground.

Control loop—a collection of instruments that work together to automatically control a process; usually consists of a sensing device, a transmitter, a controller, a transducer, and an automatic valve.

Control valves—automated valves used to regulate and throttle flow; typically provide the final control element of a control loop.

Controller—an instrument used to compare a process variable with a set point and initiate a change to return the process to a set point if a variance exists.

Convection section—the upper area of a furnace in which heat transfer is primarily through convection.

Convection tubes—tubes located above the shock bank of a furnace or away from the radiant section where heat transfer is through convection. The first pass of tubes directly above the radiant section is referred to as the *shock bank*.

Convection—the means of heat transfer in fluids resulting from currents.

Converter—see *Fixed bed reactor*.

Cooling range—the temperature difference between the hot and cold water in a cooling tower.

Cooling tower types—typically classified as induced (requires fan), forced (requires fan), atmospheric, and natural draft.

Cooling towers—evaporative coolers specifically designed to cool water or other mediums to the ambient wet-bulb air temperature.

Corrosion—electrochemical reactions between metal surfaces and fluids that result in the gradual wearing away of the metal.

Countercurrent—see *Counterflow*.

Counterflow—the movement of two flow streams in opposite directions; also called *countercurrent flow*.

Coupling—a device that attaches the drive shaft of a motor or steam turbine to a pump, compressor, or generator.

Cracking—decomposition of a chemical or mixture of chemicals by the use of heat, a catalyst, or both; thermal cracking uses heat only; catalytic cracking uses a catalyst.

Crossflow—refers to the movement of two flow streams perpendicular to each other.

Cryogenic distillation—often used to separate oxygen and nitrogen from our atmosphere by using low-temperature drying and distillation methods.

Cryogenic tank—has been designed to store liquids below −100°F.

Cylindrical furnace—a cylindrical, vertical furnace, primarily designed to transfer radiant heat to a process stream.

Damper—a device used to regulate airflow.

Datum plate—a reference point on the bottom of a tank used to measure liquid level.

Demister—a cyclone-type device used to swirl and remove moisture from a gas.

Desiccant dryer—used to remove moisture from compressor gases as they are passed over a chemical desiccant, which adsorbs the water.

Desuperheating—a process applied to remove heat from superheated steam.

Dew-point—the temperature at which a mixture produces the first drop of liquid on a vapor-liquid-equilibrium curve.

Diaphragm compressor—utilizes a hydraulically pulsed diaphragm that moves or flexes to positively displace gases.

Diaphragm pump—a reciprocating pump that uses a flexible diaphragm to positively displace fluids.

Diaphragm valve—a device that uses a flexible membrane to regulate flow.

Differential pressure (DP) cell—measures the difference in pressure between two points.

Differential pressure—the difference between inlet and outlet pressures; represented as ΔP or delta *p*.

Differential temperature—the difference between inlet and outlet temperature; represented as ΔT, or delta *t*.

Dike—a containment wall or ditch that extends around a tank to prevent product loss.

Disc—a device made of metal or ceramic that fits snugly in the seat of a valve to control flow.

Discharge head—the resistance or pressure on the outlet side of a pump.

Distillate—the condensate taken from a distillation column.

Distillation column—a cylindrical tower consisting of a series of trays or packing that provide a contact point for the vapor and liquid. The contact between the vapor and liquid in the column results in a separation of components in the mixture based on differences in boiling points.

Distillation—the separation of components in a mixture by their boiling points.

Distributed control system (DCS)—a computer-based system that controls and monitors process variables.

Diverter valve—an automatic valve used to divert the flow of solids in plastics manufacturing.

Double-acting compressor—a reciprocating compressor that compresses gas on both sides of the piston.

Downcomer flooding—occurs when the liquid flow rate in the tower is so great that liquid backs up in the downcomer and overflows to the upper tray. Liquid accumulates in the tower, differential pressure increases, and product separation is reduced.

Downcomers—the inlet tubes from the upper to lower drum of a water-tube boiler; these tubes contain hot water.

Draft—negative pressure of air and gas at different elevations in a furnace.

Drift eliminators—devices used in a cooling tower to keep water from blowing out.

Drift loss—entrained water lost from a cooling tower in the exiting air; also called *windage loss.*

Driven equipment—a device such as a compressor, pump, or generator that receives rotational energy from a driver.

Driver—a device designed to provide rotational energy to driven equipment.

Dry additive—a dry chemical additive mixed with polymer to match customer requirements.

Dry-bulb temperature (DBT)—the air temperature as measured without taking relative humidity into account.

Dryer—removes moisture from gas.

Dynamic—class of equipment such as pumps and compressors that convert kinetic energy to pressure; can be axial or centrifugal.

Economizer—a section of a fired boiler used to heat feedwater before it enters the steam drums.

Electric actuated valves—valves that utilize electricity to actuate or move the flow control device. An example of this type of valve is a solenoid.

Electrical drawings—symbols and diagrams that depict an electrical process.

Elevation drawings—a graphical representation that shows the location of process equipment in relation to existing structures and ground level.

Endothermic reaction—a chemical reaction that must have heat added to make the reactants combine to form the product.

Entrainment—the upward movement of high-velocity, large vapor droplets in a distillation column.

Equipment location drawings—show the exact floor plan for location of equipment in relation to the plan's physical boundaries.

Evaporate—to turn to vapor; evaporation removes heat energy from hot water.

Exhaust valve—valve used to block steam turbine outlet steam.

Exothermic reaction—a chemical reaction that gives off heat.

Extraction—a process for separating two materials in a mixture or solution by introducing a third material that will dissolve one of the first two materials but not the other.

Extract—the second solution that is formed when a solvent dissolves a solute.

Extruder—a complex device composed of a heated jacket, a set of screws or one screw, a heated die, large motor, gearbox, and pelletizer in plastics manufacturing.

Feed composition—the composition of the fuel entering a furnace, which must remain uniform or furnace operation will be affected.

Feed distributor—a device used in a packed distillation column to evenly distribute the liquid feed.

Feed tray—point of entry of process fluid in a distillation column, under the feed line.

Feed—the original solution to be separated in liquid—liquid extraction.

Field-mounted equipment—instruments or controllers that are mounted near the equipment in the field.

Fill—plastic or wood surfaces that direct airflow and provide for contact of water and air in a cooling tower. See also *Splash bar*.

Filter—a porous medium used to separate solid particles from a fluid by passage through it.

Final boiling point—the temperature at which the heaviest component boils.

Final control element—the device in a control loop that actually adjusts the process; typically, a control valve.

Fire-tube boiler—a type of boiler that passes hot gases through tubes to heat and vaporize water.

Firebox—the area in a furnace that contains the burners and open flames; the area of radiant heat transfer.

Fired heater—a high-temperature furnace used to heat large volumes of raw materials.

Fixed bed reactor—a vessel that contains a mass of small particles through which the reaction mixture passes; also called a *converter*.

Fixed head—a term applied to a shell-and-tube heat exchanger that has the tube sheet firmly attached to the shell.

Flame impingement—direct flame impingement occurs when the visible flame hits the tubes. Flame impingement can be classified as periodic or sustained.

Flameout—extinguishing of a burner flame during furnace operation.

Flanges—used to connect piping to equipment or where piping may have to be disconnected; consist of two mating plates fastened with bolts to compress a gasket between them.

Flare header—a pipe that connects the plant to the flare.

Flare—a device to safely burn excess hydrocarbon vapor.

Flashback—intermittent ignition of gas vapors, which then burn back in the burner; can be caused by fuel composition change.

Flat face flanges—generally used to mate against cast equipment, where bending from tightening bolts might break the flange; gasket should cover the entire face of the flange.

Float steam trap—a steam trap that operates with a float that opens a valve as the condensate level rises.

Floating head—a term applied to a tube sheet on a heat exchanger that is not firmly attached to the shell on the return head and is designed to expand (float) inside the shell as temperature rises.

Floating-roof tank—has an open top and a pan-like structure that floats on top of the liquid and moves up and down inside the tank with each change in liquid level.

Flocculation—the bridging together of coagulated particles.

Flow diagram—a simplified sketch that uses symbols to identify instruments and vessels and to describe the primary flow path through a unit.

Flow-control element—the part of a valve that regulates flow; that is, the gate or the disc.

Fluid catalytic cracking—a process that uses a fluidized bed reactor to split large gas oil molecules into smaller, more useful ones; also called *catcracking*.

Fluid coking—a process that uses a fluidized bed reactor to scrape the bottom of the barrel and squeeze light products out of residue.

Fluidized bed reactor—suspends solids by countercurrent flow of gas; heavier components fall to the bottom, and lighter ones move to the top.

Fluid—of the three forms of matter—solids, liquids, and gases—liquids and gases are considered fluids.

Foaming—often the result of too much heat, impurities in the feed, or active turbulence on the trays above undersized downcomers.

Forced draft—type of mechanical-draft cooling tower that uses fans to push air into the tower.

Fouling—buildup on the internal surfaces of devices such as cooling towers and heat exchangers, resulting in reduced heat transfer and plugging.

Foundation drawings—concrete, wire mesh, and steel specifications that identify width, depth, and thickness of footings, support beams, and foundation.

Fractional distillation—separation of two or more components through distillation.

Fuel pressure control—a pressure control loop located on the natural gas fuel line to the furnace that is designed to maintain constant pressure to the furnace burners.

Furnace flow control—a critical feature in furnace operation, temperature, and pressure control that regulates fluid feed rates in and out of the process furnace.

Furnace hi/lo alarms—alarm warnings that warn when the process flow is off specification and prevent equipment damage and harm to the environment and human life.

Furnace pressure control—monitors furnace pressure in the bottom, middle, and top of the furnace with a pressure control loop connected to the stack damper. The middle pressure reading on the furnace is compared to a set point and adjustments are made at the damper if necessary.

Furnace temperature control—adjusts fuel flow to the burners and, as flow exits the process furnace, monitors process conditions. The natural gas flow controller (*slave*) is cascaded to the (*master*) temperature controller. The temperature controller adjusts fuel flow to the burners.

Gain—the ratio of the output signal from the controller to the error signal.

Gas turbine—a device that uses high-pressure gases to turn a series of turbine wheels to provide rotational energy to turn an axle or shaft.

Gate valve—a device that places a movable metal gate in the path of a process flow.

Gate—the flow-control element of a gate valve.

Gauge hatch—a door in the roof of an atmospheric tank that enables the contents to be measured and that provides some emergency pressure relief.

Gauge pressure (psig)—the pressure above atmospheric pressure; zero is equivalent to approximately 14.7 psi at sea level.

Gearbox—a power transmission mechanism consisting of interlocking toothed wheels (gears) inside a casing.

Globe valve—a device that places a disc in the path of a process flow.

Governor valve—an automatic valve that controls steam turbine speed by regulating the amount of steam admitted.

Governor—speed-control device that adjusts the governor valve.

Grounding—is described as a procedure designed to connect an object to the earth with a copper wire and a grounding rod.

Gutted piping—see *Jacketed and gutted piping*.

Handwheel—attached to the valve stem and used to control the position of the flow-control element of a valve.

Hazy firebox or smoking stack—often occurs when not enough or excess air is going into the firebox or the fuel-air mixing ratio is incorrect.

Head—is described as Pressure (at suction) $\times$ 2.31 $+$ Specific gravity; 1 psi is equal to 2.31 feet of head.

Header box doors and gaskets—provides access to the terminal penetrations or bends on the convection tubes; also called *header box doors*. The gaskets provide a positive seal between the inside and outside of the furnace.

Header box gaskets—provide access to the terminal penetrations or bends on the convection tubes; also called *header box doors.*

Heat balance—principle that heat in equals heat out.

Heat exchanger—an energy-transfer device designed to transfer energy in the form of heat from a hotter fluid to a cooler fluid without physical contact between the two fluids.

Heat soaking—a turbine warm-up procedure designed to remove condensate and warm the internal parts; includes slow rolling the turbine at low speeds between 200 and 500 RPM.

Hemispheroid tank—has a rounded or dome-shaped top and vertical walls mounted on the ground or a concrete pad.

Homogenizer—a device in plastics manufacturing that mixes the streams of granules, peroxide, and additive by stirring with a large spiral auger as the material is moved along by the auger in a trough.

Hot tubes—glow different colors when the inside or outside of the tubes foul and when there is flame impingement, reduced flow rate, and overfiring of the furnace.

Hunting—occurs when a steam turbine's speed fluctuates while the controller searches for the correct operating speed.

Hydraulic actuated valves—a valve that utilizes a hydraulic actuator to position the flow control element. Internal designs include piston or vane.

Hydraulic turbine—a device that uses high-pressure liquids to turn a turbine wheel attached to a pump generator.

Hydrocracking—use of a multistage fixed bed reactor system to boost yields of gasoline from crude oil by splitting heavy molecules into lighter ones.

Hydrodesulfurization—removes sulfur from crude mixtures.

Impeller—a device attached to the shaft of a centrifugal pump that imparts velocity and pressure to a liquid.

Impulse turbine—a steam turbine with a blading design that causes rotation of the blades and shaft when high-velocity steam from an external source pushes on it.

Indicator gauge—an instrument used to show the value of process variables such as pressure, level, temperature, and flow.

Induced draft—type of mechanical-draft cooling tower that uses fans to pull air out of the tower.

Intercooler—a heat exchange device designed to cool compressed gas between the stages of a multistage compressor.

Interlock—a device that prevents damage to equipment and personnel by stopping or preventing the start of certain equipment if a preset condition has not been met.

Inverted bucket steam trap—a mechanical steam trap that operates with an inverted bucket inside a casing; effective on condensate and noncondensing vapors.

Isolation valve—see *Block valve*.

Jacketed and gutted piping—two concentric (one inside the other) pipes used when the conveyed fluid must be kept hot. In jacketed piping, the fluid is conveyed through the inner pipe and a heating medium is conveyed through the jacket. Gutted piping is the reverse.

Jacketed tank—an insulated system designed to hold in heat or cold.

Jet flooding—occurs when the vapor velocity is so high that liquid down flow in the tower is restricted. Liquid accumulates in the tower, differential pressure increases, and product separation is reduced.

Journal bearing—see *Radial bearings*.

Kettle reboiler—a shell-and-tube heat exchanger with a vapor disengaging cavity used to supply heat for separation of lighter and heavier components in a distillation system and to maintain heat balance.

Knockout drum—a tank located between a flare header and the flare; used to separate liquid hydrocarbons from vapor.

Labyrinth seal—a shaft seal designed to stop steam flow in a steam turbine; consists of a series of ridges and intricate paths.

Laminar flow—streamline flow that is more or less unbroken; layers of liquid flowing in a parallel path.

Latent heat—heat that cannot be measured; the heat required to change a liquid to a vapor (latent heat of evaporation) or a vapor to a liquid (latent heat of condensation).

Leaching—is the loss of wood preservative chemicals in the supporting structure of the cooling tower as water washes or flows over the exposed components.

Legends—a document used to define symbols, abbreviations, prefixes, and specialized equipment.

Lobe compressor—a rotary compressor that contains kidney-bean-shaped impellers.

Lobe pump—a rotary pump that uses kidney-bean-shaped lobes to displace and transfer fluid.

Local flooding—excessive liquid flowing down a column blocks vapor flow up the column in one section.

Low burner turn-down—a condition that can result in hazy firebox.

Low NO_x burners—a type of gas burner, invented by John Joyce, that significantly reduces the formation of oxides of nitrogen. Low NO_x burners are 100% efficient as all heat energy released from the flame is converted to useful heat.

Manometer—a device used to measure pressure or vacuum.

Manual control—allows the controller to open the control valve and set it at a predetermined percent.

Manway—a hatch or port used to provide open access into a tank.

Material balance—principle that the sum of the products leaving equals the feed entering the distillation column.

Material balancing—a method for calculating reactant amounts versus product target rates.

McCabe-Thiele—the simplest method used in the analysis of binary distillation. The composition of the mixture on each theoretical tray can be determined by the mole fraction of one of the components.

Mechanical seal—provides a leak-tight seal on a pump; consists of one stationary sealing element, usually made of carbon, and one that rotates with the shaft.

Mud drum—the lower drum of a water-tube boiler.

Multipass heat exchanger—a type of shell-and-tube heat exchanger that channels the tube-side flow across the tube bundle (heating source) more than once.

Multiport valve—has multiple inlets and/or outlets in specialized piping systems to divert flow direction, allowing fluid sources to be switched.

Needle valve—a type of globe valve that has a needle-shaped element that fits snugly into the seat.

Net positive suction head (NPSH)—the head (pressure) in feet of liquid necessary to push the required amount of liquid into the impeller of a dynamic pump without causing cavitation.

Net positive suction head available ($NPSH_a$)— a term used to indicate the required pump suction pressure so the pump can operate properly. It is defined as atmospheric pressure (converted to head) + static head + surface pressure head − vapor pressure − frictional losses.

Net positive suction head required ($NPSH_r$)—the minimum NPSH necessary to avoid cavitation. The $NPSH_a$ must be greater than or equal to the $NPSH_r$, expressed as NPSHa ≥ NPSHr. It is the reduction in total head as the liquid enters the pump.

Neutralization—a chemical reaction designed to remove hydrogen ions or hydroxyl ions from a liquid.

Nozzle—a device designed to restrict flow and convert pressure into velocity.

Oil separator—removes oil from compressed gases.

Overall flooding—local flooding expands to entire column.

Overhead product—the lighter components in a distillation column, which rise through the column and go out the overhead line, where they are condensed.

Overlap—incomplete separation of a mixture.

Overloading—operating a column at maximum conditions.

Overspeed trip—a safety device used to shut down a steam turbine when it exceeds its rotational speed limit by closing the turbine trip valve.

Oxygen analyzer—an instrument specifically designed to detect the concentration of oxygen in an air sample. Oxygen flow rates are carefully controlled through a furnace.

Packed tower—a tower that is filled with specialized packing material instead of trays.

Packing—a specially designed material used to stop fluids from entering or escaping; packed around the shaft (stem) of a valve, or shaft of a pump.

Packing gland—a mechanical device that contains and compresses packing.

Parallel flow—refers to the movement of two flow streams in the same direction; for example, tube-side flow and shell-side flow in a heat exchanger; also called concurrent.

Partial pressures—the amount of pressure per volume exerted by the various fractions in a mixture of gases.

Parts per million—one PPM equals 1 pound to every 1,000,000 pounds. Typically associated with suspended solids in the basin or product stream.

Peepblocks with peepholes—refractory blocks with holes in the center provide visual access that enable operators to inspect visually the inside of the furnace.

Peepholes—holes in the side of a furnace that enable operators to inspect visually the inside of the furnace.

Pelletizer—a device with rotating knives that cut the strands of polymer as they leave the die holes in the extruder, producing pellets.

Permissive—special type of interlock that controls a set of conditions that must be satisfied before a piece of equipment can be started.

Pig—a cylindrical device used to clean out pipes. Most pigs utilize a *pig launcher* to propel it through the line and into a pig trap.

Pipe size—the nominal (named) size of a pipe; usually close to the outside and inside diameters of the pipe but identical to neither. The outside diameter of a certain size pipe is constant. The inside diameter will change with the pipe wall thickness (schedule). See also *Pipe thickness.*

Pipe thickness—thickness of pipe wall, designated by a schedule number. Schedules 10 (thin walled), 40, 80, and 160 (heavy walled) are common. The schedule indicates a specific wall thickness for one pipe size only; a 3" schedule 40 pipe will have a different thickness than a 4" schedule 40. The pipe wall thickness increases as the schedule number increases.

Piping—used to convey all kinds of fluids, liquid or gas.

Piston pump—a reciprocating pump that uses a piston and cylinder to move fluids.

Plenum—the open area sandwiched between the fill in the center of an induced-draft cooling tower.

Plugged burner tips—make flame pattern erratic, shooting out toward a tube instead of up the firebox.

Plug valve—a device that has a plug-shaped element; used for on/off service.

Pneumatic actuated valves—utilize air to actuate the flow control element. Internal designs may be piston, vane, or diaphragm.

Positive displacement (PD)—class of equipment such as pumps and compressors that moves specific amounts of fluid from one place to another; can be rotary or reciprocating.

Preheated air—a compressed air system that typically pushes the air through tubes located in the upper section of the furnace. This preheated air takes full advantage of energy flow passing out of the furnace stack.

Pressure relief valve—used to relieve excessive pressure on the discharge of a positive displacement pump.

Primary elements and sensors—the first element of a control loop. Primary elements and sensors come in a variety of shapes and designs depending on whether they are to be used with pressure, temperature, level, flow, or analytical control loops. An example of a temperature element is a thermocouple. A flow control primary element is a turbine meter or orifice plate. A level element is a displacer. A pressure element is a bourdon tube. Many different types of primary elements or sensors are used in the chemical processing industry.

Priming—becoming filled with fluid.

Process and instrument drawing (P&ID)—a complex diagram that uses process symbols to describe a process unit; also called *piping and instrumentation drawing*.

Process flow diagram (PFD)—a simplified sketch that uses symbols to identify instruments and vessels and to describe the primary flow path through a unit.

Process heaters—combustion devices that transfer convective and radiant heat energy to chemicals or chemical mixtures. Process tubes pass through the convection and radiant sections as energy is transferred to them. This transferred energy allows the liquid to be utilized in a variety of chemical processes that require higher temperatures.

Process instrumentation—devices that control and monitor process variables; transmitters, controllers, transducers, primary elements, and sensors.

Product—the result of a chemical reaction.

Programmable logic controller (PLC)—a simple stand-alone, programmable computer that could be used to control a specific process or networked with other PLCs to control a larger operation. PLCs are inexpensive, flexible, provide reliable control, and are easy to troubleshoot.

Psychrometry—the study of cooling by evaporation.

Puking—occurs when the vapor is so great that it forces liquid up the column or out the overhead line.

Pulsation dampener—a device installed close to a pump, in the suction or discharge line, to reduce pressure variations.

Pumps—devices used to move liquids from one place to another; classified as positive displacement or dynamic.

Radial bearings—devices designed to prevent up-and-down and side-to-side movement of a shaft; also called *journal bearings*.

Radiant heat transfer—conveyance of heat by electromagnetic waves from a source to receivers.

Radiant tubes—tubes located in a furnace firebox that receive heat primarily through radiant heat transfer; also called *radiant coils*.

Radiographic inspection—use of X-rays to locate defects in metals in much the same manner as an X-ray is taken of a broken bone.

Raffinate—in liquid—liquid extraction, material that is left after a solvent has removed solute.

Raised face flange—uses a gasket that fits inside the bolts.

Reactants—the raw materials in a chemical reaction.

Reaction rate—the amount of time it takes a given amount of reactants to form a product or products.

Reactive turbine—a steam turbine with a fixed nozzle and an internal steam source.

Reactor—a device used to combine raw materials, heat, pressure, and catalysts in the right proportions to form chemical bonds that create new products.

Reboiler—a heat exchanger used to add heat to a liquid that was once boiling until the liquid boils again.

Receiver—a compressed-gas storage tank.

Reciprocating pump—a positive displacement pump that uses a plunger, piston, or diaphragm moving in a back-and-forth motion to physically displace a specific amount of fluid in a chamber.

Rectification—the separation of different substances from a solution by use of a fractionating tower.

Rectifying section—the upper section of a distillation column (above the feed line), where the higher concentration of lighter molecules is located.

Reflux—condensed distillation column product that is pumped back to increase product purity and control temperature.

Reflux ratio—defined as the amount of reflux returned to the tower divided by the amount of overhead product sent to storage. Example: 120 GPM reflux ÷ 60 GPM overhead to product tank = 2:1 reflux ratio.

Refractory—the lining of a furnace firebox that reflects heat back into the furnace.

Regenerator—used to recycle or make useable again contaminated or spent catalyst.

Relative humidity—a measurement of how much water air has absorbed at a given temperature.

Relative volatility—the characteristics associated with a liquid's tendency to change state or vaporize inside a distillation system.

Replacement reaction—a reaction designed to break a bond and form a new bond by replacing one or more of the original compound's components.

Resistance temperature detector (RTD)—a device used to measure temperature changes by changes in electrical resistance in a platinum or nickel wire.

Ring joint flange—uses only a metal ring for gasketing.

Risers—the tubes from the lower drum to the upper drum of a water-tube boiler; these tubes contain steam and water.

Rotameter—a flow meter that allows fluid to move through a clear tube that has a ball or float in it; numbers on the side of the tube indicate flow rate.

Rotary equipment—industrial equipment designed to rotate or move.

Rotary pump—a positive displacement pump that uses rotating elements to move fluids.

Rotor—the shaft and moving blades of rotary equipment or the moving conductor of an electric motor.

Ruptured tubes—flames come from opening in tubes. May cause excess oxygen levels to drop and bridge wall temperatures to increase.

Safety/relief valve—device set to automatically relieve pressure in a closed system at a predetermined set point; relief valves are used for liquids; safety valves are used for gases.

Sagging or bulged tubes—occur when guides or supports break; inside of tube fouls, causing flame impingement, reduced flow rate, over-firing furnace, or outside fouling of tubes. Note: The diameter of the tube does not change when it sags; however, it does when it bulges.

Scale—the result of suspended solids adhering to internal surfaces of equipment in the form of deposits.

Screen pack—a group of 27 horizontal mesh filtering screens that traps any particles or foreign objects in a polymer stream.

Screw pump—a rotary pump that displaces fluid with a screw.

Scrubber—a device used to remove chemicals and solids from process gases.

Seals—devices that prevent leakage between internal compartments in a rotating piece of equipment.

Sensible heat—heat that can be measured or sensed by a change in temperature.

Sentinel valve—a spring-loaded automatic relief valve that makes a high-pitched noise when turbine speed approaches the design maximum.

Set point—desired value of a process variable.

Shell-and-tube heat exchanger—a heat exchanger that has a cylindrical shell surrounding a tube bundle.

Shell side—refers to flow around the outside of the tubes of a shell-and-tube heat exchanger; see *Tube side*.

Shock bank—tubes located directly above the firebox of a furnace that receive radiant and convective heat. The shock bank is part of the convection section.

SHP—super-high-pressure (SHP) steam that operates between 1,000 and 1,200 psig.

Sight glass gauge—a level-measurement device consisting of a transparent tube and gauge attached to a vessel that allows an operator to see the corresponding liquid level.

Slip—the percentage of fluid that leaks or slips past the internal clearances of a pump over a given time.

Slop tank—or off-spec tank is used to store product that does not meet customer expectations.

Slow roll—controlling turbine speed at low (200–500) RPM.

Socket-welded piping—type of piping in which the pipe is inserted into a larger fitting before being welded to another part.

Solute—the material that is dissolved in liquid–liquid extraction.

Solution—a uniform mixture of particles that are not tied together by any chemical bond and can be separated by purely physical change.

Solvent—a chemical that will dissolve another chemical.

Soot blowers—remove soot from tubes in the convection section that consist of hollow metal rods that are inserted into the convection section and incorporate a series of timers that admit nitrogen in quick bursts.

Spalled refractory—an aging refractory that has cracked or deteriorated over time; a refractory that has not cured or dried properly; or a refractory whose anchors have failed, thus resulting in the refractory breaking loose from the sides of the furnace and falling to the furnace floor. Caused by old refractory that has cracked or deteriorated over time, or refractory that has not cured or dried properly, or broken refractory anchors.

Specific gravity—is described as the ratio between the density of a given liquid to the known density of water, or the density of gases to the density of air. The specific gravity of water or air is one.

Sphere—a circular-shaped tank with legs designed to contain high-pressure liquids or gases.

Spheroid—a circular tank with a flat bottom resting on a concrete pad or ground.

Splash bar—a device used in a cooling tower to direct the flow of falling water and increase surface area for air-water contact.

Spuds—gas-filled sections in a boiler fuel gas burner.

Stack—outlet on the top of a furnace through which hot combustion vapors escape from the furnace.

Stage—each cylinder in a compressor; specifically, the area where gas is compressed.

Star feeder—a solids feeder connected to an air system that is used to transfer plastic.

Stationary equipment—industrial equipment designed to occupy a stationary or fixed position.

Steam chest—area where steam enters a steam turbine.

Steam-generating drum—a large upper drum partially filled with feedwater. This drum is the central component of a boiler. It is connected to the lower mud drum by the downcomer and riser tubes and receives steam from the steam-generating tubes.

Steam generator—see *Boiler*.

Steam strainer—a mechanical device that removes impurities from steam.

Steam trap—a device used to separate condensate from steam and return it to the boiler to be converted to steam.

Steam turbine—an energy-conversion device that converts steam energy (kinetic energy) to useful mechanical energy; used as drivers to turn pumps, compressors, and electric generators.

Stem—a metal shaft attached to the hand-wheel and flow-control element of a valve.

Stirred reactor—a reactor designed to mix two or more components into a homogeneous mixture; also called an *autoclave*.

Stress-corrosion cracking—a mechanical-chemical type of deterioration associated with steel.

Stripping section—the section of a distillation column below the feed line, where heavier components are located.

Stuffing box—the section of a valve that contains packing.

Superheated steam—steam that is heated to a higher temperature.

Tank farm—a collection of tanks used to store and transport raw materials and products.

Tanks and pipes—vessels and tubes that store and convey fluids.

Temperature gradient—the progressively rising temperatures from the bottom of a distillation column to the top.

Terminal penetrations—provide 180° turns or pipe bends in the convection section as the pipes scroll from one side of the furnace to the other.

Ternary mixture—three components in a mixture.

Thermal shock—a form of stress resulting in metal fatigue when large temperature differences exist between a piece of equipment and the fluid in it.

Thermocouple—a temperature-measuring device composed of dissimilar metals that are connected at one end; heat applied to the connected ends causes the generation of voltage that corresponds to the temperature change, which is indicated on a temperature scale.

Thermostatic steam trap—a type of steam trap that is controlled by temperature changes.

Thermosyphon reboiler—a type of heat exchanger that generates natural circulation as a static liquid is heated to its boiling point.

Thermowell—a chamber installed in vessels or piping to hold thermocouples and RTDs.

Three-way valve—a valve with three ports (one inlet and two outlets) used to divert flow direction.

Throttling—reducing or regulating flow below the maximum output of a valve.

Thrust bearing—see *Axial bearings*.

Torque—the turning force of rotating equipment.

Total dissolved solids (TDS)—the dissolved minerals, such as magnesium and calcium, found in water, which is typically treated with sulfuric acid.

Total reflux—a process where the feed to the column is stopped and the total amount leaving the column from the top, side, or bottoms is returned to the column.

Traced piping—used when the conveyed fluid must be kept hot; usually has copper tubing containing steam or hot oil.

Transducer—a device used to convert one form of energy into another, typically electric to pneumatic or vice versa.

Transmitter—a device used to sense a process variable such as pressure, level, temperature, composition, or flow, and produce a signal that is sent to a controller, recorder, or indicator.

Tray columns—devices located on a tray in a column that allow vapors to come into contact with condensed liquids; three basic designs are bubble-cap, sieve, and valve.

Tray drying—occurs due to low feed rates, low reflux rates, or high upward vapor flow rates from tray to tray.

Trim—the flow-control element and seats in a valve.

Trip valve—a fast-closing steam inlet valve operated by an overspeed trip lever.

Tube sheet—a flat plate to which the ends of the tubes in a heat exchanger are fixed by rolling, welding, or both.

Tube side—refers to flow through the tubes of a shell-and-tube heat exchanger; see *Shell side*.

Tubular reactor—a heat exchanger in which a chemical reaction takes place; used for chemical synthesis.

Turbulent flow—random movement or mixing in swirls and eddies of a fluid.

Vacuum distillation—the process of vaporizing liquids at temperatures lower than their boiling point by reducing pressure.

Vacuum pressure—pressure below zero gauge; often expressed in inches of mercury.

Valve capacity—the total amount of fluid a valve will pass with a given pressure difference when it is fully open.

Valve—a device used to stop, start, restrict (throttle), or direct the flow of fluids.

Vane pump—a rotary pump that uses flexible or rigid vanes to displace fluids.

Vapor-liquid-equilibrium diagram—used to help a technician understand phase behavior of a two-component mixture.

Vapor lock—condition in which a pump loses liquid prime and the impellers rotate in vapor.

Vapor pressure—the outward force exerted by the molecules suspended in vapor state above a liquid at a given temperature; when the rate of liquefaction is equal to the rate of vaporization (equilibrium).

Vessel design sheets—identifies the factors entering into the selection, use, and need for periodic inspection of materials used to make vessels.

Vibrating tubes—tend to jump or move back and forth. Typically occurs in tubes outside the furnace. Vibrating tubes are often caused by two-phase slug-type flow inside the tubes. May be stopped by changing flow rates.

Viscosity—a measure of a fluid's resistance to flow.

Volute—the discharge chute of a centrifugal pump; a widening cavity that converts velocity to pressure.

Warping—a term used to describe temperature changes and pipe expansion that cause a valve to seize, or "warp." Closing a valve too quickly can cause warping, and warping can cause a valve to stick.

Water distribution header—a pipe that evenly disperses hot water over the fill of a cooling tower.

Water distribution system—typically consists of a deep pan with holes equipped with nozzles that distribute the water across the fill using gravity. Some systems utilize a pipe and spray nozzle design.

Water hammer—a condition in a boiler in which slugs of condensate (water), flowing with steam, damage equipment.

Water-tube boiler—a type of boiler that passes water-filled tubes through a heated firebox.

Weeping—occurs when the vapor velocity is too low to prevent liquid from flowing through the holes in the tray instead of across the tray. Differential pressure is reduced and product separation is reduced.

Wet-bulb temperature (WBT)—the air temperature as measured by a thermometer that takes into account the relative humidity.

Wind turbine—commonly referred to as windmill; uses air pressure to pump water, grind grain, and operate small generators.

Windage or drift—small water droplets that are carried out of the cooling tower by flowing air. See also *Drift loss*.

index

E

K

L

M